An Introduction to Ultracold Atoms with Analytical and Numerical Methods

Online at: https://doi.org/10.1088/978-0-7503-5447-9

IOP Series in Advances in Optics, Photonics and Optoelectronics

SERIES EDITOR

Professor Rajpal S Sirohi Consultant Scientist

About the Editor

Rajpal S Sirohi is currently working as a faculty member in the Department of Physics, Alabama A&M University, Huntsville, AL, USA. Prior to this, he was a consultant scientist at the Indian Institute of Science, Bangalore, and before that he was Chair Professor in the Department of Physics, Tezpur University, Assam. During 2000–2011, he was an academic administrator, being vice-chancellor to a couple of universities and the director of the Indian Institute of Technology, Delhi. He is the recipient of many international and national awards and the author of more than 400 papers. Dr Sirohi is involved with research concerning optical metrology, optical instrumentation, holography, and the speckle phenomena.

About the series

Optics, photonics, and optoelectronics are enabling technologies in many branches of science, engineering, medicine, and agriculture. These technologies have reshaped our outlook and our ways of interacting with each other, and have brought people closer together. They help us to understand many phenomena better and provide deeper insight into the functioning of nature. Further, these technologies themselves are evolving at a rapid rate. Their applications encompass very large spatial scales, from nanometres to the astronomical scale, and a very large temporal range, from picoseconds to billions of years. This series on advances in optics, photonics, and optoelectronics aims to cover topics that are of interest to both academia and industry. Some of the topics to be covered by the books in this series include biophotonics and medical imaging, devices, electromagnetics, fibre optics, information storage, instrumentation, light sources, charge-coupled devices (CCDs) and complementary metal oxide semiconductor (CMOS) imagers, metamaterials, optical metrology, optical networks, photovoltaics, free-form optics and its evaluation, singular optics, cryptography, and sensors.

About IOP ebooks

The authors are encouraged to take advantage of the features made possible by electronic publication to enhance the reader experience through the use of color, animation, and video and by incorporating supplementary files in their work.

A list of recent titles published in this series can be found here: https://iopscience.iop.org/bookListInfo/series-on-advances-in-optics-photonics-and-optoelectronics.

An Introduction to Ultracold Atoms with Analytical and Numerical Methods

Paulsamy Muruganandam
Department of Physics, Bharathidasan University, Tiruchirapalli, Tamil Nadu, India

Ramaswamy Radha
Government College for Women (Autonomous), Kumbakonam, India

IOP Publishing, Bristol, UK

ISBN 978-0-7503-5447-9 (ebook)
ISBN 978-0-7503-5445-5 (print)
ISBN 978-0-7503-5448-6 (myPrint)
ISBN 978-0-7503-5446-2 (mobi)

DOI 10.1088/978-0-7503-5447-9

Version: 20251101

IOP ebooks

British Library Cataloguing-in-Publication Data: A catalogue record for this book is available from the British Library.

Published by IOP Publishing, wholly owned by The Institute of Physics, London

IOP Publishing, No.2 The Distillery, Glassfields, Avon Street, Bristol, BS2 0GR, UK

US Office: IOP Publishing, Inc., 190 North Independence Mall West, Suite 601, Philadelphia, PA 19106, USA

In loving memory of my parents, whose enduring wisdom guides me, and to my beloved wife and children, my unwavering source of strength and joy

—Paulsamy Muruganandam.

In memory of my beloved parents for their endless love and support, to my siblings for their constant inspiration, faith and encouragement and Lord Gnanananda whose divine grace has guided every thought, word and all my endeavors through out my academic journey

—Ramaswamy Radha.

Contents

Preface

For over three decades, ultracold quantum gases have provided a versatile platform for studying quantum many-body phenomena, with Bose–Einstein condensates (BECs) at the forefront due to their tunable properties. BECs enable precise control of short-range contact interactions via Feshbach resonances, exhibit long-range anisotropic tunable dipole–dipole interactions in certain atomic species, and support two-component (vectorial) condensates with adjustable intra- and inter-species interactions. Flexible trapping geometries, from one-dimensional systems to three-dimensional lattices, further enhance their experimental utility, fostering exploration of nonlinear dynamics and exotic quantum phases.

The Gross–Pitaevskii (GP) equation, a nonlinear partial differential equation, underpins the mean-field description of BEC dynamics, capturing the effects of harmonic trapping potentials and interatomic interactions. Analytical solutions, derived under specific correlations between trap frequency and interaction strength using methods like inverse scattering transform, gauge transformation and Darboux transformation, offer insights, while numerical techniques, such as split-step Crank–Nicolson and Bayesian optimization, address general dynamics in scalar and vectorial BECs. These methods enable the study of nonlinear excitations like solitons (both bright and dark), rogue waves, vortices, Faraday waves and the dynamics of advanced systems, including spin–orbit-coupled, dipolar, and exciton–polariton condensates.

This textbook guides students and early-career researchers through the theoretical and computational tools for investigating BEC dynamics, bridging physics, applied mathematics, and computational science. In particular, an attempt has been made to give an ideal level playing field to analytical and numerical physicists. The first part (chapters 1 and 2) introduces the theory of BECs and analytical methods, deriving soliton solutions for GP-type equations. The second part (chapters 3 and 4) covers numerical techniques and analogies with optical solitons. Subsequent chapters (5–11) explore scalar, vectorial, spin–orbit-coupled, dipolar, spatially inhomogenous, and polariton BECs, focusing on the associated nonlinear excitations and their stability. Chapter 12 outlines future directions, including supersolids, quantum droplets, and quantum turbulence, highlighting their potential to advance ultracold atomic research.

Acknowledgements

The authors thank colleagues and students whose contributions have enriched this textbook on Bose–Einstein condensates. Special thanks are extended to Dr R Kishor Kumar, Dr C Senthil Kumar, Dr S Sabari, Dr P Sakthi Vinayagam, Dr T Sriraman, Dr S Bhuvaneswari, Dr R Ravisankar, Dr B Tamilarasan, Dr V Ramesh Kumar, Dr S Rajendran, Mr Anirudh Sivakumar, Ms K Rajaswathi, Mr P Raman, Ms N Rasha Shanaz, and Ms S Nirmala Jenifer for their critical feedback, numerical simulations, and editorial contributions. Their collective efforts, spanning theoretical, computational, and editorial contributions, made this textbook a valuable resource for advancing the study of ultracold atomic systems.

We sincerely thank Muthusamy Lakshmanan, Murugaian Senthilvelan, Sadhan Kumar Adhikari, Boris Malomed, Dumitru Mihalache, Lauro Tomio, Alexandru Nicolin, Sen-Yue Lou, Wu Ming Liu, Usama Al Khawaja, Antun Balaž, Arnaldo Gammal, Pankaj Kumar Mishra, Prasanta Panigrahi, Dilip Kumar Angom, and Sandeep Gautam for their invaluable contributions as mentors and collaborators in our research on ultracold atomic physics. We also acknowledge the collaboration of the late Miki Wadati and the late Kuppuswamy Porsezian during the formative years of research on Bose–Einstein condensates.

Paulsamy Muruganandam acknowledges the generous financial support received from various funding agencies over the past two decades through research projects, visiting fellowships, travel grants, and exchange programs. These include the Department of Science and Technology (DST, India), Anusandhan National Research Foundation (ANRF, India)—formerly Science and Engineering Research Board (SERB, India), Council of Scientific and Industrial Research (CSIR, India), Fundação de Amparo à Pesquisa do Estado de São Paulo (FAPESP, Brazil), TWAS-UNESCO (Italy), German Academic Exchange Service (DAAD, Germany), and ICTP South American Institute for Fundamental Research (ICTP-SAIFR, Brazil).

Ramaswamy Radha sincerely acknowledges the generous financial support extended by various national and international funding agencies over the past two decades through sponsored research projects, visiting fellowships and exchange programs. These include Department of Science and Technology (DST), Government of India, Anusandhan National Research Foundation (ANRF), formerly Science Engineering and Research Board (SERB), Government of India, University Grants Commission (UGC), Government of India, DST-Consolidation of University Research for Innovation and Excellence (DST-CURIE), Government of India, Department of Atomic Energy—National Board of Higher Mathematics (DA-NBHM), Government of India, Council of Scientific and Industrial Research (CSIR), Government of India, Indian National Science Academy (INSA)—Royal Society of London Visiting Fellowship, INSA-Polish Academy of Sciences Visiting Fellowship, Chinese Academy of Sciences Visiting Fellowship and TWAS-UNESCO Associateship.

We express our appreciation to the eBook Publishing Coordinators at IOP Publishing, particularly Ms Ashley Gasque, Ms Erika Radzvilaite, Ms Bethany Hext and Ms Mia Foulkes, for their invaluable editorial and technical support.

Author biographies

Paulsamy Muruganandam

Dr Paulsamy Muruganandam is a Professor in the Department of Physics at Bharathidasan University, Tiruchirappalli, India, where he advanced research in ultracold systems, nonlinear dynamics, and machine learning. He earned his BSc (1989) and MSc (1991) in physics from Madurai Kamaraj University, followed by an MPhil (1993) and PhD (2000) in theoretical physics from Bharathidasan University, building expertise in computational physics. His research focused on Bose–Einstein condensates (BECs), using the Gross–Pitaevskii equation to study conventional, dipolar, spin–orbit-coupled, and spinor BECs. He developed numerical methods (e.g. splitstep Crank–Nicolson, pseudo-spectral) to explore quantum vortices, lattice dynamics, and turbulent mergers, with applications in quantum simulations and analog gravity. Furthermore, from nonlinear dynamics and machine learning perspectives, his works include chaos, synchronization, time series analysis, spatiotemporal patterns, higher-order complex networks, and reservoir computing. Dr Muruganandam has authored more than 100 papers in international journals.

Ramaswamy Radha

Dr Ramaswamy Radha is an Associate Professor and Dean of Sciences at Government College for Women (Autonomous), Kumbakonam, India. She obtained her BSc (1986), MSc (1988) and M Phil (1990) from Bharathidasan University. Later, she went onto earn her PhD working on localized coherent excitations in (2+1) Dimensions and their collisional dynamics in 1997 from the same University. She was instrumental in developing an algorithm using Hirota method to construct 'Dromions' way back in 1994. She has also developed a unified approach entitled 'truncated Painlevé method' to generate localized excitations in (2 + 1)-dimensional nonlinear partial differential equations in collaboration with Professor S Y Lou, Shanghai Jiao Tong University, China. She has been serving as the Director of the Center for Nonlinear Science (CeNSc) affiliated to Government College for Women (Autonomous), Kumbakonam for the last 20 years. She started to work on ultracold atoms only from 2005. Her exploits, employing gauge transformation approach to identify several integrable models in BECs, an area dominated by numerical physicists, are considered to be an important contribution. Her notable contributions include taming of rogue waves in BECs, identification of signatures of EIT in the collision of solitons, integrability of spin orbit and Rabi-coupled BECs etc. She has authored 86 papers in peer reviewed international journals. She is a Life Member of the Indian Association of Physics Teachers (IAPT) and a Fellow of the Academy of Sciences (FASCh), Chennai. Some of the accolades bestowed on her include Third World Academy of Sciences-UNESCO Associateship Award, INSA Royal Society of London Visiting Fellow, INSA-Polish Academy of Sciences Visiting Fellow, etc.

IOP Publishing

An Introduction to Ultracold Atoms with Analytical and Numerical Methods

Paulsamy Muruganandam and Ramaswamy Radha

Chapter 1

Introduction

This chapter introduces the fundamental concepts of Bose–Einstein condensates (BECs), providing a comprehensive theoretical foundation for subsequent investigation on their localization and numerical simulation. It begins with a historical overview of BECs, tracing their theoretical origins and experimental realization. The chapter then focuses on the theoretical perspective of scalar BECs and how it can be modelled by a single-component Gross–Pitaevskii (GP) equation in the mean-field description, a variable-coefficient nonlinear Schrödinger (NLS) equation encompassing a trapping potential and interatomic interaction. The mean-field description is then extended to govern the dynamics of vectorial condensates described by a coupled GP equation with intra and inter-species (atomic) interactions. Subsequently, the chapter discusses the origin of fermionic condensates, highlighting their distinct pairing mechanisms. Beyond mean-field approaches are then introduced to address the limitations of the GP framework, incorporating quantum fluctuations and correlations. The chapter concludes with an outline of the textbook, setting the stage for analytical and numerical investigations of ultracold atoms in ensuing chapters.

1.1 History of Bose–Einstein condensates

This section has been reproduced with permission from [32].

Matter pervades the entire universe but exists in only a few forms, such as solid, liquid, and gas. Phase transitions between these states can be initiated by increasing temperature or pressure. In 1879, Sir William Crookes exploited this principle by raising the temperature to create plasma, a gas with a significant number of charge carriers [1]. Notably, only the physical state changes during phase transitions, while the chemical composition remains unchanged. Can cooling to ultracold temperatures produce a new state of matter (figure 1.1)? This question is driven by the need

doi:10.1088/978-0-7503-5447-9ch1

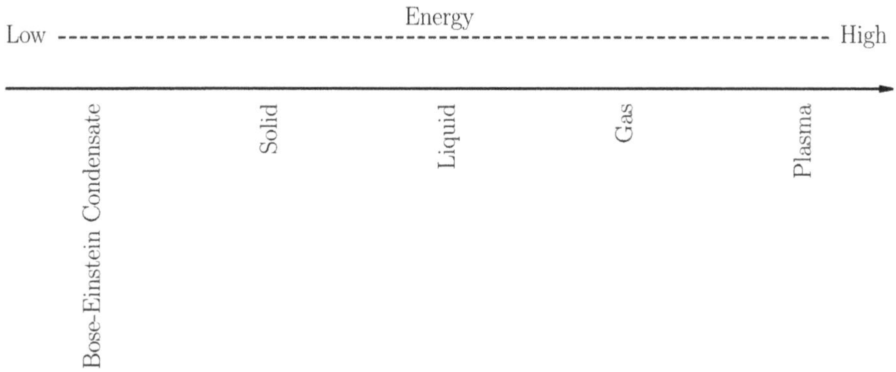

Figure 1.1. The energy levels of different physical states of matter.

to study matter wave dynamics at low temperatures. In 1908, Kamerlingh Onnes liquefied He-$_4$ at 4.2 K [2], paving the way for the discovery of superconductors. Further cooling of He-$_4$ to 2.17 K resulted in a superfluid with zero viscosity. In 1938, Fritz London proposed that superconductors and superfluids could be described by a single macroscopic wave function [3, 4], a concept first envisioned by Albert Einstein and Satyendra Nath Bose in 1924 [5, 6]. This idea was then experimentally realized in 1995 as a superatom when Eric Cornell and Carl Wieman generated BECs in rubidium atoms [7, 8], followed by Wolfgang Ketterle in sodium atoms [9]. At such ultracold temperatures, many atoms occupy the ground state or a long-lived metastable state, merging into a giant matter wave, known as BEC. The discovery of BECs spurred research in atom optics [10], condensed matter physics [11], and quantum information processing [12]. BECs were subsequently identified in other alkali atoms [13] and in atoms with long-range interactions, such as chromium [14], erbium [15], and dysprosium [16]. The success of BECs inspired experiments at the Large Hadron Collider (LHC) [17], where high-energy subatomic particles from accelerators recreated a mini Big Bang-like environment, allowing condensation. These experiments led to the discovery of the Higgs boson [18], the long elusive God particle. The identification of BECs has generated tremendous enthusiasm in atomic physics, with several investigations still in progress.

1.2 Bose–Einstein condensation: An experimental and theoretical perspective

To explore BECs from a statistical perspective, one begins by considering an ideal gas of atoms obeying Bose–Einstein statistics. The mean occupation number of atoms n with energy ε in equilibrium at temperature T is given by

$$n = \frac{1}{\exp\left(\dfrac{\varepsilon - \mu}{K_{\rm B}T}\right) - 1}, \tag{1.1}$$

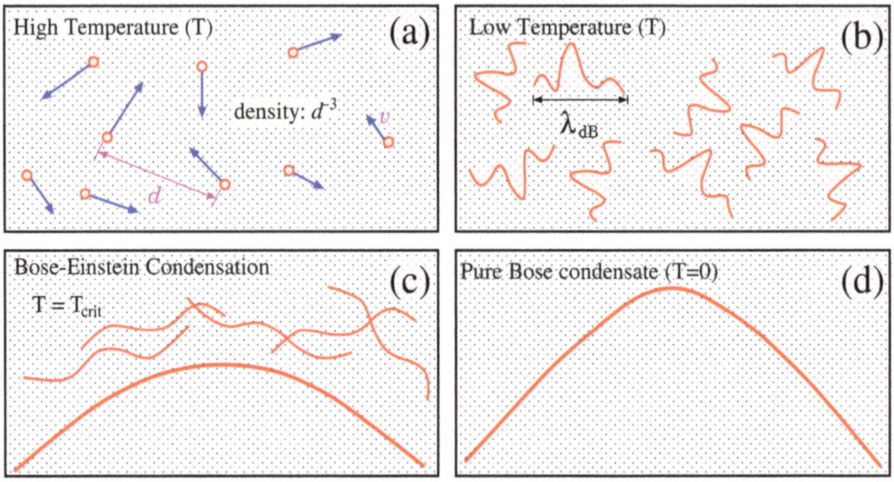

Figure 1.2. Schematic illustration of BEC formation across temperature regimes: (a) high-temperature disordered state with low particle density (d^3), (b) cooling toward critical temperature (T_{crit}) showing increasing de Broglie wavelength (λ_{dB}), (c) phase transition at $T = T_{crit}$ with BEC onset, and (d) pure BEC at $T = 0$ with macroscopic occupation of the ground state.

where μ represents the chemical potential and K_B denotes the Boltzmann constant. The thermal de Broglie wavelength associated with atomic density n is expressed as

$$\lambda_{dB} = \frac{h}{\sqrt{2\pi m K_B T}}, \tag{1.2}$$

where h is Planck's constant and m is the atomic mass. At room temperature, the de Broglie wavelength typically remains shorter than the interparticle spacing ($n^{-1/3}$). However, as the gas temperature decreases, the de Broglie wavelength increases, and at a critical temperature T_c, it coincides with the interatomic spacing. This condition marks the onset of wave function overlap, where the gas transitions into a quantum degenerate state of indistinguishable particles known as a BEC (figure 1.2). In contrast, for fermionic atoms, progressive cooling instead leads to the formation of a Fermi sea, where each low-energy state becomes occupied by exactly one particle (see figure 1.3).

Although the creation of BECs appears straightforward in principle, requiring cooling of an atomic gas until the wavepackets overlap, experimental realization proves significantly challenging. In most cases, quantum degeneracy is preempted by conventional phase transitions to liquid or solid states. These classical condensation processes can only be avoided at extremely low densities, typically of the order of a hundred-thousandth the density of ambient air.

The realization of Bose–Einstein condensation required two key developments: first, the identification of suitable gaseous systems that could remain in the gaseous phase throughout the BEC transition, with alkali atoms such as cesium, rubidium, and sodium emerging as optimal candidates, and second, the advancement of cooling and

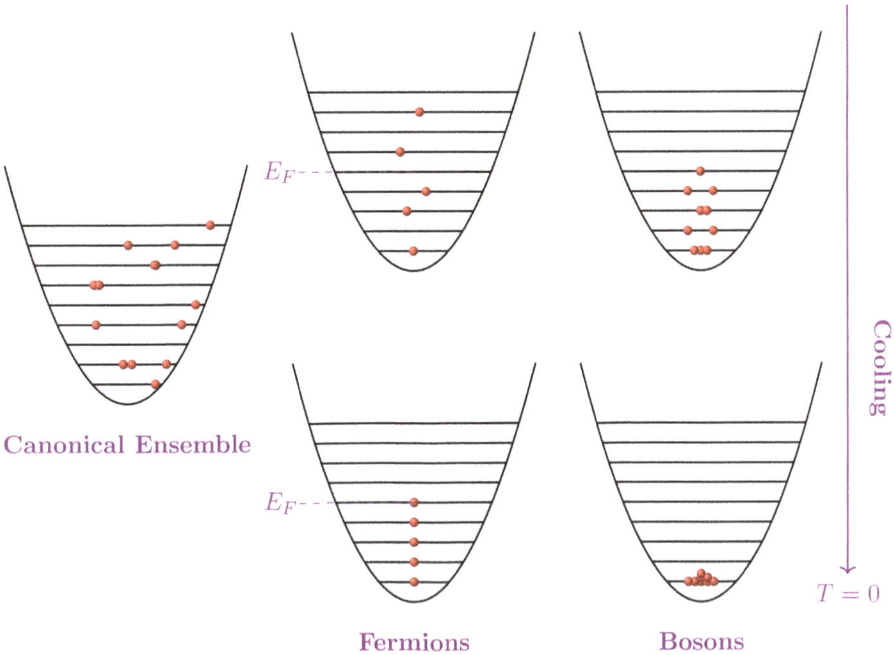

Figure 1.3. Difference between Fermions and Bosons at ultracold temperature.

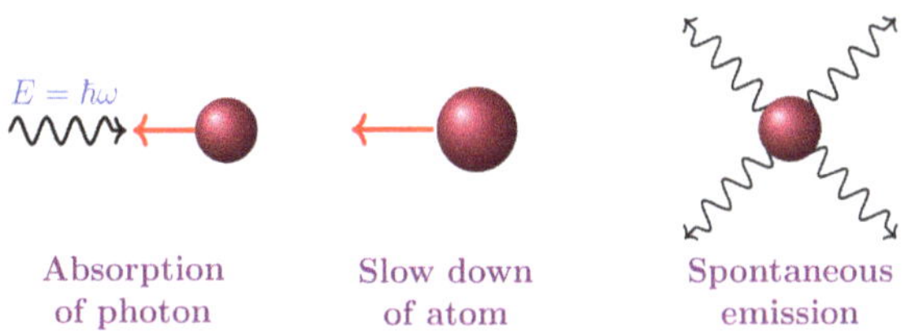

Figure 1.4. Schematic representation of the laser cooling process involved in the creation of BEC.

trapping techniques capable of achieving the necessary temperature and density regimes. It became obvious that BEC in alkali gases could only be attained through a combination of laser cooling and magnetic evaporative cooling techniques.

The main idea of laser cooling involves inducing atoms (or ions) to predominantly absorb photons propagating in the direction opposite to their motion, thereby reducing their velocity and achieving cooling as illustrated in figure 1.4. This technique [19, 20] not only cools the atoms but also confines them and keeps them away from the room-temperature wall of the enclosure.

Following initial laser cooling (precooling), subsequent cooling stages employ magnetic trapping and evaporative cooling techniques. The magnetic confinement isolates the atomic ensemble, after which forced evaporative cooling [9, 21, 22] selectively removes higher-energy atoms from the trap, resulting in further thermal

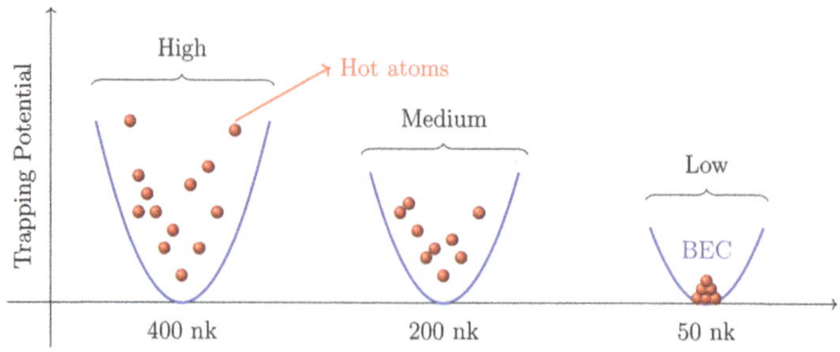

Figure 1.5. A schematic representation of magnetic evaporative cooling.

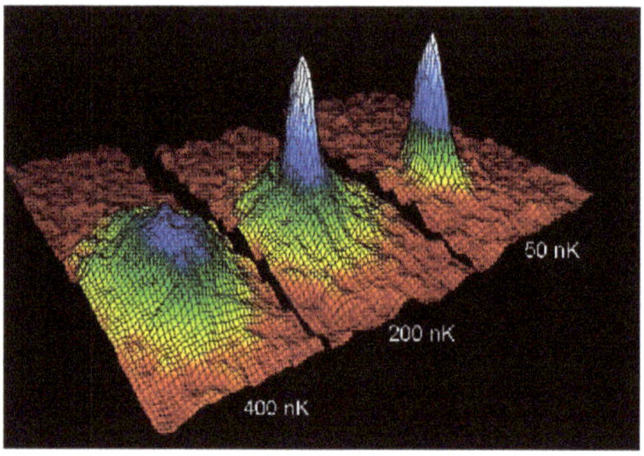

Figure 1.6. Velocity distribution data of a gas of rubidium atoms, as illustrated in reference [23], depicts three-dimensional snapshots taken sequentially over time. In these images, the atoms can be observed condensing from less dense red, yellow and green areas into very dense blue to white areas. Reprinted from [23], with permission from NIST.

reduction of the remaining atomic population. In this scheme, as shown in figure 1.5, gradual reduction of the trap potential allows the most energetic atoms to escape while the remainder rethermalize steadily at lower temperatures. The entire cooling cycle required to produce a condensate typically ranges from a few seconds to several minutes.

The first pure BEC was achieved in 1995 by Eric Cornell, Carl Wieman, and their research team at JILA. Their experiment involved cooling a dilute vapour of approximately two thousand rubidium-87 atoms to temperatures below 170 nK through a combination of laser and magnetic evaporative cooling techniques (illustrated in figure 1.6). Approximately 4 months later, a separate research group led by Wolfgang Ketterle at MIT successfully created a sodium-23 condensate [6]. Ketterle's condensate had a hundred times more atoms than the JILA experiment, enabling significant observations, including quantum mechanical interference between distinct condensates. While BECs remain extremely delicate compared to other states of matter, with even minor tinkering capable of destroying their unique quantum properties, they have nevertheless become invaluable tools for investigating fundamental physics. Following these pioneering experiments by the JILA and MIT groups, the field has witnessed substantial growth in both experimental and theoretical research, particularly in areas such as superfluid-Mott insulator transitions [24], atom lasers [25], electromagnetically induced transparency [26], optical lattice systems [27], and so on. It is important to mention that the ultralow temperature requirement of BECs of alkali metals does not generalize to all types of BECs. In 2006, a Bose–Einstein condensation of magnons (i.e. quantized spinwaves) at room temperature [28] was created admittedly by the application of pump processes.

From the above, one understands the rich variety of contexts in which the physics of Bose–Einstein condensation plays a pivotal role, thereby underscoring the importance of understanding this phenomenon to develop concrete applications.

1.3 Mean field description and Gross–Pitaevskii equation

1.3.1 Scalar Bose–Einstein condensates

This subsection has been reproduced with permission from [32].

At ultracold temperatures, where the lowest energy level is macroscopically occupied, and the gas is sufficiently dilute enough to keep particle interactions weak, BECs are well described by a variable-coefficient NLS equation, commonly known as the Gross–Pitaevskii (GP) equation. The GP model accurately characterizes both static and dynamic properties of single-component (scalar) and two-component (vector) condensates. The following section presents a detailed derivation and physical interpretation of the GP equation.

BECs consist of N interacting bosons, described by the N-particle wave function $\Psi(\mathbf{r}_1, \mathbf{r}_2, \ldots, \mathbf{r}_N)$, where $\mathbf{r}_i$ denotes the position vector of the ith atom:

$$\Psi(\mathbf{r}_1, \mathbf{r}_2, \ldots, \mathbf{r}_N) = \prod_{i=1}^{N} \phi(\mathbf{r}_i). \tag{1.3}$$

The normalized single-particle wave function $\phi(\mathbf{r}_i)$ satisfies

$$\int \mathrm{d}\mathbf{r}\, |\phi(\mathbf{r})|^2 = 1. \tag{1.4}$$

The effect of correlations due to interactions is accounted for using the effective interaction $U_0\delta(\mathbf{r} - \mathbf{r}')$. The effective Hamiltonian, within the mean-field theory, is given by

$$H = \sum_{i=1}^{N}\left[\frac{\mathbf{p}_i^2}{2m} + V(\mathbf{r}_i)\right] + U_0\sum_{i<j}\delta(\mathbf{r}_i - \mathbf{r}_j), \tag{1.5}$$

where $V(\mathbf{r}_i)$ is the external trapping potential. The energy of the state given by equation (1.3) is

$$E = N\int \mathrm{d}\mathbf{r}\left[\frac{\hbar^2}{2m}|\nabla\, \phi(\mathbf{r})|^2 + V(\mathbf{r})|\phi(\mathbf{r})|^2 + \frac{N-1}{2}U_0|\phi(\mathbf{r})|^4\right]. \tag{1.6}$$

In a uniform Bose gas, the relative reduction in the number of condensate particles is of the order $\sqrt{na^3}$, where n denotes the particle density. Introducing the condensate wave function as

$$\psi(\mathbf{r}) = N^{1/2}\phi(\mathbf{r}), \tag{1.7}$$

the particle density is expressed as

$$n(\mathbf{r}) = |\psi(\mathbf{r})|^2. \tag{1.8}$$

For $N \gg 1$, the energy of the system takes the form

$$E = \int \left[\frac{\hbar^2}{2m} |\nabla\psi(\mathbf{r})|^2 + V(\mathbf{r})|\psi(\mathbf{r})|^2 + \frac{1}{2} U_0 |\psi(\mathbf{r})|^4 \right] \mathrm{d}\mathbf{r}. \tag{1.9}$$

To find the optimal form for ψ, the energy is minimized with respect to independent variations of $\psi(\mathbf{r}, t)$ and its complex conjugate $\psi^*(\mathbf{r}, t)$, subject to the constraint that the total number of particles

$$N = \int \mathrm{d}\mathbf{r}\, |\psi(\mathbf{r}, t)|^2 \tag{1.10}$$

remains constant. This is achieved by setting $\delta E - \mu\delta N = 0$, where the chemical potential μ is the Lagrange multiplier ensuring particle conservation. The variations of ψ and ψ^* are taken as arbitrary. Equating the variation of $E - \mu N$ with respect to $\psi^*(\mathbf{r})$ to zero yields the time-independent GP equation [29–32] of the following form

$$-\frac{\hbar^2}{2m} \nabla^2 \psi(\mathbf{r}) + V(\mathbf{r})\psi(\mathbf{r}) + U_0|\psi(\mathbf{r})|^2\psi(\mathbf{r}) = \mu\psi(\mathbf{r}). \tag{1.11}$$

The above equation (1.11) is often referred to as the time-independent (stationary) GP equation. It resembles a time-independent Schrödinger equation, with the potential acting on particles being the sum of the external potential V and a nonlinear term $U_0|\psi(\mathbf{r})|^2$, which accounts for the mean-field produced by the binary interactions among bosons.

To study the dynamics of condensates, the time-dependent generalization of the Schrödinger equation, incorporating the same nonlinear interaction term, yields the time-dependent GP equation:

$$-\frac{\hbar^2}{2m} \nabla^2 \psi(\mathbf{r}, t) + V(\mathbf{r})\psi(\mathbf{r}, t) + U_0|\psi(\mathbf{r}, t)|^2\psi(\mathbf{r}, t) = \mathrm{i}\hbar\frac{\partial\psi(\mathbf{r}, t)}{\partial t}. \tag{1.12}$$

To ensure consistency between the time-dependent GP equation (1.12) and the time-independent GP equation (1.11), under stationary conditions, $\psi(\mathbf{r}, t)$ must evolve as $\exp(-\mathrm{i}\mu t/\hbar)$.

In equation (1.12), $\psi(\mathbf{r}, t)$, where $\mathbf{r} = (x, y, z)$, represents the condensate wave function, ∇^2 is the Laplacian operator, and $V(\mathbf{r})$ is the external trapping potential, assumed to be $V(\mathbf{r}) = m(\omega_r^2 r^2 + \omega_x^2 x^2)$, where $r^2 = y^2 + z^2$, and ω_r and ω_x are the confinement frequencies in the radial and axial directions, respectively.

The interaction strength is $U_0 = 4\pi\hbar^2 a/m$, where a is the s-wave scattering length and m is the atom mass.

Equation (1.12) is an inhomogeneous (3 + 1)-dimensional NLS equation. The inhomogeneity arises from the trapping potential V, which confines the atoms in the ground state, and the nonlinearity coefficient U_0, which represents the interatomic interaction governed by the scattering length a. The scattering length can be positive (repulsive interaction) or negative (attractive interaction). Recent studies have shown that the scattering length $a(t)$ can be periodically varied using Feshbach resonance [33]. Thus, understanding the dynamics of BEC reduces to solving a variable coefficient (3 + 1)-dimensional NLS equation for appropriate choices of trapping potentials V and scattering lengths $a(t)$.

Notably, while (3 + 1)-dimensional condensates exhibit intriguing geometrical structures, such as vortex tori [34, 35], Skyrmions [36], and twisted toroids [37], the (3 + 1)-dimensional GP equation (or variable-coefficient NLS) is generally non-integrable for arbitrary trapping potentials and interatomic interactions. Therefore, one must explore whether the (3 + 1)-dimensional GP equation becomes integrable in one or two spatial dimensions for specific choices of trapping potential and interatomic interaction. This involves exploring analytical approaches to identify localized nonlinear excitations in (1 + 1)- and (2 + 1)-dimensional GP equations, which would indicate the integrability of the associated dynamical system.

In a three-dimensional BEC, when the transverse trapping frequency ω_r (for $r = x, y$) is much higher than the longitudinal trapping frequency ω_z, the transverse confinement is sufficiently strong enough to ensure that atoms cannot scatter into excited states of the transverse harmonic oscillator. This leads to a cigar-shaped BEC, where the three-dimensional GP equation effectively reduces to a quasi-one-dimensional form. Conversely, when the atoms are free to move in the transverse directions ($\omega_r \ll \omega_z$), the confinement along the longitudinal axis is dominant. This results in a pancake-shaped condensate, and the (3 + 1)-dimensional GP equation reduces to a (2 + 1)-dimensional equation. A detailed derivation of this dimensionality reduction in the GP equation is presented in chapter 3, section 3.1.

1.3.2 Vectorial Bose–Einstein condensates

This subsection has been reproduced with permission from [32].

For cigar-shaped BECs, the density of quasi-one-dimensional BECs increases with time, and once it exceed a critical value, they tend to collapse. Although Feshbach resonance can marginally extend their lifetime, an alternative approach is needed to significantly enhance the stability of quasi-one-dimensional BECs. In this context, vector quasi-one-dimensional BECs assume great significance. The experimental realization of two overlapping BECs, comprising two hyperfine states of ^{87}Rb [38], has motivated research into multi-component BECs. Unlike scalar BECs, two-component BECs exhibit rich dynamics due to intra-species and inter-species interactions. These interactions provide vector BECs with unique and complex properties, such as soliton trains [39, 40], multidomain walls [41], spin switching

[40], and multimode collective excitations [42, 43], which are nonexistent in scalar BECs. Additionally, this additional degree of freedom enhances the lifetime of BECs.

To describe vectorial condensates, the scalar (single-component) GP equation can be generalized to two coupled GP equations (CGPEs). For a two-component BEC prepared in two hyperfine states, the behavior at sufficiently low temperatures is described by the following coupled GP equations [32]:

$$i\hbar\frac{\partial\psi_1}{\partial t} = \left(-\frac{\hbar^2}{2m_1}\nabla^2 + U_{11}|\psi_1|^2 + U_{12}|\psi_2|^2 + V_1\right)\psi_1, \tag{1.13a}$$

$$i\hbar\frac{\partial\psi_2}{\partial t} = \left(-\frac{\hbar^2}{2m_2}\nabla^2 + U_{21}|\psi_1|^2 + U_{22}|\psi_2|^2 + V_2\right)\psi_2, \tag{1.13b}$$

where the condensate wave functions are normalized by particle numbers $\int_{-\infty}^{\infty}|\psi_1|^2\,\mathrm{d}^3\mathbf{r} = 1$ and $\int_{-\infty}^{\infty}|\psi_2|^2\,\mathrm{d}^3\mathbf{r} = N_2/N_1$. The intra-species and inter-species interaction strengths are $U_{ii} = 4\pi\hbar^2 a_{ii}/m$ and $U_{ij} = 2\pi\hbar^2 a_{ij}/m$, respectively, with a_{ij} being the corresponding scattering lengths and m the reduced mass. The trapping potentials are $V_i = m_i[\omega_{ix}^2 x^2 + \omega_{i\perp}^2(y^2 + z^2)]/2$. Assuming $\omega_{i\perp} \gg \omega_{ix}$, such that the transverse motions of the condensates are confined to the ground state of the transverse harmonic trapping potential, the system becomes quasi-one-dimensional. By integrating out the transverse coordinates, the two coupled GP equations for binary symmetric interactions in a transient harmonic trap, in dimensionless form, become

$$i\frac{\partial\psi_1}{\partial t} = \left(-\frac{1}{2}\frac{\partial^2}{\partial x^2} + b_{11}|\psi_1|^2 + b_{12}|\psi_2|^2 + \frac{\lambda_1^2}{2}x^2\right)\psi_1, \tag{1.14a}$$

$$i\frac{\partial\psi_2}{\partial t} = \left(-\frac{k}{2}\frac{\partial^2}{\partial x^2} + b_{21}|\psi_1|^2 + b_{22}|\psi_2|^2 + \frac{\lambda_2^2}{2k}x^2\right)\psi_2, \tag{1.14b}$$

where the units for length and time are $\sqrt{\hbar/(m_1\omega_{1\perp})}$ and $2\pi/\omega_{1\perp}$, respectively, and $\psi_{1,2}$ are normalized such that $\int|\psi_1|^2\,\mathrm{d}x = 1$ and $\int|\psi_2|^2\,\mathrm{d}x = N_2/N_1$. The parameters are defined as $b_{11} = 2a_{11}N_1$, $b_{12} = 2m_1a_{12}N_1/[(1 + \omega_{2\perp}/\omega_{1\perp})m]$, $b_{21} = 2m_1a_{21}N_1/[(1 + \omega_{2\perp}/\omega_{1\perp})m]$, $b_{22} = 2a_{22}kN_1\omega_{2\perp}/\omega_{1\perp}$, $\lambda_1 = \omega_{1x}/\omega_{1\perp}$, $\lambda_2 = \omega_{2x}/\omega_{1\perp}$, and $k = m_1/m_2$.

Assuming $k = 1$ (i.e., $m_1 = m_2$) and $\omega_{1\perp} = \omega_{2\perp} = \omega_{1x} = \omega_{2x} = \omega$ (i.e., $\lambda_1 = \lambda_2$), and allowing the scattering lengths a_{ij} and the trapping potential strength λ^2

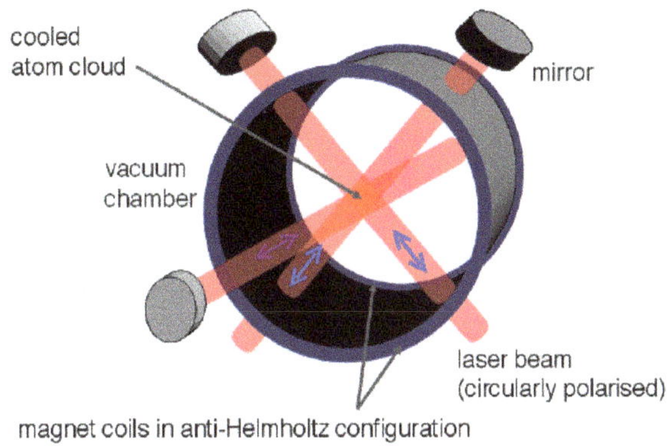

Figure 1.7. Magneto-optical trap apparatus used for sympathetic cooling in the creation of mixtures of BECs. This MOT setup image has been obtained by the author(s) from the Wikimedia website, where it is stated to have been released into the public domain. It is included within this article on that basis https://en.wikipedia.org/wiki/Magneto-optical_trap.

(where $\lambda = \omega_x/\omega_\perp$) to vary with time, the equations with $\tilde{t} = t/2$ take the following dimensionless form (omitting the tilde):

$$\mathrm{i}\frac{\partial\psi_1}{\partial t} + \frac{\partial^2\psi_1}{\partial x^2} + 2(b_{11}(t)|\psi_1|^2 + b_{12}(t)|\psi_2|^2)\psi_1 + \lambda(t)^2x^2\psi_1 = 0, \tag{1.15a}$$

$$\mathrm{i}\frac{\partial\psi_2}{\partial t} + \frac{\partial^2\psi_2}{\partial x^2} + 2(b_{21}(t)|\psi_1|^2 + b_{22}(t)|\psi_2|^2)\psi_2 + \lambda(t)^2x^2\psi_2 = 0, \tag{1.15b}$$

In equations (1.15a) and (1.15b), time t and coordinate x are measured in units $2/\omega_\perp$ and $l_\perp = \sqrt{\hbar/(m\omega_\perp)}$, respectively, where $l_\perp$ is the transverse harmonic oscillator length. The interactions are described by self-interaction $b_{ii} = 4a_{ii}N_i/l_\perp$ and inter-component interaction $b_{12} = b_{21} = 4a_{ij}N_i/l_\perp$.

The experimental realization of condensate mixtures was first achieved by Myatt *et al* at the JILA laboratory, led by Nobel laureates Cornell and Wieman [38]. A novel apparatus featuring a double magneto-optical trap and an Ioffe-type magnetic trap was used to create condensates of 2×10^6 atoms in either the $|F = 2, m = 2\rangle$ or $|F = 1, m = -1\rangle$ spin states of ^{87}Rb. Overlapping condensates of the two states were also created using nearly lossless sympathetic cooling of one state through thermal contact with the other evaporatively cooled state. The apparatus employing sympathetic cooling is shown in figure 1.7.

1.4 Fermionic condensates

Fermions have half-integer spin and constitute one-half of the fundamental particle family. To determine whether an atom is a boson or a fermion, one can examine the

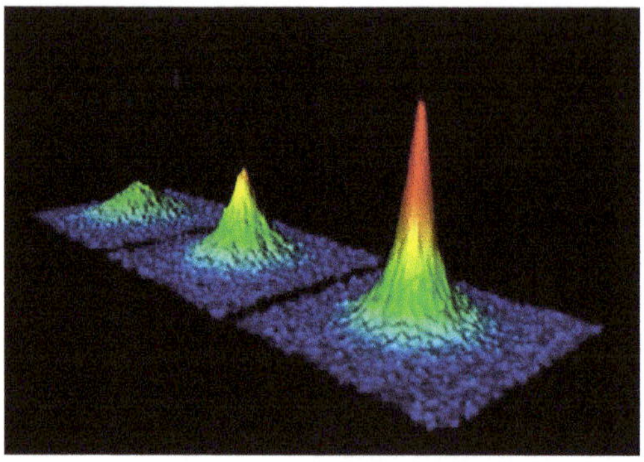

Figure 1.8. Velocity distribution of fermionic condensate. Adapted figure with permission from [45], Copyright (2004) by the American Physical Society.

number of protons, neutrons, and electrons that make up the atom. Since each of these subatomic particles has a spin of $\frac{1}{2}$, an atom composed of an **odd** total number of these particles is a **fermion** (with half-integer spin), whereas an atom with an **even** total number is a **boson** (with integer spin). It is well known that fermions obey the Pauli exclusion principle, which states that two indistinguishable fermions cannot occupy the same quantum state. As a result, in the limit of absolute zero temperature, fermions fill the lowest available energy levels of a trap one by one, with exactly one particle per state. This configuration is known as a *Fermi sea*. The key to realizing a fermionic quantum state lies in the precise control of interactions between the atoms. This tunability is essential for exploring the fascinating connection between superconductivity and Bose–Einstein condensation, a crossover regime known as the Bardeen–Cooper–Schrieffer (BCS)–BEC crossover. In this context, *Feshbach resonance* provides a powerful tool to tune the interaction strength between atoms, enabling the study of a wide range of many-body quantum phenomena (figure 1.8).

The BEC–BCS crossover describes the smooth transition between two different pairing regimes of fermions. BEC regime occurs when fermions form tightly bound pairs behaving like bosons and undergo Bose–Einstein condensation. BCS regime occurs in the case of weakly bound Cooper pairs in a Fermi sea as seen in conventional superconductors. The BEC–BCS crossover occurs when the interaction between the fermions is tuned by employing the Feshbach resonance in ultracold atomic gases. By manouvring the interaction strength, one can initiate a transition from a gas of weakly bound Cooper pairs (BCS side) to a gas of tightly bound molecules that undergo Bose–Einstein condensation (BEC side). In other words, at weak attractions (negative scattering lengths), the system behaves like a BCS superfluid while at strong attractions, fermions form bound pairs and condense as a BEC.

Even though both BECs and fermionic condensates require cooling to near absolute zero, one encounters extra challenges in fermionic condensates such as pairing fermions into Cooper pairs before being cooled. Another criteria involved is that when being paired, the size should be smaller than the interparticle spacing.

1.5 Beyond mean-field description

Describing many-body systems is a challenging task. The above statement stands valid no matter whether one considers a classical or a quantum mechanical system. The difficulty stems from the complex behavior that a given particle has due to its dependence on several others if not all particles in the system. Due to this dependence, the Schrödinger equation cannot be separated into a single-particle Hamiltonian and is harder to solve. To resolve this issue, statistical methods and models have been employed as was the case in the discovery of BECs. The statistical model often used to describe many-body problems is based on the mean-field description. The atom–atom interactions are replaced by a potential created by all the particles collectively such that all particles experience the same force. Accordingly, the dynamics of BECs can be modeled by the GP equation which is a variant of the NLS equation. This approach is found to be accurate in describing dilute and weakly interacting BECs. A large number of effects related to the phenomenon of Bose–Einstein condensation can be understood in terms of mean-field theory whereby the entire system is assumed to be condensed with thermal cloud and quantum fluctuations ignored. Although this theory works remarkably well for a broad range of experimental parameters, a more complete treatment based on an alternative model is required for understanding various experiments. Such a model should include the dynamical coupling of the condensate to the thermal cloud, the role of quantum fluctuations, the effect of dimensionality, etc.

From the mean-field theory, a BEC in two and three dimensions is expected to collapse with the attraction between the atoms. Petrov [44] suggested that quantum fluctuations can stabilize localized excitations in a two-component BEC which consists of two different kinds of atoms or the same atomic species in different states. Quantum fluctuations arise from the influence of thermal clouds (non-condensed atoms) on a coherent Bose gas. This influence is taken into account by the first-order correction to the condensate energy known as the Lee–Huang–Yang (LHY) term.

In a single-component BEC, the LHY term is negligible with respect to mean-field term. In a two-component BEC, there are intra-species repulsion and inter-species attraction and the scattering parameters can be manipulated via Feshbach resonance technique. Thus, in the absence of quantum fluctuations, such a BEC collapses. This residual attraction can be balanced by the repulsive interaction due to quantum fluctuations. As a result, the density distribution of a BEC assumes a localized profile leading to so called 'quantum droplets (QDs)' because of its liquid-like properties. It has been recently found that beyond mean-field description has also led to many interesting phenomena like 'supersolids', etc

1.6 Outline

The contents of the textbook have been organized as follows.

In chapter 2, the analytical methods which have been employed for solving GP-type equations (variants of NLS-type equations) have been elaborated. In particular, one focuses on the inverse scattering transform, gauge transformation, modified gauge transformation, Darboux transformation, and Hirota methods and how one can generate bright/dark solitons of GP-type equations. In particular, it is shown how one can derive bright and dark solitons explicitly for focussing and defocussing NLS-type equations (both scalar and vector equations).

In chapter 3, the numerical techniques which have been exploited for solving GP-type equations have been explored in detail. In particular, the algorithmic approaches employed in split-step Crank–Nicolson method, split-step Fourier transform method and Newton conjugate gradient method have been outlined. A glimpse of generating ground states of BECs using data driven approaches employing Gaussian processes as an alternative to numerical methods is shown.

In chapter 4, an analogy between BECs (matter) and optical solitons(waves) is presented. While optical solitons are generated by solving NLS-type equations, BECs are obtained by solving GP-type equations. In this chapter, physically important variable coefficient coupled NLS-type equations are studied as a precursor to the investigation of vector BECs governed by coupled GP-type equations. The signatures of electromagnetically induced transparency such as quantum storage and slow light are shown through the collisional dynamics of bright solitons.

Chapter 5 is devoted to the investigation of scalar BECs with short-range interactions using quasi-one-dimensional GP-type equations in time-independent and transient harmonic trapping potentials. Matter wave interference pattern obtained by virtue of the correlation between transient trap frequency and temporal binary interaction in the collisional dynamics of bright solitons is brought out. In addition, the impact arising by virtue of the reinforcement of three-body interaction with binary interaction is also studied.

In chapter 6, one explores the dynamics of vectorial (two-component) condensates described by coupled GP-type equations endowed with intra- and inter-species interactions. It is shown that the condensates in the temporal harmonic trap are long lived compared to their counterparts in time-independent traps. The dynamics of temporally and spatially coupled BECs are also studied. It is also demonstrated that rogue waves, which are inherently unstable can be stabilized through Feshbach resonance by manipulating scattering lengths. In other words, it is shown that one can enhance the lifetime of rogue waves through Feshbach resonance management.

In chapter 7, the focus shifts to the dynamics of spin–orbit (SO) and Rabi-coupled BECs for spin-$F = 1/2$. Employing Darboux transformation, nonlinear excitations like bright solitons, rogue waves, breathers, mixed (dark–bright) bound state solutions are explicitly derived assuming a suitable correlation between trapping frequency and temporal scattering length. For a general choice of trapping potentials and interatomic interactions, numerical techniques like split-step Fourier transform method and Newton conjugate gradient method are employed to numerically solve the dynamics of spin–orbit and Rabi-coupled BECs and the stability of the nonlinear excitations is studied.

Chapter 8 is devoted to the analysis of dipolar BECs with long-range anisotropic tunable dipole–dipole interactions, using mean-field GP equations in one-, two-, and three-dimensional traps. Split-step Crank–Nicolson method, Gaussian variational approach, and Thomas–Fermi approximations are applied to study ground-state properties, nonlinear excitations, and stability across trap geometries.

In chapter 9, the dynamics of collisionally inhomogenous BECs, focusing on Faraday and resonant waves in scalar and binary condensates is investigated. The dynamics of stable vortices arising out of spatially inhomogenous interactions in attractive BECs is analyzed. Further, the impact of cubic–quintic–septimal non-linearities arising out of collisionally inhomogenous interactions on BECs is also examined.

Chapter 10 explores the dynamics of quantum vortices in BECs, covering single vortex dynamics, vortex lattices, and their behavior in Thomas–Fermi and lowest Landau level regimes. The vortex stability via Bogoliubov equations is addressed. The dynamics of vortices in dipolar BECs and vortex formation through merging condensates are also looked into.

Chapter 11 is devoted to impart a nascent outlook into the dynamics of exciton–polariton condensates, emphasizing their non-equilibrium dynamics. The associated theoretical model is derived and the interaction of nonlinear excitations like bright solitons in polariton condensates with the impurity is studied. A theoretical model based on non resonant pumping is also derived to bring out the stability window of trapless polariton condensates.

Chapter 12 outlines future directions in store in the investigation of ultracold atoms, focusing on supersolids, quantum droplets with Lee–Huang–Yang corrections, and quantum turbulence. It discusses classical turbulence, spectral analysis, and specific turbulence regimes (Kolmogorov, Vinen, small-scale, and strong), highlighting their potential to shape ultracold atomic physics. A glimpse of quantum turbulence in self-gravitating BECs is also shown.

References

[1] Crookes W 1879 On radiant matter a lecture delivered to the British Association for the Advancement of Science, at Sheffield, Friday, August 22, 1879 *Am. J. Sci.* **s3-18** 241–62

[2] Onnes H K 1911 *Further Experiments with Liquid Helium* **vol 13** (Royal Netherlands Academy of Arts & Sciences)

[3] London F 1938 The λ-phenomenon of liquid helium and the Bose-Einstein degeneracy *Nature* **141** 643–4

[4] London F 1938 On the Bose-Einstein condensation *Phys. Rev.* **54** 947–54

[5] Bose 1924 Plancks gesetz und lichtquantenhypothese *Z. Phys.* **26** 178–81

[6] Schay G 1924 Zur quantentheorie der einatomigen idealen gase *Z. Phys.* **25** 37–41

[7] Anderson M H, Ensher J R, Matthews M R, Wieman C E and Cornell E A 1995 Observation of Bose-Einstein condensation in a dilute atomic vapor *Science* **269** 198–201

[8] Davis K B, Mewes M O, Andrews M R, van Druten N J, Durfee D S, Kurn D M and Ketterle W 1995 Bose-Einstein condensation in a gas of sodium atoms *Phys. Rev. Lett.* **75** 3969–73

[9] Ketterle W 2002 When atoms behave as waves: Bose-Einstein condensation and the atom laser *Int. J. Mod. Phys.* B **16** 4537–76

[10] Jin D S, Ensher J R, Matthews M R, Wieman C E and Cornell E A 1996 Collective excitations of a Bose-Einstein condensate in a dilute gas *Phys. Rev. Lett.* **77** 420

[11] Denschlag J, Cassettari D and Schmiedmayer J 1999 Guiding neutral atoms with a wire *Phys. Rev. Lett.* **82** 2014

[12] Bradley C C, Sackett C A and Hulet R G 1997 Bose-Einstein condensation of lithium: observation of limited condensate number *Phys. Rev. Lett.* **78** 985

[13] Ohmi T and Machida K 1998 Bose–Einstein condensation with internal degrees of freedom in alkali atom gases *J. Phys. Soc. Japan* **67** 1822–5

[14] Griesmaier A, Werner J, Hensler S, Stuhler J and Pfau T 2005 Bose-Einstein condensation of chromium *Phys. Rev. Lett.* **94** 160401

[15] Aikawa K, Frisch A, Mark M, Baier S, Rietzler A, Grimm R and Ferlaino F 2012 Bose-Einstein condensation of erbium *Phys. Rev. Lett.* **108** 210401S

[16] Tang Y, Burdick N Q, Baumann K and Lev B L 2015 Bose-Einstein condensation of 162Dy and 160Dy *New J. Phys.* **17** 045006

[17] Allen R E 2013 The Higgs bridge *Phys. Scr.* **89** 018001

[18] Higgs P W 1966 Spontaneous symmetry breakdown without massless bosons *Phys. Rev.* **145** 1156

[19] Cohen-Tannoudji C N 1998 Nobel lecture: manipulating atoms with photons *Rev. Mod. Phys.* **70** 707

[20] Phillips W D 1998 Nobel lecture: laser cooling and trapping of neutral atoms *Rev. Mod. Phys.* **70** 721

[21] Masuhara N, Doyle J M, Sandberg J C, Kleppner D, Greytak T J, Hess H F and Kochanski G P 1988 Evaporative cooling of spin-polarized atomic hydrogen *Phys. Rev. Lett.* **61** 935

[22] Andrews D L 1996 Quantum dynamics of simple systems *J. Opt. B: Quantum Semiclass. Opt.* **8** 010

[23] Cornell E 1996 Very cold indeed: the Nanokelvin physics of Bose-Einstein condensation *J. Res. Natl. Inst. Stand. Technol.* **101** 419

[24] Levi B G 2002 From superfluid to insulator: Bose-Einstein condensate undergoes a quantum phase transition *Phys. Today* **55** 18–20

[25] Ketterle W and Miesner H J 1997 Coherence properties of Bose-Einstein condensates and atom lasers *Phys. Rev.* A **56** 3291

[26] Dutton Z, Ginsberg N S, Slowe C and Hau L V 2004 The art of taming light: ultra-slow and stopped light *Europhys. News* **35** 33–9

[27] Bloch I 2005 Ultracold quantum gases in optical lattices *Nat. Phys.* **1** 23–30

[28] Demokritov S O, Demidov V E, Dzyapko O, Melkov G A, Serga A A, Hillebrands B and Slavin A N 2006 Bose-Einstein condensation of quasi-equilibrium magnons at room temperature under pumping *Nature* **443** 430–3

[29] Gross E P 1961 Structure of a quantized vortex in boson systems *Il Nuovo Cimento (1955-1965)* 454–77

[30] Gross E P 1963 Hydrodynamics of a superfluid condensate *J. Math. Phys.* **4** 195–207

[31] Pitaevskii L P 1961 Vortex lines in an imperfect bose gas *Sov. Phys. JETP* **13** 451–4

[32] Radha R and Vinayagam P S 2015 An analytical window into the world of ultracold atoms *Rom. Rep. Phys.* **67** 89

[33] Inouye S, Andrews M R, Stenger J, Miesner H J, Stamper-Kurn D M and Ketterle W 1998 Observation of Feshbach resonances in a Bose-Einstein condensate *Nature* **392** 151–4

[34] Matthews M R, Anderson B P, Haljan P C, Hall D S, Holland M J, Williams J E, Wieman C E and Cornell E A 1999 Watching a superfluid untwist itself: recurrence of Rabi oscillations in a Bose-Einstein condensate *Phys. Rev. Lett.* **83** 3358–61

[35] Madison K W, Chevy F, Wohlleben W and Dalibard J 2000 Vortex formation in a Stirred Bose-Einstein condensate *Phys. Rev. Lett.* **84** 806–9
[36] Liu Y K, Zhang C and Yang S -J 2013 3D skyrmion and knot in two-component Bose-Einstein condensates *Phys. Lett.* A **377** 3300–3
[37] Kartashov Y V, Konotop V V and Zezyulin D A 2014 Bose-Einstein condensates with localized spin-orbit coupling: soliton complexes and spinor dynamics *Phys. Rev.* A **90** 063621
[38] Myatt C J, Burt E A, Ghrist R W, Cornell E A and Wieman C E 1997 Production of two overlapping Bose-Einstein condensates by sympathetic cooling *Phys. Rev. Lett.* **78** 586–9
[39] Strecker K E, Partridge G B, Truscott A G and Hulet R G 2002 Formation and propagation of matter-wave soliton trains *Nature* **417** 150–3
[40] Son D T and Stephanov M A 2002 Domain walls of relative phase in two-component Bose-Einstein condensates *Phys. Rev.* A **65** 063621
[41] Ieda J, Miyakawa T and Wadati M 2004 Exact analysis of soliton dynamics in spinor Bose-Einstein condensates *Phys. Rev. Lett.* **93** 194102
[42] Uchiyama M, Ieda J and Wadati M 2006 Dark solitons in $F = 1$ Spinor Bose–Einstein condensate *J. Phys. Soc. Japan* **75** 064002
[43] Brazhnyi V A and Konotop V V 2005 Stable and unstable vector dark solitons of coupled nonlinear Schrödinger equations: Application to two-component Bose-Einstein condensates *Phys. Rev.* E **72** 026616
[44] Petrov D S 2015 Quantum mechanical stabilization of a collapsing Bose-Bose mixture *Phys. Rev. Lett.* **115** 155302
[45] Regal C A, Greiner M and Jin D S 2004 Observation of resonance condensation of fermionic atom pairs *Phys. Rev. Lett.* **92** 040403

Chapter 2

Analytical methods

Even though one cannot precisely define the concept of integrability of a dynamical system governed by a nonlinear partial differential equation (PDE), one can look for the possible signatures of integrability, namely, Painlevé (P-) property [1–3], Lax pair [4], soliton solutions, etc. A given nonlinear dynamical system governed by a nonlinear PDE is said to admit the P- property if the corresponding solution can be locally given in terms of a Laurent series expansion in the neighbourhood of a movable singular point/manifold. The existence of a Lax pair of a given nonlinear PDE implies that one can somehow linearize the nonlinear dynamical system and subsequently exploit it to generate soliton solutions, thereby consolidating its integrability. The present chapter [5] is completely devoted to the analytical techniques (or methods) employed for the construction of soliton solutions of the given nonlinear PDE. In particular, one focuses on the inverse scattering transform (IST), Darboux transformation approach, Gauge transformation approach, and Hirota method. The present chapter culminates with the development of a new analytical approach entitled 'Modified Gauge transformation approach' to construct dark solitons of scalar and vector NLS-type equations.

2.1 Inverse scattering transform

This section has been reproduced with permission from [5].

The IST is a nonlinear analogue of the Fourier transform, used to solve certain nonlinear PDEs. Given the initial condition $\psi(x, 0)$ and boundary conditions, two linear differential operators L and B are identified to transform a $(1 + 1)$-dimensional nonlinear PDE into two linear equations: a linear eigenvalue problem

$$L\phi = \lambda\phi, \tag{2.1}$$

doi:10.1088/978-0-7503-5447-9ch2

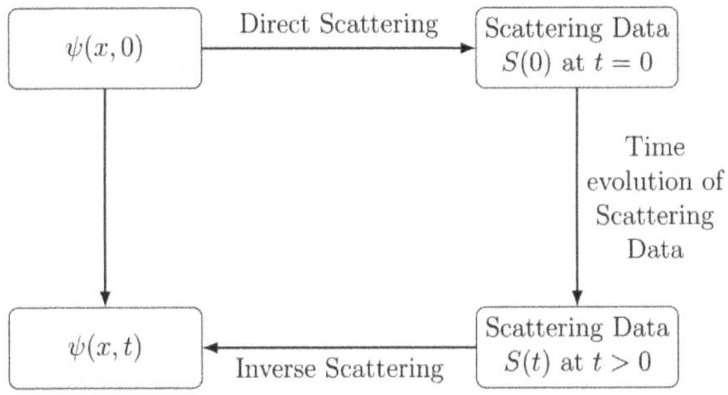

Figure 2.1. Schematic diagram of the inverse scattering transform method for solving nonlinear PDEs.

and a linear time-evolution equation

$$\phi_t = B\phi, \tag{2.2}$$

such that their compatibility condition, $L_t = [B, L]$, recovers the original nonlinear PDE. For a nonlinear dispersive system $\psi_t = K(\psi)$, where $K(\psi)$ is a nonlinear functional of ψ and its spatial derivatives, this linearization enables one to solve the Cauchy initial value problem with the boundary condition $\psi \to 0$ as $|x| \to \infty$. The process of obtaining the solution involves three steps, as outlined in figure 2.1. The IST method comprises the following steps:

(1) **Direct scattering transform analysis:** Using the initial condition $\psi(x, 0)$ as the potential, the linear eigenvalue problem (2.1) is analyzed to obtain the scattering data $S(0)$. For example, for the Korteweg–de Vries (KdV) equation

$$\psi_t + 6\psi\psi_x + \psi_{xxx} = 0, \tag{2.3}$$

the scattering data at $t = 0$ are

$$S(0) = \{\lambda_n(0),\ n = 1, 2, \ldots, N;\ C_n(0);\ R(x, 0),\ -\infty < x < \infty\}, \tag{2.4}$$

where N is the number of bound states with eigenvalues λ_n, $C_n(0)$ is the normalization constant of the bound-state eigenfunctions, and $R(x, 0)$ is the reflection coefficient.

(2) **Time evolution of scattering data:** The asymptotic form of the time-evolution equation (2.2) for the eigenfunctions is used to determine the time evolution of the scattering data $S(t)$.

(3) **Inverse scattering transform analysis:** The Gelfand–Levitan–Marchenko integral equations [6] corresponding to the scattering data $S(t)$ are constructed and solved. The solution typically consists of N localized, exponentially decaying solutions as $t \to \pm\infty$, thereby solving the initial value problem for the nonlinear PDE.

Thus, solving the initial value problem for a nonlinear PDE reduces to solving an integral equation, which can be quite complex and intricate.

2.2 Gauge transformation approach

2.2.1 Scalar nonlinear PDEs

This subsection has been reproduced with permission from [5].

This iterative method generates soliton solutions starting from a seed solution. The method begins with the Zakharov–Shabat (ZS)–Ablowitz–Kaup–Newell–Segur (AKNS) type linear systems [7–10], described by

$$\Phi_x = U\Phi, \tag{2.5a}$$

$$\Phi_t = V\Phi, \tag{2.5b}$$

where

$$U = \begin{pmatrix} -i\lambda & \psi \\ \psi^* & i\lambda \end{pmatrix}, \quad V = \begin{pmatrix} A & B \\ C & D \end{pmatrix}. \tag{2.6}$$

In equation (2.6), λ is the spectral parameter, ψ is the potential or field variable of the nonlinear PDE, and A, B, C, D are undetermined functions of λ, x, t, ψ, and their spatial and temporal derivatives. The compatibility condition $(\Phi_x)_t = (\Phi_t)_x$ yields $U_t - V_x + [U, V] = 0$, which corresponds to the nonlinear PDE. The matrices U and V, known as the Lax pair, encode the linearization of the dynamical system. For scalar nonlinear PDEs, U and V are 2×2 matrices, and the eigenfunction Φ is a 2×1 column vector.

A gauge transformation of the eigenfunction Φ yields the iterated eigenfunction $\Phi^{(1)} = g\Phi$, where $g = g(x, t, \lambda)$ is a matrix function, satisfying

$$\Phi_x^{(1)} = U^{(1)}\Phi, \tag{2.7}$$

$$\Phi_t^{(1)} = V^{(1)}\Phi, \tag{2.8}$$

with

$$U^{(1)} = gUg^{-1} + g_x g^{-1}, \tag{2.9}$$

$$V^{(1)} = gVg^{-1} + g_t g^{-1}. \tag{2.10}$$

For scalar nonlinear PDEs, the gauge function g is a 2×2 matrix. The transformation function g is derived from the solutions of the associated Riemann problems in the complex λ-plane and must be meromorphic. A simple form of g satisfying these criteria is given by

$$g = \left[I + \frac{\lambda_1 - \mu_1}{\lambda - \lambda_1} P(x, t)\right] \begin{pmatrix} 1 & 0 \\ 0 & -1 \end{pmatrix}, \tag{2.11}$$

where λ_1 and μ_1 are arbitrary complex numbers, and P is a projection matrix ($P^2 = P$). The inverse g^{-1} is given by

$$g^{-1} = \begin{pmatrix} 1 & 0 \\ 0 & -1 \end{pmatrix} \left[I - \frac{\lambda_1 - \mu_1}{\lambda - \mu_1} P(x, t)\right]. \tag{2.12}$$

The vanishing of the apparent residues at $\lambda = \lambda_1$ and $\lambda = \mu_1$ imposes the constraints

$$P_x = (I - P)\sigma_2 U(\mu_1)\sigma_2 P - P\sigma_2 U(\lambda_1)\sigma_2(I - P), \tag{2.13}$$

$$P_t = (I - P)\sigma_2 V(\mu_1)\sigma_2 P - P\sigma_2 V(\lambda_1)\sigma_2(I - P), \tag{2.14}$$

where

$$\sigma_2 = \begin{pmatrix} 1 & 0 \\ 0 & -1 \end{pmatrix}. \tag{2.15}$$

To generate a new solution from a seed solution (e.g., the vacuum solution $\psi^{(0)}$ and $\psi^{(0)*}$), the eigenvalue problem is given by

$$\Phi_x^0 = U^0\Phi^0, \quad \Phi_t^0 = V^0\Phi^0, \tag{2.16}$$

where $U^0 = U|_{\text{seed}}$ and $V^0 = V|_{\text{seed}}$. Equations (2.13) and (2.14) are solved using the vacuum eigenfunction Φ^0, with

$$P = \sigma_2 \tilde{P} \sigma_2, \tag{2.17}$$

where

$$\tilde{P} = \frac{M^{(1)}}{\mathrm{Tr}[M^{(1)}]}, \tag{2.18}$$

and $M^{(1)}$ is a 2×2 matrix defined as

$$M^{(1)} = \Phi^0(x, t; \mu_1) \begin{pmatrix} m_1 & 1/n_1 \\ n_1 & 1/m_1 \end{pmatrix} \Phi^0(x, t; \lambda_1)^{-1}, \tag{2.19}$$

with m_1 and n_1 as arbitrary complex constants, and Φ^0 satisfying equation (2.16). Substituting the meromorphic function g with the projection matrix P from equation (2.17) into equation (2.9) yields

$$U^{(1)}(\lambda) = \begin{pmatrix} -\mathrm{i}\lambda & -\psi^{(0)} \\ -\psi^{(0)*} & \mathrm{i}\lambda \end{pmatrix} - 2\mathrm{i}(\lambda_1 - \mu_1) \begin{pmatrix} 0 & \tilde{P}_{12} \\ -\tilde{P}_{21} & 0 \end{pmatrix}. \tag{2.20}$$

Comparing the eigenvalue problems for the vacuum eigenfunction Φ^0 and the transformed eigenfunction $\Phi^{(1)}$ relates the vacuum solution to the new solution, giving [11]

$$\psi^{(1)} = -\psi^{(0)} - 2\mathrm{i}(\lambda_1 - \mu_1)\tilde{P}_{12}, \tag{2.21}$$

$$\psi^{(1)*} = -\psi^{(0)*} + 2\mathrm{i}(\lambda_1 - \mu_1)\tilde{P}_{21}, \tag{2.22}$$

where

$$\tilde{P}_{12} = \frac{M_{12}^{(1)}}{M_{11}^{(1)} + M_{22}^{(1)}}, \qquad \tilde{P}_{21} = \frac{M_{21}^{(1)}}{M_{11}^{(1)} + M_{22}^{(1)}}. \tag{2.23}$$

Assuming $\psi^{(1)}$ and $\psi^{(1)*}$ as the seed solution, the above process can be repeated to generate another solution $\psi^{(2)}$ and $\psi^{(2)*}$. To construct $\psi^{(2)}$ and $\psi^{(2)*}$, one can solve the linear system given by equations (2.7) and (2.8) with matrices U^1 and V^1 associated with $\psi^{(1)}$ and $\psi^{(1)*}$. Alternatively, the solution is

$$\Phi^{(1)} = g\Phi^0. \tag{2.24}$$

The iterated solution, analogous to equations (2.21) and (2.22), is given by

$$\psi^{(2)} = -\psi^{(1)} - 2\mathrm{i}(\lambda_2 - \mu_2)\tilde{P}_{12}^{(1)}, \tag{2.25}$$

$$\psi^{(2)*} = -\psi^{(1)*} + 2\mathrm{i}(\lambda_2 - \mu_2)\tilde{P}_{21}^{(1)}. \tag{2.26}$$

This procedure can be iterated N times to obtain the Nth solution

$$\psi^{(N)} = -\psi^{(N-1)} - 2\mathrm{i}(\lambda_N - \mu_N)\tilde{P}_{12}^{(N-1)}, \tag{2.27}$$

$$\psi^{(N)*} = -\psi^{(N-1)*} + 2\mathrm{i}(\lambda_N - \mu_N)\tilde{P}_{21}^{(N-1)}. \tag{2.28}$$

This iteration is particularly useful for generating multi-soliton solutions, enabling the derivation of an N-soliton solution from the vacuum eigenfunction Φ^0 of the linear system.

2.2.2 Vector nonlinear PDEs

To generate a soliton solution for coupled (vector) nonlinear PDEs, particularly for coupled NLS-type equations, one again begins with the same AKNS type linear system given by equation (2.5), where the eigenfunction Φ is a 3×1 column vector, and U and V being 3×3 matrices of the following form:

$$U = \begin{pmatrix} -\mathrm{i}\lambda & \psi_1 & \psi_2 \\ \psi_1^* & \mathrm{i}\lambda & 0 \\ \psi_2^* & 0 & \mathrm{i}\lambda \end{pmatrix}, \qquad V = \begin{pmatrix} A & B & C \\ D & E & F \\ G & H & J \end{pmatrix}. \tag{2.29}$$

In the above equation (2.29), λ again represents the spectral parameter, ψ_j $(j = 1, 2)$ denotes the potential/field variable of the given nonlinear PDE, and A, B, C, D, E, F, G, H, and J correspond to undetermined functions of λ, x, t, ψ_j, and their spatial or time derivatives. The compatibility condition $(\Phi_x)_t = (\Phi_t)_x$ yields $U_t - V_x + [U, V] = 0$, which is equivalent to the given vector nonlinear PDE.

Now, one can gauge transform the eigenfunction Φ using the transformation function

$$g = \left[I + \frac{\lambda_1 - \mu_1}{\lambda - \lambda_1}P(x, t)\right](\sigma_3), \tag{2.30}$$

where λ_1 and μ_1 are again arbitrary complex numbers and P is an undetermined (3×3) projection matrix with

$$\sigma_3 = \begin{pmatrix} 1 & 0 & 0 \\ 0 & -1 & 0 \\ 0 & 0 & -1 \end{pmatrix}. \tag{2.31}$$

Hence, the inverse gauge function g^{-1} is now given by

$$g^{-1} = (\sigma_3) \cdot \left[I - \frac{\lambda_1 - \mu_1}{\lambda - \mu_1}P(x, t)\right]. \tag{2.32}$$

with the projection matrix P defined as

$$P = \sigma_3 \tilde{P} \sigma_3, \tag{2.33}$$

where

$$\tilde{P} = \frac{M^{(1)}}{\mathrm{Tr}[M^{(1)}]}. \tag{2.34}$$

and $M^{(1)}$ is a (3×3) matrix of the following form

$$M^{(1)} = \Phi^{(0)}(x, t, \bar{\zeta}_1) \cdot \hat{m}^{(1)} \cdot \Phi^{(0)}(x, t, \zeta_1)^{-1}. \tag{2.35}$$

In the above equation, $\hat{m}^{(1)}$ is a (3×3) arbitrary matrix taking the following form

$$\hat{m}^{(1)} = \begin{pmatrix} \mathrm{e}^{2\delta_1}\sqrt{2} & \varepsilon_1^{(1)}\mathrm{e}^{2\mathrm{i}\zeta_1} & \varepsilon_2^{(1)}\mathrm{e}^{2\mathrm{i}\zeta_1} \\ \varepsilon_1^{*(1)}\mathrm{e}^{-2\mathrm{i}\zeta_1} & \mathrm{e}^{-2\delta_1}/\sqrt{2} & 0 \\ \varepsilon_2^{*(1)}\mathrm{e}^{-2\mathrm{i}\zeta_1} & 0 & \mathrm{e}^{-2\delta_1}/\sqrt{2} \end{pmatrix}, \tag{2.36}$$

where, δ_1 and ζ_1 are arbitrary real parameters and $\varepsilon_j^{(1)}(j = 1, 2)$ are coupling parameters.

Hence, one can write down the explicit form of one soliton solution as,

$$\psi_1^{(1)} = \psi_1^{(0)} - 2\mathrm{i}(\lambda_1 - \bar{\mu}_1)P_{12}, \tag{2.37}$$

$$\psi_2^{(1)} = \psi_2^{(0)} - 2\mathrm{i}(\lambda_1 - \bar{\mu}_1)P_{13}, \tag{2.38}$$

where $\psi_i^{(0)}, i = 1, 2$, represents the zero seed solution while $\psi_i^{(1)}$, $i = 1, 2$, denotes the iterated soliton solution of the corresponding vector nonlinear PDE with

$$P_{12} = \frac{M_{12}^{(1)}}{M_{11}^{(1)} + M_{22}^{(1)} + M_{33}^{(1)}} \tag{2.39a}$$

$$P_{13} = \frac{M_{13}^{(1)}}{M_{11}^{(1)} + M_{22}^{(1)} + M_{33}^{(1)}}. \tag{2.39b}$$

The gauge transformation approach can be extended to generate multi-soliton solutions. For example, the general form of Nth order soliton solution can be written as

$$\psi_1^{(N)} = \psi_1^{(N-1)} - 2\mathrm{i}(\lambda_N - \bar{\mu}_N)\frac{\tilde{P}_{12}}{R}, \tag{2.40a}$$

$$\psi_2^{(N)} = \psi_2^{(N-1)} - 2\mathrm{i}(\lambda_N - \bar{\mu}_N)\frac{\tilde{P}_{13}}{R}, \tag{2.40b}$$

where $R = M_{11}^{(1)} + M_{22}^{(1)} + M_{33}^{(1)}$.

2.3 Darboux transformation approach

This section has been reproduced with permission from [5].

In 1882, Darboux [12] investigated the eigenvalue problem of the one-dimensional Schrödinger equation

$$-\Phi_{xx} - \psi(x)\Phi = \lambda\Phi, \tag{2.41}$$

where $\psi(x)$ is the potential function and λ is a constant spectral parameter. He demonstrated that if $\psi(x)$ and $\Phi(x, \lambda)$ satisfy equation (2.41), and $f(x) = \Phi(x, \lambda_0)$ is a solution to equation (2.41) for a fixed $\lambda = \lambda_0$, then, the transformed functions ψ' and Φ' defined by

$$\psi' = \psi + 2(\ln f)_{xx}, \quad \Phi'(x, \lambda) = \Phi_x(x, \lambda) - \frac{f_x}{f}\Phi(x, \lambda), \tag{2.42}$$

satisfy

$$-\Phi'_{xx} - \psi'\Phi' = \lambda\Phi'. \tag{2.43}$$

Since equations (2.41) and (2.43) have the same form, the transformation in equation (2.42), which maps (ψ, Φ) to (ψ', Φ'), is known as the Darboux transformation, valid for $f \neq 0$.

This approach can be applied to the ZS–AKNS linear eigenvalue problem defined by equations (2.5a) and (2.5b), where the eigenfunction Φ is a 2×2 matrix:

$$\Phi(x, t, \lambda) = \begin{pmatrix} \Phi_{11}(x, t, \lambda) & \Phi_{12}(x, t, \lambda) \\ \Phi_{21}(x, t, \lambda) & \Phi_{22}(x, t, \lambda) \end{pmatrix}. \tag{2.44}$$

Applying the Darboux transformation to a known eigenfunction yields the transformed eigenfunction [13]

$$\Phi^{(1)}(x, t, \lambda) = D(x, t, \lambda)\Phi(x, t, \lambda), \tag{2.45}$$

where $D(x, t, \lambda)$, the Darboux matrix, given by $\lambda I - S$. Here, I is the identity matrix, and the matrix S is constructed as

$$S = H\Lambda H^{-1}, \quad \det H \neq 0, \tag{2.46}$$

with $H = (h_1, \ldots, h_N)$, where h_i are column solutions of the linear eigenvalue problem in equations (2.5a) and (2.5b), and $\Lambda = \mathrm{diag}(\lambda_1, \ldots, \lambda_N)$. The transformed eigenfunction $\Phi^{(1)}$ satisfies equations (2.7) and (2.8), with the Darboux matrix D serving as the transformation function.

The transformed potentials/field variables can be obtained from

$$U^{(1)} = DUD^{-1} + D_x D^{-1}, \tag{2.47}$$

$$V^{(1)} = DVD^{-1} + D_t D^{-1}. \tag{2.48}$$

To verify the form of $U^{(1)}$ in equation (2.47), one can substitute equations (2.5a) and (2.7) into the x-derivative of equation (2.45), yielding

$$U^{(1)}\Phi^{(1)} = D_x\Phi + DU\Phi. \tag{2.49}$$

Using equation (2.45), one can confirm equation (2.47). Similarly, the form of $V^{(1)}$ in equation (2.48) can also be verified.

Starting from a seed solution $U^{(0)}$ of a nonlinear PDE, the Darboux transformation generates $U^{(1)}$. This process can be iterated to produce multi-soliton solutions. Although the Darboux transformation resembles a gauge transformation and is applicable to vector nonlinear PDEs [14], it requires nonzero seed solutions, which complicates solving the vacuum linear system in equation (2.16).

2.4 Hirota (direct) method

This section has been reproduced with permission from [5].

Although the IST was the first and a primitive analytical approach developed to address initial value problems for nonlinear PDEs, it relies on complex mathematical techniques, such as solving integral equations, rendering it intricate and highly demanding. Furthermore, this method necessitates prior knowledge of the potential

$\psi(x, t)$ at $t = 0$, specifically the initial condition $\psi(x, 0)$, along with specified boundary conditions. In contrast, the Darboux and gauge transformation methods, while being iterative and algebraic, thus bypassing the need for advanced mathematical tools, require identification of the Lax pair associated with the dynamical system. Given these challenges, an alternative approach for obtaining localized solutions, such as solitons, is essential. The Hirota direct method [15] offers a compelling solution. This technique does not demand prior information about the potential or the Lax pair of the nonlinear PDE. With its elegant algebraic and geometric framework, the Hirota method provides a straightforward and efficient means to derive soliton solutions.

The key features of the Hirota direct method are as follows:

(1) The nonlinear PDE must be converted into a bilinear form through a transformation determined via Painlevé analysis by truncating the Laurent series at the constant level term. Each term in the resulting bilinear equation has a degree of two.

The Hirota bilinear operators are defined as

$$D_t^m D_x^n(G \cdot F) = \left(\frac{\partial}{\partial t} - \frac{\partial}{\partial t'}\right)^m \left(\frac{\partial}{\partial x} - \frac{\partial}{\partial x'}\right)^n G(t, x)F(t', x')|_{t'=t, x'=x}. \tag{2.50}$$

(2) The dependent variables G and F in the bilinear equation are expressed as power series in a small parameter ε:

$$G = \varepsilon g^{(1)} + \varepsilon^3 g^{(3)} + \varepsilon^5 g^{(5)} + \cdots, \tag{2.51}$$

$$F = 1 + \varepsilon^2 f^{(2)} + \varepsilon^4 f^{(4)} + \cdots, \tag{2.52}$$

where $g^{(i)}$ and $f^{(i)}$ are functions of the independent variables, such as t and x.

(3) Substituting the power series for G and F into the bilinear equation and equating coefficients of like powers of ε produces a system of linear PDEs.

(4) Solving this system of linear PDEs enables the construction of soliton solutions.

The effectiveness of the Hirota method hinges on selecting an appropriate transformation for the dependent variables and choosing a suitable power series expansion to linearize the nonlinear PDE.

2.5 Modified gauge transformation method

The gauge transformation technique, as developed by Chau [11], can be utilized to derive bright soliton solutions from a zero seed solution and kink soliton solutions from nonzero (constant) seed solutions for nonlinear PDEs. By selecting a plane wave as the seed solution, this method can be adapted to produce dark soliton solutions for both scalar and coupled NLS-type equations [16].

The use of a nonzero plane wave as the seed solution is essential for generating dark solitons, which are characterized by dips in the density profile $|\psi|^2$. A nonzero background density, provided by the plane wave seed, is necessary to create such dips. The nonzero plane wave solution is substituted into the 3×3 Lax pair matrices U and V, as defined in equation (2.5), yielding the linear system

$$\Phi^0_x = U^0\Phi^0, \tag{2.53a}$$

$$\Phi^0_t = V^0\Phi^0, \tag{2.53b}$$

where

$$\Phi^0 = \Phi|_{\text{nonzero plane wave seed}}. \tag{2.54}$$

A gauge transformation is applied to the eigenfunction Φ^0, defined as $\hat{\Phi} = g\Phi^0$, where g is a 3×3 matrix representing the gauge function, and $\hat{\Phi}$ is the transformed eigenfunction. The Lax representation for the transformed eigenfunction $\hat{\Phi}$ is expressed as

$$\hat{\Phi}_x + U\hat{\Phi} = 0, \tag{2.55a}$$

$$\hat{\Phi}_t + V\hat{\Phi} = 0, \tag{2.55b}$$

with

$$\begin{aligned} U &= gU^0g^{-1} + g_xg^{-1}, \\ V &= gV^0g^{-1} + g_tg^{-1}. \end{aligned}$$

The gauge function $g(x, t)$ has the form specified in equation (2.30). Using the vacuum eigenfunction Φ^0, a projection matrix P is constructed, and equations (2.33)–(2.35) are employed with the matrix $\hat{m}^{(1)}$ defined as

$$\hat{m}^{(1)} = \begin{pmatrix} e^{2\delta_1}\sqrt{2} & \varepsilon_1^{(1)}e^{2i(\zeta_1+\theta_1)} & \varepsilon_2^{(1)}e^{2i(\zeta_1+\theta_2)} \\ \varepsilon_1^{*(1)}e^{-2i(\zeta_1+\theta_1)} & e^{-2\delta_1}/\sqrt{2} & 0 \\ \varepsilon_2^{*(1)}e^{-2i(\zeta_1+\theta_2)} & 0 & e^{-2\delta_1}/\sqrt{2} \end{pmatrix}. \tag{2.56}$$

Here, ζ_1, δ_1, θ_1, and θ_2 are arbitrary functions of (x, t), determined by the dispersion relation of the associated nonlinear PDEs. The structure of the $\hat{m}^{(1)}$ matrix dictates whether the soliton solutions of vector NLS-type equations are bright or dark. The dark soliton solutions for these equations are given by

$$\psi_1^{(1)} = \psi_1^{(0)} - 2i(\lambda_1 - \bar{\mu}_1)P_{12}, \tag{2.57}$$

$$\psi_2^{(1)} = \psi_2^{(0)} - 2i(\lambda_1 - \bar{\mu}_1)P_{13}, \tag{2.58}$$

where $\psi_i^{(0)}$ $(i = 1, 2)$ denotes the seed solution, $\psi_i^{(1)}$ $(i = 1, 2)$ represents the dark soliton solutions of the coupled NLS-type equation, and P_{12}, P_{13} are defined as in equation (2.39a) and (2.39b).

This modified gauge transformation method can be extended to generate multi-dark soliton solutions. For scalar NLS-type equations, the same nonzero plane wave seed solution is used, following the procedure above, but with the $\hat{m}^{(1)}$ matrix chosen as

$$\hat{m}^{(1)} = \begin{pmatrix} m_1 & 1/n_1 \\ n_1 & 1/m_1 \end{pmatrix}, \tag{2.59}$$

where

$$m_1 = \frac{k + i\lambda}{|c_i|}, \qquad n_1 = i. \tag{2.60}$$

This leads to the dark soliton solution for the corresponding integrable scalar NLS-type equation, as given by equation (2.37).

This chapter demonstrates how explicit soliton (or localized) solutions of quasi-one-dimensional Gross–Pitaevskii-type equations, with experimentally realizable potentials and scattering lengths, can be derived using the inverse scattering transform, gauge transformation, Darboux transformation, modified gauge transformation, or Hirota method. As Gross–Pitaevskii equations are variable-coefficient nonlinear PDEs, applying the IST or Hirota method can be complex. Therefore, the gauge or Darboux transformation approach is often employed to obtain bright soliton solutions (or condensates) and dark solitons for Gross–Pitaevskii-type equations.

2.6 Derivation of soliton solutions—illustration

2.6.1 Bright solitons for the focussing NLS equation

Let us consider the focussing NLS equation of the following form

$$\mathrm{i}\psi_t + \frac{1}{2}\psi_{xx} + \sigma|\psi|^2\Psi = 0, \quad \psi = \psi(x, t). \tag{2.61}$$

Lax pair matrices for the above NLS equation are given by

$$U = \begin{bmatrix} -\mathrm{i}\lambda & \psi(x, t) \\ \psi^*(x, t) & i\lambda \end{bmatrix} \tag{2.62}$$

$$V = \begin{bmatrix} -\mathrm{i}\psi\psi^* - 2\mathrm{i}\lambda^2 & \mathrm{i}\psi_x + 2\psi\lambda \\ -\mathrm{i}\psi_x^* + 2\psi^*\lambda & i\psi\psi^* + 2\mathrm{i}\lambda^2 \end{bmatrix} \tag{2.63}$$

To generate bright solitons, one considers a vacuum solution $\psi=0$ to obtain the following vacuum linear systems

$$\Phi_t^{(0)} = \begin{pmatrix} -\mathrm{i}\lambda & 0 \\ 0 & \mathrm{i}\lambda \end{pmatrix} \Phi^0 = U^{(0)}\Phi^{(0)}, \tag{2.64a}$$

$$\Phi_z^{(0)} = \begin{pmatrix} -2\mathrm{i}\lambda^2 & 0 \\ 0 & 2\mathrm{i}\lambda^2 \end{pmatrix} \Phi^{(0)} = V^{(0)}\Phi^{(0)}. \tag{2.64b}$$

Solving the above vacuum linear systems, one obtains the vacuum eigenfunction given by

$$\Phi^{(0)}(x,\, t) = \begin{pmatrix} \mathrm{e}^{\lambda t + 4\lambda^3 \int \alpha_3 \mathrm{d}x + 2\mathrm{i}\lambda^2 \int \alpha_1 dx} & 0 \\ 0 & \mathrm{e}^{-\lambda t - 4\lambda^3 \int \alpha_3 \mathrm{d}x - 2\mathrm{i}\lambda^2 \int \alpha_1 dx} \end{pmatrix}. \tag{2.65}$$

One can now gauge transform the vacuum eigenfunction $\phi^{(0)}(x,\, t)$ by a meromorphic (regular) function g and invoke gauge transformation approach to obtain the Projection matrix

$$P_1(t,\, z) = \begin{pmatrix} \frac{1}{2}\mathrm{sech}(\theta_1)\mathrm{e}^{\theta_1} & -\frac{1}{2}\mathrm{sech}(\theta_1)\mathrm{e}^{\mathrm{i}\xi_1} \\ -\frac{1}{2}\mathrm{sech}(\theta_1)\mathrm{e}^{-\mathrm{i}\xi_1} & \frac{1}{2}\mathrm{sech}(\theta_1)\mathrm{e}^{-\theta_1} \end{pmatrix}, \tag{2.66}$$

where

$$\theta_1 = 2\beta x + 8t\alpha\beta - 2\delta_1, \tag{2.67a}$$

$$\xi_1 = -2\mathrm{i}[\alpha x + 2t(\alpha^2 - \beta^2) + \phi_1] \tag{2.67b}$$

Now, substituting the meromorphic function along with the projection matrix given by (2.66) into (2.9), one obtains

$$U^{(1)}(\lambda) = \begin{pmatrix} -\mathrm{i}\lambda & 0 \\ 0 & \mathrm{i}\lambda \end{pmatrix} - 2\mathrm{i}(\lambda_1 - \mu_1)\begin{pmatrix} 0 & P_{1(12)} \\ -P_{1(21)} & 0 \end{pmatrix}. \tag{2.68}$$

Thus, the bright solitons at $\sigma = 1$ can be written as

$$\psi = 2\beta \, \mathrm{e}^{\mathrm{i}(-2\mathrm{i}[\alpha x + 2t(\alpha^2 - \beta^2) + \phi_1])} \mathrm{sech}(2\beta x + 8t\alpha\beta - 2\delta_1), \tag{2.69}$$

where δ_1, ϕ_1 are arbitrary real constants.

2.6.2 Bright solitons for a vector (two-coupled) NLS equation/focussing Manakov model

The focussing coupled NLS equation is of the following form,

$$\mathrm{i}\psi_{1t} + \psi_{1xx} + 2(g_{11}|\psi_1|^2 + g_{12}|\psi_2|^2)\psi_1 = 0 \tag{2.70a}$$

$$\mathrm{i}\psi_{2t} + \psi_{2xx} + 2(g_{21}|\psi_1|^2 + g_{22}|\psi_2|^2)\psi_2 = 0 \tag{2.70b}$$

In the above equation, g_{11} and g_{22} account for the strengths of self-phase modulation while g_{12} and g_{21} represent the strengths of cross-phase modulation. It has been found that the equation is integrable if either (i) $g_{11} = g_{12} = g_{21} = g_{22}$ or (ii) $g_{11} = g_{21} = -g_{12} = -g_{22}$. The first choice $g_{11} = g_{12} = g_{21} = g_{22} = 1$ represents the focussing Manakov model while the choice $g_{11} = g_{12} = g_{21} = g_{22} = -1$ represents the defocussing Manakov model. Equation (2.70a) admits the linear eigenvalue problem given by equation (2.5) where $\Phi = (\phi_1, \phi_2, \phi_3)^{\mathrm{T}}$ and

$$U = \begin{pmatrix} -\mathrm{i}\lambda & \psi_1 & \psi_2 \\ -\psi_1^* & \mathrm{i}\lambda & 0 \\ -\psi_2^* & 0 & \mathrm{i}\lambda \end{pmatrix},$$

$$V = \begin{pmatrix} -2\mathrm{i}\lambda^2 + \mathrm{i}(|\psi_1|^2 + |\psi_2|^2) & 2\lambda\psi_1 + \psi_{1x} & 2\lambda\psi_2 + \psi_{2x} \\ -2\lambda\psi_1^* + \psi_{1x}^* & 2\mathrm{i}\lambda^2 + \mathrm{i}|\psi_1|^2 & -\mathrm{i}\psi_1^*\psi_2 \\ -2\lambda\psi_2^* + \psi_{2x}^* & -\mathrm{i}\psi_1\psi_2^* & 2\mathrm{i}\lambda^2 + \mathrm{i}|\psi_2|^2 \end{pmatrix}$$

To generate bright solitons, one can substitute $\psi_j = \psi_j^* = 0$, where, $j = 1, 2$ to obtain vacuum linear systems

$$\Phi_x^0 = U^0\Phi^0, \quad U^0 = \begin{pmatrix} -\mathrm{i}\lambda & 0 & 0 \\ 0 & \mathrm{i}\lambda & 0 \\ 0 & 0 & \mathrm{i}\lambda \end{pmatrix} \tag{2.71}$$

$$\Phi_t^0 = V^0\Phi^0, \quad V^0 = \begin{pmatrix} -2\mathrm{i}\lambda^2 & 0 & 0 \\ 0 & 2\mathrm{i}\lambda^2 & 0 \\ 0 & 0 & 2\mathrm{i}\lambda^2 \end{pmatrix} \tag{2.72}$$

Substituting the vacuum linear system given by (2.71) and (2.72) and solving them, one obtains the following:

$$\Phi^0 = \begin{pmatrix} e^{-\int 2i\lambda dx - \int 6i\lambda^2 dt} & 0 & 0 \\ 0 & e^{\int i\lambda dx + \int 3i\lambda^2 dt} & 0 \\ 0 & 0 & e^{\int i\lambda dx + \int 3i\lambda^2 dt} \end{pmatrix} \tag{2.73}$$

Gauge transforming the vacuum eigenfunction $\Phi^{(0)}(x, t)$ by a meromorphic (regular) function g to obtain a new eigenfunction and following the procedure given by (2.30) to (2.39), one can generate soliton solutions as

$$\psi_1^{(1)} = -\psi_1^{(0)} - 2i(\lambda - \mu)P_{12} \tag{2.74}$$

$$\psi_2^{(1)} = -\psi_2^{(0)} - 2i(\lambda - \mu)P_{13} \tag{2.75}$$

Substituting the values of the projection matrix, and taking $\mu = \alpha - i\beta$ and $\lambda = \alpha + i\beta$ in the above equation leads to the bright soliton solutions of the following form.

$$\psi_1^{(1)} = 2i\beta[-e^{-3ix\alpha - 9it\alpha^2 + x\beta + 6t\alpha\beta + 9it\beta^2 + i\delta_1}]\mathrm{sech}[4(x + 6t\alpha)\beta + 2\delta_1]\epsilon_{11} \tag{2.76}$$

$$\psi_2^{(2)} = 2i\beta[-e^{-3ix\alpha - 9it\alpha^2 + x\beta + 6t\alpha\beta + 9it\beta^2 + i\delta_1}]\mathrm{sech}[4(x + 6t\alpha)\beta + 2\delta_1]\epsilon_{21} \tag{2.77}$$

subject to $|\varepsilon_1^{(j)}|^2 + |\varepsilon_2^{(j)}|^2 = 1, (j = 1, 2)$ where $\varepsilon_{1,2}$ represent coupling parameters.

2.6.3 Dark soliton solutions for the defocussing nonlinear Schrödinger equation

The defocussing NLS equation takes the following form

$$i\psi_t + \frac{1}{2}\psi_{xx} + \sigma|\psi|^2\psi = 0, \quad \text{where} \quad \psi = \psi(x, t) \quad \text{and} \quad \sigma = -1. \tag{2.78}$$

Lax pair matrices for the above NLSE are given by

$$U = \begin{bmatrix} -i\lambda & \psi(x, t) \\ -\psi^*(x, t) & i\lambda \end{bmatrix} \tag{2.79}$$

$$V = \begin{bmatrix} i\sigma\psi(x, t)\psi^*(x, t) - 2i\lambda^2 & \psi(x, t)_x + 2i\lambda\psi(x, t) \\ \psi^*(x, t)_x - 2i\lambda\psi^*(x, t) & i\sigma\psi(x, t)\psi^*(x, t) + 2i\lambda^2 \end{bmatrix} \tag{2.80}$$

To generate dark solitons, a seed solution is considered in the form $\psi^{(0)} = a_1 e^{i(k_1 x + w_1 t)}$ and $\psi^{(0)*} = a_2 e^{-i(k_1 x + w_1 t)}$, where $w_1 = 2a_1 a_2 + k_1^2$. Solving the corresponding vacuum linear systems yields the vacuum eigenfunction given by

$$\Phi^{(0)} = \begin{pmatrix} \Phi_{11}^{(0)} & 0 \\ 0 & \Phi_{22}^{(0)} \end{pmatrix} \tag{2.81}$$

where

$$\Phi_{11}^{(0)} = \exp[-\sqrt{a_1a_2 - \lambda^2}(2\lambda t + x) + ia_1a_2t](c_1e^{2\sqrt{a_1a_2 - \lambda^2}(2\lambda t+x)} + c_2),$$
$$\Phi_{22}^{(0)} = \exp[-\sqrt{a_1a_2 - \lambda^2}(2\lambda t + x) + ia_1a_2t](c_4e^{2\sqrt{a_1a_2 - \lambda^2}(2\lambda t+x)} + c_3),$$

One can gauge transform the vacuum eigenfunction $\Phi^{(0)}(x, t)$ by a meromorphic (regular) function g to obtain a new eigenfunction $\Phi^{(1)}(x, t)$ and follow the procedure given by equations (2.55) to (2.59) in section 2.5 to construct the dark solitons of the following form

$$\psi = \beta_1\sqrt{\frac{-2a1}{a2}}\tanh[\alpha_1\beta_1x - 2\delta_1]e^{-i(2a_1\beta_1^2(t-2\chi_1)+\phi_0)}, \tag{2.82}$$

where α_1, β_1, δ_1, χ_1 and ϕ_0 are arbitrary real constants.

2.7 Defocussing Manakov model

The defocussing Manakov model admits the linear eigenvalue problem given by equation (2.5) where $\Phi = (\Phi_1, \Phi_2, \Phi_3)^T$ and U, V matrices are given by

$$U = \begin{pmatrix} -i\lambda & \psi_1 & \psi_2 \\ \psi_1^* & i\lambda & 0 \\ \psi_2^* & 0 & i\lambda \end{pmatrix},$$
$$V = \begin{pmatrix} -2i\lambda^2 + i(|\psi_1|^2 + |\psi_2|^2) & 2\lambda\psi_1 + \psi_{1x} & 2\lambda\psi_2 + \psi_{2x} \\ 2\lambda\psi_1^* - \psi_{1x}^* & 2i\lambda^2 - i|\psi_1|^2 & -i\psi_1^*\psi_2 \\ 2\lambda\psi_2^* - \psi_{2x}^* & -i\psi_1\psi_2^* & 2i\lambda^2 - i|\psi_2|^2 \end{pmatrix} \tag{2.83}$$

To generate dark solitons, one can substitute the seed solution as $\psi_i^0 = c_ie^{[i(a_ix-(\frac{a_i^2}{2}+\Sigma_{l=1}^2\sigma_lc_l^2))]}$ and solve the respective linear eigenvalue problem to arrive at

$$\Phi^{(0)} = \begin{pmatrix} \Phi_{11}^{(0)} & 0 & 0 \\ 0 & \Phi_{22}^{(0)} & 0 \\ 0 & 0 & \Phi_{33}^{(0)} \end{pmatrix} e^{i\left[\mu x+\left(\zeta\mu-\frac{1}{2}\zeta^2+\frac{1}{2}\mu^2+\sigma_lc_l^2\right)t\right]} \tag{2.84}$$

where

$$\Phi_{11}^{(0)} = e^{-2i\lambda x-6i\lambda^2t}, \quad \Phi_{22}^{(0)} = e^{i\lambda x+3i\lambda^2t}, \quad \Phi_{33}^{(0)} = e^{i\lambda x+3i\lambda^2t},$$

One now gauge transforms the eigenfunction Φ^0 such that $\hat{\Phi} = g\Phi^0$ where g is a gauge function represented by a 3×3 matrix while $\hat{\Phi}$ is an iterated eigenfunction. Again following the procedure given by equations (2.30) to (2.39a), one obtains

$$\psi_1^{(1)} = -\psi_1^0 - 2i(\lambda - \mu)P_{12} \tag{2.85}$$

$$\psi_2^{(1)} = -\psi_2^0 - 2i(\lambda - \mu)P_{13} \tag{2.86}$$

where, P_{12} and P_{12} are the elements of the projection matrix given by

$$\tilde{P} = \frac{M^{(1)}}{\mathrm{Tr}[M^{(1)}]}, \tag{2.87}$$

and

$$M^{(1)} = \Phi^{(0)}(x, t, \bar{\zeta}_1) \cdot \hat{m}^{(1)} \cdot \Phi^{(0)}(x, t, \zeta_1)^{-1}, \; P = \sigma_3 \tilde{P} \sigma_3 \tag{2.88}$$

where $\Phi^{(0)}$ is given by equation (2.84). In the above equation, $\hat{m}^{(1)}$ is a 3×3 arbitrary matrix of the form

$$\hat{m}^{(1)} = \begin{pmatrix} e^{2\delta_1}\sqrt{2} & \varepsilon_1^{(1)} e^{2i(\chi_1 + \xi_1)} & \varepsilon_2^{(1)} e^{2i(\chi_1 + \xi_2)} \\ \varepsilon_1^{*(1)} e^{-2i(\chi_1 + \xi_1)} & e^{-2\delta_1}/\sqrt{2} & 0 \\ \varepsilon_2^{*(1)} e^{-2i(\chi_1 + \xi_2)} & 0 & e^{-2\delta_1}/\sqrt{2} \end{pmatrix}, \tag{2.89}$$

where $\chi_1, \delta_1, \xi_1, \xi_2$, are arbitrary functions of (x, t) and their choice is governed by the dispersion relation of the associated nonlinear PDEs while $\varepsilon_1^{(1)}$ and $\varepsilon_2^{(1)}$ are the coupling parameters. The dark solitons can now be written as

$$\psi_1^{(1)} = a_1 \varepsilon_1^{(1)} [\alpha_1 + i\beta_1 \tanh(\theta_1)] e^{i(-\xi_1)}, \tag{2.90a}$$

$$\psi_2^{(1)} = a_2 \varepsilon_2^{(1)} [\alpha_1 + i\beta_1 \tanh(\theta_1)] e^{i(-\xi_1)}, \tag{2.90b}$$

where

$$\begin{aligned} \theta_1 &= \sigma(x - vt) - 2\delta_1, \\ \xi_1 &= 2i(a_1^2 - a_2^2)t, \end{aligned}$$

subject to $|\varepsilon_1^{(j)}|^2 + |\varepsilon_2^{(j)}|^2 = 1$, $(j = 1, 2)$ where $\varepsilon_{1,2}$ represent coupling parameters. $\alpha_1 = \cos(\phi)$ and $\beta_1 = \sin(\phi)$ while $\phi = v/2\sqrt{a_1^2 + a_2^2}$ and μ is the hidden spectral parameter. It should be pointed out that one can extend the Gauge transformation approach to generate multi-dark solitons.

2.8 Summary and future challenges

In this chapter, analytical methods/algorithms to derive exact soliton solutions (bright and dark) have been extensively reviewed. In particular, the derivation of bright and dark solitons employing gauge transformation approach has been demonstrated in an attempt to draw more researchers into the analytical domain of ultracold atoms. It should be pointed out that the IST and gauge/Darboux transformation approaches can be employed either for solving the initial value problem or generating soliton solutions of the associated nonlinear PDEs in (1 + 1)-dimensions, respectively, provided the associated linear eigenvalue problem exists. Even though Hirota method can be employed directly to nonlinear PDEs whose Lax pair is not known, it suffers from the following limitations. It heavily depends on the exponential ansatz and it does not always work. It suits only for generating soliton solutions and struggles in generating other excitations like rogue waves. The expression for the multi-soliton solution becomes extremely messy. In addition, not all integrable nonlinear PDEs can be bilinearized. This scenario warrants the identification of a unified approach to analytically solve nonlinear PDEs and it would be a huge challenge if researchers in future can come up with such an approach. To help the readers implement either gauge/Darboux transformation approach, some of the open problems are listed below.

2.9 Problems

Exercise 2.1

1. Employ gauge transformation and generate soliton solutions of the following coherently coupled NLS equation given by

   Case 1: If $q_{-1} = q_1$, i.e., $\mathbf{Q} = \begin{pmatrix} q_1 & q_0 \\ q_0 & q_{-1} \end{pmatrix}$.

$$iq_{1x} + q_{1tt} + 2(|q_1|^2 + |q_2|^2)q_1 + 2q_2^2 q_1^* = 0$$
$$iq_{2x} + q_{2tt} + 2(|q_1|^2 + |q_2|^2)q_2 + 2q_1^2 q_2^* = 0$$

   Case 2: If $q_{-1} = -q_1$, i.e., $\mathbf{Q} = \begin{pmatrix} q_1 & q_0 \\ q_0 & q_{-1} \end{pmatrix}$.

$$iq_{1x} + q_{1tt} + 2(|q_1|^2 + |q_2|^2)q_1 - 2q_2^2 q_1^* = 0$$
$$iq_{2x} + q_{2tt} + 2(|q_1|^2 + |q_2|^2)q_2 - 2q_1^2 q_2^* = 0$$

   The above equation admits the following Lax pair

$$\Phi_x = (\lambda\mathbf{J} + \mathbf{U}_0)\Phi \equiv \mathbf{U}\Phi$$
$$\Phi_t = (2\lambda^2\mathbf{J} + \lambda V_1 + V_0\mathbf{U}_0)\Phi \equiv \mathbf{V}\Phi$$

   where $\Phi = (\phi_1, \phi_2, \phi_3, \phi_4)^{\mathrm{T}}$

$$J = i\begin{pmatrix} -\mathbf{I}_{2\times 2} & 0 \\ 0 & \mathbf{I}_{2\times 2} \end{pmatrix}, \quad U_0 = \begin{pmatrix} 0 & \mathbf{Q} \\ -\mathbf{Q}^\dagger & 0 \end{pmatrix},$$

$$V_1 = 2\begin{pmatrix} 0 & \mathbf{Q} \\ -\mathbf{Q}^\dagger & 0 \end{pmatrix}, \quad V_0 = \begin{pmatrix} \mathbf{Q}\mathbf{Q}^\dagger & \mathbf{Q}_t \\ -\mathbf{Q}_t^\dagger & -\mathbf{Q}^\dagger\mathbf{Q} \end{pmatrix},$$

2. Generate the soliton solutions by employing Darboux transformation of the following coupled NLS-type equation of the following form

$$
\begin{aligned}
&\mathrm{i}q_t(x,t) + q_{xx}(x,t) + 2q(x,t)(a|q(x,t)|^2) + a|r(x,t)|^2\\
&+ bq(x,t)r^*(x,t) + bq^*(x,t)r(x,t) = 0\\
&\mathrm{i}r_t(x,t) + r_{xx}(x,t) + 2r(x,t)(a|q(x,t)|^2) + a|r(x,t)|^2\\
&+ bq(x,t)r^*(x,t) + bq^*(x,t)r(x,t) = 0
\end{aligned}
$$

The above equation admits the following Lax pair given by

$$
\begin{aligned}
\Phi_x &= U\Phi\\
\Phi_t &= V\Phi
\end{aligned}
$$

in which $\Phi = \Phi(x, t; \lambda)$ is a matrix function of complex spectral parameter λ, and

$$
\begin{aligned}
U &= \mathrm{i}\lambda\Lambda + Q,\\
V &= -2\mathrm{i}\lambda^2\Lambda - 2\lambda Q - \mathrm{i}(Q^2 + Q_x)\Lambda
\end{aligned}
$$

$$
Q \equiv Q(x,t) = \begin{pmatrix} 0 & 0 & q(x,t)\\ 0 & 0 & q^*(x,-t)\\ -aq^*(x,t) - bq(x,-t) & -aq(x,-t) - b^*q^*(x,t) & 0 \end{pmatrix}.
$$

3. Employing gauge transformation approach, construct the soliton solutions of the following (2 + 1)-dimensional Hirota–Maxwell–Bloch system

$$
\begin{aligned}
\mathrm{i}q_t(x,y,t) + \epsilon_1 q_{xy}(x,y,t) + \mathrm{i}\epsilon_2 q_{xxy}(x,y,t) + \mathrm{i}[w(x,y,t)q(x,y,t)]_x&\\
- v(x,y,t)q(x,y,t) - 2\mathrm{i}p(x,y,t) &= 0\\
v_x(x,y,t) + 2\epsilon_1\sigma[q_y(x,y,t)q^*(x,-y,t) - q(x,y,t)q_y^*(x,-y,t)]&\\
+ 2\mathrm{i}\epsilon_2\sigma[q(x,y,t)q^*(x,-y,t) + q^*(x,y,t)q_{xy}(x,-y,t)] &= 0\\
\omega_x(x,y,t) - 2\epsilon_2\sigma[q_y(x,y,t)q^*(x,-y,t)] &= 0\\
p_x(x,y,t) - 2\mathrm{i}\omega p(x,y,t) - 2q(x,y,t)\eta(x,y,t) &= 0\\
\eta_x(x,y,t) + \sigma p(x,y,t)q^*(x,-y,t) + \delta p^*(x,-y,t)q(x,y,t) &= 0
\end{aligned}
$$

The above (2 + 1)-dimensional nonlinear PDE admits the following linear eigenvalue problem

$$
\begin{aligned}
&\varphi_x(x,y,t;\lambda) = A(x,y,t;\lambda)\varphi(x,y,t;\lambda),\\
&\varphi_t(x,y,t;\lambda) = (2\epsilon_1\lambda + 4\epsilon_2\lambda^2)\varphi_y(x,y,t;\lambda) + B(x,y,t;\lambda)\varphi(x,y,t;\lambda)
\end{aligned}
$$

The matrix $A(x, y, t; \lambda)$ has the form:

$$
A = -\mathrm{i}\lambda\sigma_3 + A_0
$$

$$\sigma_3 = \begin{pmatrix} 1 & 0 \\ 0 & -1 \end{pmatrix}, \quad A_0 = \begin{pmatrix} 0 & q(x, y, t) \\ -\sigma q^*(x, -y, t) & 0 \end{pmatrix}.$$

$$B = \lambda B_1 + B_0 + \frac{\mathrm{i}}{\lambda + \omega} B_{-1},$$

$$B_1 = \mathrm{i}\omega\sigma_3 + 2\mathrm{i}\epsilon_2\sigma_3 A_{0y} = \begin{pmatrix} \mathrm{i}\omega(x, y, t) & 2\mathrm{i}\epsilon_2 q_y(x, y, t) \\ -2\mathrm{i}\epsilon_1\sigma q_y^*(x, y, t) & -\mathrm{i}\omega(x, y, t) \end{pmatrix},$$

$$B_0 = \begin{pmatrix} -\frac{\mathrm{i}}{2}v & \mathrm{i}\epsilon_1 q_y - \epsilon_2 q_{xy} - wq \\ -\mathrm{i}\epsilon_1\sigma q_y^* - \epsilon_2\sigma q_{xy}^* + \sigma w q^* & \frac{\mathrm{i}}{2}v \end{pmatrix},$$

$$B_{-1} = \begin{pmatrix} \eta(x, y, t) & -p(x, y, t) \\ -2\delta p^*(x, -y, t) & -\eta(x, y, t) \end{pmatrix}.$$

References

[1] Weiss J, Tabor M and Carnevale G 1983 The Painlevé property for partial differential equations *J. Math. Phys.* **24** 522

[2] Lakshmanan M and Sahadevan R 1993 Painlevé analysis, lie symmetries, and integrability of coupled nonlinear oscillators of polynomial type *Phys. Rep.* **224** 1

[3] Gramaticos B and Ramani A 1997 *Integrability of Nonlinear Systems* (Springer)

[4] Lax P D 1968 Integrals of nonlinear equations of evolution *Phys. Rep.* **21** 467

[5] Radha R and Vinayagam P S 2015 An analytical window into the world of ultracold atoms *Rom. Rep. Phys.* **67** 89

[6] Lakshmanan M and Rajasekar S 2015 *Nonlinear Dynamics: Integrability, Chaos and Patterns* (Berlin: Springer)

[7] Zakharov V E and Shabat A B 1973 Interaction between solitons in a stable medium *Sov. Phys. JETP* **37** 823

[8] Ablowitz M J, Kaup D J, Newell A C and Segur H 1973 The inverse scattering transform-Fourier analysis for nonlinear problems *Appl. Math.* **53** 249–315

[9] Ablowitz M J, Kaup D J, Newell A C and Segur H 1973 Method for solving the sine-Gordon equation *Phys. Rev. Lett.* **30** 1262

[10] Ablowitz M J, Kaup D J, Newell A C and Segur H 1973 Nonlinear-evolution equations of physical significance *Phys. Rev. Lett.* **31** 125

[11] Chau L -L, Shaw J C and Yen H C 1991 An alternative explicit construction of dimensions *J. Math. Phys.* **32** 1737–43

[12] Darboux G 1882 Sur une proposition relative auxé quations liné aires *C. R. Hebd. Seances Acad. Sci.* **94** 1456–9

[13] Matveev V B and Salle M 1991 *Darboux Transformations and Solitons* (Springer)

[14] Degasperis A and Lombardo S 2007 Multicomponent integrable wave equations: I. darboux-dressing transformation *J. Math. Phys.* **40** 961

[15] Hirota R 2004 *The Direct Method in Soliton Theory* (Cambridge University Press)

[16] Vinayagam P S, Radha R, Vyas V M and Porsezian K 2015 Generalized gauge transformation approach to construct dark solitons of coupled nonlinear Schrödinger type equations *Rom. Rep. Phys.* **67** 737

Chapter 3

Numerical techniques

The study of Bose–Einstein condensates (BECs) within the mean-field approximation is fundamentally centred around solving the Gross–Pitaevskii (GP) equation, a nonlinear partial differential equation (PDE) describing the evolution of the condensate wavefunction in three-dimensional space. The GP equation accounts for the interplay between kinetic energy, external trapping potentials, and interatomic interactions modelled by a mean-field nonlinearity. While the equation is analytically tractable in certain idealized scenarios, such as specific combinations of harmonic trapping potentials and scattering lengths in quasi-one-dimensional geometries, it lacks general analytical solutions in higher dimensions or for arbitrary potentials. Consequently, numerical methods play a crucial role in determining the ground-state and dynamical properties of BECs beyond these restricted cases. This chapter systematically presents a few key numerical techniques employed to solve the GP type equations, including their theoretical foundations, computational implementations, and applications to realistic physical systems.

3.1 The Gross–Pitaevskii (GP) equation

The time-dependent nonlinear GP equation in the mean-field approximation describing the evolution of the condensate wavefunction $\Psi(\mathbf{r}, \tau)$ at position $\mathbf{r}$ and time τ at absolute zero temperature is given by

$$\mathrm{i}\hbar\frac{\partial\Psi(\mathbf{r};\tau)}{\partial\tau} = \left[-\frac{\hbar^2}{2m}\nabla^2 + V(\mathbf{r}) + gN\,|\Psi(\mathbf{r};\tau)|^2\right]\Psi(\mathbf{r};\tau) = 0, \tag{3.1}$$

where m is the atomic mass and N the number of atoms in the condensate, $g = 4\pi\hbar^2 a/m$ the strength of interatomic interaction, with a the atomic scattering length, $V(\mathbf{r})$ corresponds to the trapping potential with $\mathbf{r} \in (x, y, z)$, and ∇^2 is the Laplacian operator, given by

doi:10.1088/978-0-7503-5447-9ch3

$$\nabla^2 = \frac{\partial^2}{\partial x^2} + \frac{\partial^2}{\partial y^2} + \frac{\partial^2}{\partial z^2}.$$

The condensate wavefunction is subjected to the following normalization condition

$$\int |\Psi(\mathbf{r};\,\tau)|^2\,\mathrm{d}\mathbf{r} = 1. \tag{3.2}$$

The above GP equation (3.1) is a nonlinear PDE in three space and one time dimensions, and in general, it does not possesses analytical or closed-form solutions.

3.1.1 Dimensionless Gross–Pitaevskii equation

For numerical convenience, it is advantageous to express the GP equation (3.1) in dimensionless form. Consider a BEC confined in an anisotropic harmonic potential

$$V(\mathbf{r}) = \frac{1}{2}m\omega^2(\nu^2x^2 + \kappa^2y^2 + \lambda^2z^2), \tag{3.3}$$

where the trap frequencies along each axis are $\omega_x = \nu\omega$, $\omega_y = \kappa\omega$, and $\omega_z = \lambda\omega$ for the x-, y-, and z-directions, respectively.

Defining the dimensionless quantities $x' = \ell^{-1}x$, $y' = \ell^{-1}y$, $z' = \ell^{-1}z$, $g' = \ell^{-1}g$, and $\Psi'(x',\,y',\,z';\,t') = \ell^{3/2}\Psi(\mathbf{r};\,\tau)$ with $\ell = \sqrt{\hbar/(m\omega)}$ and $t' = \omega t$, the GP equation can be expressed in dimensionless form as:

$$\mathrm{i}\frac{\partial\Psi(x,\,y,\,z;\,t)}{\partial t'} = \left[-\frac{1}{2}\nabla^2 + V(x,\,y,\,z) + gN\,|\Psi(x,\,y,\,z;\,t)|^2\right]\Psi(x,\,y,\,z;\,t) = 0, \tag{3.4}$$

where the primes have been omitted for notational convenience. Here, $\ell = \sqrt{\hbar/(m\omega)}$ represents the oscillator length, which establishes the unit of length, while ω^{-1} serves as the unit for time.

3.1.2 Bose–Einstein condensate in a spherically symmetric trap: effective one-dimensional Gross–Pitaevskii equation

The GP equation (3.1) can be reduced to an effective one-dimensional form when describing a BEC confined in a spherically symmetric harmonic potential $V(\mathbf{r}) = \frac{1}{2}m\omega^2r^2$, where $r = \sqrt{x^2 + y^2 + z^2}$ represents the radial coordinate. This particular trapping potential configuration corresponds to the case where the parameters satisfy $\nu = \kappa = \lambda = 1$ in equation (3.3). Through partial-wave projection, the full wavefunction $\Psi(\mathbf{r};\,\tau)$ simplifies to its radial component $\psi(r,\,\tau)$, yielding the reduced form $\Psi(\mathbf{r};\,\tau) = \psi(r,\,\tau)/r$. After a transformation of variables to dimensionless quantities defined by $r = \sqrt{2}\,\tilde{r}/\ell$, $t = \tau\omega$ and $\phi(r;\,t) \equiv \varphi(r;\,t)/r = \psi(\tilde{r},\,\tau)[\ell^3/\sqrt{2\sqrt{2}}\,]$, the GP equation (3.1) in this case becomes

$$\mathrm{i}\frac{\partial}{\partial t}\varphi(r;\,t) = \left[-\frac{\partial^2}{\partial r^2} + \frac{r^2}{4} + g\left|\frac{\varphi(r;\,t)}{r}\right|^2\right]\varphi(r;\,t), \tag{3.5}$$

where $g = 8\sqrt{2}\pi Na/\mathfrak{l}$. The transformation of the wavefunction from ψ to $\varphi = r\psi$ is primarily motivated by analytical convenience and offers several advantages. First, this substitution eliminates the first-order derivative $\partial/\partial r$ in the differential equation (3.5), yielding a simplified form [1]. Second, while the original wavefunction ψ remains constant (with $\partial\psi/\partial r = 0$) at the origin $r = 0$, the transformed variable φ satisfies $\varphi(0, t) = 0$. Consequently, when solving the differential equation (3.5), one may impose straightforward boundary conditions: $\varphi \to 0$ as $r \to 0$ or $r \to \infty$. In contrast, the boundary conditions for ψ will be mixed, requiring ψ to vanish at infinity while its first spatial derivative vanishes at the origin. In this case, the normalization condition, equation (3.2), becomes

$$4\pi \int_0^\infty dr|\varphi(r;\, t)|^2 = 1. \tag{3.6}$$

However, equation (3.5) does not represent the only possible dimensionless form of the GP equation in this context. Various alternative dimensionless formulations have been derived and employed by different researchers. For example, using the transformations $r = \tilde{r}/\mathfrak{l}$, $t = \tau\omega$, and $\phi(r;\, t) \equiv \varphi(r;\, t)/r = \psi(\tilde{r}, \tau)\mathfrak{l}^{3/2}$, the GP equation (3.1) becomes

$$\mathrm{i}\frac{\partial}{\partial t}\varphi(r;\, t) = \left[-\frac{1}{2}\frac{\partial^2}{\partial r^2} + \frac{1}{2}r^2 + g\left|\frac{\varphi(r;\, t)}{r}\right|^2\right]\varphi(r;\, t), \tag{3.7}$$

where $g = 4\pi Na/\mathfrak{l}$ with normalization (3.6). Finally, using the transformations $r = \tilde{r}/\mathfrak{l}$, $t = \tau\omega/2$, and $\phi(r;\, t) \equiv \varphi(r;\, t)/r = \psi(\tilde{r}, \tau)\mathfrak{l}^{3/2}$, the GP equation (3.1) becomes

$$\mathrm{i}\frac{\partial}{\partial t}\varphi(r;\, t) = \left[-\frac{\partial^2}{\partial r^2} + r^2 + g\left|\frac{\varphi(r;\, t)}{r}\right|^2\right]\varphi(r;\, t), \tag{3.8}$$

where $g = 8\pi Na/\mathfrak{l}$ with normalization (3.6). These three sets of dimensionless GP equations have been widely used in the literature and will be considered here. Equations (3.5), (3.7), and (3.8) allow stationary solutions $\varphi(r;\, t) \equiv \varphi(r)\exp(-\mathrm{i}\mu t)$ where μ is the chemical potential. The boundary conditions for the solution of these equations are $\varphi(0, t) = 0$ and $\lim_{r\to\infty} \varphi(r, t) = 0$ [2].

3.1.3 Bose–Einstein condensate in an axially symmetric trap: effective two-dimensional Gross–Pitaevskii equation

For a BEC confined in an axially symmetric trap (where $\nu = \kappa \neq \lambda$ in equation (3.3)), the dimensionless GP equation (3.4) can be simplified by adopting cylindrical coordinates $\mathbf{r} \equiv (\rho, z)$. Here, $\rho = \sqrt{x^2 + y^2}$ represents the radial coordinate, and z corresponds to the axial coordinate. The resulting equation is expressed as:

$$\mathrm{i}\frac{\partial}{\partial t}\varphi(\rho, z;\, t) = \left[-\frac{1}{2}\frac{\partial^2}{\partial \rho^2} - \frac{1}{2\rho}\frac{\partial}{\partial \rho} - \frac{1}{2}\frac{\partial^2}{\partial z^2} + \frac{1}{2}(\kappa^2\rho^2 + \lambda^2 z^2) + g\,|\varphi(\rho, z;\, t)|^2\right]\varphi(\rho, z;\, t), \tag{3.9}$$

where $g = 4\pi aN/\mathfrak{l}$ and the function $\varphi(\rho, z;\, t)$ is subject to the following normalization condition

$$2\pi \int_0^\infty \rho \, d\rho \int_{-\infty}^\infty |\varphi(\rho, z; t)|^2 \, dz = 1. \tag{3.10}$$

The appropriate boundary conditions in this case are [3]

$$\lim_{z \to \pm\infty} \varphi(\rho, z; t) = 0, \quad \lim_{\rho \to \infty} \varphi(\rho, z; t) = 0, \quad \text{and} \quad \left. \frac{\partial \varphi(\rho, z; t)}{\partial \rho} \right|_{\rho=0} = 0.$$

3.1.4 Bose–Einstein condensate in a strong transverse and axial confinements: quasi-one- and two-dimensional Gross–Pitaevskii equations

It is also instructive to examine scenarios where the BEC is subject to (i) strong axial confinement,and (ii) strong transverse confinement. Under these conditions, the GP equation (3.1) can be reduced to quasi-two-dimensional and quasi-one-dimensional forms, respectively.

3.1.4.1 Strong axial trap: anisotropic GP equation in quasi-2D

In the case of a disk-shaped trap, which represents an anisotropic trap in two dimensions with strong axial confinement, equation (3.4) reduces to a two-dimensional form. This reduction is achieved by assuming that the system remains confined to the ground state in the axial direction. Under this condition, the wavefunction of equation (3.4) can be expressed as

$$\Psi(x, y, z; t) = \tilde{\varphi}(x, y; t)\phi_0(z) \exp\left(-\frac{i\lambda}{2}t\right), \tag{3.11}$$

where

$$\phi_0(z) = \left(\frac{\lambda}{2\pi}\right)^{1/4} \exp\left(-\frac{\lambda}{4}z^2\right)$$

is the ground state wavefunction in the z-direction. Substituting the above ansatz into equation (3.4), multiplying by $\phi_0(z)$, integrating over z, and simplifying (by dropping the tilde notation for φ and setting $\nu = 1$) yields

$$i\frac{\partial}{\partial t}\varphi(x, y; t) = \left[-\frac{1}{2}\frac{\partial^2}{\partial x^2} - \frac{1}{2}\frac{\partial^2}{\partial y^2} + \frac{1}{2}(x^2 + \kappa y^2) + g\,|\varphi(x, y; t)|^2\right]\varphi(x, y; t) = 0, \tag{3.12}$$

where $g = \frac{4aN\sqrt{2\pi\lambda}}{l}$, and the normalization condition is given by

$$\int_{-\infty}^\infty dx \int_{-\infty}^\infty dy \, |\varphi(x, y; t)|^2 = 1. \tag{3.13}$$

3.1.4.2 Strong transverse trap: GP equation in quasi-1D

In the case of an elongated cigar-shaped trap which is essentially an axially symmetric trap with strong transverse confinement, equation (3.4) reduces to a quasi-one-dimensional form. This reduction is achieved by assuming that the system remains confined to the ground state in the transverse direction with the wave-function of the following form

$$\Psi(x, y, z; t) = \tilde{\varphi}(x; t)\phi_0(y)\, \phi_0(z)\exp\left[-\frac{\mathrm{i}}{2}(\lambda + \kappa)t\right], \tag{3.14}$$

where the ground state wavefunctions in the y- and z-directions are given by

$$\phi_0(y) = \left(\frac{\kappa}{\pi}\right)^{1/4} \exp\left(-\frac{\kappa}{2}y^2\right) \quad \text{and} \quad \phi_0(z) = \left(\frac{\lambda}{\pi}\right)^{1/4} \exp\left(-\frac{\lambda}{2}z^2\right),$$

respectively. Through substitution of this ansatz into equation (3.4) followed by multiplication with $\phi_0(y)\phi_0(z)$, integration over the transverse coordinates y and z, removal of the tilde notation from φ, and setting $\nu = 1$, the quasi-one-dimensional GP equation is obtained as

$$\mathrm{i}\frac{\partial}{\partial t}\varphi(x; t) = \left[-\frac{1}{2}\frac{\partial^2}{\partial x^2} + V(x) + g\, |\varphi(x; t)|^2\right]\varphi(x; t). \tag{3.15}$$

with $V(x) = \frac{1}{2}x^2$, $g = 2aN\sqrt{\lambda\kappa}/l$ and normalization

$$\int_{-\infty}^{\infty} dx\, |\varphi(x; t)|^2 = 1. \tag{3.16}$$

3.2 Split-step Crank–Nicolson method

Several numerical methods are available for solving the time-dependent GP equation. These methods typically involve a time-stepping approach, but their specific implementations can vary. The GP equation is often discretized in both space and time and then solved iteratively, beginning with an initial guess. A commonly used discretization scheme for the GP equation is the semi-implicit Crank–Nicolson scheme, which has certain advantages.

The split-step method is a numerical technique for solving nonlinear PDEs. The fundamental idea is to decompose the equation into simpler sub-problems, each representing either the linear or nonlinear component, and then solve these sub-problems sequentially over small time steps. This approach can substantially enhance computational efficiency and accuracy compared to traditional methods for certain types of PDEs. In the split-step method, the original PDE is split into two or more simpler equations, each incorporating a specific part of the original equation. For instance, one can divide the GP equation into a linear derivative term and terms representing the trap and nonlinear interactions. The solution is then advanced in time by iteratively applying these simplified equations over small time intervals. Each equation is solved independently using appropriate numerical methods. For example, the solution to the linear part might be obtained using standard linear solvers like finite difference, spectral, or pseudospectral methods, while the nonlinear part is solved in the time domain.

The numerical solution procedure for solving the GP equation described in the following section is adapted from reference [4], utilizing split-step Crank–Nicolson (SSCN) method.

3.2.1 Numerical solution to one-dimensional Gross–Pitaevskii equation

This subsection has been reproduced with permission from [4].

The one-dimensional GP equation (3.15) can be rewritten through the substitution $\varphi(x; t) = \psi(x, t)$, introduced for notational simplicity, yielding

$$\mathrm{i}\frac{\partial\psi(x, t)}{\partial t} = \left[-\frac{1}{2}\frac{\partial^2}{\partial x^2} + V(x) + g\,|\psi(x, t)|^2\right]\psi(x, t) \equiv H\psi(x, t), \tag{3.17}$$

where H denotes the Hamiltonian operator encompassing both linear terms (kinetic energy and harmonic potential) and the nonlinear term (mean-field interaction). One solves this equation by iteration [2, 4–7]. A given trial input solution is propagated in time over small time steps until a stable final solution is reached. The GP equation is discretized in space and time using the finite difference scheme. This procedure results in a set of algebraic equations which can be solved by time iteration using an input solution consistent with the known boundary condition. In the present split-step method [7], this iteration is conveniently done in several steps by breaking up the full Hamiltonian into different derivative and non-derivative parts, i.e., $H = H_1 + H_2 + H_3$, where

$$H_1 = \frac{1}{2}[V(x) + g\,|\psi(x, t)|^2] \tag{3.18a}$$

$$H_2 = -\frac{1}{2}\frac{\partial^2}{\partial x^2}, \tag{3.18b}$$

$$H_3 = H_1. \tag{3.18c}$$

To numerically solve the above set of equations, the time variable has to be discretized as $t_n = n\Delta$, where Δ is the time step. Let ψ^n represent the discretized wavefunction at time t_n. The wavefunction at $t_{n+1} = t_n + \Delta$ is obtained by solving (3.17) successfully with H_1, H_2 and H_3 as described below.

The solution ψ^n is advanced first over the time step Δ at time t_n by solving equation (3.18a) to produce an intermediate solution $\psi^{n+1/3}$ from ψ^n. Since there is no derivative in H_1, this propagation is performed essentially exactly for small Δ through the operation

$$\psi^{n+1/3} = \mathcal{O}_{\mathrm{nd}}(H_1)\psi^n \equiv \mathrm{e}^{-\mathrm{i}\Delta H_1}\psi^n, \tag{3.19}$$

where $\mathcal{O}_{\rm nd}(H_1)$ denotes time-evolution operation with H_1 and the suffix 'nd' denotes non-derivative. Next, the time propagation corresponding to the operator H_2 is performed numerically by the following semi-implicit Crank–Nicolson scheme [5]:

$$\frac{\psi^{n+2/3} - \varphi^{n+1/3}}{-{\rm i}\Delta} = \frac{1}{2}H_2(\psi^{n+2/3} + \psi^{n+1/3}). \tag{3.20}$$

The formal solution to (3.20) is

$$\psi^{n+2/3} = \mathcal{O}_{\rm CN}(H_2)\ \psi^{n+1/3} \equiv \frac{1 - {\rm i}\Delta H_2/2}{1 + {\rm i}\Delta H_2/2}\psi^{n+1/3}, \tag{3.21}$$

where $\mathcal{O}_{\rm CN}$ denotes time-evolution operation with H_2. The suffix 'CN' refers to the Crank–Nicolson algorithm. The operation $\mathcal{O}_{\rm CN}$ is used to propagate the intermediate solution $\psi^{n+1/3}$ by a time step Δ to generate the solution $\psi^{n+2/3}$ at time $t_n + \Delta$. The final solution is obtained by repeating the procedure described for solving (3.17) with $H = H_1$. That is,

$$\psi^{n+1} = \mathcal{O}_{\rm nd}(H_1)\psi^{n+2/3} \equiv {\rm e}^{-{\rm i}\Delta H_1}\psi^{n+2/3}. \tag{3.22}$$

Consequently, a single time iteration from t_n to t_{n+1} is performed via the following three-step operation:

$$\psi^{n+1} = \mathcal{O}_{\rm nd}(H_3)\mathcal{O}_{\rm CN}(H_2)\mathcal{O}_{\rm nd}(H_1)\psi^n. \tag{3.23}$$

The break-up of the non-derivative term in two parts H_1 and H_3 (with half time steps each) symmetrically around the derivative term H_2, increases enormously the stability of the method and reduces the numerical error.

The advantage of the above split-step method with a small time step Δ is due to the following three factors [5, 7]. First, all iterations conserve normalization of the wavefunction. Second, the error involved in splitting the Hamiltonian is proportional to Δ^2 and can be neglected. Moreover, the method preserves the symplectic structure of the Hamiltonian formulation. Finally, a major part of the Hamiltonian, including the nonlinear term, is treated fairly accurately without mixing with the delicate Crank–Nicolson propagation. The method can handle an arbitrarily large nonlinear term and yield stable and reasonably accurate results.

This section describes the Crank–Nicolson algorithm for solving the discrete equation (3.20). The GP equation is mapped onto a one-dimensional spatial grid of N_x points in x. Equation (3.17) is discretized taking $H = H_2$ of (3.18b) using the following Crank–Nicolson scheme [2, 5, 7]:

$$\frac{{\rm i}(\psi_j^{n+1} - \psi_j^n)}{\Delta} = -\frac{1}{4h^2}\Big[(\psi_{j+1}^{n+1} - 2\psi_j^{n+1} + \psi_{j-1}^{n+1}) + (\psi_{j+1}^n - 2\varphi_j^n + \psi_{j-1}^n)\Big], \tag{3.24}$$

where $\psi_j^n = \psi(x_j, t_n)$ represents the value of ψ at $x = x_j = -\frac{1}{2}N_x h + jh$, $j = 0, 1, 2, \ldots, N_x$ with N_x being an even number and h the spatial step size. Note that the total number of spatial grid points in this case is $N_x + 1$. This scheme is constructed by approximating the time derivative $\partial/\partial t$ using a two-point formula and the spatial second derivative $\partial^2/\partial x^2$ by a three-point formula. This process yields a set of tridiagonal equations of the form (3.24) involving ψ_{j+1}^{n+1}, ψ_j^{n+1}, and ψ_{j-1}^{n+1} at time t_{n+1}, which are solved subject to appropriate boundary conditions. These tridiagonal equations emerging from (3.24) can be explicitly written as

$$A_j^- \psi_{j-1}^{n+1} + A_j^0 \psi_j^{n+1} + A_j^+ \psi_{j+1}^{n+1} = b_j, \tag{3.25a}$$

where

$$b_j = \frac{\mathrm{i}\Delta}{4h^2}\left(\psi_{j+1}^n - 2\psi_j^n + \psi_{j-1}^n\right) + \psi_j^n, \quad j = 0, 1, 2, \ldots, N, \tag{3.25b}$$

and $A_k^0 = 1 + \mathrm{i}\Delta/(2h^2)$, $A_k^- = A_k^+ = -\mathrm{i}\Delta/(4h^2)$. All quantities in b_j refer to time step t_n and are considered known. The only unknowns in equation (3.25a) are the values of $\psi_{i\pm1}^{n+1}$, and ψ_i^{n+1} at time t_{n+1}.

To solve equation (3.24), one assumes the one-term forward recursion relation [2]

$$\psi_{j+1}^{n+1} = \alpha_j \psi_j^{n+1} + \beta_j \tag{3.26}$$

where α_i and β_i are coefficients to be determined. Substituting equation (3.26) into equation (3.25a), one obtains

$$A_j^- \psi_{j-1}^{n+1} + A_j^0 \psi_j^{n+1} + A_j^+\left(\alpha_j \psi_j^{n+1} + \beta_j\right) = b_j. \tag{3.27}$$

Solving the above equation for ψ_j^{n+1} leads to

$$\psi_j^{n+1} = \gamma_j\left(A_j^- \psi_{j-1}^{n+1} + A_j^+ \beta_j - b_j\right), \tag{3.28a}$$

where

$$\gamma_j = -\frac{1}{A_j^0 + A_j^+ \alpha_j}. \tag{3.28b}$$

The recursion relations for the coefficients α_j and β_j are obtained using equations (3.26) and (3.28a) as

$$\alpha_{j-1} = \gamma_j A_j^- \quad \text{and} \quad \beta_{j-1} = \gamma_j\left(A_j^+ \beta_j - b_j\right). \tag{3.29}$$

The recursion relations (3.28a), (3.28b) and (3.29) are used in a backward sweep to determine the coefficients α_j, β_j and γ_j for j running from $N_x - 2$ down to 0 with initial values $\alpha_{N_x-1} = 0$ and $\beta_{N_x-1} = \psi_{N_x}^{n+1}$. Once α_j and β_j and γ_j are determined, equation (3.26) can be used to compute the solution for the entire spatial range from $j = 1$ to $N_x - 1$, starting with the value $\psi_0^{n+1}(=0)$ given by the boundary condition $\psi = 0$ at $x = \pm\infty$. Therefore, the solution is determined by applying two sets of recursion relations across the spatial grid points, each involving approximately N_x operations.

The Crank–Nicolson scheme described above has certain properties worth mentioning [5, 7]. The error in this scheme is second order in both space and time steps, so that for small Δ and h, the error is negligible. This scheme is also unconditionally stable. The boundary condition at infinity is preserved for small values of Δ/h^2 [5].

3.2.2 Imaginary time propagation

While the SSCN scheme outlined above offers several advantages, it necessitates addressing the time evolution of the wavefunction, or simply, the condensate's dynamics. For stationary ground states, the imaginary-time propagation method proves more convenient. This approach indirectly solves the time-independent GP equation.

The time-independent GP equation can be obtained by explicitly expressing the time dependence of the one-dimensional wavefunction as $\psi(x, t) = \psi(x)\mathrm{e}^{-\mathrm{i}\mu t}$, where $\psi(x) \in \mathbb{R}^1$ and μ is the chemical potential. The GP equation (3.1) becomes

$$\mu\psi(x) = \left[-\frac{1}{2}\frac{\partial^2}{\partial x^2} + V(x) + g\psi(x)^2\right]\psi(x), \tag{3.30}$$

and the chemical potential μ is given by

$$\mu = \int_{-\infty}^{\infty}\left(\frac{1}{2}\left[\frac{\partial\psi(x)}{\partial x}\right]^2 + [V(x) + g\psi(x)^2]\psi(x)^2\right)\mathrm{d}x, \tag{3.31}$$

and the energy parameter E is expressed as

$$E = \int_{-\infty}^{\infty}\left(\frac{1}{2}\left[\frac{\partial\psi(x)}{\partial x}\right]^2 + \left[V(x) + \frac{g}{2}\psi(x)^2\right]\psi(x)^2\right)\mathrm{d}x. \tag{3.32}$$

The wavefuntion $\psi(x)$ is subjected to the following normalization condition

$$\int_{-\infty}^{\infty}\psi^2(x)\mathrm{d}x = 1. \tag{3.33}$$

The root-mean-squared distance, denoted by $\langle x\rangle_{\mathrm{rms}}$, is another quantity of interest. It is defined as

$$\langle x\rangle_{\mathrm{rms}} = \left[\int_{-\infty}^{+\infty} x^2\psi(x)^2\mathrm{d}x\right]^{\frac{1}{2}}. \tag{3.34}$$

[1] The choice of real values for the wavefunction is valid only for the ground state, excluding phase considerations, and cannot be generalized to Hamiltonians incorporating angular momentum, spin–orbit coupling, complex potentials, etc. Nevertheless, the imaginary time propagation method detailed below can be employed to calculate the stationary states of systems governed by such Hamiltonians.

Solution to the time-independent GP equation (3.30) yields the stationary state $\psi(x)$. However, without knowing the chemical potential μ, solving (3.30) is not straightforward. Alternatively, one can numerically solve the time-dependent GP equation (3.17) by replacing the time variable t with its imaginary counterpart, that is, $t \to -it$ and $\psi(x, t)$ with its time-independent counterpart $\psi(x)$. With the above substitution, the GP equation (3.17) becomes

$$-\frac{\partial\psi(x)}{\partial t} = \left[-\frac{1}{2}\frac{\partial^2}{\partial x^2} + V(x) + g\psi(x)^2\right]\psi(x) \equiv H\psi(x), \tag{3.35}$$

Note that the variable t in the resulting equation (3.35) is merely a mathematical parameter and does not carry any physical significance.

Equation (3.35) can be solved iteratively using the SSCN method. For this purpose, let us split (3.35) as

$$H_1 = \frac{1}{2}[V(x) + g\psi(x)^2] = H_3 \tag{3.36a}$$

$$H_2 = -\frac{1}{2}\frac{\partial^2}{\partial x^2}, \tag{3.36b}$$

As discussed earlier, one needs to discretize the time variable as $t_n = n\Delta$, where Δ is the time step. Suppose ψ^n is the discretized wavefunction at t_n, the next approximation to the wavefunction at $t_{n+1} = t_n + \Delta$ is obtained by solving the GP equation (3.35) successfully with H replaced by H_1, H_2 and H_3.

The solution ψ^n is advanced first over the time step Δ at time t_n by solving equation (3.36a) to produce an intermediate solution $\psi^{n+1/3}$ from ψ^n. Since there is no derivative in H_1, this propagation is performed through the operation

$$\psi^{n+1/3} = \mathcal{O}_{\rm nd}(H_1)\psi^n \equiv \mathrm{e}^{-\Delta H_1}\psi^n. \tag{3.37}$$

Next, one performs the imaginary time propagation corresponding to the operator H_2 by the semi-implicit Crank–Nicolson scheme. As before, one maps the GP equation (3.35) with $H = H_2$ onto a one-dimensional spatial grid of N_x points in x and discretize using the Crank–Nicolson scheme as follows

$$-\frac{(\psi_j^{n+1} - \psi_j^n)}{\Delta} = -\frac{1}{4h^2}\left[(\psi_{j+1}^{n+1} - 2\psi_j^{n+1} + \psi_{j-1}^{n+1}) + (\psi_{j+1}^n - 2\varphi_j^n + \psi_{j-1}^n)\right]. \tag{3.38}$$

The tridiagonal system of equations (3.25a) become

$$A_j^-\psi_{j-1}^{n+1} + A_j^0\psi_j^{n+1} + A_j^+\psi_{j+1}^{n+1} = b_j, \tag{3.39a}$$

with

$$b_j = \frac{\Delta}{4h^2}\left(\psi_{j+1}^n - 2\psi_j^n + \psi_{j-1}^n\right) + \psi_j^n, \quad j = 0, 1, 2, \ldots, N, \tag{3.39b}$$

and $A_k^0 = 1 - \Delta/(2h^2)$, $A_k^- = A_k^+ = \Delta/(4h^2)$. The remaining procedures are identical to those used in real-time propagation, as discussed earlier. Alternatively, the imaginary time propagation can be achieved by replacing i by 1 in equations (3.19) and (3.22), and by -1 in the remaining equations in section 3.2.1. Subsequently, the entire analysis outlined for real-time propagation in section 3.2.1 is applicable.

However, there is a caveat that remains to be addressed. In equation (3.35), one can see that an eigenstate ψ_j of a eigenvalue E_j, satisfying $H\psi_j = E_j\psi_j$ behaves under imaginary-time propagation as

$$\frac{\partial \psi_j}{\partial t} = -E_j\psi_j. \tag{3.40}$$

When starting with an arbitrary initial wavefunction $\psi(x)$, expressed as a linear combination of all eigenfunctions of the Hamiltonian H, imaginary-time propagation causes each eigenfunction to decay exponentially. Excited states, with higher eigenvalues E_j, decay faster than the ground state, which has the smallest eigenvalue. As a result, after sufficient time, the ground state dominates the solution. Since all states decay during imaginary-time propagation, the wavefunction must be renormalized at each iteration to prevent it from approaching zero.

3.3 Split-step Fourier transform method

Another popularly known numerical solution procedure often used to solve the cubic nonlinear Schrödinger equation is the split-step Fourier transform (SSFT) (or simply Fourier) method [8–11]. This method is a pseudospectral numerical method that involves forward and backward Fourier transforms because the numerical solution of the derivative part is performed in the frequency domain.

The SSFT methods also employ a similar splitting of the Hamiltonian (3.18) as the SSCN method discussed in the previous sections. In the SSFT method, the non-derivative parts of the Hamiltonian are solved identically to the SSCN method. The linear derivative part is solved in the frequency domain using the forward and backward Fourier transforms employed through fast Fourier transform tools.

In the SSFT method, the numerical solution to the one-dimensional GP equation (3.17) can be obtained by discretizing it like that described in the previous section, i.e., the wavefunction at time $t = t_n$ is represented by ψ^n and advanced in small time steps Δ. The entire procedure for advancing the wavefunction over one time step, i.e., from ψ^n at time t_n to ψ^{n+1} at t_{n+1}, using SSFT, can be symbolically represented as

$$\psi^{n+1} = \mathcal{O}_{\rm nd}(H_3)\ \mathcal{O}_{\rm FT}(H_2)\ \mathcal{O}_{\rm nd}(H_1)\ \psi^n. \tag{3.41}$$

where $\mathcal{O}_{\rm nd}$ represents the propagation of non-derivative part given by equation (3.19) and $\mathcal{O}_{\rm FT}$ denotes the solution procedure in the frequency domain using Fourier and inverse Fourier transforms To perform the time propagation using the Fourier transform method $\mathcal{O}_{\rm FT}$, the wavefunction is spatially discretized on a one-dimensional grid with N_x points[2], $x_j = -\frac{1}{2}N_x h, +jh, j = 0, 1, 2, \ldots, N_x - 1$.

3.3.1 Computation of ground state wavefunction for the one-dimensional GP equation

The numerical solution of the GP equation (3.17) is demonstrated using both the SSCN and Fourier spectral methods. The analysis focusses on the one-dimensional case with a harmonic trapping potential given by $V(x) = \frac{1}{2}\omega^2 x^2$. To obtain the stationary wavefunction profile, imaginary time propagation is employed. This approach requires an initial seed solution, for which the non-interacting ground state wavefunction (corresponding to $g = 0$ in the GP equation) serves as the starting configuration, that is,

$$\psi_0(x) = \pi^{-\frac{1}{4}} \exp\left(-\frac{1}{2}\omega x^2\right). \tag{3.42}$$

Figure 3.1 illustrates the numerical solution to the one-dimensional GP equation (3.17) for a BEC with harmonic trap frequency $\omega = 1$ and interaction strength $g = 1$, as obtained using the split-step Fourier transform and SSCN methods, employing a spatial step $\mathrm{d}x = 0.005$ and time step $\mathrm{d}t = 10^{-5}$ using imaginary time propagation. Figure 3.1(a) displays the wavefunction $\psi(x)$ as a function of position x, showing the converged spatial profile of the BEC ground state for both SSFT and SSCN approaches. Figure 3.1(b) presents the residue, defined as $(H - \mu)\psi(x)$, where H is the Hamiltonian and μ is the chemical potential, quantifying the accuracy of the numerical solutions.

The provided tables 3.1 and 3.2 evaluate the performance of two numerical methods, namely, SSCN and SSFT, with imaginary time propagation for solving the one-dimensional GP equation (3.17) to obtain the ground state wavefunction of a BEC. The simulations use a harmonic trap with frequency $\omega = 1$ and interaction strength $g = 1$. Both tables list results for varying spatial steps $\mathrm{d}x$ (0.001 to 0.1) and time steps $\mathrm{d}t$, reporting the chemical potential μ, energy E, root-mean-square

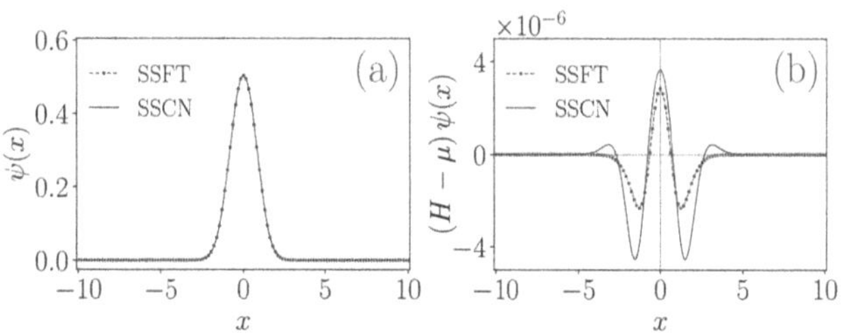

Figure 3.1. Illustration of numerical solution to one-dimensional GP equation (3.17) with $\omega = 1$ and $g = 1$ as obtained using the SSFT and SSCN methods with $\mathrm{d}x = 0.005$ and $\mathrm{d}t = 10^{-5}$. (a) Plot of the wavefunction $\psi(x)$ against x as obtained using imaginary time propagation and (b) the residue, i.e., $(H - \mu)\psi(x)$.

[2] Choosing N_x as a power of 2 typically makes the fast Fourier transform computational routine more efficient.

Table 3.1. Table of dx, dt, chemical potential μ, energy E, root-mean-squared distance $\langle x \rangle_{rms}$, and the absolute maximum of the residue $(H - \mu)\psi(x)$ illustrating the performance of the SSCN method with imaginary time propagation for computing the ground state wavefunction of the one-dimensional GP equation (3.17) with $\omega = 1$ and $g = 1$.

dx	dt	μ	E	$\langle x \rangle_{rms}$	$\psi(0)$	Res. (abs. max)
0.1	1.00×10^{-3}	0.870 034	0.689 484	0.773 74	0.709 22	1.172×10^{-3}
0.05	2.50×10^{-4}	0.869 967	0.689 487	0.774 02	0.709 09	3.021×10^{-4}
0.025	6.25×10^{-5}	0.869 950	0.689 487	0.774 09	0.709 06	8.338×10^{-5}
0.01	1.00×10^{-5}	0.869 945	0.689 487	0.774 11	0.709 05	1.220×10^{-5}
0.005	2.50×10^{-6}	0.869 944	0.689 487	0.774 12	0.709 05	1.378×10^{-4}
0.0025	6.25×10^{-7}	0.869 944	0.689 487	0.774 12	0.709 05	1.370×10^{-4}
0.001	10^{-5}	0.869 944	0.689 487	0.774 12	0.709 05	2.864×10^{-6}

Table 3.2. Table of dx, dt, chemical potential μ, energy E, root-mean-squared distance $\langle x \rangle_{rms}$, and the absolute maximum of the residue $(H - \mu)\psi(x)$ illustrating the performance of the SSFT method with imaginary time propagation for computing the ground state wavefunction of the one-dimensional GP equation (3.17) with $\omega = 1$ and $g = 1$.

dx	dt	μ	E	$\langle x \rangle_{rms}$	$\psi(0)$	Res. (abs. max)
0.1	1.00×10^{-3}	0.869 963	0.689 484	0.780 47	0.709 10	2.881×10^{-4}
0.05	2.50×10^{-4}	0.869 949	0.689 487	0.775 71	0.709 06	7.109×10^{-5}
0.025	6.25×10^{-5}	0.869 945	0.689 487	0.774 52	0.709 05	1.771×10^{-5}
0.01	1.00×10^{-5}	0.869 944	0.689 487	0.774 18	0.709 05	2.832×10^{-6}
0.005	2.50×10^{-6}	0.869 944	0.689 487	0.774 13	0.709 05	7.078×10^{-7}
0.0025	6.25×10^{-7}	0.869 944	0.689 487	0.774 12	0.709 05	1.012×10^{-6}
0.001	10^{-5}	0.869 944	0.689 487	0.774 12	0.709 05	2.831×10^{-6}

distance $\langle x \rangle_{rms}$, central wavefunction value $\psi(0)$, and absolute maximum residual $|(H - \mu)\psi(x)|$.

For the SSCN method (table 3.1), as dx decreases from 0.1 to 0.001, μ converges to $\sim$0.869 944, E stabilizes at 0.689 487, and $\langle x \rangle_{rms}$ approaches 0.774 12, with $\psi(0) \approx 0.709\,05$. Residuals decrease significantly, reaching 2.864×10^{-6} at $dx = 0.001$, though some smaller dx values (e.g., 0.005, 0.0025) show slightly higher residuals ($\sim 10^{-4}$), possibly due to numerical instabilities at very small dt.

The SSFT method (table 3.2) shows similar convergence, with $\mu \approx 0.8699\,44$, $E = 0.689\,487$, and $\psi(0) \approx 0.709\,05$ at $dx = 0.001$. However, $\langle x \rangle_{rms}$ is slightly higher (e.g., 0.774 52 at $dx = 0.025$) for larger dx, indicating sensitivity to spatial discretization. Residuals are generally lower, reaching 7.078×10^{-7} at $dx = 0.005$, suggesting better accuracy than SSCN for finer grids.

Both methods achieve high accuracy, with residuals $\sim 10^{-6}$–10^{-7} for $dx \leqslant 0.01$, confirming their effectiveness for stationary BEC solutions (tables 3.3 and 3.4).

Table 3.3. Table of g, chemical potential μ, energy E, root-mean-squared distance $\langle x \rangle_{\text{rms}}$, and the absolute maximum of the residue $(H - \mu)\psi(x)$ demonstrates the performance of the SSCN method with imaginary time propagation for calculating the ground state wavefunction of the one-dimensional GP equation (3.17) with $\omega = 1$, $\mathrm{d}x = 0.005$, $\mathrm{d}t = 10^{-5}$, and various values of g.

g	μ	E	$\langle x \rangle_{\text{rms}}$	$\psi(0)$	Residue (abs. max)
1	0.869 944	0.689 487	0.774 12	0.709 05	4.555×10^{-6}
10	3.107 254	1.947 127	1.169 21	0.549 33	2.727×10^{-5}
50	8.919 921	5.391 355	1.904 45	0.421 71	1.799×10^{-4}
100	14.134 549	8.508 527	2.386 44	0.375 73	4.081×10^{-4}
500	41.283 856	24.781 020	4.065 24	0.287 34	2.692×10^{-3}
1000	65.529 273	39.322 424	5.119 61	0.256 00	6.049×10^{-3}

Table 3.4. Table of g, chemical potential μ, energy E, root-mean-squared distance $\langle x \rangle_{\text{rms}}$, and the absolute maximum of the residue $(H - \mu)\psi(x)$ demonstrates the performance of the SSFT with imaginary time propagation for calculating the ground state wavefunction of the one-dimensional GP equation (3.17) with $\omega = 1$, $\mathrm{d}x = 0.005$, $\mathrm{d}t = 10^{-5}$, and various values of g.

g	μ	E	$\langle x \rangle_{\text{rms}}$	$\psi(0)$	Residue (abs. max)
1	0.869 944	0.689 487	0.774 12	0.709 05	2.831×10^{-6}
10	3.107 254	1.947 127	1.169 21	0.549 33	2.798×10^{-5}
50	8.919 921	5.391 355	1.904 45	0.421 71	1.814×10^{-4}
100	14.134 549	8.508 527	2.386 44	0.375 73	4.089×10^{-4}
500	41.283 856	24.781 020	4.065 24	0.287 34	2.692×10^{-3}
1000	65.529 273	39.322 424	5.119 61	0.256 00	6.049×10^{-3}

The SSFT method appears more stable for small $\mathrm{d}x$ and $\mathrm{d}t$, aligning with its spectral accuracy in momentum space, while SSCN remains robust but slightly less precise at the finest grids. Furthermore, it is worth noting that the SSCN and SSFT require approximately the same amount of time to solve the above one-dimensional problem with imaginary time propagation.

Figures 3.2 and 3.3 illustrate numerical solutions to the two-dimensional GP equation, as given by equation (3.17), for a BEC with parameters $\nu = 1$, $\gamma = 2$, and interaction strength $g = 12.5484$. The simulations employ a spatial grid with $\mathrm{d}x = \mathrm{d}y = 0.1$ and a time step $\mathrm{d}t = 10^{-4}$, using imaginary time propagation to obtain the ground state. Figure 3.2 presents results from the SSCN method, while figure 3.3 shows results from the SSFT method. Each figure includes surface plots of (a) the two-dimensional density profile $|\psi(x, y)|^2$, depicting the condensate's spatial distribution, and figure 3.3(b) the residual $(H - \mu)\psi(x, y)$. These results again demonstrate the effectiveness of SSCN and SSFT methods in accurately solving the two-dimensional GP equation.

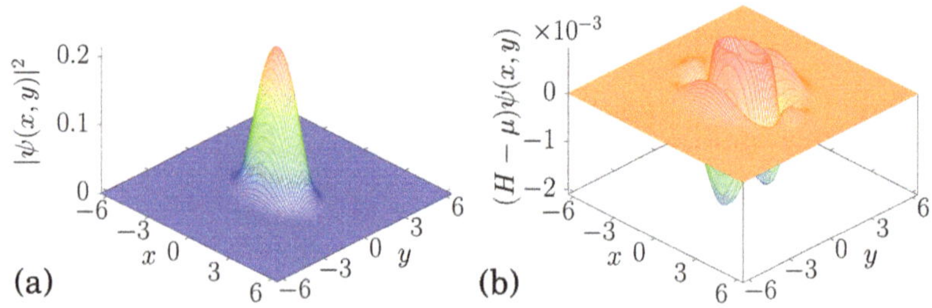

Figure 3.2. Illustration of numerical solution to the two-dimensional GP equation (3.17) with $\nu = 1, \gamma = 2$ and $g = 12.5484$, as obtained using SSCN method with $\mathrm{d}x = \mathrm{d}y = 0.1$ and $\mathrm{d}t = 10^{-4}$. Surface plots of (a) the two-dimensional density profile $|\psi(x, y)|^2 x$ as obtained using imaginary time propagation and (b) the residue $(H - \mu)\psi(x, y)$.

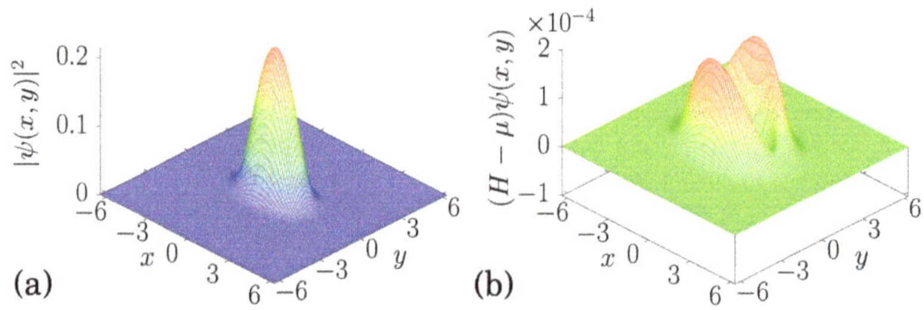

Figure 3.3. Illustration of numerical solution to two-dimensional GP equation (3.17) with $\nu = 1, \gamma = 2$ and $g = 12.5484$ as obtained using SSFT method with $\mathrm{d}x = \mathrm{d}y = 0.1$ and $\mathrm{d}t = 10^{-4}$. Surface plots of (a) two-dimensional density profile $|\psi(x, y)|^2 x$ as obtained using imaginary time propagation and (b) residue - $(H - \mu)\psi(x, y)$.

3.4 Newton conjugate-gradient methods

The Newton conjugate-gradient (NCG) method combines Newton iterations with conjugate-gradient iterations to solve the linear Newton correction equation [11, 12]. When the linearization operator is self-adjoint, the preconditioned conjugate-gradient method becomes applicable for solving the linear equation.

Consider a general real-valued nonlinear wave system in arbitrary spatial dimensions, described by:

$$\mathbf{L}_0\mathbf{u}(\mathbf{x}) = 0. \tag{3.43}$$

Here, $\mathbf{x}$ represents a vector spatial variable, $\mathbf{u}(\mathbf{x})$ denotes a real-valued vector solution satisfying equation (3.43), with the boundary condition $\mathbf{u} \to 0$ as $|\mathbf{x}| \to \infty$.

Let $\mathbf{u}_n(\mathbf{x})$ represent an approximate solution that is close to the exact solution $\mathbf{u}(\mathbf{x})$. The exact solution can be expressed through the iterative formulation

$$\mathbf{u}(\mathbf{x}) = \mathbf{u}_n(\mathbf{x}) + \mathbf{e}_n(\mathbf{x}), \tag{3.44}$$

where $\|\mathbf{e}_n(\mathbf{x})\| \ll 1$ represents the correction term. Substitution of this expression into (3.43) and expansion about $\mathbf{u}_n(\mathbf{x})$ yields

$$\mathbf{L}_0\mathbf{u}_n + \mathbf{L}_{1n}\mathbf{e}_n + \mathcal{O}(\mathbf{e}_n^2) = 0. \tag{3.45}$$

Here, $\mathbf{L}_{1n}$ denotes the linearization operator (Jacobian) $\mathbf{L}_1$ evaluated at $\mathbf{u}_n(\mathbf{x})$. Neglecting higher-order terms in $\mathbf{e}_n$ reduces equation (3.45) to a linear inhomogeneous equation for the error term $\mathbf{e}_n$.

The approximate solution is updated according to

$$\mathbf{u}_{n+1}(\mathbf{x}) = \mathbf{u}_n(\mathbf{x}) + \Delta\mathbf{u}_n(\mathbf{x}), \tag{3.46}$$

where $\Delta\mathbf{u}_n(\mathbf{x})$ represents the solution to the linearized equation for $\mathbf{e}_n$, expressed as

$$\mathbf{L}_{1n}\Delta\mathbf{u}_n(\mathbf{x}) = -\mathbf{L}_0\mathbf{u}_n(\mathbf{x}). \tag{3.47}$$

When the Newton correction equation (3.47) is solved exactly or with precision significantly exceeding that of the size of $\Delta\mathbf{u}_n$, the iterations (3.46) exhibit quadratic convergence to the exact solution $\mathbf{u}(\mathbf{x})$. In practical implementations, the numerical solution of equation (3.47) requires the conjugate-gradient method, as described in the following.

The application of conjugate-gradient methods for solving the linear equation (3.47) depends on whether the linearization operator $\mathbf{L}_1$ is self-adjoint. Two approaches are available:

- The preconditioned conjugate-gradient method, applicable when $\mathbf{L}_1$ is self-adjoint.
- The preconditioned biconjugate-gradient (BiCG) method, used when $\mathbf{L}_1$ is non-self-adjoint.

3.4.1 The preconditioned conjugate-gradient method for self-adjoint linearization operators $\mathbf{L}_1$

In a conservative Hamiltonian system, the linearization operator $\mathbf{L}_1$ is self-adjoint, meaning it is Hermitian. This property carries significant implications. In computational practice, the linearization operator is typically represented by a matrix. When $\mathbf{L}_1$ is self-adjoint, its corresponding matrix is symmetric. For equations involving symmetric and positive-definite matrices, the preconditioned conjugate-gradient method is generally the most efficient solver.

The preconditioned conjugate-gradient method, as applied to the linear operator in equation (3.47), is described below. For simplicity, the subscripts 'n' in (3.47) are omitted, and the initial guess $\Delta\mathbf{u}^{(0)}$ is assumed to be zero. The preconditioned conjugate-gradient algorithm for solving the linear Newton correction system (3.47) proceeds as follows:

$$\Delta\mathbf{u}^{(0)} = 0, \tag{3.48a}$$

$$\mathbf{R}^{(0)} = -\mathbf{L}_0\mathbf{u}, \tag{3.48b}$$

$$\mathbf{D}^{(0)} = \mathbf{M}^{-1}\mathbf{R}^{(0)}, \tag{3.48c}$$

$$a^{(i)} = \frac{\langle \mathbf{R}^{(i)}, \mathbf{M}^{-1}\mathbf{R}^{(i)} \rangle}{\langle \mathbf{D}^{(0)}, \mathbf{L}_1\mathbf{D}^{(i)} \rangle}, \tag{3.48d}$$

$$\Delta\mathbf{u}^{(i+1)} = \Delta\mathbf{u}^{(i)} + a^{(i)}\mathbf{D}^{(i)} \tag{3.48e}$$

$$\mathbf{R}^{(i+1)} = \mathbf{R}^{(i)} - a^{(i)}\mathbf{L}_1\mathbf{D}^{(i)}, \tag{3.48f}$$

$$b^{(i+1)} = \frac{\langle \mathbf{R}^{(i+1)}, \mathbf{M}^{-1}\mathbf{R}^{(i+1)} \rangle}{\langle \mathbf{R}^{(i)}, \mathbf{M}^{-1}\mathbf{R}^{(i)} \rangle}, \tag{3.48g}$$

$$\mathbf{D}^{(i+1)} = \mathbf{M}^{-1}\mathbf{R}^{(i+1)} + b^{(i+1)}\mathbf{D}^{(i)}, \tag{3.48h}$$

where $i = 0, 1, 2, \ldots$ is the index of conjugate-gradient (CG) iterations, $\langle \cdot, \cdot \rangle$ represents the standard inner product. $\mathbf{M}$ is the preconditioning operator, assumed to be self-adjoint and positive-definite. The preconditioning operator $\mathbf{M}$ accelerates the convergence of the conjugate-gradient iterations. It is a wise choice to choose operator $\mathbf{M}$ such that it is easily invertible. In practice, the linear derivative part of $\mathbf{L}_0$ is the best choice of $\mathbf{M}$. The above conjugate-gradient iterations are embedded within Newton iteration (3.46) and called the NCG method

The linearization operator $\mathbf{L}_1$ typically possesses both positive and negative eigenvalues, along with a zero eigenvalue. In the case of the ground state, $\mathbf{L}_1$ usually has one eigenvalue whose sign differs from all others, whereas, for an excited state, it generally exhibits two or more eigenvalues with signs opposite to the rest. Due to this property, the CG iterations in the Newton-CG method may fail, as the denominator in the $a^{(i)}$ formula can become zero. However, even if such a breakdown occurs, modifying the initial guess function $\mathbf{u}_0(x)$ can often resolve the issue by altering the eigenvalue distribution and preventing the denominator from vanishing.

In the case of a zero eigenvalue for the linearization operator $\mathbf{L}_1$, the solution to the linear equation (3.47) is non-unique, as any linear combination of its eigenfunctions can be added without altering the result. When these eigenfunctions arise from the invariances of the wave solutions, such as $\mathbf{u}_{x_j}$ for a solution $\mathbf{u}(x)$ that is translation-invariant in x_j, this non-uniqueness is inconsequential, as it merely corresponds to a shift in the free parameter of the wave. This behaviour is analogous to that observed in other iterative methods. To enforce uniqueness, such as when converging to a symmetric localized wave solution centred at $x = 0$, the initial guess $\mathbf{u}_0(x)$ can be restricted to symmetric functions in x. Under this constraint, the initial error function $\mathbf{e}_0(x)$ excludes the position-shifting eigenmode $\mathbf{u}_x$, ensuring the peak remains fixed at $x = 0$ throughout the iteration.

Interestingly, even when the kernel of $\mathbf{L}_1$ contains eigenfunctions not induced by invariances of the wave solutions, the Newton-CG method remains convergent. This behaviour contrasts with several other iterative methods, which typically fail to converge under such conditions. The underlying mechanism can be illustrated through an analogy with Newton's method applied to the algebraic equation

$$f(x) = (x - \alpha)^2 = 0, \tag{3.49}$$

which possesses a degenerate root at $x = \alpha$. The Newton correction step for this equation takes the form

$$\Delta x_n = -\frac{f(x_n)}{f'(x_n)}. \tag{3.50}$$

At the root $x = \alpha$, the derivative vanishes ($f'(\alpha) = 0$), resulting in a non-empty kernel of $f'(\alpha)$. This mirrors the non-empty kernel of $\mathbf{L}_1$ in the original problem. Despite this singularity, Newton's iteration for (3.49) retains convergence, albeit with a reduction in convergence rate from quadratic to linear.

The reason lies in the fact that as x_n approaches the root α, the function $f(x)$ converges to zero more rapidly than $f'(x_n)$. This behaviour ensures that the correction term Δx_n remains non-singular. Similarly, when the kernel of $\mathbf{L}_1$ is non-empty, the Newton-CG method retains its convergence. Such convergence, despite a non-empty kernel, represents one of the key advantages of the Newton-CG method compared to alternative approaches.

A termination criterion must be established for the Newton-CG iterations. Typically, the CG iterations are terminated once the approximate solution $\Delta\mathbf{u}^{(i)}$ reaches a desired level of accuracy. The error in this solution can be quantified using the residual $\mathbf{R}^{(i)}$ from the CG iterations, which corresponds to the residual of the linear equation (3.47), given by:

$$\mathbf{R}^{(i)} = -\mathbf{L}_0\mathbf{u}_n - \mathbf{L}_{1n}\Delta\mathbf{u}_n^{(i)}. \tag{3.51}$$

This error can be more conveniently measured using the $\mathbf{M}^{-1}$-weighted 2-norm[3] of $\mathbf{R}^{(i)}$:

$$\|\mathbf{R}^{(i)}\|_M \equiv \langle \mathbf{R}^{(i)}, \mathbf{M}^{-1}\mathbf{R}^{(i)} \rangle^{1/2}, \tag{3.52}$$

a quantity that naturally arises in the CG iterations. The accuracy with which the linear Newton correction equation (3.47) is solved plays a crucial role in the overall performance of the Newton-CG method.

The approximate solution $\mathbf{u}_{n+1}$ is updated using the formula $\mathbf{u}_n + \Delta\mathbf{u}_n$. If the accuracy of $\mathbf{u}_n$ (compared to the exact solution $\mathbf{u}$) is poor, requiring excessive accuracy in solving for $\Delta\mathbf{u}_n$ from the linear equation (3.47) becomes inefficient, as it does not improve the overall accuracy of $\mathbf{u}_{n+1}$. This phenomenon is referred to as *oversolving*. However, when $\mathbf{u}_n$ approaches the exact solution $\mathbf{u}$, higher accuracy in solving for $\Delta\mathbf{u}_n$ from (3.47) becomes necessary to sustain the rapid convergence of Newton's method.

[3] The weighted 2-norm (also called the weighted Euclidean norm) of a vector $\mathbf{x} \in \mathbb{R}^n$ is defined as $\|\mathbf{x}\|_\mathbf{W} = \sqrt{\mathbf{x}^T\mathbf{W}\mathbf{x}} = (\sum_{i=1}^n w_i x_i^2)^{1/2}$, where $\mathbf{W} = \mathrm{diag}(w_1, w_2, \ldots, w_n)$ is a diagonal weight matrix with positive weights $w_i > 0$, and $\mathbf{x}^\mathrm{T}$ denotes the transpose of $\mathbf{x}$.

To minimize oversolving, an effective strategy involves adaptively determining the accuracy for solving the linear Newton correction (3.47) based on the accuracy of $\mathbf{u}_n$. This accuracy can be quantified using the $\mathbf{M}^{-1}$-weighted 2-norm of the residual of equation (3.43):

$$\|\mathbf{L}_0\mathbf{u}_n\|_M = \langle \mathbf{L}_0\mathbf{u}_n, \mathbf{M}^{-1}\mathbf{L}_0\mathbf{u}_n\rangle^{\frac{1}{2}}. \tag{3.53}$$

A practical stopping criterion for the CG iterations when solving (3.47) is to ensure that the error in $\Delta\mathbf{u}_n^{(i)}$ falls below a specified fraction of the error in $\mathbf{u}_n$ itself. Accordingly, the CG iterations may be terminated when

$$\|\mathbf{R}^{(i)}\|_M \leqslant \epsilon_{\text{cg}} \|\mathbf{L}_0\mathbf{u}_n\|_M, \tag{3.54}$$

where ϵ_{cg} is a small positive tolerance parameter. Note that the residual $\mathbf{L}_0\mathbf{u}_n$ corresponds to the inhomogeneous term in (3.47). With zero initial conditions for the CG iterations, $\mathbf{R}^{(0)} = -\mathbf{L}_0\mathbf{u}_n$, the stopping criterion given by (3.54) simplifies to

$$\|\mathbf{R}^{(i)}\|_M \leqslant \epsilon_{\text{cg}} \|\mathbf{R}^{(0)}\|_M. \tag{3.55}$$

The choice of ϵ_{cg} is critical: if too small, oversolving occurs, rendering the process inefficient; If too large, the inaccurately solved linear system (3.47) may prevent convergence of Newton's method. Numerical experiments suggest that the optimal ϵ_{cg} typically lies between 10^{-1} and 10^{-3}.

3.4.2 The preconditioned biconjugate-gradient method for non-self-adjoint linearization operators $\mathbf{L}_1$

When the linearization operator $\mathbf{L}_1$ is non-self-adjoint, the conjugate-gradient iterations applied to equation (3.47) typically fail to converge. In the context of matrix computations, several extensions of the CG method have been developed to handle non-symmetric matrices. The BiCG method stands out due to its computational efficiency and straightforward implementation. This approach employs two sequences of residuals and search directions, updated using both the matrix and its transpose. The application of the BiCG method to equation (3.47) with incorporating preconditioning leads to the following preconditioned BiCG iterations:

$$\Delta\mathbf{u}^{(0)} = 0, \tag{3.56a}$$

$$\mathbf{R}^{(0)} = \tilde{\mathbf{R}}^{(0)} = -\mathbf{L}_0\mathbf{u}, \tag{3.56b}$$

$$\mathbf{D}^{(0)} = \tilde{\mathbf{D}}^{(0)} = \mathbf{M}^{-1}\mathbf{R}^{(0)}, \tag{3.56c}$$

$$a^{(i)} = \frac{\langle \tilde{\mathbf{R}}^{(i)}, \mathbf{M}^{-1}\mathbf{R}^{(i)}\rangle}{\langle \tilde{\mathbf{D}}^{(0)}, \mathbf{L}_1\mathbf{D}^{(i)}\rangle}, \tag{3.56d}$$

$$\Delta\mathbf{u}^{(i+1)} = \Delta\mathbf{u}^{(i)} + a^{(i)}\mathbf{D}^{(i)} \tag{3.56e}$$

$$\mathbf{R}^{(i+1)} = \mathbf{R}^{(i)} - a^{(i)}\mathbf{L}_1\mathbf{D}^{(i)}, \tag{3.56f}$$

$$\tilde{\mathbf{R}}^{(i+1)} = \tilde{\mathbf{R}}^{(i)} - a^{(i)}\mathbf{L}_1^{\dagger}\tilde{\mathbf{D}}^{(i)}, \tag{3.56g}$$

$$b^{(i+1)} = \frac{\langle \tilde{\mathbf{R}}^{(i+1)}, \mathbf{M}^{-1}\mathbf{R}^{(i+1)} \rangle}{\langle \tilde{\mathbf{R}}^{(i)}, \mathbf{M}^{-1}\mathbf{R}^{(i)} \rangle}, \tag{3.56h}$$

$$\mathbf{D}^{(i+1)} = \mathbf{M}^{-1}\mathbf{R}^{(i+1)} + b^{(i)}\mathbf{D}^{(i)}, \tag{3.56i}$$

$$\tilde{\mathbf{D}}^{(i+1)} = \mathbf{M}^{-1}\tilde{\mathbf{R}}^{(i+1)} + b^{(i+1)}\tilde{\mathbf{D}}^{(i)}, \tag{3.56j}$$

where $\mathbf{L}_1^{\dagger}$ is the adjoint operator (or, more generally, the Hermitian conjugate to $\mathbf{L}_1$) of $\mathbf{L}_1$, and $\mathbf{M}$ is a self-adjoint and positive-definite preconditioning operator. When these BiG iterations are used within the Newton iterations (3.46), the resulting method is known as the Newton-BCG method. Note that a single BiCG iteration requires four operator evaluations: $\mathbf{L}_1$, $\mathbf{L}_1^{\dagger}$, and two $\mathbf{M}^{-1}$, which is double the cost of one CG iteration. If the linear operator $\mathbf{L}_1$ is self-adjoint, these BiCG iterations become equivalent to the CG iterations discussed in section 3.4.1, but with twice the cost per iteration.

Example 3.4.1 Solution of the time-independent GP equation using CG methods.

The time-independent GP equation (3.30) is solved numerically using CG methods, with the parameters $g = 1$, $\mu = 1$, $V(x) = \frac{1}{2}\omega x^2$, $\omega = 1$, and the initial wavefunction $\psi_0(x) = \pi^{-1/4}\exp\left(-\frac{1}{2}x^2\right)$.

The time-independent GP equation is rewritten as

$$\mathbf{L}\psi(x) \equiv \left[-\frac{1}{2}\frac{\partial^2}{\partial x^2} + V(x) + g|\psi(x)|^2 - \mu\right]\psi(x) = 0. \tag{3.57}$$

The corresponding linearization operator for this equation is

$$\mathbf{L}_1 = -\frac{1}{2}\frac{\partial^2}{\partial x^2} + V(x) + 3g|\psi(x)|^2 - \mu. \tag{3.58}$$

An alternative expression for the operator $\mathbf{L}_1$ is

$$\mathbf{L}_1 = \left[\frac{1}{2}\frac{\partial^2}{\partial x^2} + V(x) + g|\psi(x)|^2 - \mu\right] - 2g|\psi(x)|^2. \tag{3.59}$$

To implement the CG method, an easily invertible positive-definite self-adjoint operator (the acceleration operator, $\mathbf{M}$) is introduced:

$$\mathbf{M} = c - \frac{1}{2}\frac{\partial^2}{\partial x^2}, \tag{3.60}$$

where c is a positive constant.

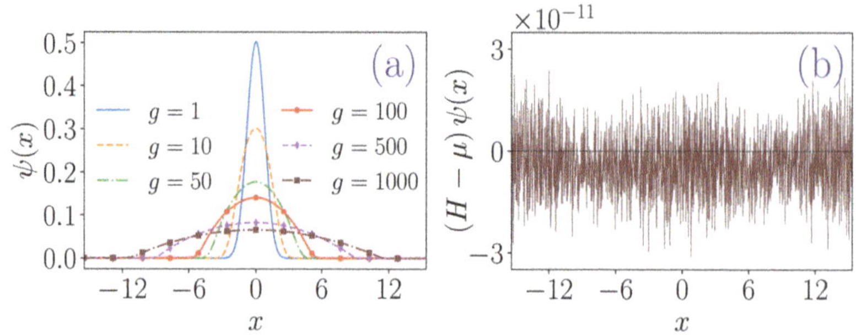

Figure 3.4. Illustration of improved numerical solutions to the one-dimensional GP equation with $V(x) = x^2/2$, obtained using the NCG method: (a) plots of the wavefunction $\psi(x)$ against x for various interaction strengths and (b) residuals of the solutions to $(H - \mu)\psi(x) = 0$ for case with $g = 1000$.

Table 3.5. Table of g, chemical potential μ, energy E, root-mean-squared distance $\langle x\rangle_{\mathrm{rms}}$, and the absolute maximum of the residue $(H - \mu)\psi(x)$ displaying the performance of NCG method for computing the ground state wavefunction of one-dimensional GP equation (3.17) with $\omega = 1$, $\mathrm{d}x = 0.005$ and for different interaction strengths (g).

g	μ	E	$\langle x\rangle_{\mathrm{rms}}$	$\psi(0)$	Residue (abs. max)
1	0.869 944	0.689 487	0.774 12	0.709 05	0.8965×10^{-12}
10	3.107 243	1.947 127	1.169 22	0.549 33	0.5508×10^{-12}
50	8.919 819	5.391 355	1.904 50	0.421 70	0.8091×10^{-12}
100	14.134 287	8.508 527	2.386 56	0.375 72	0.3535×10^{-12}
500	41.281 591	24.781 019	4.065 80	0.287 32	0.4060×10^{-12}
1000	65.523 555	39.322 422	5.120 73	0.255 97	0.3255×10^{-12}

Figure 3.4 illustrates the numerical solutions to the one-dimensional GP equation for a BEC in a harmonic trap $V(x) = x^2/2$, obtained using the NCG method. Figure 3.4(a) displays the ground state wavefunction $\psi(x)$ as a function of position x for various interaction strengths g, highlighting the effect of nonlinear interactions on the condensate's spatial profile. Figure 3.4(b) shows the residuals of the solutions to the stationary GP equation $(H - \mu)\psi(x) = 0$ for $g = 1000$, demonstrating the convergence of the NCG method. Table 3.5 summarizes the performance of the NCG method for computing the ground state, listing the chemical potential μ, energy E, root-mean-squared distance $\langle x\rangle_{\mathrm{rms}}$, central wavefunction value $\psi(0)$, and absolute maximum residual for interaction strengths $g = 1$, 10, 50, 100, 500, 1000. The simulations use a trap frequency $\omega = 1$ and spatial step $dx = 0.005$. Residuals of the order of 10^{-12} across all cases confirm the high accuracy and robustness of the NCG method for solving the GP equation (3.17). A note on the computational tools for solving GP equation for conventional BECs, dipolar BECs and spin–orbit coupled BECs is given in appendix A.

3.5 Bayesian optimization of BECs—data-driven approach

This section has been reproduced from [35].

In recent years, the scientific computing community has begun embracing machine learning techniques as efficient and scalable alternatives to conventional numerical simulation methods. This emerging field, often referred to as data-driven scientific computing, has shown considerable potential by effectively modelling complex physical systems using sparse and noisy data derived from coarse-grained numerical simulations. Unlike traditional solvers that scale poorly with increasing grid resolution and often suffer from instability due to numerical differentiation, machine learning models offer fixed computational costs and benefit from the accuracy and robustness of automatic differentiation. As a result, these models are capable of capturing the underlying dynamics of physical processes with remarkable efficiency and fidelity.

The rapid progress in machine learning has driven groundbreaking achievements across a variety of domains, including computer vision [13], natural language processing [14], and computational biology [15], often surpassing human-level performance in specific tasks [16–19]. These advances have begun to significantly influence computational sciences, where machine learning methods are being adopted to complement or enhance traditional simulation-based techniques. At the heart of this revolution are artificial neural networks (ANNs), the foundational models of modern machine learning and deep learning which excel at automatically learning abstract representations from complex data. In physical sciences, ANNs have been successfully employed to identify order parameters and detect phase transitions without prior knowledge of the underlying physics [20–23]. Greitemann *et al* [24] utilized a kernel-based learning approach to identify phases in magnetic materials and uncover intricate order parameters. In the study of nonlinear dynamical systems, Jaeger *et al* [25] demonstrated that ANNs could predict chaotic time-series trajectories with a dramatic increase in accuracy outperforming earlier methods by a factor of up to 2400. Machine learning has also made a substantial impact on quantum many-body physics, where the exponential complexity of quantum states poses significant computational challenges. Carleo and Troyer [26] introduced an ANN-based variational representation of quantum states that enabled efficient modelling of strongly correlated systems. Expanding on this idea, Nomura *et al* [27] developed machine learning techniques for accurately constructing ground-state wavefunctions of entangled quantum spin systems and fermionic lattice models. Gao and Duan [28] further combined restricted Boltzmann machines with traditional variational Monte Carlo methods, yielding a powerful solver for quantum many-body systems. These developments collectively demonstrate that ANNs are capable of representing large-scale quantum states, including ground states of complex Hamiltonians and dynamically evolving quantum systems [29–32], establishing their role as indispensable tools in modern computational physics.

ANNs have emerged as powerful tools in physical sciences, particularly for tasks traditionally reliant on domain-specific knowledge. Notably, ANNs have been employed to identify order parameters and detect phase transitions without requiring prior understanding of the system's underlying physics [20–23]. Greitemann *et al* [24] applied a kernel-based learning framework to classify phases in magnetic materials and to unearth intricate order parameters that are otherwise difficult to identify. In the context of nonlinear dynamical systems, Jaeger and Haas [25] demonstrated the effectiveness of ANNs in forecasting chaotic time-series data, achieving a dramatic improvement in prediction accuracy up to 2400 times greater than that of previous methods.

Machine learning has also emerged as a powerful tool in quantum many-body physics, a field where the exponential growth of complexity with system size poses major computational hurdles. Carleo and Troyer [26] introduced a novel approach using ANNs as variational representations of quantum states, enabling the efficient modelling of strongly correlated quantum systems. Building on this framework, Nomura *et al* [27] developed methods to accurately approximate ground-state wavefunctions for entangled quantum spin systems and fermionic models on lattices. In a complementary effort, Gao and Duan [28] combined restricted Boltzmann machines with variational Monte Carlo techniques, producing a robust and scalable solver for quantum many-body problems.

These advancements clearly demonstrate that ANNs are capable of representing large-scale quantum states, including the ground states of complex Hamiltonians and quantum systems undergoing dynamic evolution [29–32]. As such, ANNs have become indispensable tools in modern computational physics.

Despite their versatility and expressive power, ANNs have several inherent limitations. They typically require large amounts of data to perform well, making them less suitable for problems with limited observations. Moreover, they tend to produce overconfident predictions and are known to be susceptible to adversarial perturbations. A further drawback is that learning in ANNs is generally based on maximum likelihood estimation, which yields point estimates of model parameters rather than full probability distributions, limiting their ability to quantify uncertainty in predictions.

On the other hand, Gaussian processes offer a principled alternative in this regard. As non-parametric, probabilistic models, Gaussian processes define distributions over functions and are particularly effective in learning from small datasets. Their Bayesian nature allows them to capture uncertainty directly by producing a posterior distribution over possible functions conditioned on observed data. This enables not only accurate interpolation but also a coherent estimation of uncertainty, making Gaussian processes well-suited for tasks that require robust predictions with quantified confidence, especially in data-scarce regimes.

Machine learning methods have been applied to model BECs, with Liang *et al* [33] notably employing convolutional neural networks (CNNs) to approximate ground-state wavefunctions. Although their study showed that neural networks can replicate wavefunctions, these models do not consistently outperform traditional numerical methods [34] in accuracy or computational efficiency. This challenge can be addressed by leveraging Gaussian processes as a computationally efficient and statistically robust alternative for ground-state wavefunction modelling in BECs. Gaussian processes offer enhanced interpretability, scalability, and inherent uncertainty quantification, providing clear advantages over neural network-based approaches.

The machine learning approach for solving the GP equation in the following section is adapted from [35] utilizing the potential of Gaussian processes to model the ground state wavefunction of BECs more accurately with minimum data.

One now demonstrates the potential of Gaussian processes as surrogate models for data-driven emulation of one-dimensional scalar and vector BECs. The present validation framework evaluates Gaussian processes based on the following criteria:

1. *Correctness*: Can the Gaussian process accurately model the ground state of BECs?
2. *Versatility*: Can the Gaussian process adapt to different settings of BECs?
3. *Data efficiency*: How many data points are necessary to model a wavefunction?
4. *Compute efficiency*: How do they fare against numerical techniques when used as a simulator, after training?

To answer these questions, the performance of Gaussian processes on different settings of BECs are studied. The Trotter–Suzuki approximation, a numerical technique for simulating the ground state wavefunctions ψ has been employed. One runs simulations by setting up a one-dimensional grid and varies the coupling strength g, the parameter that controls the interaction between the atoms besides varying the trapping potential. Similarly for two-component BECs, one runs simulations by varying interaction parameters $\{g_{11}, g_{12}, g_{22}\}$ which results in two wavefunctions $\{\psi_1, \psi_2\}$. One models the simulated wavefunctions using a Gaussian process with radial basis function (RBF) kernel [36]. The Gaussian process models the wavefunction as a function of space x and coupling strength g. The results reveal the versatility of Gaussian processes in modelling different kinds of wavefunctions, efficiently and accurately, with uncertainty estimates. In addition to modelling uncertainty from data, this method performs better than Liang *et al* [33] in terms of efficiency and model complexity. The present model being simpler in terms of model complexity, uses just a small fraction ($\frac{1}{50}$th) of the data points used by Liang *et al* to achieve similar accuracy. Furthermore, comparing its efficiency of the method in predicting wavefunctions, with Trotter–Suzuki approximation, one finds that the above method performs 36× faster.

3.6 Background

3.6.1 Trotter–Suzuki approximation

The behaviour of any physical system can be studied by solving the PDEs which represent the dynamics of that physical phenomenon. In practice, most PDEs with any real application are nonlinear in nature and are hard to solve analytically. This is particularly true for complex dynamical systems which are quite difficult to solve and necessitate high computational resources to arrive at highly accurate solutions. Trotter–Suzuki decomposition implemented by Wittek and Cucchietti [37], exploits optimized kernels to solve the GP equation of a free particle. The exponential operators in PDEs are notoriously hard to approximate. Trotter–Suzuki decomposes the Hamiltonian into the sum of diagonal matrices which eases the task of computing the exponential. The evolution operator is calculated using the Trotter–Suzuki approximation. Given a Hamiltonian as a sum of Hermitian operators, for instance $H = H_1 + H_2 + H_3$, the evolution is approximated as [37]

$$e^{-i\Delta t H} = e^{-i\frac{\Delta t}{2}H_1}e^{-i\frac{\Delta t}{2}H_2}e^{-i\frac{\Delta t}{2}H_3}e^{-i\frac{\Delta t}{2}H_3}e^{-i\frac{\Delta t}{2}H_2}e^{-i\frac{\Delta t}{2}H_1}. \tag{3.61}$$

3.6.2 Gaussian processes

Gaussian processes [38] are probabilistic machine learning models that define a distribution over functions. They are infinite dimensional realizations of a multivariate Gaussian distribution. Given a set of points in the input space $\{x_1, x_2, \ldots x_n\}$ and the function f evaluated at those points $\{f_e^1, f_e^2, \ldots f_e^n\}$, a GP can be formally defined as follows.

Definition: $p(f)$ is a Gaussian process if for any subset $\{x_1, x_2, \ldots x_n\} \subset \mathcal{X}$, the marginal distribution over the subset $p(f_e)$ has a multivariate Gaussian distribution.

Considering a multivariate Gaussian distribution given by,

$$z_i \sim \mathcal{N}(\mu, \Sigma),$$

where $\mu \in \mathbb{R}^k$ is a k-dimensional zero vector and $\Sigma \in \mathbb{R}^{k\times k}$ is the covariance matrix that captures the correlation between dimensions $\{X_1, X_2, ..X_k\}$. Let us sample four points $\{z_1, z_2, z_3, z_4\}$ from the distribution and plot it sequentially, as shown in figure 3.5(a). From figure 3.5(a), one can make two inferences: (1) points closer to each other on the x-axis behave similarly on the y-axis; and (2) points further apart from each other behave differently on the y-axis. That is, closer points are highly correlated with each other while further points are not. This intuition leads to the understanding that by connecting these points sequentially, one can realize smooth functions like the ones shown in figure 3.5(b). The functions being realized by connecting the points sequentially can be considered as samples from a Gaussian process by a zero-mean vector μ and the hitherto unknown covariance matrix Σ.

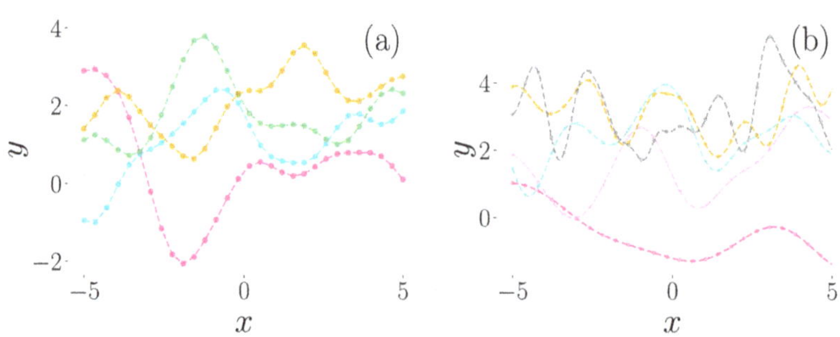

Figure 3.5. (a) Four samples from a multivariate normal distribution $\mathcal{N}(0, \Sigma)$ plotted sequentially, and (b) samples from Gaussian process prior conditioned on different values of length scale l_k. Reproduced from [35]. CC BY 4.0.

Yet another inference that can be made from figure 3.5(a) is that the nature of the functions realized are dependent on the covariance matrix Σ. By manipulating the covariance matrix Σ, one can manipulate the kind of functions realized from k-dimensional samples $\{z_i\}$. The mechanism to manipulate the covariance matrix is through an algebraic function named *covariance function* or *kernel*. The kernel $k(x, x') \to \mathcal{R}$ takes two points from the input space $\mathcal{X}$ as inputs and returns a scalar value $c \in [0, 1]$. This value represents the similarity between the two input points. The covariance matrix Σ is constructed by calculating the similarity between k equally spaced points in the input space $\{x_1, \ldots, x_k\}$. By controlling the kernel function, one can control the covariance matrix that governs the GP, thereby controlling the nature of functions generated. The kernel is usually parameterized by one or more tunable variables which act as knobs for controlling the shape of the functions that Gaussian process generates.

Assuming that for didactic purposes the x and y are scalars and the function f is given by $f: \mathcal{X} \to R$, one considers RBF parameterized by length scale k_l as the kernel.

$$k(x_i, x_j) = \exp\left(-\frac{1}{2k_l^2}(x_i - x_j)^2\right) \tag{3.62}$$

Length scale k_l controls the order of the functions generated, as evident from figure 3.5(b).

3.6.2.1 Inference

Inference in Gaussian processes consists of two main steps: *kernel parameter Search* and *posterior estimation*. In the kernel parameter search phase, the kernel parameters are optimized to achieve the best fit with the observed data. The optimal kernel parameters are determined through maximum likelihood estimation (MLE) by minimizing the marginal negative log-likelihood [39], which is given by:

$$\log p(y|X) = \log \mathcal{N}(y|0, K_y) = -\frac{1}{2}y^T K_y^{-1} y - \frac{1}{2}\log|K_y| - \frac{N}{2}\log(2\pi) \tag{3.63}$$

$$K_y = k(X, X) \tag{3.64}$$

where N is the number of data points, (X, y) represent the data points.

Then, one estimates Gaussian process posterior conditioned on data, i.e., mean vector and covariance matrix conditioned on data. The definition of Gaussian process states that the joint distribution of observed data y and predictions f_* has a multivariate Gaussian distribution. Given data X and new input X_*, one can write the joint distribution as,

$$\begin{bmatrix} y \\ f_* \end{bmatrix} \sim \mathcal{N}\left(\mathbf{0}, \begin{pmatrix} K_y & K_* \\ K_*^T & K_{**} \end{pmatrix}\right) \tag{3.65}$$

$$K_* = k(X, X_*), \quad K_{**} = k(X_*, X_*)$$

Using the multivariate Gaussian theorem [40], one arrives at the parameters of the posterior,

$$\mu_* = K_*^T K_y^{-1} y, \quad \Sigma_* = K_{**} - K_*^T K_y^{-1} K_*.$$

3.6.2.2 Motivation

In our experiments, one considers the ground state wavefunction of one-dimensional BECs [34]. The simulation takes as input multiple parameters including grid parameters like radius, length, potential function (V_{ext}), coupling strength (g), etc, and generates a one-dimensional wavefunction using imaginary time evolution method. The time period to complete a simulation depends on the dimensions and size of the grid. The promise of data-driven scientific computing is the ability to model any process without closed-form solutions using experimental data and building a predictive model that can provide predictions at any point in the grid with absolute guarantee. As discussed earlier, a data-driven model trained on scattered data points from simulations is capable of predicting the wavefunction at any point within the grid in a few fixed number of central processing unit cycles irrespective of grid dimensions or size.

Gaussian processes model the wavefunction as a probability distribution over functions supported by a kernel with optimal parameters. Being a Bayesian model, Gaussian process is data-efficient and makes uncertainty-aware predictions. The ability to model uncertainty is especially important in modelling physical phenomenon from a few scattered noisy observations. Bayesian methods rely on Bayesian Inference which estimates posterior over unknowns in contrast to non-probabilistic models which result in point estimates of unknowns. As a consequence, it is possible to generate entire wavefunctions from the Gaussian processes that are subject to constraints on the input parameters like coupling strength, omega (Rabi coupling), etc.

3.7 Methods

One employs Trotter–Suzuki-mpi [41], a massively parallel implementation of the Trotter–Suzuki approximation to simulate the evolution of BECs. The simulation setup consists of a one-dimensional grid which represents the physical system. This discretized space is defined by a lattice of 512 nodes within a physical length of 24 units. The physics of BECs is described by the Hamiltonian which represents the gravitational potential energy equation. The Hamiltonian requires a grid and a trapping potential. The trapping potential is defined as a function which takes x and y as arguments and returns a scalar value as output. In this case, one considers a harmonic trapping potential given by $\frac{x^2}{2}$. The state of the system is initialized in Gaussian form. Imaginary time evolution evolves the system state using the dynamics defined by the Hamiltonian, in 10^3 iterations with time step being $\Delta_t = 10^{-4}$ for each iteration. To collect data, M simulations are conducted by varying the interaction parameter g. The sklearn Gaussian process API [42, 43] with an RBF kernel is employed as a surrogate to model the wavefunction ψ as a simpler, continuous function of space x and the interaction parameter g.

The prior on the ground state wavefunction of BEC is given by

$$\psi_{\text{prior}} \sim GP(0, k(x, x'; \theta)).$$

N data points of the form (x, g, ψ) are sampled randomly from M simulation outcomes. The RBF kernel given in equation (3.62) is used throughout the numerical experiments.

The hyper-parameters θ of the kernel are tuned using maximum likelihood estimation which essentially amounts to finding the parameters that minimize the expression (3.63), given a list of data points.

The RBF kernel consists of the following parameters: length scale k_l and variance k_σ. $\psi_{\text{posterior}}$ is estimated by conditioning ψ_{prior} on the sampled data points. The expression for $\psi_{\text{posterior}}$ reduces to a mathematically tractable form given by the mean vector and covariance matrix presented in equation (3.65). Inference consists of calculating the marginal mean $\mu_{2|1}$ and covariance $\Sigma_{2|1}$ by substituting the training points X and new test points X_* into the analytical forms presented in equation (3.65). The marginal mean $\mu_{2|1}$ constitutes the predicted wavefunction. The diagonal elements of the covariance matrix form the variance σ on each prediction. Together, they make the Gaussian process posterior parameterized by $[\mu_{2|1}, \Sigma_{2|1}]$. The fidelity of the predicted wavefunction is measured by calculating the mean-squared error (MSE) metric against the ground truth data. Figure 3.6 portrays the results of an illustrative experiment on one-dimensional BECs with a harmonic trapping potential.

The experimental setup described above can be extended to different settings of simulation and data-driven approximation, to include two-component BECs and different kinds of trapping potentials. A detailed report of the experiments conducted is presented in the next section.

3.8 Experiments

All the experiments to be covered in this section will follow along similar lines as the illustrative experiment discussed in the previous section: simulation, data-driven approximation and evaluation. In the previous experiment, 300 simulations shown in figure 3.6(a) for equally spaced values of g ranging from 1 to 100 were run. One randomly sampled 500 samples of the form (x, g, ψ) from the simulation results, which were used to fit the Gaussian process. Evaluation leads to an MSE score of 1.23×10^{-7} which indicates that the trained Gaussian process can accurately predict ground state wavefunctions for 100 different values of the interaction parameter g within the range (1, 100).

$$i\hbar\frac{\partial\psi_1}{\partial t} = \left(-\frac{\hbar^2}{2m}\nabla^2 + V_{ext}(x) + g_{11}|\psi_1|^2 + g_{12}|\psi_2|^2\right)\psi_1 + \Omega\psi_2 \tag{3.66}$$

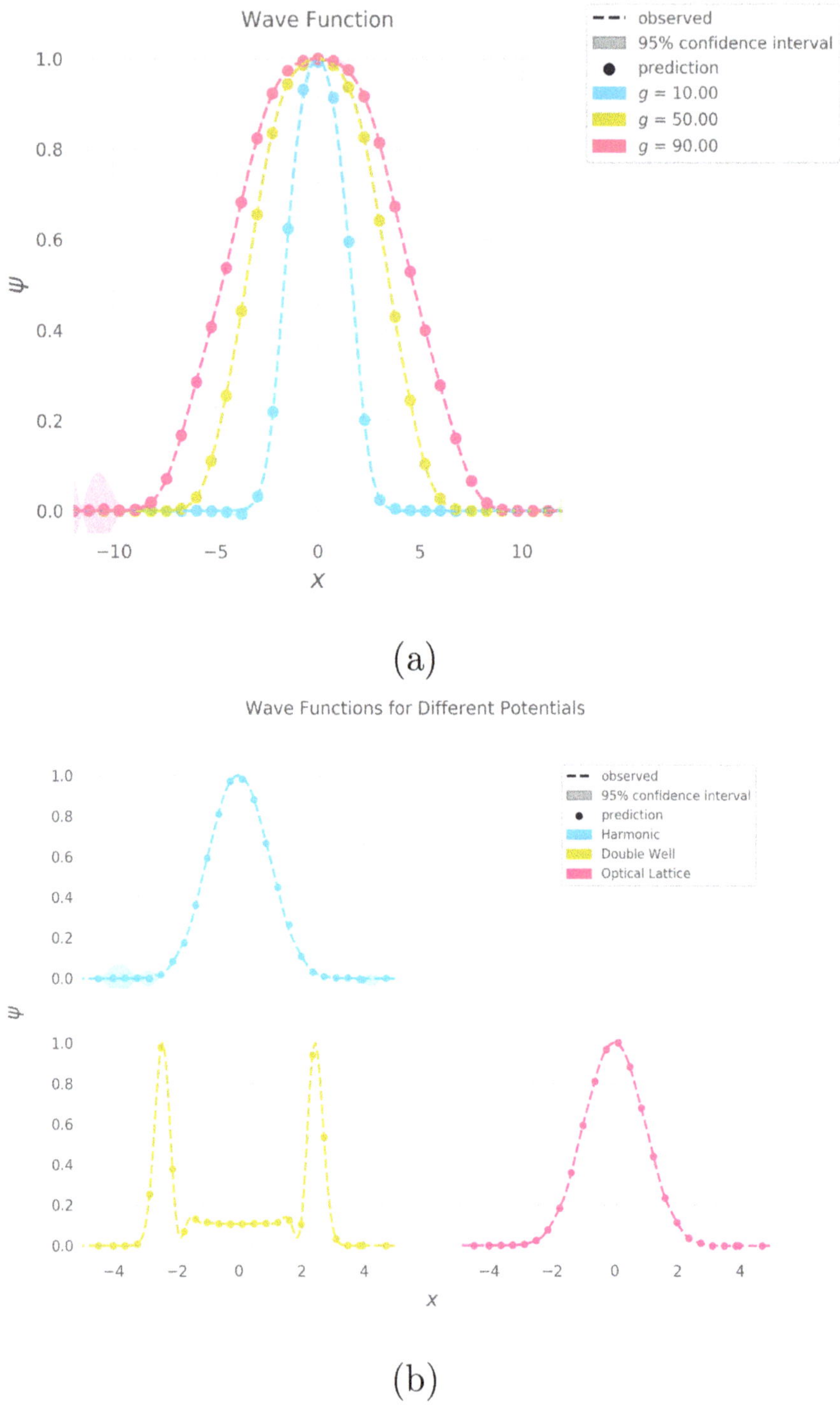

Figure 3.6. (a) Simulated and predicted ground state wavefunctions for different values of the interaction parameter $g = 10, 50, 90$, and (b) prediction of wavefunctions conditioned on different trapping potentials—harmonic, optical lattice, double-well potentials. Reproduced from [35]. CC BY 4.0.

$$i\hbar\frac{\partial\psi_2}{\partial t} = \left(-\frac{\hbar^2}{2m}\nabla^2 + V_{ext}(x) + g_{21}|\psi_1|^2 + g_{22}|\psi_2|^2\right)\psi_2 + \Omega\psi_1 \tag{3.67}$$

To follow up, one can ask the question, how many data points are necessary to build a reliable predictive model of ground state wavefunction? In order to answer this question, one varies the number of training samples and observes the effect it has on MSE score of GP trained on those samples. One reuses the experimental setup from the previous experiment and vary the number of data points N sampled from $M = 300$ simulations. The results are presented in figure 3.7. Figure 3.7 depicts the relationship between variance and number of samples N and it shows the rate of decrease of MSE w.r.t. N. The MSE curve shows a trend of saturation close to 0 as the number of samples are increased, as shown in table 3.6. Decrease in variance and MSE is evidence of confidence and accuracy in prediction, respectively. The next experiment tests the versatility of the Gaussian process by modelling the ground states of BECs for different choices of trapping potentials. One considers harmonic, double-well and optical lattice potentials in figure 3.8. It is better to restrict the range of g to (0, 2) in order to ensure the stability of generated wavefunction for different potentials. 100 simulations were run and 500 samples were collected for each potential. A Guassian process for each potential is fitted using collected data points. Evaluating the models results in MSE scores of 1.23×10^{-7}, 1.57×10^{-6} and 4.30×10^{-5}, for harmonic, double-well and optical lattice potentials [44],

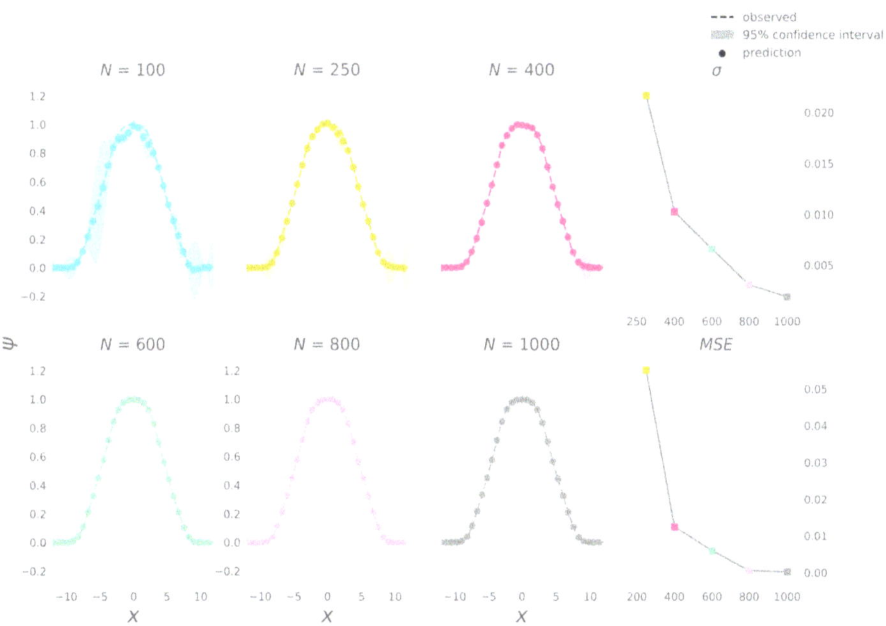

Figure 3.7. Gaussian processes fit on training samples of different sizes, $N = \{100, 250, 400, 600, 800, 1000\}$, variance σ and MSE plotted against training sample size. Reproduced from [35]. CC BY 4.0.

Table 3.6. Error estimates from experiments. Reproduced from [35]. CC BY 4.0.

Experiment	Number of training samples	Mean-squared error
One-component BEC with harmonic potential	500	1.23×10^{-7}
One-component BEC with double-well potential	500	1.57×10^{-6}
One-component BEC with optical lattice potential	500	4.30×10^{-5}
Two-component BEC with harmonic potential	1000	2.52×10^{-5}
Two-component BEC with double-well potential	1000	2.43×10^{-4}
Two-component BEC with optical lattice potential	1000	3.27×10^{-4}
Rabi-coupled two-component BEC with cosine potential	1200	4.20×10^{-4}

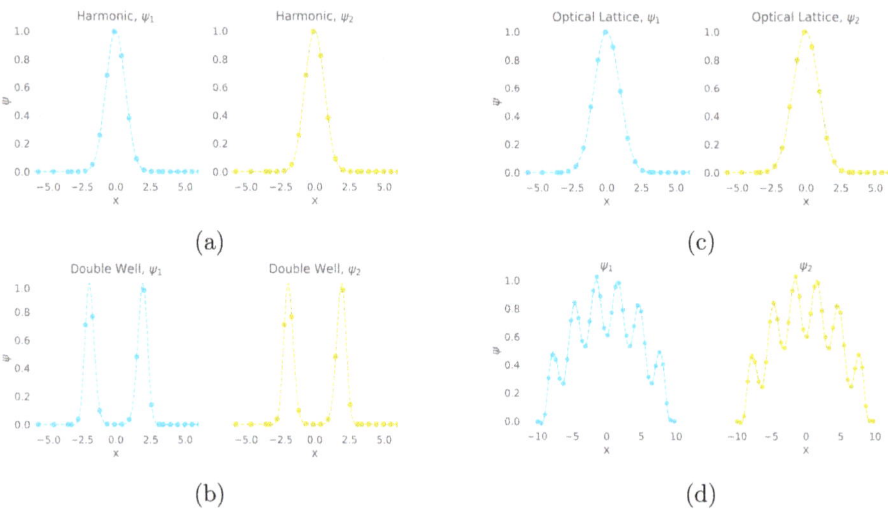

Figure 3.8. (a, b, c) Prediction of two-component BECs ground-state wavefunctions for different trapping potentials: (a) harmonic, (b) double-well, (c) optical lattice, (d) emulation of Rabi-coupled two-component BECs conditioned on cosine potential. Reproduced from [35]. CC BY 4.0.

respectively. The predictive wavefunctions for a value of $g = 1$ are plotted against the ground truth in figures 3.8(a–c).

The ground state wavefunctions of two-component BECs are given by ψ_1 and ψ_2. The interaction strength is defined by parameters g_{11}, g_{12} and g_{22}. One models the system using a Gaussian process which learns the relationship between (ψ_1, ψ_2) and the interaction parameters. A total of 500 data points are sampled across 200 simulations with values of interaction parameters lying within the range $(-1, 1)$. The choice of this range is dictated by the zone of stability of the system. Predictions and ground truth wavefunctions for different potentials conditioned on values of

Table 3.7. Trotter–Suzuki versus Gaussian processes. Reproduced from [35]. CC BY 4.0.

Setting	Trotter–Suzuki (ms)	Gaussian processes (ms)
One-component BEC with grid size $d = 512$	160	11
One-component BEC with grid size $d = 1024$	330	22.6
Two-component BEC with grid size $d = 512$	930	14
Two-component BEC with grid size $d = 1024$	1150	23

interaction parameters given by $\{g_{11} = 0.1, g_{12} = -0.1, g_{22} = 0.1\}$, are presented in figures 3.8(a–c). Evaluation on a test set within the range $g \in (-1, 1)$ results in MSE in the range of 10^{-4}. One reproduces an experiment conducted by Liang *et al* [33] using Gaussian process—emulation of Rabi-coupled two-component BECs. Figure 3.8(d) shows the predicted ground state wavefunction of Rabi-coupled two-component BECs described by equations (3.66) and (3.67), conditioned on a cosine potential given by $\frac{x^2}{2} + 24\cos^2 x$ with interaction parameters $\{g_{11} = 103, g_{12} = 100, g_{22} = 97\}$. Finally, one employs trained Gaussian processes as simulators to generate wavefunctions and compares their performance against Trotter–Suzuki. One experiments with one-component and two-component BECs in 512 and 1024 grid spaces. The results tabulated in table 3.7 indicate an average of $36\times$ speed up in wave generation with reasonable accuracy.

3.9 Summary and future challenges

This chapter presented numerical techniques for solving the GP equation to model BECs across various trapping potentials and interaction strengths. Numerical solutions were obtained for one-dimensional, two-dimensional, and quasi-one/two-dimensional GP equations in spherically and axially symmetric traps, employing the SSCN, SSFT, and NCG gradient methods. These approaches enabled the computation of ground-state wavefunctions via imaginary time propagation and time evolution of BEC dynamics. Additionally, data-driven methods employing Gaussian processes were explored as alternatives to traditional numerical simulations, leveraging the Trotter–Suzuki approximation for enhanced efficiency. These techniques provided robust tools for studying BEC ground states and dynamics in diverse experimental configurations.

Even though it looks as if substantial progress has been made in this direction, several challenges still need to be overcome, giving scope for future research. By exploring the application of physics-informed neural networks (PINNs) to model the dynamics of scalar (single-component) and vector (multi-component) BECs, one can develop innovative solutions to complex problems, thereby offering potential improvements in computational efficiency over SSCN, SSFT, and NCG methods. Developing PINNs to handle complex trapping geometries and nonlinear interactions while maintaining physical accuracy warranted further exploration. Additionally, integrating data-driven approaches with high-precision numerical methods to optimize BEC simulations in real-time experimental settings poses an open question, aiming to bridge theoretical models with practical implementations.

3.10 Problems

Exercise 3.1

1. A BEC consists of approximately 6×10^6 ^{87}Rb atoms confined in an axially symmetric harmonic trap with trapping frequencies $\omega_x = 2\pi \times 210$ Hz, $\omega_y = 2\pi \times 210$ Hz, and $\omega_z = 2\pi \times 33$ Hz along the x, y, and z directions, respectively, at a temperature of 500 nK [45]. Calculate the characteristic time and length scales of the trap and the interaction parameter g for the BEC.
2. Solve the time-independent 1D GP equation for a BEC of 10^5 ^{87}Rb atoms ($m = 1.443 \times 10^{-25}$ kg, $a = 100a_0 \approx 5.29 \times 10^{-9}$ m) confined in a harmonic trap with potential $V(x) = \frac{1}{2}m\omega^2x^2$, where $\omega = 2\pi \times 200$ Hz. Use the imaginary-time propagation method with an SSCN scheme to compute the ground state and the first two excited states. Compare the chemical potential and rms size of the ground state with the Thomas–Fermi approximation.
3. Simulate the dynamics of a 1D BEC of 5×10^4 ^{23}Na atoms ($m = 3.817 \times 10^{-26}$ kg, $a = 2.75$ nm) in an optical lattice potential $V(x) = V_0 \sin^2(kx)$, where $V_0 = 10E_r$ (recoil energy $E_r = \hbar^2k^2/2m$, $k = 2\pi/\lambda$, $\lambda = 1064$ nm) and a harmonic trap $V_h(x) = \frac{1}{2}m\omega^2x^2$ with $\omega = 2\pi \times 50$ Hz. Use a split-step Fourier method to solve the time-dependent GP equation with an initial Gaussian wavefunction displaced by $x_0 = 5$ μm. Analyze the onset of Bloch oscillations and the effect of interaction strength on oscillation amplitude [46].
4. Compute the ground state of a 2D BEC of 10^4 ^{7}Li atoms ($m = 1.165 \times 10^{-26}$ kg, $a = -27a_0 \approx -1.43 \times 10^{-9}$ m) in an isotropic harmonic trap $V(x, y) = \frac{1}{2}m\omega^2(x^2 + y^2)$, with $\omega = 2\pi \times 150$ Hz. Use the NCG method to solve the time-independent GP equation. Investigate the critical number of atoms for collapse by incrementally increasing N until the solution becomes unstable [47].
5. Simulate vortex formation in a 2D BEC of 2×10^5 ^{87}Rb atoms in a harmonic trap $V(x, y) = \frac{1}{2}m\omega^2(x^2 + y^2)$, $\omega = 2\pi \times 100$ Hz, under rotation at angular frequency $\Omega = 0.7\omega$. Use the time-dependent GP equation with an SSCN method to evolve an initial Thomas–Fermi profile. Introduce a small random perturbation to break symmetry. Analyze the number and arrangement of vortices as a function of time [48].
6. Study the dynamics of a dark soliton in a 1D BEC of 3×10^4 ^{23}Na atoms in a double-well potential $V(x) = \frac{1}{2}m\omega^2x^2 + V_0e^{-x^2/\sigma^2}$, with $\omega = 2\pi \times 80$ Hz, $V_0 = 5\hbar\omega$, and $\sigma = 2$ μm. Use an SSCN method to solve the time-dependent GP equation, starting with a dark soliton solution in one well. Analyze the soliton's oscillation between wells and its stability against perturbations [49].

7. Simulate transcritical flow of a 1D BEC of 10^5 ^{87}Rb atoms through a repulsive Gaussian barrier $V(x) = V_0 e^{-x^2/\sigma^2}$, with $V_0 = 10\hbar\omega$, $\sigma = 1\ \mu\text{m}$, and $\omega = 2\pi \times 50$ Hz, in a box potential ($V = \infty$ for $|x| > L$, $L = 20\ \mu\text{m}$). Use a time-dependent GP equation with a split-step Fourier method to model an incoming flow (velocity $v_0 = 0.5$ mm s^{-1}). Analyze the formation of dispersive shock waves and compare with the Whitham modulation theory [50].
8. Compute the ground and excited states of a 2D BEC of 10^5 ^{87}Rb atoms in a square optical lattice $V(x, y) = V_0[\sin^2(kx) + \sin^2(ky)]$, with $V_0 = 8E_r$, $k = 2\pi/780$ nm. Use a split-step Fourier method with imaginary-time propagation to solve the time-independent GP equation. Analyze the band structure by computing the chemical potential for different lattice depths ($V_0 = 4, 8, 12E_r$) [51].
9. Simulate a 1D BEC of 2×10^4 ^{23}Na atoms in a harmonic trap $V(x, t) = \frac{1}{2}m\omega^2(x - x_0(t))^2$, where $\omega = 2\pi \times 100$ Hz and the trap centre moves as $x_0(t) = A\sin(\omega_d t)$, with $A = 5\ \mu\text{m}$, $\omega_d = 2\pi \times 20$ Hz. Use a stochastic GP equation with phenomenological damping ($\gamma = 0.01\omega$) to model dissipation [52, 53]. Analyze the condensate's breathing mode and energy dissipation over time [54].
10. Solve for the stationary states of a 2D BEC of 8×10^4 ^{87}Rb atoms in a double-well potential $V(x, y) = \frac{1}{2}m\omega_x^2 x^2 + \frac{1}{2}m\omega_y^2 y^2 + V_0 e^{-x^2/\sigma^2}$, with $\omega_x = \omega_y = 2\pi \times 120$ Hz, $V_0 = 8\hbar\omega_x$, $\sigma = 1.5\ \mu\text{m}$. Use the BiCG method to solve the time-independent GP equation.

References

[1] Gammal A, Frederico T and Tomio L 1999 Improved numerical approach for the time-independent Gross-Pitaevskii nonlinear Schrödinger equation *Phys. Rev.* E **60** 2421–4

[2] Koonin S E 2018 *Computational Physics: Fortran Version* (CRC Press)

[3] Brtka M, Gammal A and Tomio L 2006 Relaxation algorithm to hyperbolic states in Gross-Pitaevskii equation *Phys. Lett.* A **359** 339–44

[4] Adhikari S K and Muruganandam P 2002 Bose-Einstein condensation dynamics from the numerical solution of the Gross-Pitaevskii equation *J. Phys. B: At. Mol. Opt. Phys.* **35** 2831–43

[5] Dautray R 2000 *Mathematical Analysis and Numerical Methods for Science and Technology* (Springer)

[6] Muruganandam P and Adhikari S K 2009 Fortran programs for the time-dependent Gross-Pitaevskii equation in a fully anisotropic trap *Comput. Phys. Commun.* **180** 1888–912

[7] Ames W F 2014 *Numerical Methods for Partial Differential Equations* 3rd edn (Academic) (description based upon print version of record)

[8] Weideman J A C and Herbst B M 1986 Split-step methods for the solution of the nonlinear Schrödinger equation *SIAM J. Numer. Anal.* **23** 485–507

[9] Trefethen L N 2000 *Spectral Methods in MATLAB* (Society for Industrial and Applied Mathematics)

[10] Agrawal G 2010 *Nonlinear Fiber Optics* 4th edn (Academic)

[11] Yang J 2010 *Nonlinear Waves in Integrable and Nonintegrable Systems* (Society for Industrial and Applied Mathematics)

[12] Yang J 2009 Newton-conjugate-gradient methods for solitary wave computations *J. Comput. Phys.* **228** 7007–24

[13] Tan M, Pang R and Le Q V 2020 EfficientDet: scalable and efficient object detection *2020 IEEE/CVF Conf. on Computer Vision and Pattern Recognition (CVPR)* (IEEE)

[14] Ng N, Yee K, Baevski A, Ott M, Auli M and Edunov S 2019 Facebook FAIR's WMT19 News translation task submission *Proc. of the Fourth Conf. on Machine Translation (Volume 2: Shared Task Papers, Day 1)* (Association for Computational Linguistics)

[15] Christiansen E M *et al* 2018 In silico labeling: predicting fluorescent labels in unlabeled images *Cell* **173** 792–803.e19

[16] Weyand T, Kostrikov I and Philbin J 2016 Planet-photo geolocation with convolutional neural networks *Proc. Computer Vision–ECCV 2016: 14th Eur. Conf. (Amsterdam, The Netherlands, 11–14 October 2016) Part VIII 14* (Springer) pp 37–55

[17] Brown N and Sandholm T 2019 *Superhuman AI for multiplayer poker Science* **365** 885–90

[18] Ibarz B, Leike J, Pohlen T, Irving G, Legg S and Amodei D 2018 Reward learning from human preferences and demonstrations in atari *NIPS'18: Proc. of the 32nd Int. Conf. on Neural Information Processing Systems* vol 3 8022–34

[19] Borowiec S 2016 Alphago seals 4-1 victory over go grandmaster lee sedol *The Guardian* https://www.theguardian.com/technology/2016/mar/15/googles-alphago-seals-4-1-victory-over-grandmaster-lee-sedol

[20] Carrasquilla J and Melko R G 2017 Machine learning phases of matter *Nat. Phys.* **13** 431–4

[21] Morningstar A and Melko R G 2018 Deep learning the Ising model near criticality *J. Mach. Learn. Res.* **18** 1–17

[22] Tanaka A and Tomiya A 2017 Detection of phase transition via convolutional neural networks *J. Phys. Soc. Japan* **86** 063001

[23] Zdeborová L 2017 New tool in the box *Nat. Phys.* **13** 420–1

[24] Greitemann J, Liu K, Jaubert L D C, Yan H, Shannon N and Pollet L 2019 Identification of emergent constraints and hidden order in frustrated magnets using tensorial kernel methods of machine learning *Phys. Rev.* B **100** 174408

[25] Jaeger H and Haas H 2004 Harnessing nonlinearity: predicting chaotic systems and saving energy in wireless communication *Science* **304** 78–80

[26] Carleo G and Troyer M 2017 Solving the quantum many-body problem with artificial neural networks *Science* **355** 602–6

[27] Nomura Y, Darmawan A S, Yamaji Y and Imada M 2017 Restricted Boltzmann machine learning for solving strongly correlated quantum systems *Phys. Rev.* B **96** 205152

[28] Gao X and Duan L -U 2017 Efficient representation of quantum many-body states with deep neural networks *Nat. Commun.* **8** 662

[29] Carleo G, Nomura Y and Imada M 2018 Constructing exact representations of quantum many-body systems with deep neural networks *Nat. Commun.* **9** 5322

[30] Czischek S, Gärttner M and Gasenzer T 2018 Quenches near Ising quantum criticality as a challenge for artificial neural networks *Phys. Rev.* B **98** 024311

[31] Schmitt M and Heyl M 2018 Quantum dynamics in transverse-field Ising models from classical networks *SciPost Phys.* **4** 013

[32] Fabiani G and Mentink J 2019 Investigating ultrafast quantum magnetism with machine learning *SciPost Phys.* **7** 004

[33] Liang X, Zhang H, Liu S, Li Y and Zhang Y -S 2018 Generation of Bose-Einstein condensates' ground state through machine learning *Sci. Rep.* **8** 16337
[34] Bao W, Jaksch D and Markowich P A 2003 Numerical solution of the Gross-Pitaevskii equation for Bose-Einstein condensation *J. Comput. Phys.* **187** 318–42
[35] Bakthavatchalam T A, Ramamoorthy S, Sankarasubbu M, Ramaswamy R and Sethuraman V 2021 Bayesian Optimization of Bose-Einstein Condensates *Sci. Rep.* **11** 5054
[36] Vert J -P, Tsuda K and Schölkopf B 2004 *A Primer on Kernel Methods* (The MIT Press) pp 35–70
[37] Wittek P and Cucchietti F M 2013 A second-order distributed Trotter-Suzuki solver with a hybrid CPU-GPU kernel *Comput. Phys. Commun.* **184** 1165–71
[38] MacKay D J 2006 *The Humble Gaussian Distribution* (Cavendish Laboratory)
[39] Rasmussen C E and Williams C K I 2005 *Gaussian Processes for Machine Learning* (The MIT Press)
[40] Murphy K P 2012 *Machine Learning: a Probabilistic Perspective* (MIT Press)
[41] Wittek P and Calderaro L 2015 Extended computational kernels in a massively parallel implementation of the Trotter-Suzuki approximation *Comput. Phys. Commun.* **197** 339–40
[42] Pedregosa F *et al* 2011 Scikit-learn: Machine learning in python *J. Mach. Learn. Res.* **12** 2825–30
[43] Buitinck L *et al* 2013 API design for machine learning software: experiences from the scikit-learn project *ECML PKDD Workshop: Languages for Data Mining and Machine Learning* 108–22
[44] Bao W and Cai Y 2013 Mathematical theory and numerical methods for Bose-Einstein condensation *Kinet. Relat. Mod.* **6** 1–135
[45] Greiner M, Bloch I, Hänsch T W and Esslinger T 2001 Magnetic transport of trapped cold atoms over a large distance *Phys. Rev.* A **63** 031401
[46] Witthaut D, Werder M, Mossmann S and Korsch H J 2005 Bloch oscillations of Bose-Einstein condensates: Breakdown and revival *Phys. Rev.* E **71** 036625
[47] Gammal A, Tomio L and Frederico T 2002 Critical numbers of attractive Bose-Einstein condensed atoms in asymmetric traps *Phys. Rev.* A **66** 043619
[48] Cozzini M, Stringari S and Tozzo C 2006 Vortex lattices in Bose-Einstein condensates: From the Thomas-Fermi regime to the lowest-Landau-level regime *Phys. Rev.* A **73** 023615
[49] Gubeskys A and Malomed B A 2007 Symmetric and asymmetric solitons in linearly coupled Bose-Einstein condensates trapped in optical lattices *Phys. Rev.* A **75** 063602
[50] Leszczyszyn A M, El G A, Gladush Y G and Kamchatnov A M 2009 Transcritical flow of a Bose-Einstein condensate through a penetrable barrier *Phys. Rev.* A **79** 063608
[51] Bao W and Du Q 2004 Computing the ground state solution of Bose-Einstein condensates by a normalized gradient flow *SIAM J. Sci. Comput.* **25** 1674–97
[52] Gardiner C W, Anglin J R and Fudge T I A 2002 The stochastic Gross-Pitaevskii equation *J. Phys. B: At. Mol. Opt. Phys.* **35** 1555–82
[53] Gardiner C W and Davis M J 2003 The stochastic Gross-Pitaevskii equation: II *J. Phys. B: At. Mol. Opt. Phys.* **36** 4731–53
[54] Cockburn S P, Gallucci D and Proukakis N P 2011 Quantitative study of quasi-one-dimensional Bose gas experiments via the stochastic Gross-Pitaevskii equation *Phys. Rev.* A **84** 023613

IOP Publishing

An Introduction to Ultracold Atoms with Analytical and Numerical Methods

Paulsamy Muruganandam and Ramaswamy Radha

Chapter 4

Bose–Einstein condensates and optical solitons

Bose–Einstein condensates (BECs) and optical solitons exhibit several fundamental similarities. Both systems are characterized by a macroscopic wave function that describes the collective behaviour of the entire dynamical system. A key shared property is coherence, where phase relationships between individual components are maintained, leading to unified dynamics. The dynamics of optical solitons are governed by the nonlinear Schrödinger equation (NLS), while BECs are described by the Gross–Pitaevskii (GP) equation, which represents a modified form of the NLS equation. In BECs, nonlinear interatomic interactions dictate the system's evolution, whereas, for optical solitons, the nonlinear interaction with the propagating medium plays an analogous role. Matter-wave solitons in BECs demonstrate striking parallels with optical solitons, as both maintain stable propagation profiles resistant to perturbations over extended durations. Mathematically, the descriptions of BECs and optical solitons share a common structure, differing primarily in the physical origin of their nonlinear terms: interatomic interactions for BECs versus light–matter interactions for optical solitons.

The NLS equation is known to be completely integrable, and the GP equation has also been shown to possess integrable solutions for specific trapping potentials and interaction parameters. This integrability implies that both systems conserve fundamental quantities such as energy, momentum, and norm (representing either particle number in BECs or optical intensity in solitons). Both NLS and GP equations employ a mean-field approximation, treating the respective systems as single macroscopic quantum wave functions. Historically, developments in optical soliton research provided crucial insights that later facilitated advancements in BEC studies. This chapter reviews the foundational progress in optical soliton physics that inspired subsequent investigations into analogous phenomena in BEC systems.

doi:10.1088/978-0-7503-5447-9ch4

4.1 Coupled nonlinear Schrödinger equations

This chapter lays the foundation for studying coupled GP equations by examining a class of physically significant, variable-coefficient coupled NLS equations, also referred to as inhomogeneous NLS systems. The analysis begins with the well-known Manakov model and progresses to the derivation of bright soliton solutions, complemented by a novel perspective on their collision dynamics.

The chapter introduces a specific system of coupled NLS equations characterized by opposing signs for the kinetic and gradient terms in the two equations. This system incorporates time-varying nonlinearity coefficients and an external harmonic trapping potential. By carefully selecting a time-dependent trapping potential, the system sustains stable soliton solutions, which are derived analytically through the gauge transformation approach. These exact solutions are corroborated by numerical simulations, which verify their robustness and stability over time.

Subsequently, the discussion extends to a generalized coupled NLS (GCNLS) equation that encompasses self-phase modulation (SPM), cross-phase modulation (XPM), and four-wave mixing (FWM), governed by four arbitrary real parameters. The chapter elaborates on the formulation of the corresponding Lax pair and the construction of bright soliton solutions for this generalized system. An in-depth analysis of soliton collision dynamics reveals a distinctive feature: the ability to redistribute intensity (or energy) not only between the two modes but also within a single mode. This phenomenon, absent in the standard Manakov model, highlights the enhanced flexibility of the generalized system in manipulating soliton interactions.

4.2 Coupled nonlinear Schrödinger equations (the Manakov model): new signatures

The concept of solitons as information carriers in optical fibres was theoretically proposed by Hasegawa and Tappert [1, 2] in 1973 and experimentally validated by Mollenauer *et al* [3] in 1980. Subsequent research has extensively explored the dynamics of temporal optical solitons in applications such as long-distance optical fibre communication and optical switching devices [4]. The propagation of electromagnetic waves in single-mode optical fibres is mathematically modelled by the integrable NLS equation, which emerges when the Kerr nonlinearity, specifically SPM, exactly counteracts group velocity dispersion [1, 4].

Single-mode optical fibres can sustain two orthogonal polarization modes. In an ideal fibre with perfect cylindrical symmetry and isotropic material properties, these polarization states would remain independent. However, practical imperfections, such as deviations in fibre geometry or stress-induced material anisotropy, disrupt this symmetry, leading to polarization mode coupling [5, 6]. In real optical fibres, birefringence varies along the fibre length due to fluctuations in core shape and mechanical stresses, further complicating polarization dynamics.

When multiple optical waves propagate simultaneously within a fibre, their interactions are governed by the fibre's nonlinear properties. A key interaction mechanism is cross-phase modulation, where the effective refractive index experienced by one wave depends not only on its own intensity but also on the intensity of co-propagating waves. Cross-phase modulation occurs concurrently with SPM and, in the case of orthogonally polarized waves, induces nonlinear birefringence within the fibre. Consequently, the propagation of solitons in birefringent nonlinear fibres is described by a system of coupled NLS equations [7–11], expressed as:

$$i\psi_{1,t} + \psi_{1,xx} + 2(g_{11}|\psi_1|^2 + g_{12}|\psi_2|^2)\psi_1 = 0, \tag{4.1a}$$

$$i\psi_{2,t} + \psi_{2,xx} + 2(g_{21}|\psi_1|^2 + g_{22}|\psi_2|^2)\psi_2 = 0, \tag{4.1b}$$

where $\psi_i(x, t)$, for $i = 1, 2$, represent the complex amplitudes of the two polarization components, subscripts t and x denote partial derivatives with respect to time and space, respectively, g_{11} and g_{22} quantify the SPM strengths, and g_{12} and g_{21} describe the cross-phase modulation strengths.

Prior studies have identified that the system of equations (4.1) is integrable under two specific conditions: (i) $g_{11} = g_{12} = g_{21} = g_{22}$, corresponding to the focusing Manakov model [12], or (ii) $g_{11} = g_{21} = -g_{12} = -g_{22}$, known as the modified Manakov model [13, 14]. The focusing Manakov model has been widely studied [15–20], with significant emphasis on the redistribution of bright soliton intensities. The modified Manakov model has also been thoroughly investigated for its unique soliton interactions. Bright solitons, localized solutions of equations (4.1), remain a focal point of research in nonlinear optics [21] and BECs [22]. In the focussing Manakov model, cross-phase modulation facilitates soliton radiation trapping, whereas in the modified model, vector-soliton outcoupling arises due to differences in intra- and interspecies scattering lengths. Notably, quantum superposition is not feasible in these systems, as solitons operate in a high kinetic energy regime [23].

To support high-bit-rate communication, wavelength division multiplexing is employed, utilizing coupled NLS equations to enable soliton transmission across multiple channels with distinct carrier frequencies [4]. In such scenarios, the dynamics of multiple co-propagating fields are governed by equations (4.1), which are generally non-integrable. Additionally, higher-order effects, such as third-order dispersion, Kerr dispersion, and stimulated Raman scattering, have been incorporated into extended coupled NLS models [4]. Beyond optical fibres, coupling phenomena are observed in systems such as parallel waveguides coupled via evanescent field overlap, polarization mode interactions in uniform waveguides, nonlinear optical waveguide arrays, and distributed feedback structures [4]. Nonlinear couplers, in particular, leverage solitons for all-optical switching applications [24, 25].

A hallmark of coupled NLS equations is the intensity redistribution among solitons, a feature with significant implications for optical fibre communications, including the development of intensity pump sources and soliton switching mechanisms [24, 25]. However, identifying additional dynamical signatures of these equations could further enhance the performance of soliton-based communication systems. Exploring such novel properties is critical for advancing the efficiency and functionality of optical soliton propagation in practical applications.

4.3 Lax pair and bright solitons

This section has been reproduced with permission from [42].

For particular choices of coupling coefficients, specifically $g_{11} = g_{12} = g_{21} = g_{22}$ or $g_{11} = g_{21} = -g_{12} = -g_{22}$, the system governed by equation (4.1) supports a linear eigenvalue problem, as defined in equation (2.5). The eigenvector is given by $\Phi = (\phi_1, \phi_2, \phi_3)^T$, with the matrices

$$U = \begin{pmatrix} -2i\zeta & \sqrt{a}\psi_1 & \sqrt{b}\psi_2 \\ \sqrt{a}\psi_1^* & i\zeta & 0 \\ \sqrt{b}\psi_2^* & 0 & i\zeta \end{pmatrix}, \quad V = \begin{pmatrix} -(B+J) & A & K \\ A^* & B & G \\ K^* & H & J \end{pmatrix}, \tag{4.2}$$

where the components are

$$A = i\sqrt{a}\psi_{1x} + 3\sqrt{a}\zeta\psi_1, \ K = i\sqrt{b}\psi_{2x} + 3\sqrt{b}\zeta\psi_2, \ A^* = -i\sqrt{a}\psi_{1x}^* + 3\sqrt{a}\zeta\psi_1^*,$$
$$K^* = -i\sqrt{b}\psi_{2x}^* + 3\sqrt{b}\zeta\psi_2^*, \ B = 3i\zeta^2 + ia\psi_1\psi_1^*, \ J = 3i\zeta^2 + i\psi_2\psi_2^*,$$
$$G = i\sqrt{a}\sqrt{b}\psi_2\psi_1^*, \ H = i\sqrt{a}\sqrt{b}\psi_1\psi_2^*, \ \zeta = \mu ab$$

In these expressions, μ denotes a complex isospectral parameter, and a and b are real constants. The compatibility condition $U_t - V_x + [U, V] = 0$ leads to the coupled nonlinear partial differential equations

$$i\psi_{1t} + \psi_{1xx} + 2(a|\psi_1|^2 + b|\psi_2|^2)\psi_1 = 0, \tag{4.3a}$$

$$i\psi_{2t} + \psi_{2xx} + 2(a|\psi_1|^2 + b|\psi_2|^2)\psi_2 = 0. \tag{4.3b}$$

When the condition $a = b$ holds, these equations correspond to the focussing Manakov model, as analyzed in references [15–20]. Alternatively, when $a = -b$, the system represents the modified Manakov model, as discussed in references [13, 26].

To obtain bright vector-soliton solutions for equations (4.3), one starts with the vacuum solution where $\psi_1^{(0)} = \psi_2^{(0)} = 0$. Under this condition. The eigenvalue problem simplifies to

$$\Phi_x^{(0)} = U^{(0)}\Phi^{(0)}, \tag{4.4a}$$

$$\Phi_t^{(0)} = V^{(0)}\Phi^{(0)}, \tag{4.4b}$$

with the matrices defined as

$$U^{(0)} = \begin{pmatrix} -i\zeta(t) & 0 & 0 \\ 0 & i\zeta(t) & 0 \\ 0 & 0 & i\zeta(t) \end{pmatrix}, \quad V^{(0)} = \begin{pmatrix} -3i\zeta(t)^2 & 0 & 0 \\ 0 & 3i\zeta(t)^2 & 0 \\ 0 & 0 & 3i\zeta(t)^2 \end{pmatrix}. \tag{4.5}$$

The solution to this linear system yields

$$\Phi^{(0)} = \begin{pmatrix} e^{-i\zeta(t)x-3i\int\zeta(t)^2\,dt} & 0 & 0 \\ 0 & e^{i\zeta(t)x+3i\int\zeta(t)^2\,dt} & 0 \\ 0 & 0 & e^{i\zeta(t)x+3i\int\zeta(t)^2\,dt} \end{pmatrix}. \tag{4.6a}$$

Applying gauge transformation approach, one derives the bright soliton solutions [27]

$$\psi_1^{(1)} = \varepsilon_1^{(1)}\beta_1 \mathrm{sech}(\theta_1)e^{i(-\xi_1)}, \tag{4.7a}$$

$$\psi_2^{(1)} = \varepsilon_2^{(1)}\beta_1 \mathrm{sech}(\theta_1)e^{i(-\xi_1)}, \tag{4.7b}$$

where the phase and argument are

$$\theta_1 = 2\beta_1 x + 8\alpha_1\beta_1 t - 2\delta_1, \tag{4.8a}$$

$$\xi_1 = 2\alpha_1 x + 4(\alpha_1^2 - \beta_1^2)t - 2\chi_1, \tag{4.8b}$$

with parameters $\alpha_1 = \alpha_{10}ab$, $\beta_1 = \beta_{10}ab$, and arbitrary constants δ_1 and χ_1. The coefficients $\varepsilon_1^{(1)}$ and $\varepsilon_2^{(1)}$ are coupling parameters that modulate the soliton amplitudes.

4.3.1 Collisional dynamics of bright solitons

The bright soliton solution reveals that its amplitude is influenced not only by the coupling parameters $\varepsilon_1^{(1)}$ and $\varepsilon_2^{(1)}$, but also by the SPM and XPM parameters a and b. Consequently, the effects of SPM and XPM can be effectively incorporated into the collisional dynamics of bright solitons.

To investigate the roles of SPM and XPM in the coupled NLS equation, the two-soliton solution, derived via the gauge transformation approach described in section 2.2.2, is considered in the form

$$\psi_1^{(2)} = \psi_1^{(1)} - 2i(\zeta_2 - \bar{\zeta}_2)\frac{\tilde{P}_{12}}{R}, \tag{4.9a}$$

$$\psi_2^{(2)} = \psi_2^{(1)} - 2i(\zeta_2 - \bar{\zeta}_2)\frac{\tilde{P}_{13}}{R}, \tag{4.9b}$$

where $\zeta_2 = \bar{\zeta}_2 = \alpha_2 + i\beta_2$. The explicit expressions for $\tilde{P}_{12}$ and $\tilde{P}_{13}$ are

$$\begin{aligned} \tilde{P}_{12} = -\Bigg[& M_{12}^{(1)}\left(\left(\tau + \gamma M_{11}^{(1)}\right)M_{11}^{(2)} + \gamma\left(\frac{\gamma^* M_{12}^{(1)}M_{21}^{(2)}}{\tau^2} + M_{13}^{(1)}M_{31}^{(2)}\right)\right) \\ & + M_{32}^{(1)}\left(\left(\tau + \gamma M_{11}^{(1)}\right)M_{13}^{(2)} + \gamma\left(M_{12}^{(1)}M_{23}^{(2)} + M_{13}^{(1)}M_{33}^{(2)}\right)\right)\frac{\gamma^*}{\tau^2} \\ & + \left(\left(\tau + \gamma M_{11}^{(1)}\right)M_{12}^{(2)} + \gamma\left(M_{12}^{(1)}M_{22}^{(2)} + M_{13}^{(1)}M_{32}^{(2)}\right)\right)\frac{\tau + \gamma^* M_{22}^{(1)}}{\tau^2}\Bigg], \end{aligned} \tag{4.10a}$$

$$\begin{aligned}\tilde{P}_{13} = -\Bigg[& M_{13}^{(1)}\Big(\big(\tau + \gamma M_{11}^{(1)}\big)M_{11}^{(2)} + \gamma\big(M_{12}^{(1)}M_{21}^{(2)} + M_{13}^{(1)}M_{31}^{(2)}\big)\Big)\frac{\gamma^*}{\tau^2} \\ &+ M_{23}^{(1)}\Big(\big(\tau + \gamma M_{11}^{(1)}\big)M_{12}^{(2)} + \gamma\big(M_{12}^{(1)}M_{22}^{(2)} + M_{13}^{(1)}M_{32}^{(2)}\big)\Big)\frac{\gamma^*}{\tau^2} \\ &+ \Big(\big(\tau + \gamma M_{11}^{(1)}\big)M_{13}^{(2)} + \gamma\big(M_{12}^{(1)}M_{23}^{(2)} + M_{13}^{(1)}M_{33}^{(2)}\big)\Big)\frac{\tau + \gamma^* M_{33}^{(1)}}{\tau^2}\Bigg],\end{aligned} \tag{4.10b}$$

with

$$\tau = M_{11}^{(1)} + M_{22}^{(1)} + M_{33}^{(1)}, \quad R = \tilde{P}_{11} + \tilde{P}_{22} + \tilde{P}_{33}, \ \gamma = \frac{\lambda_1 - \mu_1}{\mu_2 - \lambda_1}, \ \gamma^* = -\frac{\lambda_1 - \mu_1}{\lambda_2 - \mu_1},$$

and

$$\begin{aligned}\tilde{P}_{11} &= \frac{\gamma^*}{\tau^2}\Big[M_{21}^{(1)}\Big(\big(\tau + \gamma M_{11}^{(1)}\big)M_{12}^{(2)} + \gamma\big(M_{12}^{(1)}M_{22}^{(2)} + M_{13}^{(1)}M_{32}^{(2)}\big)\Big) \\ &\quad + M_{31}^{(1)}\Big(\big(\tau + \gamma M_{11}^{(1)}\big)M_{13}^{(2)} + \gamma\big(M_{12}^{(1)}M_{23}^{(2)} + M_{13}^{(1)}M_{33}^{(2)}\big)\Big)\Big] \\ &\quad + \frac{\tau + \gamma^* M_{11}^{(1)}}{\tau^2}\Big(\big(\tau + \gamma M_{11}^{(1)}\big)M_{11}^{(2)} + \gamma\big(M_{12}^{(1)}M_{21}^{(2)} + M_{13}^{(1)}M_{31}^{(2)}\big)\Big),\end{aligned} \tag{4.10c}$$

$$\begin{aligned}\tilde{P}_{22} &= \frac{\gamma^*}{\tau^2}\Big[M_{12}^{(1)}\Big(\gamma M_{11}^{(2)}M_{21}^{(1)} + \big(\tau + \gamma M_{22}^{(1)}\big)M_{21}^{(2)} + \gamma M_{23}^{(1)}M_{31}^{(2)}\Big) \\ &\quad + M_{32}^{(1)}\Big(\gamma M_{13}^{(2)}M_{21}^{(1)} + \big(\tau + \gamma M_{22}^{(1)}\big)M_{23}^{(2)} + \gamma M_{23}^{(1)}M_{33}^{(2)}\Big)\Big] \\ &\quad + \Big(\gamma M_{12}^{(2)}M_{21}^{(1)} + \big(\tau + \gamma M_{22}^{(1)}\big)M_{22}^{(2)} + \gamma M_{23}^{(1)}M_{32}^{(2)}\Big)\frac{\tau + \gamma^* M_{22}^{(1)}}{\tau^2},\end{aligned} \tag{4.10d}$$

$$\tilde{P}_{33} = \frac{\gamma^*}{\tau^2}\Big[M_{13}^{(1)}\Big(\gamma M_{11}^{(2)}M_{31}^{(1)} + \gamma M_{21}^{(2)}M_{32}^{(1)} + \big(\tau + \gamma M_{33}^{(1)}\big)M_{31}^{(2)}\Big) \tag{4.10e}$$

$$\begin{aligned}&+ M_{23}^{(1)}\Big(\gamma M_{12}^{(2)}M_{31}^{(1)} + \gamma M_{22}^{(2)}M_{32}^{(1)} + \big(\tau + \gamma M_{33}^{(1)}\big)M_{32}^{(2)}\Big)\Big] \\ &+ \Big(\gamma M_{13}^{(2)}M_{31}^{(1)} + \gamma M_{23}^{(2)}M_{32}^{(1)} + \big(\tau + \gamma M_{33}^{(1)}\big)M_{33}^{(2)}\Big)\frac{\tau + \gamma^* M_{33}^{(1)}}{\tau^2},\end{aligned} \tag{4.10f}$$

with

$$\begin{pmatrix} M_{11}^{(j)} & M_{12}^{(j)} & M_{13}^{(j)} \\ M_{21}^{(j)} & M_{22}^{(j)} & M_{23}^{(j)} \\ M_{31}^{(j)} & M_{32}^{(j)} & M_{33}^{(j)} \end{pmatrix} = \begin{pmatrix} e^{-\theta_j}\sqrt{2} & e^{-i\xi_j} & e^{-i\xi_j}\varepsilon_2^{(j)} \\ e^{i\xi_j} & e^{\theta_j}/\sqrt{2} & 0 \\ e^{i\xi_j}\varepsilon_2^{*(j)} & 0 & e^{\theta_j}/\sqrt{2} \end{pmatrix} \equiv \mathbf{M}^{(j)}, \tag{4.11}$$

where $j = 1, 2$, and

$$\theta_2 = 8\alpha_2\beta_2 t + 2x\beta_2 - 2\delta_2, \ \xi_2 = 4(\alpha_2^2 - \beta_2^2)t + 2x\alpha_2 - 2\chi_2. \tag{4.12}$$

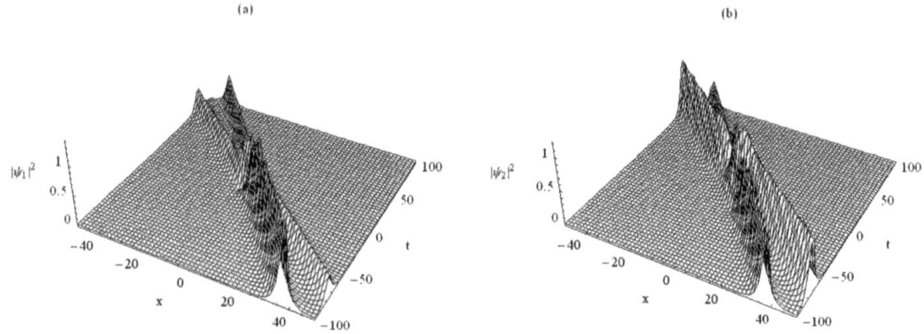

Figure 4.1. Intensity distribution in the coupled NLS equation for the parameters $a = 1$, $b = 1$, $\varepsilon_1^{(1)} = 0.85i$, $\varepsilon_1^{(2)} = 0.5$. Reprinted with permission from [27], Copyright (2013) by the American Physical Society.

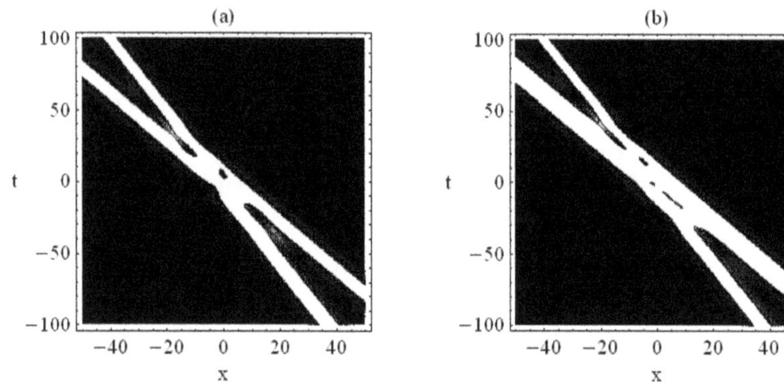

Figure 4.2. Trajectories of bright solitons in the two modes. Reprinted with permission from [27], Copyright (2013) by the American Physical Society.

The densities of the two modes satisfy the relation $|\varepsilon_1^{(j)}|^2 + |\varepsilon_2^{(j)}|^2 = 1$, for $j = 1, 2$.

Figure 4.1 illustrates the intensity distribution in the coupled NLS equation, while the contour plot in figure 4.2 depicts the soliton trajectories. Adjusting the strengths of SPM and XPM (with SPM equal to XPM) results in a rotation of the bright soliton trajectories and a realignment of the intensity distribution between the two modes, as shown in figure 4.3. The contour plot in figure 4.4 confirms this rotation. Further variations in the parameters a and b enhance the angle of rotation, as depicted in figure 4.5, with the corresponding contour plot in figure 4.6. Comparing the density profiles in figures 4.1, 4.3, and 4.5 reveals both the rotation of soliton trajectories and a realignment of intensity distribution between the modes ψ_1 and ψ_2. This rotation and realignment stem from additional energy introduced into the system by varying the SPM and XPM parameters, which influences both the trajectory dynamics and the intensity distribution.

The modified coupled NLS equation (Manakov model) also exhibits rotation of bright soliton trajectories. The intensity profile in figure 4.7 shows shape-changing collisional dynamics similar to those in the standard coupled NLS equation.

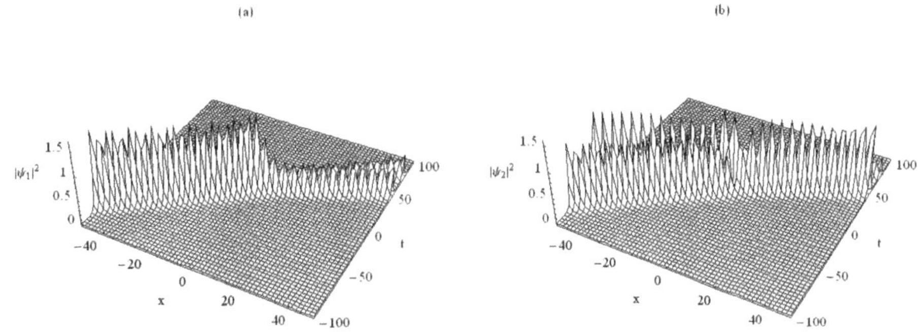

Figure 4.3. Realignment of intensity distribution for the parameters $a = 1.5$, $b = 1.5$, $\varepsilon_1^{(1)} = 0.85i$, $\varepsilon_1^{(2)} = 0.5$. Reprinted with permission from [27], Copyright (2013) by the American Physical Society.

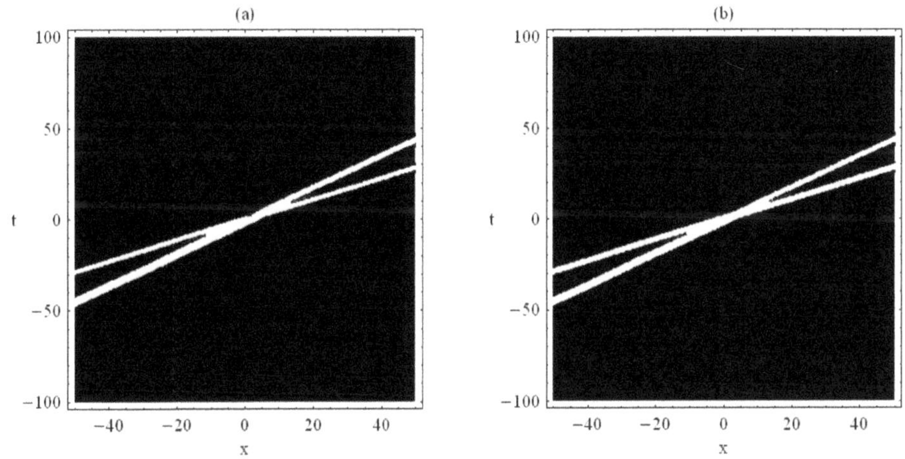

Figure 4.4. Rotation of bright soliton trajectories. Reprinted with permission from [27], Copyright (2013) by the American Physical Society.

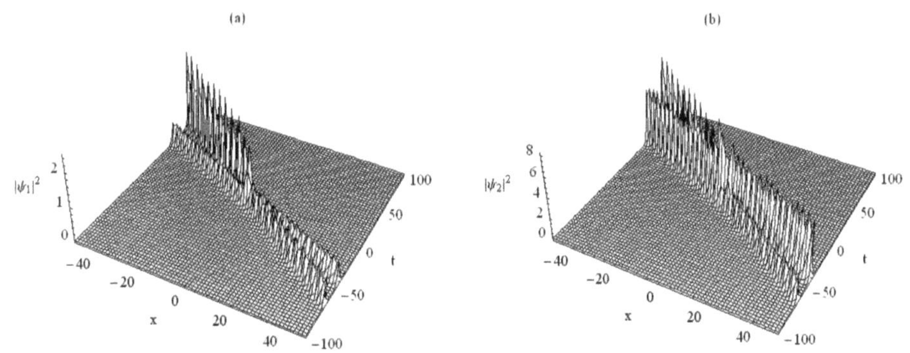

Figure 4.5. Further realignment of intensity distribution for the parameters $a = 2.5$, $b = 2.5$, $\varepsilon_1^{(1)} = 0.85i$, $\varepsilon_1^{(2)} = 0.5$. Reprinted with permission from [27], Copyright (2013) by the American Physical Society.

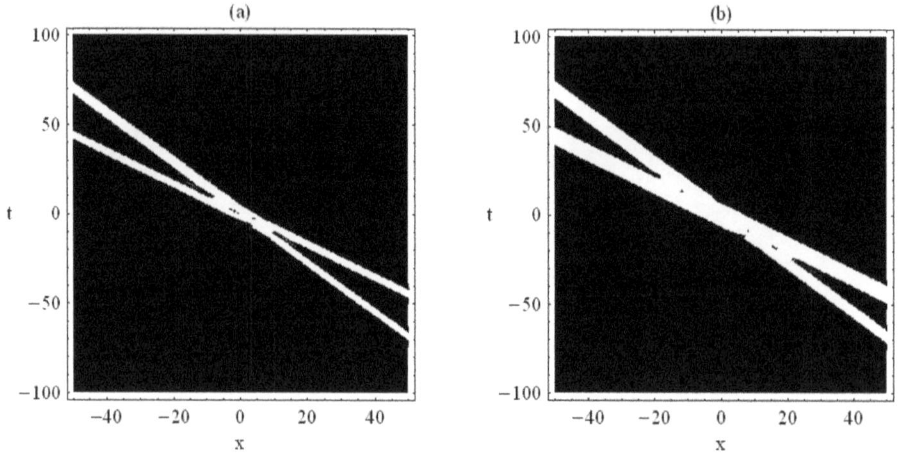

Figure 4.6. Enhanced rotation of bright soliton trajectories. Reprinted with permission from [27], Copyright (2013) by the American Physical Society.

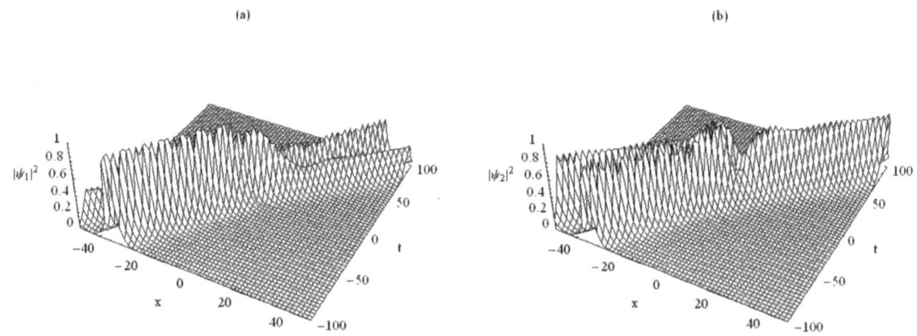

Figure 4.7. Collisional dynamics of bright solitons in the modified coupled NLS equation for $a = -b = 0.9$, $\varepsilon_1^{(1)} = 0.85i$, $\varepsilon_1^{(2)} = 0.5$. Reprinted with permission from [27], Copyright (2013) by the American Physical Society.

However, the soliton trajectories in figure 4.8 are diagonally opposite to those observed in figure 4.2. The angular separation between solitons can be adjusted by tuning the complex spectral parameter μ, as illustrated in figures 4.9 and 4.10. Notably, this variation in angular separation occurs within the coupled NLS equation itself, a phenomenon not previously reported.

Thus, variations in SPM or XPM parameters inject additional energy into the system, leading to both the rotation of bright soliton trajectories and the realignment of their intensity distribution.

In integrable systems associated with coupled NLS-type equations, the ratio of SPM to XPM coefficients, which govern interactions within and between compo-

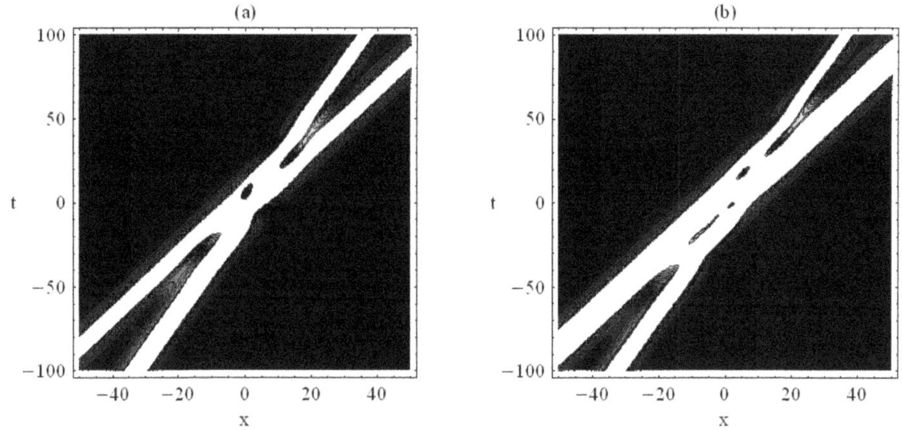

Figure 4.8. Diagonally opposite trajectories of bright solitons in the modified coupled NLS equation. Reprinted with permission from [27], Copyright (2013) by the American Physical Society.

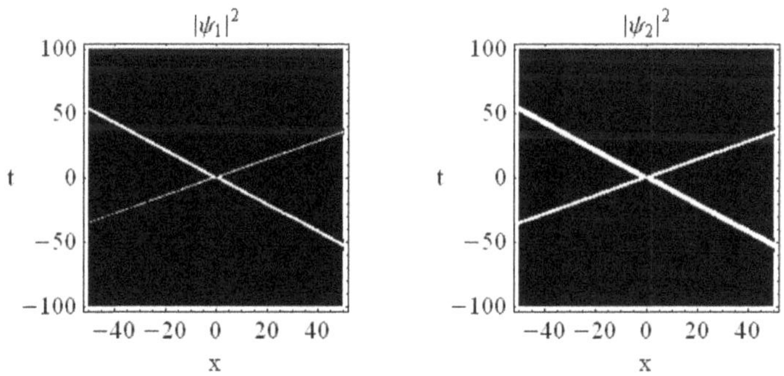

Figure 4.9. Enhanced angular separation between solitons by varying μ_i, $i = 1, 2$, for $\alpha_{10} = -0.1$, $\beta_{10} = -0.2$, $\alpha_{20} = 0.15$, $\beta_{20} = 0.3$, with other parameters as in figure 4.3. Reprinted with permission from [27], Copyright (2013) by the American Physical Society.

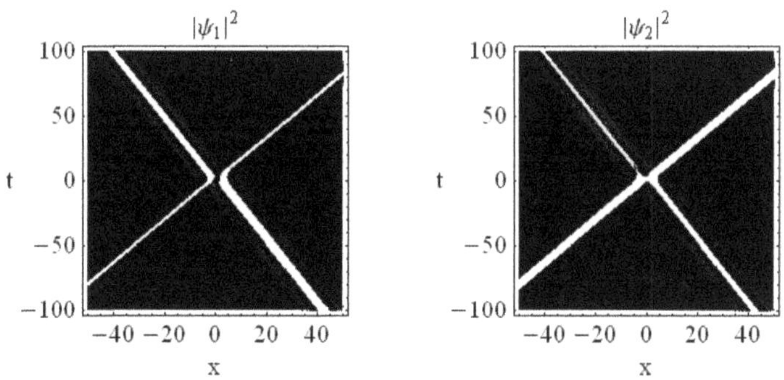

Figure 4.10. Further enhancement of angular separation between solitons for $\alpha_{10} = -0.1$, $\beta_{10} = -0.25$, $\alpha_{20} = 0.2$, $\beta_{20} = 0.3$, with other parameters as in figure 4.3. Reprinted with permission from [27], Copyright (2013) by the American Physical Society.

nents, is typically equal. However, physically realistic systems often deviate from this constraint. Consequently, the search for new solvable models of coupled NLS equations continues. In this context, Park and Shin [28] introduced novel integrable NLS-type equations that extend beyond the conventional Manakov model by incorporating FWM terms.

4.4 Coherently coupled nonlinear Schrödinger equations

Parts of this section have beenreproduced with permission from [37].

Models of bimodal light propagation in nonlinear birefringent optical fibres incorporate not only XPM but also FWM terms, specifically $q_1^2 q_2^*$ and $q_2^2 q_1^*$, which describe coherent nonlinear interactions between two orthogonal polarizations of electromagnetic waves [4, 29]. This study introduces a novel system of coupled NLS equations that include SPM, XPM, and FWM terms with a time-dependent coefficient alongside a time-dependent anti-trapping parabolic potential. The notation follows conventions used in BEC models derived from GP equations [22, 30]:

$$i\frac{\partial q_1}{\partial t} + \frac{1}{2}\frac{\partial^2 q_1}{\partial x^2} + \gamma(t)(|q_1|^2 - 2|q_2|^2)q_1 - \gamma(t)q_2^2 q_1^* + \frac{1}{2}\lambda^2(t)x^2 q_1 = 0, \quad (4.13a)$$

$$i\frac{\partial q_2}{\partial t} + \frac{1}{2}\frac{\partial^2 q_2}{\partial x^2} + \gamma(t)(2|q_1|^2 - |q_2|^2)q_2 + \gamma(t)q_1^2 q_2^* + \frac{1}{2}\lambda^2(t)x^2 q_2 = 0, \quad (4.13b)$$

where $\gamma(t)$ represents the strength of the FWM terms, and $\lambda^2(t)$ denotes the strength of the anti-trapping (expulsive) potential. Such potentials arise in physical scenarios, including interactions of optical and matter-wave solitons with barriers and wave packet splitting in interferometers [31].

The system of equations (4.13) can be derived from the Lagrangian

$$\begin{aligned} L = &\frac{i}{2}\left(q_1^*\frac{\partial q_1}{\partial t} - q_1\frac{\partial q_1^*}{\partial t}\right) - \frac{1}{2}\left|\frac{\partial q_1}{\partial x}\right|^2 + \frac{1}{2}\gamma(t)|q_1|^4 + \frac{1}{2}\lambda^2(t)x^2|q_1|^2 \\ &- 2\gamma(t)|q_1|^2|q_2|^2 - \frac{1}{2}\gamma(t)\left[q_2^2(q_1^*)^2 + q_1^2(q_2^*)^2\right] - \frac{i}{2}\left(q_2^*\frac{\partial q_2}{\partial t} - q_2\frac{\partial q_2^*}{\partial t}\right) \quad (4.14) \\ &+ \frac{1}{2}\left|\frac{\partial q_2}{\partial x}\right|^2 + \frac{1}{2}\gamma(t)|q_2|^4 - \frac{1}{2}\lambda^2(t)x^2|q_2|^2, \end{aligned}$$

where * denotes the complex conjugate. A distinctive feature of this Lagrangian is its sign-indefinite nature, as the kinetic and gradient terms, involving time and spatial derivatives, respectively, have opposite signs for the components q_1 and q_2. Consequently, the system exhibits ‘opposite time directions’ for the two subsystems, making it distinct from typical physical systems, though it bears some resemblance to an idealized optical coupler with normal and negative-refractive-index cores [32]. Despite its non-standard nature, the system is of theoretical interest as a novel nonlinear wave model.

The opposite time directions in the subsystems lead to the non-conservation of the total norm, defined as

$$N = \int_{-\infty}^{\infty} (|q_1(x)|^2 + |q_2(x)|^2)\, \mathrm{d}x. \tag{4.15}$$

From equations (4.13), the evolution of the norm is governed by

$$\frac{\mathrm{d}N}{\mathrm{d}t} = 4\gamma(t) \int_{-\infty}^{\infty} \mathrm{Im}\left\{(q_1^*(x))^2 q_2^2(x)\right\} \mathrm{d}x. \tag{4.16}$$

However, the system conserves the difference between the norms of the two subsystems

$$\frac{\mathrm{d}}{\mathrm{d}t} \int_{-\infty}^{\infty} (|q_1(x)|^2 - |q_2(x)|^2)\, \mathrm{d}x = 0, \tag{4.17}$$

reflecting the conservative nature of the system with opposite time directions. Equation (4.17) corresponds to the conservation of energy in the dynamical system described by equations (4.13). Additional conserved quantities, as discussed in [33], further support the integrability of equations (4.13). Unlike typical systems where coherent nonlinear coupling facilitates norm exchange between subsystems while conserving the total norm, the opposite time directions here allow the coupling to generate or absorb the norm.

4.4.1 Linear eigenvalue problem and integrability condition

Equations (4.13) admit a linear eigenvalue problem associated with the Lax pair matrices $\mathcal{U}$ and $\mathcal{V}$, defined as

$$\mathcal{U} = \begin{pmatrix} \mathrm{i}\zeta(t) & \sqrt{\gamma(t)}\,\mathrm{e}^{\mathrm{i}\phi(x,\,t)} q_1(x,\,t) & \sqrt{\gamma(t)}\,\mathrm{e}^{\mathrm{i}\phi(x,\,t)} q_2(x,\,t) \\ -\sqrt{\gamma(t)}\,\mathrm{e}^{-\mathrm{i}\phi(x,\,t)} r_1(x,\,t) & -\mathrm{i}\zeta(t) & 0 \\ -\sqrt{\gamma(t)}\,\mathrm{e}^{-\mathrm{i}\phi(x,\,t)} r_2(x,\,t) & 0 & -\mathrm{i}\zeta(t) \end{pmatrix}, \tag{4.18}$$

$$\mathcal{V} = \begin{pmatrix} V_{11} & V_{12} & V_{13} \\ V_{21} & V_{22} & V_{23} \\ V_{31} & V_{32} & V_{33} \end{pmatrix}, \tag{4.19}$$

where

$$
\begin{aligned}
V_{11} =& - \mathrm{i}\zeta(t)^2 + \mathrm{i}\Omega(t)x\zeta(t) + \frac{\mathrm{i}}{2}\mathrm{e}^{\mathrm{i}\phi(x,\,t)}q_1(x,\,t)\mathrm{e}^{-\mathrm{i}\phi(x,\,t)}r_1(x,\,t)\gamma(t) \\
&+ \frac{\mathrm{i}}{2}\mathrm{e}^{\mathrm{i}\phi(x,\,t)}q_2(x,\,t)\mathrm{e}^{-\mathrm{i}\phi(x,\,t)}r_2(x,\,t)\gamma(t), \\
V_{12} =& (\Omega(t)x - \zeta(t))\mathrm{e}^{\mathrm{i}\phi(x,\,t)}q_1(x,\,t)\sqrt{\gamma(t)} + \frac{\mathrm{i}}{2}\sqrt{\gamma(t)}\frac{\partial}{\partial x}(\mathrm{e}^{\mathrm{i}\phi(x,\,t)}q_1(x,\,t)), \\
V_{13} =& (\Omega(t)x - \zeta(t))\mathrm{e}^{\mathrm{i}\phi(x,\,t)}q_2(x,\,t)\sqrt{\gamma(t)} + \frac{\mathrm{i}}{2}\sqrt{\gamma(t)}\frac{\partial}{\partial x}(\mathrm{e}^{\mathrm{i}\phi(x,\,t)}q_2(x,\,t)), \\
V_{21} =& - (\Omega(t)x - \zeta(t))\mathrm{e}^{-\mathrm{i}\phi(x,\,t)}r_1(x,\,t)\sqrt{\gamma(t)} + \frac{\mathrm{i}}{2}\sqrt{\gamma(t)}\frac{\partial}{\partial x}(\mathrm{e}^{-\mathrm{i}\phi(x,\,t)}r_1(x,\,t)), \\
V_{22} =& \mathrm{i}\zeta(t)^2 - \mathrm{i}\Omega(t)x\zeta(t) - \frac{\mathrm{i}}{2}\mathrm{e}^{\mathrm{i}\phi(x,\,t)}q_1(x,\,t)\mathrm{e}^{-\mathrm{i}\phi(x,\,t)}r_1(x,\,t)\gamma(t), \\
V_{23} =& - \frac{\mathrm{i}}{2}\mathrm{e}^{\mathrm{i}\phi(x,\,t)}q_2(x,\,t)\mathrm{e}^{-\mathrm{i}\phi(x,\,t)}r_2(x,\,t)\gamma(t), \\
V_{31} =& - (\Omega(t)x - \zeta(t))\mathrm{e}^{-\mathrm{i}\phi(x,\,t)}r_2(x,\,t)\sqrt{\gamma(t)} + \frac{\mathrm{i}}{2}\sqrt{\gamma(t)}\frac{\partial}{\partial x}(\mathrm{e}^{-\mathrm{i}\phi(x,\,t)}r_2(x,\,t)), \\
V_{32} =& - \frac{\mathrm{i}}{2}\mathrm{e}^{\mathrm{i}\phi(x,\,t)}q_1(x,\,t)\mathrm{e}^{-\mathrm{i}\phi(x,\,t)}r_1(x,\,t)\gamma(t), \\
V_{33} =& \mathrm{i}\zeta(t)^2 - \mathrm{i}\Omega(t)x\zeta(t) - \frac{\mathrm{i}}{2}\mathrm{e}^{\mathrm{i}\phi(x,\,t)}q_2(x,\,t)\mathrm{e}^{-\mathrm{i}\phi(x,\,t)}r_2(x,\,t)\gamma(t),
\end{aligned} \tag{4.20a}
$$

with

$$r_1(x,\,t) = -aq_1^*(x,\,t) - bq_2^*(x,\,t) + d_1b_1q_1^*(x,\,t), \tag{4.20b}$$

$$r_2(x,\,t) = b_1q_1^*(x,\,t) - cq_2^*(x,\,t) + dbq_2^*(x,\,t), \tag{4.20c}$$

$$\phi(x,\,t) = \frac{1}{2}\Omega(t)x^2, \tag{4.20d}$$

where a, b, c, d, b_1, d_1 are arbitrary constants. By appropriately selecting these parameters, equations (4.13) can be obtained from the compatibility condition $\mathcal{U}_t - \mathcal{V}_x + [\mathcal{U},\,\mathcal{V}] = 0$, where the spectral parameter $\zeta(t)$ satisfies

$$\frac{\mathrm{d}\zeta(t)}{\mathrm{d}t} = \Omega(t)\zeta(t), \tag{4.21a}$$

with

$$\lambda^2(t) = \Omega^2(t) - \frac{\mathrm{d}\Omega(t)}{\mathrm{d}t}, \tag{4.21b}$$

$$\Omega(t) = -\frac{\mathrm{d}}{\mathrm{d}t}\ln\gamma(t). \tag{4.21c}$$

The Riccati equation (4.21b) has been used to solve GP-type equations [34, 35]. Notably, the identification of this Riccati-type equation (4.21b) provides an initial indication of the complete integrability of equations (4.13). Equation (4.21b), which governs the strength of the parabolic potential $\lambda^2(t)$, shows its relation to the FWM strength $\gamma(t)$ through the integrability condition. Substituting equation (4.21c) into equation (4.21b) yields

$$-\gamma''(t)\gamma(t) + 2(\gamma'(t))^2 - \lambda^2(t)\gamma^2(t) = 0. \tag{4.22}$$

Thus, the coupled GP equations (4.13), or equivalently coupled NLS equations with a time-dependent harmonic trap, are completely integrable for appropriate choices of $\lambda(t)$ and $\gamma(t)$ that satisfy equation (4.22). For a constant $\lambda(t) = c_1$, equation (4.22) gives $\gamma(t) = e^{c_1 t}$.

The integrable form of equations (4.13) can be transformed via the substitution

$$q_{1,2}(x, t) = \frac{1}{\sqrt{\gamma(t)}\,l(t)}Q_{1,2}(X, T)\exp\left(-i\frac{\Omega(t)x^2}{2}\right), \tag{4.23}$$

where $X = x/l(t)$, $T = T(t)$, $\frac{\mathrm{d}l}{\mathrm{d}t} = 2\Omega(t)l$, and $\frac{\mathrm{d}T}{\mathrm{d}t} = 1/l^2$, into a system of coupled perturbed NLS equations with constant coefficients:

$$\mathrm{i}\frac{\partial Q_1}{\partial T} + \frac{\partial^2 Q_1}{\partial X^2} + (|Q_1|^2 - 2|Q_2|^2)Q_1 - Q_2^2 Q_1^* = \mathrm{i}\epsilon(t)Q_1, \tag{4.24a}$$

$$\mathrm{i}\frac{\partial Q_2}{\partial T} + \frac{\partial^2 Q_2}{\partial X^2} + (2|Q_1|^2 - |Q_2|^2)Q_2 - Q_1^2 Q_2^* = \mathrm{i}\epsilon(t)Q_2, \tag{4.24b}$$

where $\epsilon(t) = \left(\Omega(t) + \frac{1}{\gamma(t)}\frac{\mathrm{d}\gamma(t)}{\mathrm{d}t}\right)l^2$. By setting $\Omega(t) = -\frac{1}{\gamma(t)}\frac{\mathrm{d}\gamma(t)}{\mathrm{d}t}$, which makes $\epsilon(t) = 0$, the system reduces to the coherently coupled NLS equations studied in reference [36] using the Hirota method.

4.4.2 Persistent solitons and collisional dynamics

4.4.2.1 Analytical results

To derive bright vector solitons for the coupled NLS equations (4.13), one starts with the vacuum solution ($q_1^{(0)} = q_2^{(0)} = 0$) and applies gauge transformation approach to obtain the explicit one-soliton solution [37]

$$q_1^{(1)} = 2A_1\varepsilon_1^{(1)}\beta_0\mathrm{sech}(\theta_1)\mathrm{e}^{\mathrm{i}(-\xi_1 + \phi(x, t))}, \tag{4.25a}$$

$$q_2^{(1)} = 2A_2\varepsilon_2^{(1)}\beta_0\text{sech}(\theta_1)\text{e}^{\text{i}(-\xi_1 + \phi(x,\, t))}, \tag{4.25b}$$

where

$$\theta_1 = 2x\beta_1(t) - 4\int \alpha_1(t)\beta_1(t)\ \text{d}t + 2\delta_1, \tag{4.26a}$$

$$\xi_1 = 2x\alpha_1(t) - 2\int (\alpha_1(t)^2 - \beta_1(t)^2)\ \text{d}t - 2\chi_1, \tag{4.26b}$$

$$\phi(x,\, t) = \frac{1}{2}\Omega(t)x^2, \tag{4.26c}$$

$$A_1 = A_2 = \exp\left(\frac{1}{2}\int \Omega(t)\ \text{d}t\right), \tag{4.26d}$$

$$\{\alpha_1(t),\, \beta_1(t)\} = \{\alpha_{10},\, \beta_{10}\}\exp\left(-\int \Omega(t)\ \text{d}t\right), \tag{4.26e}$$

and δ_1, χ_1, and β_0 are arbitrary parameters. Here, $\alpha_1(t)$ and $\beta_1(t)$ represent time-dependent scattering lengths, and the coupling parameters $\varepsilon_1^{(1)}$, $\varepsilon_2^{(1)}$ satisfy the constraint $|\varepsilon_1^{(1)}|^2 + |\varepsilon_2^{(1)}|^2 = 1$.

For a time-independent parabolic potential ($\Omega(t)$=constant), the bright solitons exhibit either growth or decay, depending on the potential's strength, as illustrated in figures 4.11 and 4.12. Such growth or decay is a hallmark of variable-coefficient NLS equations. For instance, in BECs with exponentially varying scattering lengths in a parabolic trap, the condensate density increases or decreases over time based on the potential's sign, yet the system remains completely integrable and conservative [38].

To stabilize the solitons, a time-dependent parabolic potential is introduced, with $\Omega(t)$ chosen as depicted in figure 4.13. The corresponding density profile, shown in

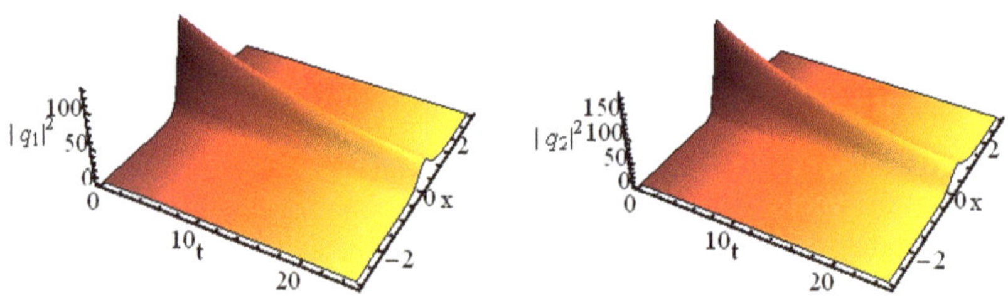

Figure 4.11. Decay of the soliton solution in a time-independent parabolic potential with $\Omega(t) = -0.02$ (or $\gamma(t) = \exp(0.02t)$), $\varepsilon_1^{(1)} = 0.3$, $\alpha_{10} = 0.1$, $\beta_{10} = 0.3$, $\chi_1 = 0.1$, $\delta_1 = 0.2$. Reprinted from [37], Copyright (2016), with permission from Elsevier.

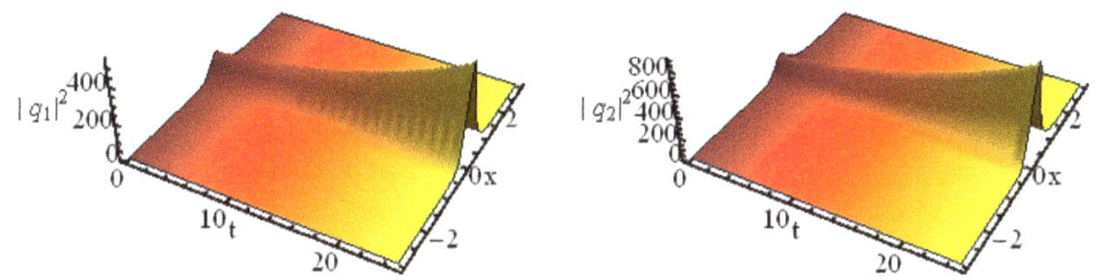

Figure 4.12. Growth of the soliton solution with $\Omega(t) = 0.02$ (or $\gamma(t) = \exp(-0.02t)$), $\varepsilon_1^{(1)} = 0.3$, $\alpha_{10} = 0.5$, $\beta_{10} = 0.5$, $\chi_1 = 0.5$, $\delta_1 = 0.2$. Reprinted from [37], Copyright (2016), with permission from Elsevier.

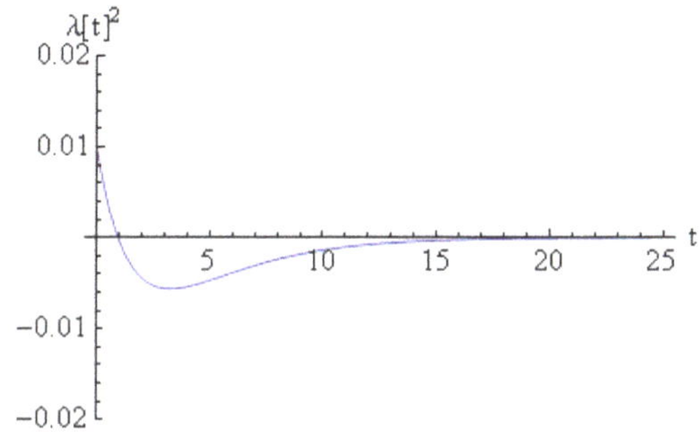

Figure 4.13. Evolution of the parabolic potential strength $\lambda^2(t)$ (which may be positive or negative), governed by equation (4.22), for $\gamma(t) = \exp\left[\frac{2}{3}(1 - e^{-0.3t})\right]$. Reprinted from [37], Copyright (2016), with permission from Elsevier.

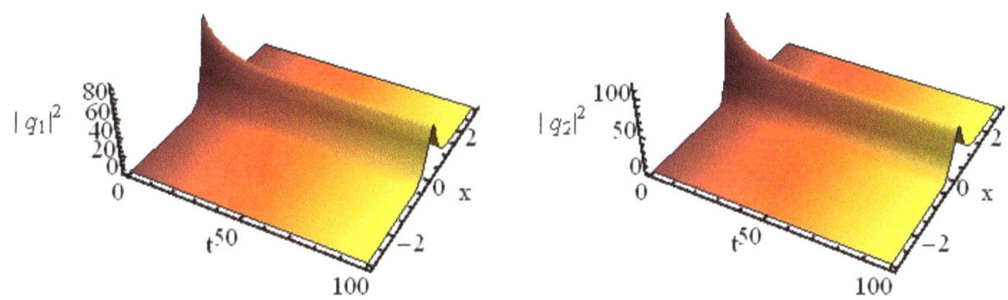

Figure 4.14. Persistent soliton with the same $\gamma(t)$ as in figure 4.13, and parameters $\varepsilon_1^{(1)} = 0.3$, $\alpha_{10} = 0.2$, $\beta_{10} = 0.5$, $\chi_1 = 0.5$, $\delta_1 = 0.2$. Reprinted from [37], Copyright (2016), with permission from Elsevier.

figure 4.14, demonstrates that the bright solitons maintain their shape. These solutions are referred to as 'persistent bright solitons' due to their stability.

4.4.2.2 Numerical simulation

The analytical results are validated by numerically solving equations (4.13) using the split-step Crank–Nicolson method [39]. Figure 4.15 compares the analytical persistent bright soliton solutions, given by equations (4.25), with their numerical

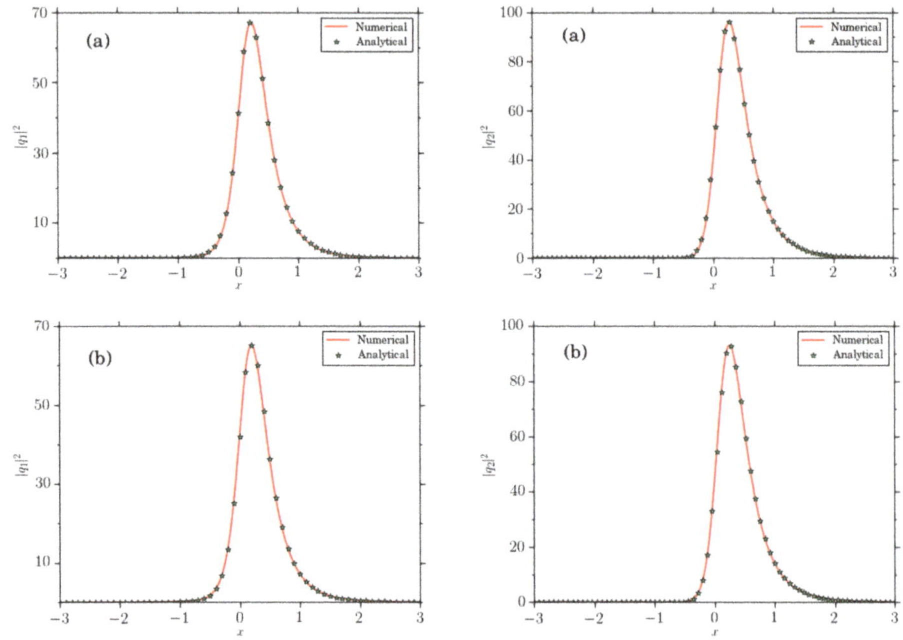

Figure 4.15. Comparison of analytical and numerical soliton solutions. (a) Upper left: q_1 component at $t = 10$; upper right: q_2 at $t = 10$. (b) Lower left: q_1 at $t = 20$; lower right: q_2 at $t = 20$. Parameters match those in figure 4.14. Reprinted from [37], Copyright (2016), with permission from Elsevier.

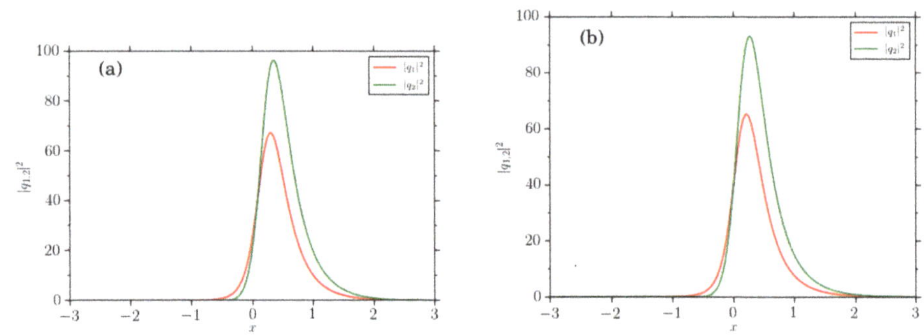

Figure 4.16. Density profiles after a sudden 10% increase in the potential strength, with $\gamma(t)= \exp\left[\frac{20}{3}(1 - e^{-0.3t})\right]$, at (a) $t = 10$, (b) $t = 20$. Reprinted from [37], Copyright (2016), with permission from Elsevier.

counterparts at $t = 10$ and $t = 20$. The figure confirms that the analytical and numerical solutions match precisely.

The stability of persistent solitons, which depends on the time-dependent parabolic potential strength (see equations (4.21b)–(4.22)), is tested by introducing a sudden 10% increase or decrease in the potential strength, as shown in figures 4.16

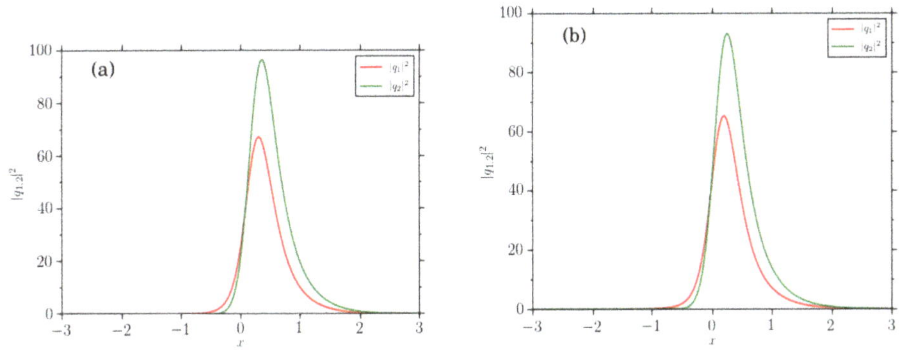

Figure 4.17. Density profiles after a sudden 10% decrease in the potential strength, with $\gamma(t) = \exp\left[\frac{1}{15}(1 - e^{-0.3t})\right]$, at (a) $t = 10$, (b) $t = 20$. Reprinted from [37], Copyright (2016), with permission from Elsevier.

and 4.17. These figures indicate that such perturbations do not compromise the solitons' stability.

4.4.3 Collisional dynamics of persistent solitons

The gauge transformation approach can be extended to derive multisoliton solutions [40]. Specifically, the two-soliton solution for the two modes, $q_1^{(2)}$ and $q_2^{(2)}$, is given by

$$q_1^{(2)} = \frac{2IA_1}{B},\ q_2^{(2)} = \frac{2IA_2}{B}, \tag{4.27}$$

where

$$\begin{aligned}
A_1 = {} & M_{12}^{(1)}M_{22}^{(2)}(\zeta_2 - \zeta_1)(\zeta_1 - \zeta_1^*)(\zeta_2 - \zeta_2^*) + M_{12}^{(2)}M_{22}^{(1)}(\zeta_2 - \zeta_1^*)(\zeta_2^* - \zeta_1)(\zeta_2 - \zeta_2^*) \\
& + M_{11}^{(1)}M_{12}^{(2)}(\zeta_2 - \zeta_1^*)(\zeta_2^* - \zeta_1^*)(\zeta_2 - \zeta_2^*) + M_{11}^{(2)}M_{12}^{(1)}(\zeta_1 - \zeta_1^*)(\zeta_2^* - \zeta_1)(\zeta_2^* - \zeta_1^*), \\
A_2 = {} & M_{11}^{(2)}M_{21}^{(1)}(\zeta_2 - \zeta_1)(\zeta_1 - \zeta_1^*)(\zeta_2 - \zeta_2^*) + M_{11}^{(1)}M_{21}^{(2)}(\zeta_2 - \zeta_1^*)(\zeta_2^* - \zeta_1)(\zeta_2 - \zeta_2^*) \\
& + M_{21}^{(2)}M_{22}^{(1)}(\zeta_2 - \zeta_1^*)(\zeta_1 - \zeta_1^*)(\zeta_2 - \zeta_2^*) + M_{21}^{(1)}M_{22}^{(2)}(\zeta_1 - \zeta_1^*)(\zeta_2^* - \zeta_1)(\zeta_2^* - \zeta_1^*), \\
B = {} & (M_{12}^{(2)}M_{21}^{(1)} + M_{12}^{(1)}M_{21}^{(2)})(\zeta_1 - \zeta_1^*)(\zeta_2 - \zeta_2^*) + (M_{11}^{(2)}M_{22}^{(1)} + M_{11}^{(1)}M_{22}^{(2)})(\zeta_2 - \zeta_1^*) \\
& \times (\zeta_2^* - \zeta_1) + (M_{11}^{(1)}M_{11}^{(2)} + M_{22}^{(1)}M_{22}^{(2)})(\zeta_2 - \zeta_1)(\zeta_2^* - \zeta_1^*),
\end{aligned} \tag{4.28}$$

and the matrix elements are given in equation (4.11) with

$$\theta_j = 2x\beta_j(t) - 4\int \alpha_j(t)\beta_j(t)\, dt + 2\delta_j, \tag{4.29a}$$

$$\xi_j = 2x\alpha_j(t) - 2\int (\alpha_j(t)^2 - \beta_j(t)^2)\, dt - 2\chi_j, \tag{4.29b}$$

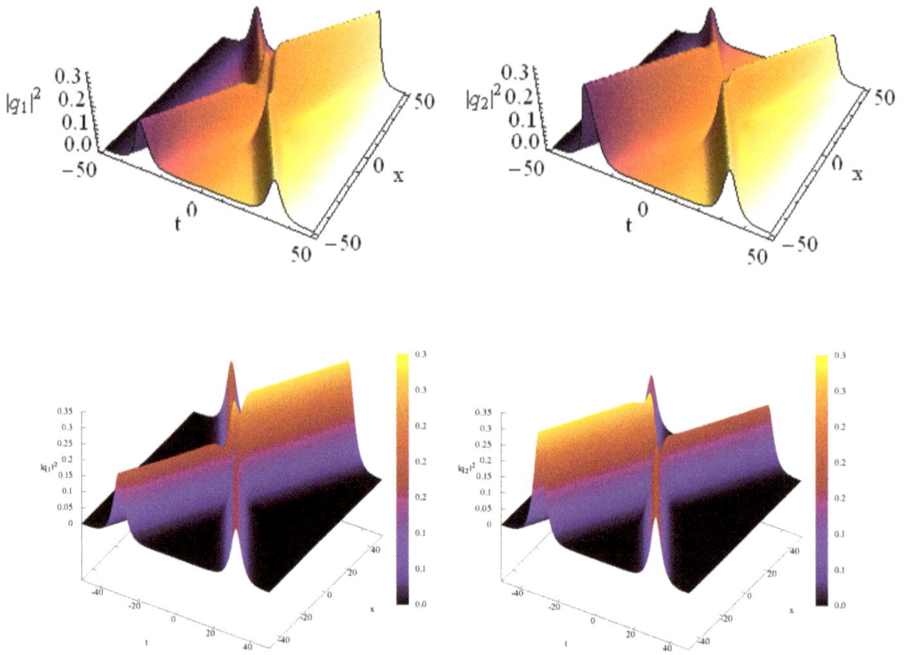

Figure 4.18. Inelastic collision of solitons with $\gamma(t) = \exp\left[\frac{2}{3}(1 - e^{-0.3t})\right]$, $\alpha_{10} = 0.1$, $\alpha_{20} = 0.25$, $\beta_{10} = 0.3$, $\beta_{20} = 0.2$, $\delta_1 = 0.1$, $\delta_2 = 0.2$, $\chi_1 = 0.3$, $\chi_2 = 0.4$, $\varepsilon_1^{(1)} = 0.85i$, $\varepsilon_1^{(2)} = 0.5$, satisfying $|\varepsilon_1^{(j)}|^2 + |\varepsilon_2^{(j)}|^2 = 1$ ($j = 1, 2$). Top panels: analytical solution; bottom panels: numerical solution. Reprinted from [37], Copyright (2016), with permission from Elsevier.

for $j = 1, 2$. Figure 4.18 illustrates the inelastic collision of persistent solitons, with the analytical solution (top panels) and numerical solution (bottom panels) showing identical collisional dynamics.

4.5 Four-wave mixing induced manipulation of light in coupled nonlinear Schrödinger equations

The shape-changing collisions of bright solitons in the Manakov model are achieved by adjusting phase-related parameters [19]. Recent studies have demonstrated that soliton trajectories can be rotated by varying system parameters, such as SPM and XPM, without compromising the integrability of the Manakov or modified Manakov model [27]. However, the coupled NLS equations generally lack exact soliton solutions for arbitrary (unequal) SPM and XPM coefficients, rendering them non-integrable due to the requirement of a specific ellipticity angle of 35 degrees [8, 18]. Despite the development of various integrable single and coupled NLS-type equations supporting bright and dark solitons, progress beyond the Manakov model in terms of intensity redistribution between optical beams remains limited. Although Park and Shin [28] explored a generalized coupled NLS equation incorporating FWM alongside SPM and XPM, identifying four integrable model classes, and Wang *et al* [41] analyzed a variant with three arbitrary parameters (two real for SPM and XPM, one complex for FWM), the role of FWM in the collisional dynamics of solitons within the Manakov model has not been fully examined. Furthermore, the potential of four arbitrary parameters in a generalized coupled NLS equation to manipulate soliton dynamics remains unexplored.

A critical question arises: is it possible to control the intensity (or energy) of a specific electromagnetic mode or bound state in optical pulses? Addressing this question is significant, as it offers the ability to selectively enhance the energy of a particular mode or bound state as optical beams propagate through a fibre, enabling advanced applications in optical communication and switching.

4.5.1 Generalized coupled nonlinear Schrödinger equations with four-wave mixing and Lax pair

The interaction between co-propagating optical beams in a nonlinear medium is fundamental to the functionality of optical fibres. The propagation of optical pulses in a nonlinear birefringent fibre is described by the GCNLS equations

$$i\frac{\partial\psi_1}{\partial t} + \frac{\partial^2\psi_1}{\partial x^2} + 2\left(a|\psi_1|^2 + c|\psi_2|^2 + b\psi_1\psi_2^* + d\psi_2\psi_1^*\right)\psi_1 = 0, \tag{4.30a}$$

$$i\frac{\partial\psi_2}{\partial t} + \frac{\partial^2\psi_2}{\partial x^2} + 2\left(a|\psi_1|^2 + c|\psi_2|^2 + b\psi_1\psi_2^* + d\psi_2\psi_1^*\right)\psi_2 = 0, \tag{4.30b}$$

where ψ_1 and ψ_2 denote the amplitudes of the electromagnetic beams. The real coefficients a and c represent SPM and XPM, respectively, while b and d, also real in this model, account for FWM. Although b and d can be complex, real values are adopted here for simplicity. When $b = d = 0$ and $a = c$, equations (4.30) reduce to the integrable Manakov model [12, 15, 16]. The system has been studied for the case $d = b^*$ [41], where the dynamics of the collisions of the solitons was analyzed.

The Lax pair matrices $\mathcal{U}$ and $\mathcal{V}$ associated with equations (4.30) are:

$$\mathcal{U} = \begin{pmatrix} i\zeta_1 & \psi_1 & \psi_2 \\ -R_1 & -i\zeta_1 & 0 \\ -R_2 & 0 & -i\zeta_1 \end{pmatrix}, \tag{4.31}$$

$$\mathcal{V} = \begin{pmatrix} V_{11} & V_{12} & V_{13} \\ V_{21} & V_{22} & V_{23} \\ V_{31} & V_{32} & V_{33} \end{pmatrix}, \tag{4.32}$$

where

$$\begin{aligned} V_{11} &= -i\zeta_1^2 + \frac{\mathrm{i}}{2}\psi_1 R_1 + \frac{\mathrm{i}}{2}\psi_2 R_2,\ V_{12} = -\zeta_1\psi_1 + \frac{\mathrm{i}}{2}\frac{\partial\psi_1}{\partial x},\ V_{13} = -\zeta_1\psi_2 + \frac{\mathrm{i}}{2}\frac{\partial\psi_2}{\partial x}, \\ V_{21} &= \zeta_1 R_1 + \frac{\mathrm{i}}{2}\frac{\partial R_1}{\partial x},\ V_{22} = i\zeta_1^2 - \frac{\mathrm{i}}{2}\psi_1 R_1,\ V_{23} = -\frac{\mathrm{i}}{2}\psi_2 R_1, \\ V_{31} &= \zeta_1 R_2 + \frac{\mathrm{i}}{2}\frac{\partial R_2}{\partial x},\ V_{32} = -\frac{\mathrm{i}}{2}\psi_1 R_2,\ V_{33} = i\zeta_1^2 - \frac{\mathrm{i}}{2}\psi_2 R_2, \end{aligned} \tag{4.33}$$

and

$$R_1 = -a\psi_1^*(x, t) - b\psi_2^*(x, t),\ R_2 = -d\psi_1^*(x, t) - c\psi_2^*(x, t). \tag{4.34}$$

The spectral parameter ζ_1 is isospectral, and the compatibility condition $(\Phi_x)_t = (\Phi_t)_x$, equivalent to the zero-curvature condition $\mathcal{U}_t - \mathcal{V}_x + [\mathcal{U}, \mathcal{V}] = 0$, yields the integrable GCNLS equations (4.30).

4.5.2 Bright solitons and collisional dynamics

To derive bright vector-soliton solutions for equations (4.30), one starts with the vacuum solution ($\psi_1(x, t) = \psi_2(x, t) = 0$) and applies the gauge transformation approach to obtain [42]

$$\psi_1^{(1)} = 2\varepsilon_1^{(1)}\beta_1 \mathrm{sech}(\theta_1)\mathrm{e}^{-\mathrm{i}\xi_1}, \tag{4.35a}$$

$$\psi_2^{(1)} = 2\varepsilon_2^{(1)}\beta_1 \mathrm{sech}(\theta_1)\mathrm{e}^{-\mathrm{i}\xi_1}, \tag{4.35b}$$

where

$$\theta_1 = 2x\beta_1 - 4\int \alpha_1\beta_1\ \mathrm{d}t + 2\delta_1, \tag{4.36a}$$

$$\xi_1 = 2x\alpha_1 - 2\int \left(\alpha_1^2 - \beta_1^2\right)\mathrm{d}t - 2\chi_1, \tag{4.36b}$$

with $\alpha_1 = \alpha_{10}(a\tau_1^2 + b\tau_1\tau_2 + d\tau_1\tau_2 + c\tau_2^2)$, $\beta_1 = \beta_{10}(a\tau_1^2 + b\tau_1\tau_2 + d\tau_1\tau_2 + c\tau_2^2)$, and δ_1, χ_1, τ_1, τ_2 as arbitrary parameters. The coupling parameters $\varepsilon_1^{(1)}$ and $\varepsilon_2^{(1)}$ satisfy $|\varepsilon_1^{(1)}|^2 + |\varepsilon_2^{(1)}|^2 = 1$.

The soliton amplitude depends on the SPM (a), XPM (c), and FWM (b, d) parameters, enabling energy switching between two optical pulses or their bound states. The gauge transformation approach [40] can be extended to obtain multi-soliton solutions. The two-soliton solution for the two modes is given by

$$\psi_1^{(2)} = \frac{2IA_1}{B},\ \psi_2^{(2)} = \frac{2IA_2}{B}, \tag{4.37}$$

where

$$\begin{aligned}
A_1 &= M_{12}^{(1)}M_{22}^{(2)}(\zeta_2 - \zeta_1)\left(\zeta_1 - \zeta_1^*\right)\left(\zeta_2 - \zeta_2^*\right) + M_{12}^{(2)}M_{22}^{(1)}\left(\zeta_2 - \zeta_1^*\right)\left(\zeta_2^* - \zeta_1\right)\left(\zeta_2 - \zeta_2^*\right)\\
&\quad + M_{11}^{(1)}M_{12}^{(2)}\left(\zeta_2 - \zeta_1^*\right)\left(\zeta_2^* - \zeta_1^*\right)\left(\zeta_2 - \zeta_2^*\right) + M_{11}^{(2)}M_{12}^{(1)}\left(\zeta_1 - \zeta_1^*\right)\left(\zeta_2^* - \zeta_1\right)\left(\zeta_2^* - \zeta_1^*\right),\\
A_2 &= M_{11}^{(2)}M_{21}^{(1)}(\zeta_2 - \zeta_1)\left(\zeta_1 - \zeta_1^*\right)\left(\zeta_2 - \zeta_2^*\right) + M_{11}^{(1)}M_{21}^{(2)}\left(\zeta_2 - \zeta_1^*\right)\left(\zeta_2^* - \zeta_1\right)\left(\zeta_2 - \zeta_2^*\right)\\
&\quad + M_{21}^{(2)}M_{22}^{(1)}\left(\zeta_2 - \zeta_1^*\right)\left(\zeta_1 - \zeta_1^*\right)\left(\zeta_2 - \zeta_2^*\right) + M_{21}^{(1)}M_{22}^{(2)}\left(\zeta_1 - \zeta_1^*\right)\left(\zeta_2^* - \zeta_1\right)\left(\zeta_2^* - \zeta_1^*\right),\\
B &= \left(M_{12}^{(1)}M_{21}^{(2)} + M_{12}^{(2)}M_{21}^{(1)}\right)\left(\zeta_1 - \zeta_1^*\right)\left(\zeta_2 - \zeta_2^*\right) + \left(M_{11}^{(1)}M_{22}^{(2)} + M_{11}^{(2)}M_{22}^{(1)}\right)\\
&\quad \times \left(\zeta_2 - \zeta_1^*\right)\left(\zeta_2^* - \zeta_1\right) + \left(M_{11}^{(1)}M_{11}^{(2)} + M_{22}^{(1)}M_{22}^{(2)}\right)(\zeta_2 - \zeta_1)\left(\zeta_2^* - \zeta_1^*\right),
\end{aligned} \tag{4.38}$$

with $\zeta_2 = \alpha_2 + i\beta_2$, and with the matrix elements given in equation (4.11) where

$$\theta_j = 2x\beta_j - 4\int \alpha_j\beta_j \, dt + 2\delta_j, \tag{4.39a}$$

$$\xi_j = 2x\alpha_j - 2\int \left(\alpha_j^2 - \beta_j^2\right) dt - 2\chi_j, j = 1, 2. \tag{4.39b}$$

4.5.3 Intramodal collision of bright solitons

To examine FWM's impact on soliton collisions in the Manakov model, consider the elastic collision depicted in figure 4.19. Introducing FWM not only rotates the soliton trajectories but also enhances their intensities, as shown in figure 4.20. To amplify the intensity of a specific bound state in a given mode, the FWM parameter d can be adjusted. Figure 4.21 illustrates that increasing d enhances the intensity I_1 at the expense of I_2 in each mode. Conversely, adjusting b reverses this effect, enhancing I_2, as shown in figure 4.22.

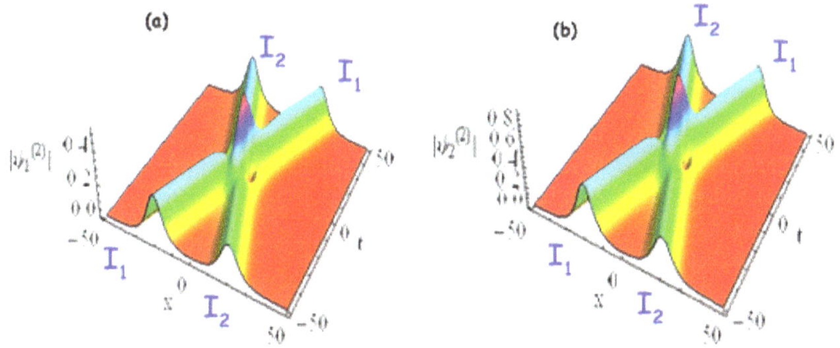

Figure 4.19. Elastic collision of solitons in the Manakov model with $a = c = 1$, $b = d = 0$, $\alpha_{10} = 0.1$, $\alpha_{20} = 0.25$, $\beta_{10} = 0.3$, $\beta_{20} = 0.1$, $\chi_1 = 0.1$, $\chi_2 = 0.3$, $\delta_1 = 0.1$, $\delta_2 = 0.2$, $\varepsilon_1^{(1)} = 0.5$, $\varepsilon_1^{(2)} = 0.8$. Reprinted from [42], Copyright (2016), with permission from Elsevier.

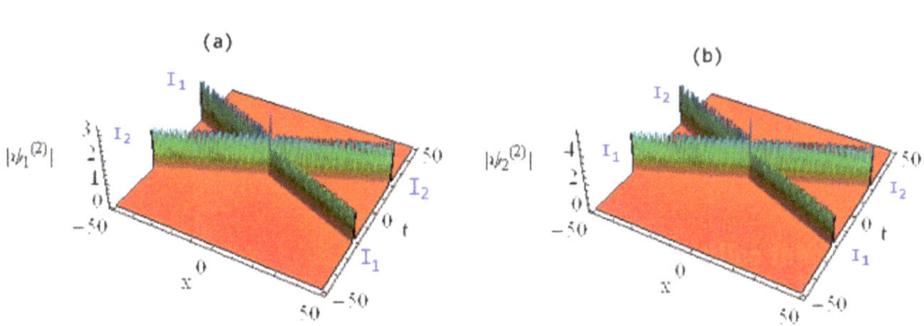

Figure 4.20. FWM-induced rotation and intensity enhancement of solitons with $a = c = 1$, $b = d = 0.5$, and other parameters as in figure 4.19. Reprinted from [42], Copyright (2016), with permission from Elsevier.

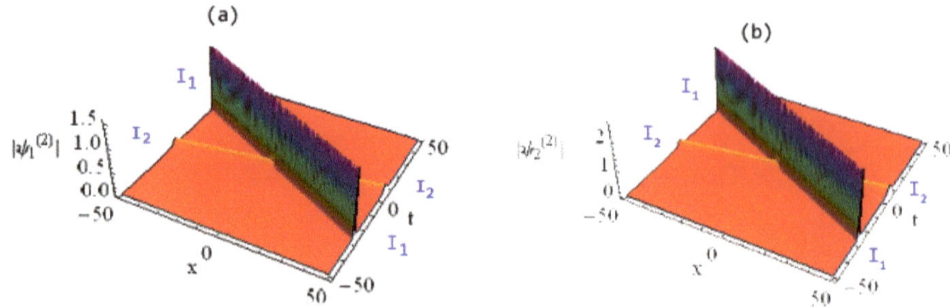

Figure 4.21. Enhanced intensity I_1 by setting $d = 3.5$, with other parameters as in figure 4.20. Reprinted from [42], Copyright (2016), with permission from Elsevier.

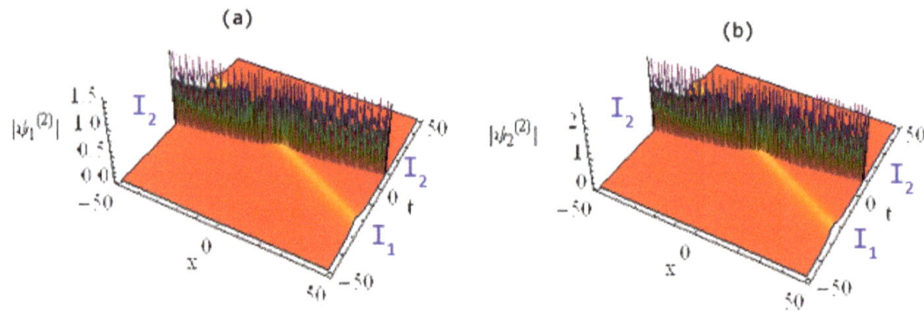

Figure 4.22. Enhanced intensity I_2 by setting $b = 3.5$, with other parameters as in figure 4.20. Reprinted from [42], Copyright (2016), with permission from Elsevier.

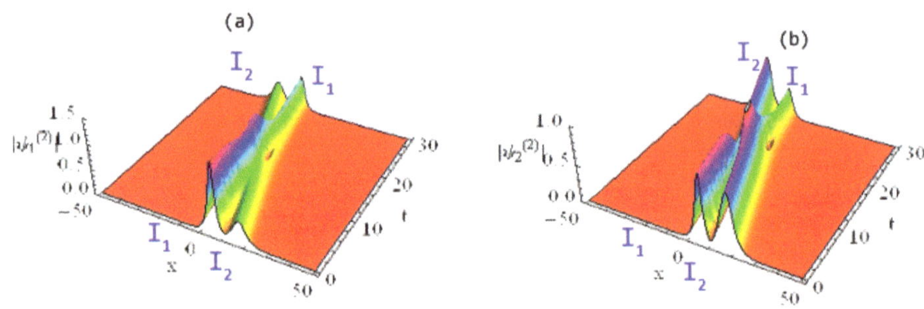

Figure 4.23. Inelastic collision of bright solitons in the Manakov model with parameters $a = c = 1$, $b = d = 0$, $\alpha_{10} = 0.1$, $\alpha_{20} = 0.25$, $\beta_{10} = 0.2$, $\beta_{20} = 0.3$, $\chi_1 = 0.2$, $\chi_2 = 0.3$, $\delta_1 = 0.1$, $\delta_2 = 0.2$, $\varepsilon_1^{(1)} = 0.5$, and $\varepsilon_1^{(2)} = 0.8$. Reprinted from [42], Copyright (2016), with permission from Elsevier.

4.5.4 Intermodal collision of bright solitons

To control the intensity of a specific optical beam, consider the inelastic collision in the Manakov model without FWM, shown in figure 4.23 [43]. Introducing FWM induces energy exchange between the beams, similar to the Manakov model, while

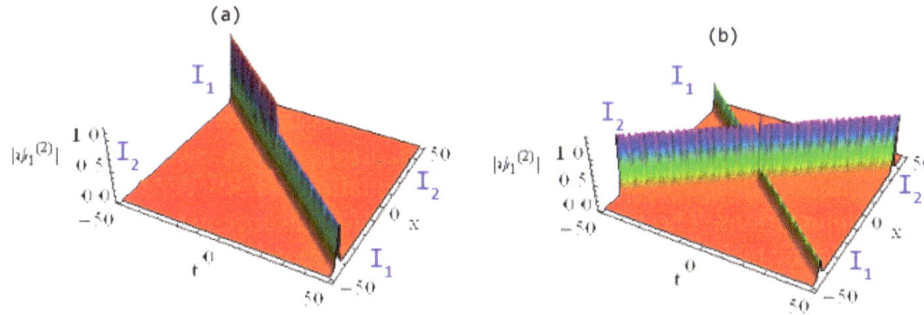

Figure 4.24. Rotation of bright solitons and energy transfer between modes with $b = 0.5$, $d = 3.5$, and other parameters identical to those in figure 4.23. Reprinted from [42], Copyright (2016), with permission from Elsevier.

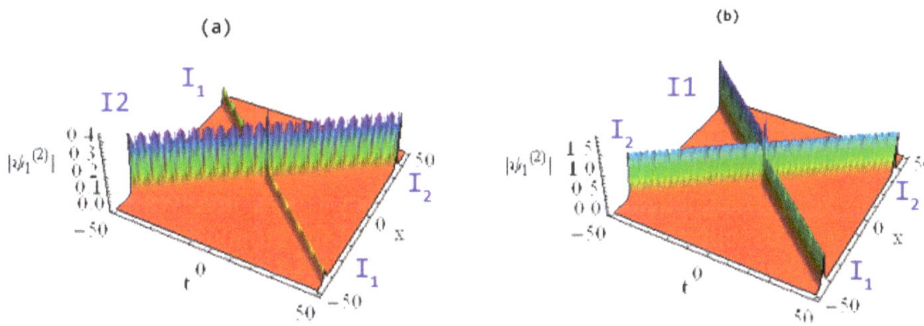

Figure 4.25. Reversed intensity enhancement between modes with $b = 2.5$, $d = 0.5$, and other parameters matching those in figure 4.23. Reprinted from [42], Copyright (2016), with permission from Elsevier.

also causing soliton rotation, as depicted in figures 4.24 and 4.25. By tuning the FWM parameters b and d, desired intensity distributions are achieved, as demonstrated by comparing figures 4.24 and 4.25. These results highlight the ability to tailor beam intensities. The combination of FWM with unequal SPM and XPM slightly increases intensities and rotates soliton trajectories.

Neglecting SPM and XPM ($a = c = 0$) in equations (4.30) yields

$$i\frac{\partial \psi_1}{\partial t} + \frac{\partial^2 \psi_1}{\partial x^2} + 2\left(b\psi_1\psi_2^* + d\psi_2\psi_1^*\right)\psi_1 = 0, \tag{4.40a}$$

$$i\frac{\partial \psi_2}{\partial t} + \frac{\partial^2 \psi_2}{\partial x^2} + 2\left(b\psi_1\psi_2^* + d\psi_2\psi_1^*\right)\psi_2 = 0, \tag{4.40b}$$

a special case of equations (4.30) where soliton reflection and non-interaction occur [41]. Under appropriate transformations, equations (4.30) map to GP equations, enabling intensity switching of matter waves in specific bound states or hyperfine states of atoms, applicable to matter-wave switching in BECs.

The collisional dynamics reveal the flexibility to maintain desired intensities in specific optical beam bound states or pulses, surpassing the capabilities of the Manakov model. Unlike the Manakov model, where energy switching relies on phase parameters, intensity manipulation in equations (4.30) depends on light-medium interactions. These findings pave the way for advancements in optical and matter-wave switching in nonlinear optics and BECs, potentially inspiring parallel developments in ultracold atom research.

4.6 Propagation of light through matter and electromagnetically induced transparency

The interaction of two laser pulses with an atomic system offers intriguing possibilities, particularly for quantum information processing and quantum computing. When a single laser beam propagates through a Λ-type atomic ensemble, the medium absorbs the light, leading to heating. However, when two laser beams, with frequencies tuned to the atomic transition frequencies, irradiate the medium, absorption is suppressed, and the medium becomes transparent. This phenomenon, known as electromagnetically induced transparency (EIT), involves two coherent light sources interacting with three quantum states of the atomic system. For EIT to occur, two of the three possible transitions must be dipole-allowed (induced by an oscillating electric field), while the third transition is dipole-forbidden.

4.6.1 Maxwell–Bloch equations and gauge transformation approach

Consider two electric fields with complex envelopes ϵ_1 and ϵ_2, and carrier frequencies ω_1 and ω_2, propagating along the z-direction through a Λ-type atomic ensemble, as depicted in figure 4.26. The ensemble has three energy levels: the ground state $|1\rangle$, the

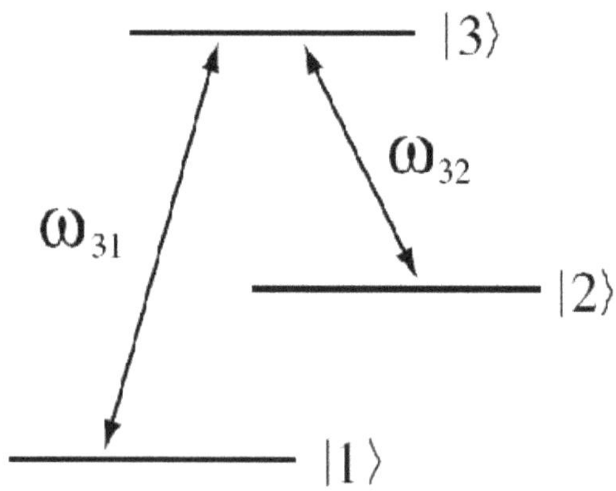

Figure 4.26. The Λ-type atomic configuration. Reprinnted with permission from [44], Copyright (2008) by the American Physical Society.

intermediate state $|2\rangle$, and the excited state $|3\rangle$. The transition frequencies between $|3\rangle$ and $|1\rangle$, and $|3\rangle$ and $|2\rangle$, match ω_1 and ω_2, respectively.

Assuming pulse durations are much shorter than irreversible relaxation times, the evolution of short pulses under the slowly varying envelope approximation is governed by the Maxwell–Bloch equations [45–47]

$$\left(\frac{\partial}{\partial z} + \frac{n}{c}\frac{\partial}{\partial t}\right)\epsilon_1 = -i\frac{2\pi N\hbar\omega_1}{cn}\mu_{13}\langle\hat{\sigma}_{13}\rangle, \tag{4.41}$$

$$\left(\frac{\partial}{\partial z} + \frac{n}{c}\frac{\partial}{\partial t}\right)\epsilon_2 = -i\frac{2\pi N\hbar\omega_2}{cn}\mu_{23}\langle\hat{\sigma}_{23}\rangle, \tag{4.42}$$

$$i\hbar\frac{\partial}{\partial t}\hat{\sigma} = [\hat{H}, \hat{\sigma}], \tag{4.43}$$

where $\mu_{ij} = -d_{ij}/\hbar$, and the Hamiltonian $\hat{H}$ for the density matrix is

$$\hat{H} = \hbar\begin{pmatrix} 0 & 0 & \mu_{13}\epsilon_1^* \\ 0 & \Delta\omega_1 - \Delta\omega_2 & \mu_{32}\epsilon_2^* \\ \mu_{31}\epsilon_1 & \mu_{32}\epsilon_2 & \Delta\omega_1 \end{pmatrix}, \tag{4.44}$$

with detunings $\Delta\omega_1 = \omega_{31} - \omega_1$, $\Delta\omega_2 = \omega_{32} - \omega_2$, and d_{ij} as the allowed transition matrix elements. Here, $\omega_p = \omega_1$ and $\omega_c = \omega_2$ are the probe and coupling laser frequencies, respectively.

Introducing new independent variables

$$x = t - \frac{nz}{c},\ T = \frac{z}{l_1}, \tag{4.45}$$

where:

$$l_v^{-1} = \frac{2\pi N\hbar\omega_v}{cn}|\mu_{3v}|^2,\ v = 1, 2, \tag{4.46}$$

the Maxwell–Bloch equations under resonance conditions ($\omega_{31} = \omega_1$, $\omega_{32} = \omega_2$) become

$$\frac{\partial\epsilon_1}{\partial T} = -i\frac{\sigma_{31}}{\mu_{31}}, \tag{4.47}$$

$$\frac{\partial\epsilon_2}{\partial T} = -i\frac{l_1}{l_2}\frac{\sigma_{32}}{\mu_{32}}, \tag{4.48}$$

$$i\hbar\frac{\partial\sigma}{\partial x} = [H_{\text{res}}, \sigma], \tag{4.49}$$

where $\sigma = (\sigma_{ij})$ is a 3×3 Hermitian matrix representing the expectation values of the operator $\hat{\sigma}_{ij}$. Redefining the variables

$$q_1 = \mu_{31}\epsilon_1, \; q_2 = \mu_{32}\epsilon_2, \tag{4.50}$$

and setting $l_1 = l_2$, equations (4.47) and (4.48) become integrable

$$\begin{aligned} \frac{\partial q_1}{\partial T} &= -\, i\sigma_{31}, \\ \frac{\partial q_2}{\partial T} &= -\, i\sigma_{32}, \\ i\hbar\frac{\partial \sigma}{\partial x} &= [H_{\text{res}}, \sigma], \end{aligned} \tag{4.51}$$

with the resonant Hamiltonian taking the following form

$$H_{\text{res}} = \hbar\begin{pmatrix} 0 & 0 & q_1^* \\ 0 & 0 & q_2^* \\ q_1 & q_2 & 0 \end{pmatrix}. \tag{4.52}$$

To identify localized excitations in the coherent light sources driving the atomic system, soliton solutions are sought, as their collisions may reveal EIT signatures, facilitating efficient information transfer. Equations (4.51) and (4.52) admit the linear eigenvalue problem

$$\Phi_x = \left[i\lambda\begin{pmatrix} 1 & 0 & 0 \\ 0 & 1 & 0 \\ 0 & 0 & -1 \end{pmatrix} - \frac{\mathrm{i}}{\hbar}H_{\text{res}} \right]\Phi, \tag{4.53}$$

$$\Phi_T = \frac{\mathrm{i}}{2\lambda}\begin{pmatrix} \sigma_{11} & \sigma_{12} & \sigma_{13} \\ \sigma_{21} & \sigma_{22} & \sigma_{23} \\ \sigma_{31} & \sigma_{32} & \sigma_{33} \end{pmatrix}\Phi, \tag{4.54}$$

where σ_{ij} (for $i, j = 1, 2, 3$) represents the atomic population density matrix.

Starting with the vacuum state

$$\begin{aligned} &q_1^{(0)} = q_2^{(0)} = 0, \\ &\sigma_{31} = \sigma_{13} = \sigma_{32} = \sigma_{23} = \sigma_{21} = \sigma_{12} = 0, \\ &\sigma_{jj} = \text{constant}, j = 1, 2, 3, \end{aligned} \tag{4.55}$$

the vacuum linear system is given by

$$\Phi_x^0 = i\lambda\begin{pmatrix} 1 & 0 & 0 \\ 0 & 1 & 0 \\ 0 & 0 & -1 \end{pmatrix}\Phi^0, \tag{4.56}$$

$$\Phi_T^0 = \frac{\mathrm{i}}{2\lambda}\begin{pmatrix} \sigma_{11} & 0 & 0 \\ 0 & \sigma_{22} & 0 \\ 0 & 0 & \sigma_{33} \end{pmatrix}\Phi^0. \tag{4.57}$$

Solving this, the trivial matrix eigenfunction becomes:

$$\psi^0(x, T) = \begin{pmatrix} \mathrm{e}^{\mathrm{i}\lambda x+\frac{\mathrm{i}}{2\lambda}\sigma_{11}T} & 0 & 0 \\ 0 & \mathrm{e}^{\mathrm{i}\lambda x+\frac{\mathrm{i}}{2\lambda}\sigma_{22}T} & 0 \\ 0 & 0 & \mathrm{e}^{\mathrm{i}\lambda x+\frac{\mathrm{i}}{2\lambda}\sigma_{33}T} \end{pmatrix}. \tag{4.58}$$

Using the gauge transformation approach, the one-soliton solution for the probe and coupling fields is written as [35, 44, 48]

$$\epsilon_1^{(1)} = \frac{q_1^{(1)}}{\tilde{\mu}_{31}} = \frac{2\sqrt{2}}{\tilde{\mu}_{31}}\beta_1\varepsilon_1^{(1)}\frac{A_1}{B_1 + B_2}, \tag{4.59}$$

$$\epsilon_2^{(1)} = \frac{q_2^{(1)}}{\tilde{\mu}_{32}} = \frac{2\sqrt{2}}{\tilde{\mu}_{32}}\beta_1\varepsilon_2^{(1)}\frac{A_2}{B_1 + B_2}, \tag{4.60}$$

where:

$$\begin{aligned} A_1 &= \mathrm{e}^{\mathrm{i}\mu_1 x+\frac{\mathrm{i}}{2\mu_1}\sigma_{11}T-2i\xi_1 + i\lambda_1 x-\frac{\mathrm{i}}{2\lambda_1}\sigma_{33}T}, \\ A_2 &= \mathrm{e}^{\mathrm{i}\mu_1 x+\frac{\mathrm{i}}{2\mu_1}\sigma_{22}T+\mathrm{i}\lambda_1 x-\frac{\mathrm{i}}{2\lambda_1}\sigma_{33}T}, \\ B_1 &= \frac{1}{2}\mathrm{e}^{\mathrm{i}\mu_1 x+2\delta_1 - i\lambda_1 x}\left(e^{\frac{\mathrm{i}}{2\mu_1}\sigma_{11}T-\frac{\mathrm{i}}{2\lambda_1}\sigma_{11}T} + e^{\frac{\mathrm{i}}{2\mu_1}\sigma_{22}T-\frac{\mathrm{i}}{2\lambda_1}\sigma_{22}T}\right), \\ B_2 &= \mathrm{e}^{-\mathrm{i}\mu_1 x+\frac{\mathrm{i}}{2\mu_1}\sigma_{33}T-2\delta_1 + i\lambda_1 x-\frac{\mathrm{i}}{2\lambda_1}\sigma_{33}T}, \end{aligned} \tag{4.61}$$

with the spectral parameter $\lambda_1 = \mu_1^* = \alpha_1 + i\beta_1$. When $\alpha_1 = 0$ and $\sigma_{11} = \sigma_{22}$, the solutions simplify to:

$$\bar{\epsilon}_1^{(1)} = \frac{\sqrt{2}}{\tilde{\mu}_{31}}\beta_1\varepsilon_1^{(1)}\mathrm{sech}\left[2\beta_1 x - \frac{1}{2\beta_1}(\sigma_{11} - \sigma_{33})T\right], \tag{4.62}$$

$$\bar{\epsilon}_2^{(1)} = \frac{\sqrt{2}}{\tilde{\mu}_{32}}\beta_1\varepsilon_2^{(1)}\mathrm{sech}\left[2\beta_1 x - \frac{1}{2\beta_1}(\sigma_{11} - \sigma_{33})T\right], \tag{4.63}$$

termed as ‘simultons’ due to their identical velocities. When the ground and intermediate state populations are equal, the complex soliton solutions (4.59) and (4.60) reduce to these real simulton solutions, consistent with [45].

4.6.2 Signatures of EIT in soliton collisions

The gauge transformation approach can be generalized to generate multisoliton solutions. The two-soliton solution is given by

$$q_1^{(2)} = -q_1^{(1)} - 2i(\lambda_2 - \mu_2)\frac{P_{13}}{R}, \tag{4.64a}$$

$$q_2^{(2)} = -q_2^{(1)} - 2i(\lambda_2 - \mu_2)\frac{P_{23}}{R}, \tag{4.64b}$$

where

$$\begin{aligned} P_{13} = &-\left(M_{11}^{(2)} + \frac{\gamma}{\tau}M_{11}^{(1)}M_{11}^{(2)} + \frac{\gamma}{\tau}M_{13}^{(1)}M_{31}^{(2)}\right)\left(\frac{\gamma^*}{\tau}M_{13}^{(1)}\right) - \left(\frac{\gamma}{\tau}M_{13}^{(1)}M_{32}^{(2)}\right)\left(\frac{\gamma^*}{\tau}M_{23}^{(1)}\right) \\ &-\left(M_{13}^{(2)} + \frac{\gamma}{\tau}M_{11}^{(1)}M_{13}^{(2)} + \frac{\gamma}{\tau}M_{13}^{(1)}M_{33}^{(2)}\right) - \left(M_{13}^{(2)} + \frac{\gamma}{\tau}M_{11}^{(1)}M_{13}^{(2)} + \frac{\gamma}{\tau}M_{13}^{(1)}M_{33}^{(2)}\right) \\ &\times\left(\frac{\gamma^*}{\tau}M_{33}^{(1)}\right), \end{aligned} \tag{4.65a}$$

$$\begin{aligned} P_{23} = &-\left(M_{22}^{(2)} + \frac{\gamma}{\tau}M_{22}^{(1)}M_{22}^{(2)} + \frac{\gamma}{\tau}M_{23}^{(1)}M_{32}^{(2)}\right)\left(\frac{\gamma^*}{\tau}M_{23}^{(1)}\right) - \left(\frac{\gamma}{\tau}M_{23}^{(1)}M_{31}^{(2)}\right)\left(\frac{\gamma^*}{\tau}M_{13}^{(1)}\right) \\ &-\left(M_{23}^{(2)} + \frac{\gamma}{\tau}M_{22}^{(1)}M_{23}^{(2)} + \frac{\gamma}{\tau}M_{23}^{(1)}M_{33}^{(2)}\right) - \left(M_{23}^{(2)} + \frac{\gamma}{\tau}M_{22}^{(1)}M_{23}^{(2)} + \frac{\gamma}{\tau}M_{23}^{(1)}M_{33}^{(2)}\right) \\ &\times\left(\frac{\gamma^*}{\tau}M_{33}^{(1)}\right), \end{aligned} \tag{4.65b}$$

$$R = P_{11} + P_{22} + P_{33}, \tag{4.65c}$$

and

$$\begin{aligned} P_{11} = &\left(M_{11}^{(2)} + \frac{\gamma}{\tau}M_{11}^{(1)}M_{11}^{(2)} + \frac{\gamma}{\tau}M_{13}^{(1)}M_{31}^{(2)}\right) + \left(M_{11}^{(2)} + \frac{\gamma}{\tau}M_{11}^{(1)}M_{11}^{(2)} + \frac{\gamma}{\tau}M_{13}^{(1)}M_{31}^{(2)}\right) \\ &\times\left(\frac{\gamma^*}{\tau}M_{11}^{(1)}\right) + \left(M_{13}^{(2)} + \frac{\gamma}{\tau}M_{11}^{(1)}M_{13}^{(2)} + \frac{\gamma}{\tau}M_{13}^{(1)}M_{33}^{(2)}\right)\left(\frac{\gamma^*}{\tau}M_{31}^{(1)}\right), \end{aligned} \tag{4.65d}$$

$$\begin{aligned} P_{22} = &\left(M_{22}^{(2)} + \frac{\gamma}{\tau}M_{22}^{(1)}M_{22}^{(2)} + \frac{\gamma}{\tau}M_{23}^{(1)}M_{32}^{(2)}\right) + \left(M_{22}^{(2)} + \frac{\gamma}{\tau}M_{22}^{(1)}M_{22}^{(2)} + \frac{\gamma}{\tau}M_{23}^{(1)}M_{32}^{(2)}\right) \\ &\times\left(\frac{\gamma^*}{\tau}M_{22}^{(1)}\right) + \left(M_{23}^{(2)} + \frac{\gamma}{\tau}M_{22}^{(1)}M_{23}^{(2)} + \frac{\gamma}{\tau}M_{23}^{(1)}M_{33}^{(2)}\right)\left(\frac{\gamma^*}{\tau}M_{32}^{(1)}\right), \end{aligned} \tag{4.65e}$$

$$\begin{aligned} P_{33} = &\left(M_{31}^{(2)} + \frac{\gamma}{\tau}M_{31}^{(1)}M_{11}^{(2)} + \frac{\gamma}{\tau}M_{33}^{(1)}M_{31}^{(2)}\right)\left(\frac{\gamma^*}{\tau}M_{13}^{(1)}\right) \\ &+\left(M_{32}^{(2)} + \frac{\gamma}{\tau}M_{32}^{(1)}M_{22}^{(2)} + \frac{\gamma}{\tau}M_{33}^{(1)}M_{32}^{(2)}\right)\left(\frac{\gamma^*}{\tau}M_{23}^{(1)}\right) \\ &+\left(M_{33}^{(2)} + \frac{\gamma}{\tau}M_{31}^{(1)}M_{13}^{(2)} + \frac{\gamma}{\tau}M_{32}^{(1)}M_{23}^{(2)} + \frac{\gamma}{\tau}M_{33}^{(1)}M_{33}^{(2)}\right)\left(1 + \frac{\gamma^*}{\tau}M_{33}^{(1)}\right), \end{aligned} \tag{4.65f}$$

with matrix elements

$$
\begin{aligned}
M_{11}^{(j)} &= \frac{1}{\sqrt{2}} \mathrm{e}^{-2\delta_j - \mathrm{i}(\lambda_j - \mu_j)x - \frac{\mathrm{i}}{2}\left(\frac{1}{\lambda_j} - \frac{1}{\mu_j}\right)\sigma_{11} T}, \ M_{12}^{(j)} = 0, \\
M_{13}^{(j)} &= \mathrm{e}^{-2\mathrm{i}\xi_j + \mathrm{i}(\lambda_j + \mu_j)x - \frac{\mathrm{i}}{2\lambda_j}\sigma_{33} T + \frac{\mathrm{i}}{2\mu_j}\sigma_{11} T} \varepsilon_1^{(j)}, \\
M_{21}^{(j)} &= 0, \ M_{22}^{(j)} = \frac{1}{\sqrt{2}} \mathrm{e}^{-2\delta_j - \mathrm{i}(\lambda_j - \mu_j)x - \frac{\mathrm{i}}{2}\left(\frac{1}{\lambda_j} - \frac{1}{\mu_j}\right)\sigma_{22} T}, \\
M_{23}^{(j)} &= \mathrm{e}^{-2\mathrm{i}\xi_j + \mathrm{i}(\lambda_j + \mu_j)x - \frac{\mathrm{i}}{2\lambda_j}\sigma_{33} T + \frac{\mathrm{i}}{2\mu_j}\sigma_{22} T} \varepsilon_2^{(j)}, \\
M_{31}^{(j)} &= \mathrm{e}^{2\mathrm{i}\xi_j - \mathrm{i}(\lambda_j + \mu_j)x - \frac{\mathrm{i}}{2\lambda_j}\sigma_{11} T + \frac{\mathrm{i}}{2\mu_j}\sigma_{33} T} \varepsilon_1^{*(j)}, \\
M_{32}^{(j)} &= \mathrm{e}^{2\mathrm{i}\xi_j - \mathrm{i}(\lambda_j + \mu_j)x - \frac{\mathrm{i}}{2\lambda_j}\sigma_{22} T + \frac{\mathrm{i}}{2\mu_j}\sigma_{33} T} \varepsilon_2^{*(j)}, \\
M_{33}^{(j)} &= \sqrt{2} \mathrm{e}^{-2\delta_j + \mathrm{i}(\lambda_j - \mu_j)x - \frac{\mathrm{i}}{2}\left(\frac{1}{\lambda_j} - \frac{1}{\mu_j}\right)\sigma_{33} T},
\end{aligned}
\tag{4.66}
$$

$$
\tau = M_{11}^{(1)} + M_{22}^{(1)} + M_{33}^{(1)}, \tag{4.67a}
$$

$$
\gamma = \frac{\lambda_1 - \mu_1}{\mu_2 - \lambda_1}, \ \gamma^* = -\frac{\lambda_1 - \mu_1}{\lambda_2 - \mu_1}, \ \lambda_n = \mu_n^* = \alpha_n + i\beta_n, \tag{4.67b}
$$

for $n, j = 1, 2$. By selecting appropriate density matrix elements σ_{11}, σ_{22}, and σ_{33}, representing atomic populations, and coupling parameters $\varepsilon_1^{(j)}$, $\varepsilon_2^{(j)}$ (for $j = 1, 2$), which control soliton intensity sharing, various collision behaviours are observed under realistic physical conditions.

Figure 4.27 illustrates energy exchange between the probe and coupling fields in a three-level atomic system when the ground and intermediate state populations are equal ($\sigma_{11} = \sigma_{22}$). Figure 4.28 shows that when the probe field (q_1) is present, the coupling field (q_2) is absent, and vice versa, indicating a switching-off of the coupling beam while the probe pulse remains in the medium. This behaviour enables quantum storage by adiabatically turning off the coupling beam. Figure 4.29 depicts elastic collisions of simultons, where probe and coupling fields propagate without energy

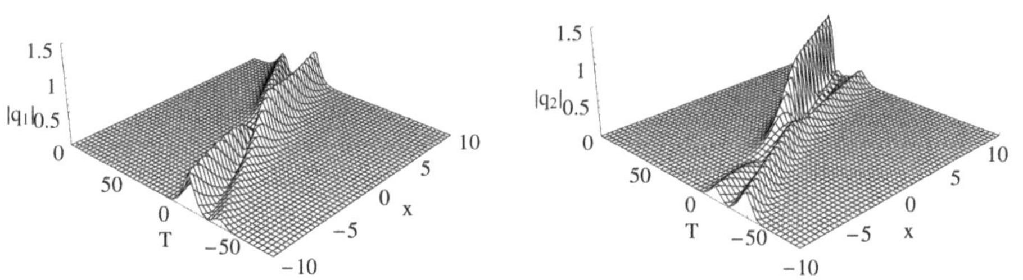

Figure 4.27. Energy exchange between probe and coupling fields in a three-level medium with $\sigma_{11} = \sigma_{22} = 0.4$, $\sigma_{33} = 0.2$, $\alpha_1 = \alpha_2 = 0$, $\beta_1 = 0.5$, $\beta_2 = 0.4$, $\varepsilon_1^{(1)} = 0.3$, $\varepsilon_1^{(2)} = 0.93$, satisfying $|\varepsilon_1^{(j)}|^2 + |\varepsilon_2^{(j)}|^2 = 1$, for $j = 1, 2$. Reprinted with permission from [44], Copyright (2008) by the American Physical Society.

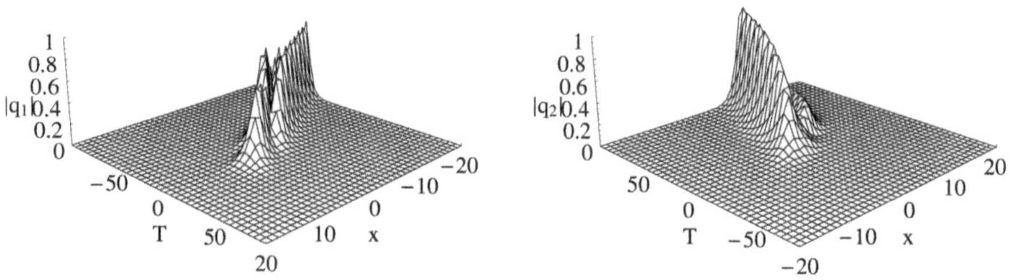

Figure 4.28. Soliton interaction with $\sigma_{11} = 0.5$, $\sigma_{22} = 0.3$, $\sigma_{33} = 0.2$, $\alpha_1 = 0.1$, $\alpha_2 = 0.2$, $\beta_1 = 0.5$, $\beta_2 = 0.4$, $\varepsilon_1^{(1)} = 0.1$, $\varepsilon_1^{(2)} = 0.99$, satisfying $|\varepsilon_1^{(j)}|^2 + |\varepsilon_2^{(j)}|^2 = 1$, for $j = 1, 2$. Reprinted with permission from [44], Copyright (2008) by the American Physical Society.

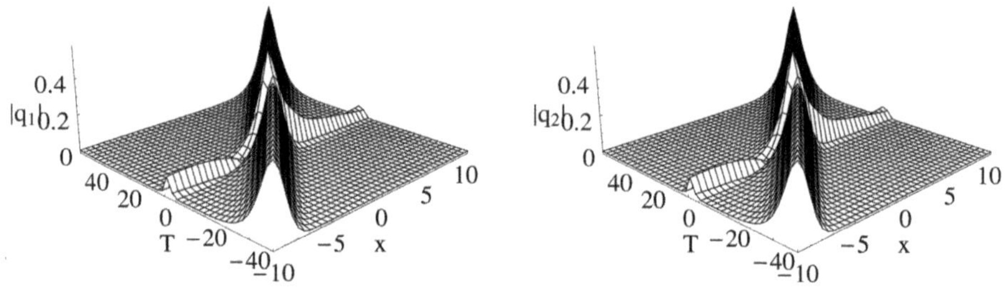

Figure 4.29. Elastic collision of simultons with $\sigma_{11} = \sigma_{22} = 0.4$, $\sigma_{33} = 0.2$, $\alpha_1 = \alpha_2 = 0$, $\beta_1 = 0.5$, $\beta_2 = 0.1$, $\varepsilon_1^{(1)} = \varepsilon_1^{(2)} = 1/\sqrt{2}$, satisfying $|\varepsilon_1^{(j)}|^2 + |\varepsilon_2^{(j)}|^2 = 1$, for $j = 1, 2$. Reprinted with permission from [44], Copyright (2008) by the American Physical Society.

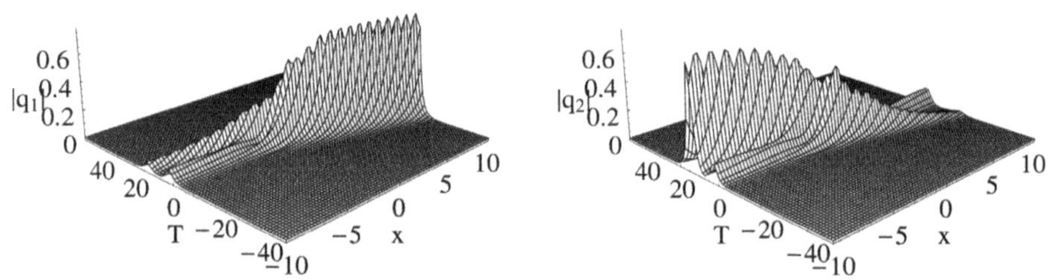

Figure 4.30. Two-soliton interaction with $\sigma_{11} = 0.2$, $\sigma_{22} = 0$, $\sigma_{33} = 0.8$, $\alpha_1 = 0.1$, $\alpha_2 = 0.2$, $\beta_1 = 0.5$, $\beta_2 = 0.1$, $\varepsilon_1^{(1)} = \varepsilon_1^{(2)} = 1/\sqrt{2}$, satisfying $|\varepsilon_1^{(j)}|^2 + |\varepsilon_2^{(j)}|^2 = 1$, for $j = 1, 2$. Reprinted with permission from [44], Copyright (2008) by the American Physical Society.

exchange. In figure 4.30, with the intermediate state unpopulated ($\sigma_{22} = 0$), the probe pulse undergoes compression, increasing its amplitude, while the coupling pulse broadens, indicating low group velocity for the probe.

These collisions, particularly in figure 4.28, suggest the potential for trapping and retrieving light pulses [49], while figure 4.30 indicates the possibility of slowing light through probe pulse compression. These results demonstrate EIT signatures in soliton collisions without relying on the dark state polariton framework [50], offering applications in quantum information processing.

4.7 Summary and future challenges

In this chapter, the dynamics of coupled NLS-type equations describing the propagation of two light beams under different settings have been reviewed, and the interaction of solitons has been brought out in detail. The signatures of EIT like quantum storage and slow light when two laser beams pass through matter/atomic system have also been brought out in detail through the collision of bright solitons. The existence of localized solutions and their stability in the counterpart of the Manakov model in two and three dimensions from the point of view of integrability continue to be elusive and challenging even today. The investigation of the above model is still relevant today, and it assumes tremendous significance in getting a deeper understanding of BECs, plasmas, planar waveguides, etc.

References

[1] Hasegawa A and Tappert F 1973 Transmission of stationary nonlinear optical pulses in dispersive dielectric fibers. I. Anomalous dispersion *Appl. Phys. Lett.* **23** 142–4

[2] Hasegawa A and Tappert F 1973 Transmission of stationary nonlinear optical pulses in dispersive dielectric fibers. II. Normal dispersion *Appl. Phys. Lett.* **23** 171–2

[3] Mollenauer L F, Stolen R H and Gordon J P 1980 Experimental observation of picosecond pulse narrowing and solitons in optical fibers *Phys. Rev. Lett.* **45** 1095–8

[4] Agrawal G P 2013 *Nonlinear Fiber Optics* 5th edn (Academic) (includes bibliographical references and index)

[5] Zakharov V and Schulman E 1982 To the integrability of the system of two coupled nonlinear Schrödinger equations *Physica* D **4** 270–4

[6] Sahadevan R, Tamizhmani K M and Lakshmanan M 1986 Painleve analysis and integrability of coupled non-linear Schrödinger equations *J. Phys.* A **19** 1783–91

[7] Akhmediev N N and Ankiewicz A 1997 Solitons *Optical and Quantum Electronics Series* 1st edn (Chapman and Hall) ch 5 pp 311–32

[8] Menyuk C 1987 Nonlinear pulse propagation in birefringent optical fibers *IEEE J. Quantum Electron.* **23** 174–76

[9] Wadati M, Iizuka T and Hisakado M 1992 A coupled nonlinear Schrödinger equation and optical solitons *J. Phys. Soc. Japan* **61** 2241–5

[10] Porsezian K and Nakkeeran K 1997 Optical solitons in birefringent fibre - Bäcklund transformation approach *J. Opt. A: Pure Appl. Opt.* **6** L7–11

[11] Hisakado M, Iizuka T and Wadati M 1994 Coupled hybrid nonlinear Schrödinger equation and optical solitons *J. Phys. Soc. Japan* **63** 2887–94

[12] Manakov S V 1974 On the theory of two-dimensional stationary self-focusing of electromagnetic waves *Sov. Phys.-JETP* **38** 248–53

[13] Makhankov V, Makhaldiani N and Pashaev O 1981 On the integrability and isotopic structure of the one-dimensional Hubbard model in the long wave approximation *Phys. Lett.* A **81** 161–4

[14] Kanna T, Tsoy E N and Akhmediev N 2004 On the solution of multicomponent nonlinear Schrödinger equations *Phys. Lett.* A **330** 224–9

[15] Kaup D J and Malomed B A 1993 Soliton trapping and daughter waves in the Manakov model *Phys. Rev.* A **48** 599–604

[16] Abdullaev F K and Tsoy E 2002 The evolution of optical beams in self-focusing media *Physica* D **161** 67–78
[17] Hisakado M and Wadati M 1995 Integrable multi-component hybrid nonlinear Schrödinger equations *J. Phys. Soc. Japan* **64** 408–13
[18] Radhakrishnan R and Lakshmanan M 1995 Bright and dark soliton solutions to coupled nonlinear Schrodinger equations *J. Phys.* A **28** 2683–92
[19] Radhakrishnan R, Lakshmanan M and Hietarinta J 1997 Inelastic collision and switching of coupled bright solitons in optical fibers *Phys. Rev.* E **56** 2213–6
[20] Radhakrishnan R and Lakshmanan M 1999 Suppression and enhancement of soliton switching during interaction in periodically twisted birefringent fibers *Phys. Rev.* E **60** 2317–21
[21] Saleh M F and Biancalana F 2013 Soliton-radiation trapping in gas-filled photonic crystal fibers *Phys. Rev.* A **87** 043807
[22] Feijoo D, Paredes A and Michinel H 2013 Outcoupling vector solitons from a Bose-Einstein condensate with time-dependent interatomic forces *Phys. Rev.* A **87** 063619
[23] Gertjerenken B, Billam T P, Khaykovich L and Weiss C 2012 Scattering bright solitons: quantum versus mean-field behavior *Phys. Rev.* A **86** 033608
[24] Enns R and Rangnekar S 1987 Bistable solitons and optical switching IEEE *J. Quantum Electron.* **23** 1199–204
[25] Cancellieri G, Chiaraluce F, Gambi E and Pierleoni P 1995 Coupled-soliton photonic logic gates: practical design procedures *J. Opt. Soc. Am.* B **12** 1300
[26] Tsoy E N and Akhmediev N 2006 Dynamics and interaction of pulses in the modified Manakov model *Opt. Commun.* **266** 660–8
[27] Radha R, Vinayagam P S and Porsezian K 2013 Rotation of the trajectories of bright solitons and realignment of intensity distribution in the coupled nonlinear Schrödinger equation *Phys. Rev.* E **88** 032903
[28] Park Q H and Shin H J 1999 Painlevé analysis of the coupled nonlinear Schrödinger equation for polarized optical waves in an isotropic medium *Phys. Rev.* E **59** 2373–9
[29] Blow K J, Doran N J and Wood D 1987 Polarization instabilities for solitons in birefringent fibers *Opt. Lett.* **12** 202
[30] Ieda J, Miyakawa T and Wadati M 2004 Exact analysis of soliton dynamics in spinor Bose-Einstein condensates *Phys. Rev. Lett.* **93** 194102
[31] Cuevas J, Kevrekidis P G, Malomed B A, Dyke P and Hulet R G 2013 Interactions of solitons with a Gaussian barrier: splitting and recombination in quasi-one-dimensional and three-dimensional settings *New J. Phys.* **15** 063006
[32] Maimistov A I and Gabitov I R 2007 Nonlinear optical effects in artificial materials *Eur. Phys. J. Spec. Top.* **147** 265–86
[33] Zhang H Q, Li J, Xu T, Zhang Y X, Hu W and Tian B 2007 Optical soliton solutions for two coupled nonlinear Schrödinger systems via Darboux transformation *Phys. Scr.* **76** 452–60
[34] Atre R, Panigrahi P K and Agarwal G S 2006 Class of solitary wave solutions of the one-dimensional Gross-Pitaevskii equation *Phys. Rev.* E **73** 056611
[35] Kumar V R, Radha R and Panigrahi P K 2008 Dynamics of Bose-Einstein condensates in a time-dependent trap *Phys. Rev.* A **77** 023611
[36] Sakkaravarthi K and Kanna T 2013 Bright solitons in coherently coupled nonlinear Schrödinger equations with alternate signs of nonlinearities *J. Math. Phys.* **54**

[37] Radha R, Vinayagam P, Sudharsan J and Malomed B A 2016 Persistent bright solitons in sign-indefinite coupled nonlinear Schrödinger equations with a time-dependent harmonic trap *Commun. Nonlinear Sci.* **31** 30–9
[38] Radha R and Ramesh Kumar V 2007 Bright matter wave solitons and their collision in Bose-Einstein condensates *Phys. Lett.* A **370** 46–50
[39] Muruganandam P and Adhikari S 2009 Fortran programs for the time-dependent Gross-Pitaevskii equation in a fully anisotropic trap *Comput. Phys. Commun.* **180** 1888–912
[40] Ramesh Kumar V, Radha R and Wadati M 2010 Collision of bright vector solitons in two-component Bose-Einstein condensates *Phys. Lett.* A **374** 3685–94
[41] Wang C, Gao C, Jian C -M and Zhai H 2010 Spin-orbit coupled spinor Bose-Einstein condensates *Phys. Rev. Lett.* **105** 160403
[42] Radha R, Vinayagam P and Porsezian K 2016 Manipulation of light in a generalized coupled nonlinear Schrödinger equation *Commun. Nonlinear Sci.* **37** 354–61
[43] Malomed B A 1992 Inelastic collisions of polarized solitons in a birefringent optical fiber *J. Opt. Soc. Am.* B **9** 2075–82
[44] Kumar V R, Radha R and Wadati M 2008 Collision of solitons in electromagnetically induced transparency *Phys. Rev.* A **78** 041803
[45] Wadati M 2008 Matter rogue wave in Bose–Einstein condensates with attractive atomic interaction *J. Phys. Soc. Japan* **77** 024003
[46] Konopnicki M J and Eberly J H 1981 Simultaneous propagation of short different-wavelength optical pulses *Phys. Rev.* A **24** 2567–83
[47] Maĭmistov A I 1984 Rigorous theory of self-induced transparency in the case of a double resonance in a three-level medium *Sov. J. Quantum Electron.* **14** 385–9
[48] Chau L L, Shaw J C and Yen H C 1991 An alternative explicit construction of N-soliton solutions in 1+1 dimensions *J. Math. Phys.* **32** 1737–43
[49] Bajcsy M, Zibrov A S and Lukin M D 2003 Stationary pulses of light in an atomic medium *Nature* **426** 638–41
[50] Liu X J, Jing H and Ge M L 2004 Solitons formed by dark-state polaritons in an electromagnetic induced transparency *Phys. Rev.* A **70** 055802

Paulsamy Muruganandam and Ramaswamy Radha

Chapter 5

Dynamics of scalar Bose–Einstein condensates with short-range interactions

This chapter provides a comprehensive review of the dynamics of Bose–Einstein condensates (BECs) in experimentally realizable simple potentials, with particular emphasis on the interplay between confinement geometry, interatomic interactions, and external driving forces [1]. The analysis begins with investigating BEC dynamics in a time-independent harmonic oscillator potential with exponentially modulated scattering lengths which supports the formation and stabilization of bright matter-wave solitons. The generation of associated bright solitons and their collision dynamics are examined through systematic analysis. The discussion then progresses to include time-dependent harmonic trapping potentials, focussing specifically on how the temporal variation of both trap frequency and scattering length generates distinct matter-wave interference patterns during bright soliton collisions. The analysis includes examining how the reinforcement of three-body interactions, both attractive and repulsive, combined with binary interactions can influence condensate behaviour.

5.1 Quasi-one-dimensional Bose–Einstein condensates in a time-independent harmonic trap

In a time-independent harmonic oscillator potential with exponentially varying scattering lengths, the dynamics of a quasi-one-dimensional BEC is described by the Gross–Pitaevskii (GP) equation [2–6]:

$$i\frac{\partial\psi(x, t)}{\partial t} + \frac{\partial^2\psi(x, t)}{\partial x^2} + 2a(t)|\psi(x, t)|^2\psi(x, t) + \frac{1}{4}\lambda^2x^2\psi(x, t) = 0, \tag{5.1}$$

doi:10.1088/978-0-7503-5447-9ch5

where time t and coordinate x are measured in units of $2/\omega_\perp$ and $a_\perp$, respectively. Here, $a_\perp = \sqrt{\hbar/(m\omega_\perp)}$ and $a_0 = \sqrt{\hbar/(m\omega_0)}$ denote the linear oscillator lengths in the transverse and cigar-axis directions, with $\omega_\perp$ and ω_0 being the corresponding harmonic oscillator frequencies. The atomic mass is m, and the trap anisotropy parameter is $\lambda = 2|\omega_0|/\omega_\perp \ll 1$. The Feshbach resonance-managed nonlinear coefficient, representing the scattering length, is $a(t) = a_0 \exp(\lambda t)$.

The Lax pair for equation (5.1), adapted from [5], is given by:

$$\frac{\partial \Phi}{\partial x} = U\Phi, \quad U = \begin{pmatrix} \mathrm{i}\zeta & Q \\ -Q^* & -\mathrm{i}\zeta \end{pmatrix}, \tag{5.2a}$$

$$\frac{\partial \Phi}{\partial t} = V\Phi, \quad V = \begin{pmatrix} -2\mathrm{i}\zeta^2 + \mathrm{i}\lambda x\zeta + \mathrm{i}|Q|^2 & (\lambda x - 2\zeta)Q + \mathrm{i}\dfrac{\partial Q}{\partial x} \\ -(\lambda x - 2\zeta)Q^* + \mathrm{i}\dfrac{\partial Q^*}{\partial x} & 2\mathrm{i}\zeta^2 - \mathrm{i}\lambda x\zeta - \mathrm{i}|Q|^2 \end{pmatrix}, \tag{5.2b}$$

where a complex nonisospectral parameter $\zeta = \alpha(t) + \mathrm{i}\beta(t)$ satisfies:

$$\frac{\partial \zeta}{\partial t} = \lambda\zeta, \tag{5.3}$$

and the macroscopic wave function $\psi(x, t)$ is related to Q via:

$$Q = \exp\left(\frac{\lambda t}{2} + \mathrm{i}\frac{\lambda x^2}{4}\right)\psi(x, t). \tag{5.4}$$

5.1.1 Gauge transformation and bright solitons

To derive bright soliton solutions for equation (5.1), one considers the vacuum solution $Q^{(0)} = 0$, yielding the vacuum linear systems

$$\frac{\partial \Phi^{(0)}}{\partial x} = U^{(0)}\Phi^{(0)}, \quad U^{(0)} = \begin{pmatrix} \mathrm{i}\zeta & 0 \\ 0 & -\mathrm{i}\zeta \end{pmatrix}, \tag{5.5a}$$

$$\frac{\partial \Phi^{(0)}}{\partial t} = V^{(0)}\Phi^{(0)}, \quad V^{(0)} = \begin{pmatrix} -2\mathrm{i}\zeta^2 + \mathrm{i}\lambda x\zeta & 0 \\ 0 & 2\mathrm{i}\zeta^2 - \mathrm{i}\lambda x\zeta \end{pmatrix}. \tag{5.5b}$$

Solving these linear systems in accordance with the time-varying spectral parameter ζ given by equation (5.3), gives

$$\Phi^{(0)}(x, t, \zeta) = \begin{pmatrix} \mathrm{e}^{\mathrm{i}x\zeta - 2\mathrm{i}\int \zeta^2\,\mathrm{d}t} & 0 \\ 0 & \mathrm{e}^{-\mathrm{i}x\zeta + 2\mathrm{i}\int \zeta^2\,\mathrm{d}t} \end{pmatrix}. \tag{5.6}$$

Using complex parameters $\zeta_1 = \zeta = \alpha_1(t) + i\beta_1(t)$, $\mu_1 = \zeta^*$, and applying the gauge transformation approach [7] as described in section 2.2.1, the bright soliton solution for the GP equation is obtained as [6]

$$\psi^{(1)}(x, t) = 2\beta_0 \exp\left(\frac{\lambda t}{2} - i\frac{\lambda x^2}{4}\right) \operatorname{sech}(\theta_1) e^{i\xi_1}, \tag{5.7}$$

where

$$\theta_1 = 2\beta_1 x - 8\int \alpha_1\beta_1 \, dt + 2\delta_1, \tag{5.8a}$$

$$\xi_1 = 2\alpha_1 x - 4\int (\alpha_1^2 - \beta_1^2) \, dt - 2\phi_1, \tag{5.8b}$$

$$\alpha_1 = \alpha_{10} e^{\lambda t}, \tag{5.8c}$$

$$\beta_1 = \beta_{10} e^{\lambda t}. \tag{5.8d}$$

This bright soliton solution, given by equations (5.7) and (5.8), matches with the result derived by Liang *et al* [4] using the Darboux transformation. Density profiles, shown in figure 5.1(a), illustrate that the matter-wave density $|\psi(x, t)|^2$ increases with the absolute value of the scattering length a_s, compressing the BEC bright soliton trains. The contour plot in figure 5.1(b), depicting the x–t plane, reveals progressively narrowing soliton widths.

Conversely, figure 5.2(a) shows that reducing the absolute value of a_s decreases the peak matter-wave density $|\psi(x, t)|^2$, broadening the soliton trains. This broadening is confirmed by the contour plot in figure 5.2(b). These findings demonstrate that tuning the scattering length enables precise compression of BEC bright solitons to a desired peak density without instabilities or expansion of localized solitons while maintaining condensate coherence.

These results align with the theoretical predictions by Liang *et al* [4]. Further studies of the quasi-one-dimensional GP equation in expulsive parabolic potentials with positive scattering lengths corroborate these observations [5, 8–12].

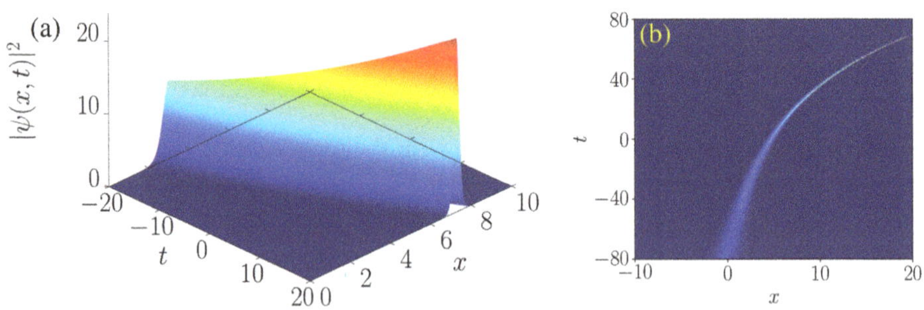

Figure 5.1. (a) Dynamics of a bright soliton and (b) density contour plot of a bright soliton with parameters $\lambda = 0.02$, $\beta_0 = 2$, $\alpha_{10} = 0.05$, $\delta_1 = 0.5$, $\phi_1 = 0.1$.

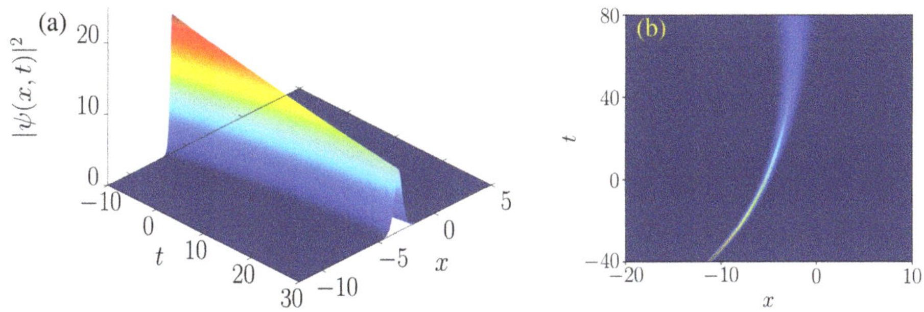

Figure 5.2. (a) Dynamics of a bright soliton and (b) density contour plot of a bright soliton with parameters $\lambda = -0.02$, $\beta_0 = 2$, $\alpha_{10} = 0.05$, $\delta_1 = 0.5$, $\phi_1 = 0.1$.

The gauge transformation method can be extended to obtain multisoliton solutions. The two-soliton solution for the matter wave is written as

$$\psi^{(2)}(x, t) = \frac{A_1 + A_2 + A_3 + A_4}{B_1 + B_2} \exp\left(-\frac{\lambda t}{2} - i\frac{\lambda x^2}{4}\right), \tag{5.9}$$

where

$$A_1 = \left[-2\beta_2\left((\alpha_2 - \alpha_1)^2 - (\beta_1^2 - \beta_2^2)\right) - 4\mathrm{i}\beta_1\beta_2(\alpha_2 - \alpha_1)\right]\mathrm{e}^{\theta_1 + \mathrm{i}\xi_2}, \tag{5.10a}$$

$$A_2 = -2\beta_2\left[(\alpha_2 - \alpha_1)^2 + (\beta_1^2 + \beta_2^2)\right]\mathrm{e}^{-\theta_1 + \mathrm{i}\xi_2}, \tag{5.10b}$$

$$A_3 = \left[-2\beta_1\left((\alpha_2 - \alpha_1)^2 + (\beta_1^2 - \beta_2^2)\right) + 4\mathrm{i}\beta_1\beta_2(\alpha_2 - \alpha_1)\right]\mathrm{e}^{\mathrm{i}\xi_1 + \theta_2}, \tag{5.10c}$$

$$A_4 = -4\mathrm{i}\beta_1\beta_2[(\alpha_2 - \alpha_1) - \mathrm{i}(\beta_1 - \beta_2)]\mathrm{e}^{\mathrm{i}\xi_1 - \theta_2}, \tag{5.10d}$$

$$B_1 = -4\beta_1\beta_2[\sinh(\theta_1)\sinh(\theta_2) + \cos(\xi_1 - \xi_2)], \tag{5.10e}$$

$$B_2 = 2\cosh(\theta_1)\cosh(\theta_2)\left[(\alpha_2 - \alpha_1)^2 + (\beta_1^2 + \beta_2^2)\right], \tag{5.10f}$$

and

$$\begin{aligned} \theta_i &= 2\beta_i x - 8\int \alpha_i\beta_i \,\mathrm{d}t + 2\delta_i, \\ \xi_i &= 2\alpha_i x - 4\int (\alpha_i^2 - \beta_i^2)\,\mathrm{d}t - 2\phi_i, \\ \alpha_i &= \alpha_{i0}\mathrm{e}^{\lambda t}, \quad \beta_i = \beta_{i0}\mathrm{e}^{\lambda t}, \quad i = 1, 2. \end{aligned} \tag{5.11}$$

Figures 5.3(a)–(d) illustrate the interaction of two bright matter-wave solitons. Despite the exponentially increasing scattering length enhancing the matter-wave density, the amplitude remains stable during propagation, indicating conservation of the total number of atoms despite atom exchange during soliton compression.

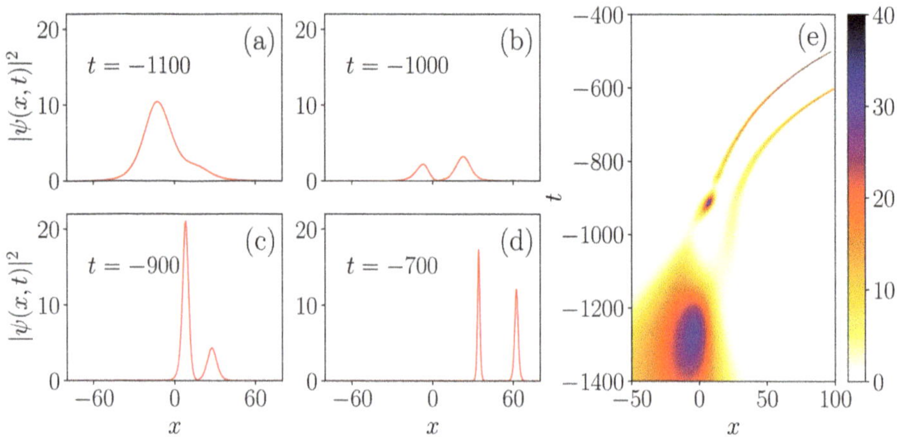

Figure 5.3. Plots illustrating the interaction of two bright solitons at different time intervals: (a) $t = -1100$, (b) $t = -1000$, (c) $t = -900$, (d) $t = -700$, and (e) contour plot of the two bright soliton interactions. The parameters in equations (5.9), (5.10a), and (5.11) are set as follows: $\lambda = 0.005$, $\beta_{10} = 15$, $\alpha_{10} = 3$, $\beta_{20} = 10$, $\alpha_{20} = 5$, $\delta_1 = 0.1$, $\delta_2 = 0.15$, $\phi_1 = 0.1$, and $\phi_2 = 0$.

The contour plot in figure 5.3(e) shows that the two soliton trains diverge along distinct longitudinal trajectories post-interaction, with their amplitudes fluctuating over time.

These observations indicate that bright solitons in an expulsive time-independent trap can be compressed or broadened by exponentially increasing or decreasing the scattering length. However, the dependence of the scattering length on trap frequency may limit the experimental relevance of these exact solutions. Investigating the ramifications of a general time-dependent scattering length and a time-dependent trap could enhance tunability, allowing precise control of the trap's effect on the condensates.

5.2 Impact of transient trap on Bose–Einstein condensates

A time-dependent trapping potential subjects a condensate to both time-varying scattering lengths and confinement strengths. For BECs modelled as weakly interacting atomic gases in a quasi-one-dimensional system, the dynamics is governed by the GP equation in dimensionless units [13]

$$i\frac{\partial\psi}{\partial t} + \frac{1}{2}\frac{\partial^2\psi}{\partial x^2} + \gamma(t)|\psi|^2\psi - \frac{M(t)}{2}x^2\psi = 0, \tag{5.12}$$

where $\gamma(t) = -2a_s(t)/a_B$, with a_B the Bohr radius, represents the time-dependent scattering length, and $M(t) = \omega_0^2(t)/\omega_\perp^2$ defines the harmonic confinement, with $M(t) > 0$ for a confining potential and $M(t) < 0$ for an expulsive potential. This GP equation can be mapped onto a linear Schrödinger eigenvalue problem, yielding one-soliton solutions expressed as doubly periodic Jacobian elliptic functions [13]. However, this mapping is inadequate for deriving analytical multisoliton solutions. The gauge transformation approach proves effective here, enabling the systematic construction of multisoliton solutions from the vacuum linear system's solutions.

To derive bright soliton solutions for equation (5.12) in both confining and expulsive potentials, a modified lens transformation is introduced [12–15]

$$\psi(x, t) = \sqrt{A(t)}\; Q(x, t)\, \mathrm{e}^{\mathrm{i}\chi(x,\, t)}, \tag{5.13}$$

with the phase defined as:

$$\chi(x, t) = -\frac{1}{2}c(t)x^2. \tag{5.14}$$

Substituting equation (5.13) into equation (5.12), the modified nonlinear Schrödinger (NLS) equation becomes

$$\mathrm{i}\frac{\partial Q}{\partial t} + \frac{1}{2}\frac{\partial^2 Q}{\partial x^2} - \mathrm{i}c(t)x\frac{\partial Q}{\partial x} - \mathrm{i}c(t)Q + \gamma(t)A(t)|Q|^2 Q = 0, \tag{5.15}$$

subject to:

$$M(t) = \frac{\mathrm{d}c(t)}{\mathrm{d}t} - c(t)^2, \tag{5.16}$$

and

$$c(t) = -\frac{\mathrm{d}}{\mathrm{d}t}\ln A(t). \tag{5.17}$$

Equation (5.15) is associated with the following linear eigenvalue problem:

$$\frac{\partial \Phi}{\partial x} = U\Phi, \quad U = \begin{pmatrix} \mathrm{i}\zeta(t) & Q \\ -Q^* & -\mathrm{i}\zeta(t) \end{pmatrix}, \tag{5.18}$$

$$\frac{\partial \Phi}{\partial t} = V\Phi,$$

$$V = \begin{pmatrix} -\mathrm{i}\zeta(t)^2 + \mathrm{i}c(t)x\zeta(t) + \frac{\mathrm{i}}{2}\gamma(t)A(t)|Q|^2 & (c(t)x - \zeta(t))Q + \frac{\mathrm{i}}{2}\frac{\partial Q}{\partial x} \\ -(c(t)x - \zeta(t))Q^* + \frac{\mathrm{i}}{2}\frac{\partial Q^*}{\partial x} & \mathrm{i}\zeta(t)^2 - \mathrm{i}c(t)x\zeta(t) - \frac{\mathrm{i}}{2}\gamma(t)A(t)|Q|^2 \end{pmatrix}, \tag{5.19}$$

where the complex spectral parameter $\zeta(t)$ is nonisospectral, satisfying:

$$\frac{\mathrm{d}\zeta(t)}{\mathrm{d}t} = c(t)\zeta(t), \tag{5.20}$$

and $\gamma(t) = 1/A(t)$. The compatibility condition $(\partial_t\partial_x\Phi) = (\partial_x\partial_t\Phi)$ yields equation (5.15).

Combining equations (5.17) and (5.16) with $\gamma(t) = 1/A(t)$, one obtains

$$\gamma''(t)\gamma(t) - 2\gamma'(t)^2 - M(t)\gamma(t)^2 = 0. \tag{5.21}$$

This condition ensures the complete integrability of equation (5.12) when the trapping potential $M(t)$ and scattering length $\gamma(t)$ are related by equation (5.21), consistent with [12]. Integrability holds only for specific choices of $M(t)$ that satisfy equation (5.16). For instance, with $M(t) = k$ (constant) and $\gamma(t) = \exp(\lambda t)$, where λ is the trap frequency, equation (5.12) describes BECs in an expulsive parabolic potential with an exponentially varying scattering length [4, 6]. This satisfies equation (5.21) when $k = -\lambda^2$, ensuring integrability, and has been experimentally realized [16].

To obtain the soliton solution for equation (5.15), one starts with the seed solution $Q^{(0)} = 0$ and solves the linear systems (5.18) and (5.19), consistent with equation (5.20), yielding

$$\Phi^{(0)}(x, t, \zeta) = \begin{pmatrix} e^{ix\zeta(t) - i\int_0^t \zeta(t')^2 \, dt'} & 0 \\ 0 & e^{-ix\zeta(t) + i\int_0^t \zeta(t')^2 \, dt'} \end{pmatrix}. \tag{5.22}$$

Using the gauge transformation approach [7] with $\zeta_1 = \zeta = \alpha_1(t) + i\beta_1(t)$ and $\mu_1 = \zeta_1^*$, the one-soliton solution for equation (5.12) is given by [17]

$$\psi^{(1)}(x, t) = \sqrt{\frac{1}{\gamma(t)}} \, 2\beta_1(t) \, \mathrm{sech}(\theta_1) \, e^{i\left(-\frac{1}{2}c(t)x^2 + \xi_1\right)}, \tag{5.23}$$

where

$$\begin{aligned} \theta_1 &= 2\beta_1(t)x - 4\int_0^t \alpha_1(t')\beta_1(t') \, dt' + 2\delta_1, \\ \xi_1 &= 2\alpha_1(t)x - 2\int_0^t (\alpha_1(t')^2 - \beta_1(t')^2) \, dt' - 2\phi_1, \\ \alpha_1(t) &= \alpha_{10} e^{\int_0^t c(t') \, dt'}, \\ \beta_1(t) &= \beta_{10} e^{\int_0^t c(t') \, dt'}, \end{aligned} \tag{5.24}$$

and ϕ_1, δ_1, α_{10}, and β_{10} are arbitrary real constants.

The amplitude of this bright soliton depends strongly on the scattering length $a_s(t)$ and the time-dependent trap $M(t)$, while its velocity is determined solely by $M(t)$.

This approach can be extended to obtain multisoliton solutions. The two-soliton solution is given by

$$\psi^{(2)}(x, t) = \sqrt{\frac{1}{\gamma(t)}} \, \frac{A_1 + A_2 + A_3 + A_4}{B_1 + B_2} \, e^{-\frac{i}{2}c(t)x^2}, \tag{5.25}$$

where:

$$
\begin{aligned}
A_1 &= \left\{-2\beta_2\left[(\alpha_2-\alpha_1)^2-(\beta_1^2-\beta_2^2)\right]-4\mathrm{i}\beta_1\beta_2(\alpha_2-\alpha_1)\right\}\mathrm{e}^{\theta_1+\mathrm{i}\xi_2},\\
A_2 &= -2\beta_2\left[(\alpha_2-\alpha_1)^2+(\beta_1^2+\beta_2^2)\right]\mathrm{e}^{-\theta_1+\mathrm{i}\xi_2},\\
A_3 &= \left\{-2\beta_1\left[(\alpha_2-\alpha_1)^2+(\beta_1^2-\beta_2^2)\right]+4\mathrm{i}\beta_1\beta_2(\alpha_2-\alpha_1)\right\}\mathrm{e}^{\mathrm{i}\xi_1+\theta_2},\\
A_4 &= -4\mathrm{i}\beta_1\beta_2[(\alpha_2-\alpha_1)-\mathrm{i}(\beta_1-\beta_2)]\mathrm{e}^{\mathrm{i}\xi_1-\theta_2},\\
B_1 &= -4\beta_1\beta_2[\sinh(\theta_1)\sinh(\theta_2)+\cos(\xi_1-\xi_2)],\\
B_2 &= 2\cosh(\theta_1)\cosh(\theta_2)\left[(\alpha_2-\alpha_1)^2+(\beta_1^2+\beta_2^2)\right],\\
\theta_i &= 2\beta_i x-4\int_0^t \alpha_i(t')\beta_i(t')\,dt'+2\delta_i,\\
\xi_i &= 2\alpha_i x-2\int_0^t (\alpha_i(t')^2-\beta_i(t')^2)\,\mathrm{d}t'-2\phi_i,\\
\alpha_i(t) &= \alpha_{i0}\mathrm{e}^{\int_0^t c(t')\,\mathrm{d}t'},\\
\beta_i(t) &= \beta_{i0}\mathrm{e}^{\int_0^t c(t')\,\mathrm{d}t'},\\
\gamma(t) &= \gamma_0\mathrm{e}^{\int_0^t c(t')\,\mathrm{d}t'},
\end{aligned}
\tag{5.26}
$$

for $i = 1, 2$. Various soliton profiles can be achieved by selecting $\gamma(t)$ and $M(t)$ consistent with equation (5.21).

Figures 5.4(a) and (b) illustrate the evolution of two-soliton solutions in an expulsive trap ($M(t) < 0$) with $\gamma(t) = 0.5\exp(-0.125t^2)$ for different initial conditions. The matter-wave density $|\psi|^2$ decreases gradually due to the diminishing absolute value of the scattering length, with soliton trajectories determined by the initial conditions. Identifying this parameter regime, where slow condensate decay occurs, allows for operation in a safer parameter range to mitigate decay.

Figure 5.5(a) depicts soliton interactions for $\gamma(t) = 0.5\exp(0.0025t^2)$. As shown in figure 5.5(b), the trap remains confining ($M(t) > 0$) for a limited time ($t < 14$),

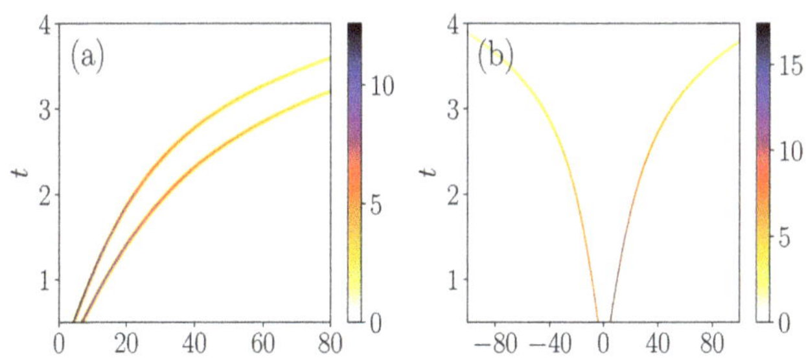

Figure 5.4. Two-soliton interaction in an expulsive trap ($M(t) < 0$) with $\gamma(t) = \gamma_0\exp(-0.125t^2)$, $\gamma_0 = 0.5$, $\alpha_{10} = 2.31$, $\beta_{10} = 1.5$, $\beta_{20} = 1.2$, and with (a) $\alpha_{20} = 3.12$, $\phi_1 = 0.005$, $\delta_1 = 0.002$, $\phi_2 = 0.002$, $\delta_2 = 0.001$ and (b) $\alpha_{20} = -2.12$, $\phi_1 = 0.05$, $\delta_1 = 0.02$, $\phi_2 = 0.02$, $\delta_2 = 0.01$.

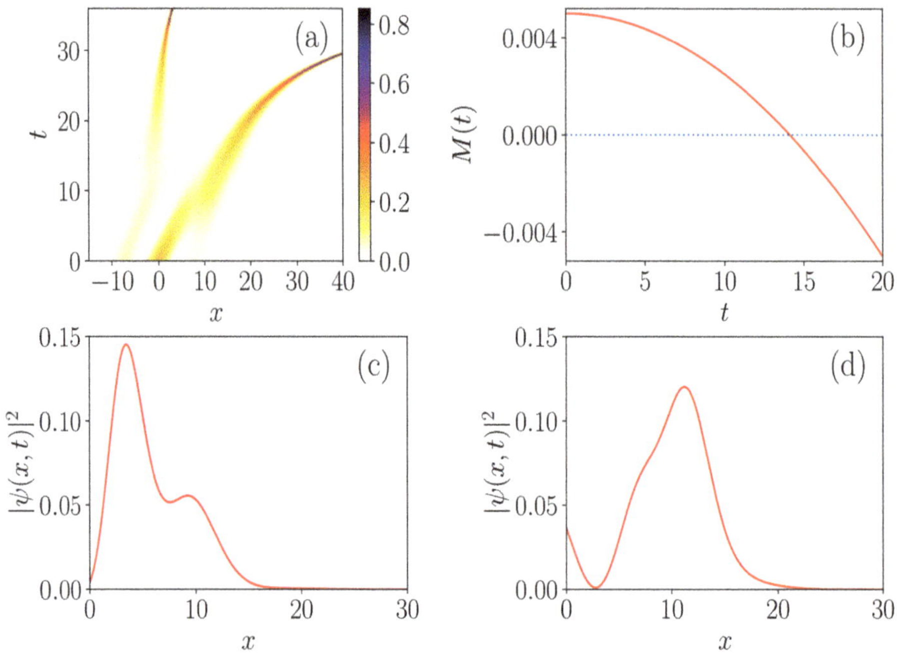

Figure 5.5. Two-soliton interaction in a confining trap ($M(t) > 0$) at different times with $\gamma(t) = \gamma_0 \exp(0.0025t^2)$, $\gamma_0 = 0.5$, $\alpha_{10} = 0.01$, $\beta_{10} = 0.1$, $\alpha_{20} = 0.28$, $\beta_{20} = 0.11$, $\phi_1 = \delta_2 = 0.1$, $\phi_2 = \delta_1 = 0.2$.

during which the solitons slide past each other, resembling liquid droplets, as seen in figures 5.5(c) and 5.5(d). Beyond this period ($t > 14$), the trap becomes expulsive, leading to soliton compression and an increase in the matter-wave density $|\psi|^2$. For $\gamma(t) = 0.5\exp(-0.125t^2)$, the decreasing scattering length causes slow condensate decay, whereas for $\gamma(t) = 0.5\exp(0.0025t^2)$, the increasing scattering length results in soliton compression after a time delay introduced by the time-dependent trap. When $M(t)$ is constant, equation (5.12) describes BECs in an expulsive potential with a time-independent scattering length, leading to immediate soliton compression without delay [4, 6]. This delay can be controlled by adjusting $M(t)$, allowing precise tuning of soliton width and amplitude, consistent with numerical results [18].

For $\gamma(t) = 0.5\exp(0.0025t^2)$, the trap becomes expulsive again at $t > 14$, as shown in figure 5.5(b). To maintain a confining trap ($M(t) > 0$), a complex scattering length is required, as in cold alkaline-earth metal atoms [19]. This leads to periodic variations in the matter-wave density $|\psi|^2$ due to the periodic modulation of the scattering length [20, 21], resembling Faraday waves observed experimentally [22]. Figure 5.6 shows continuous energy exchange between the two soliton pulses during propagation.

The time-dependent trap enables prolonged stabilization of condensates by carefully tuning $M(t)$. Future investigations could explore condensate dynamics under an external time-dependent or constant force to control evolution without a trapping potential.

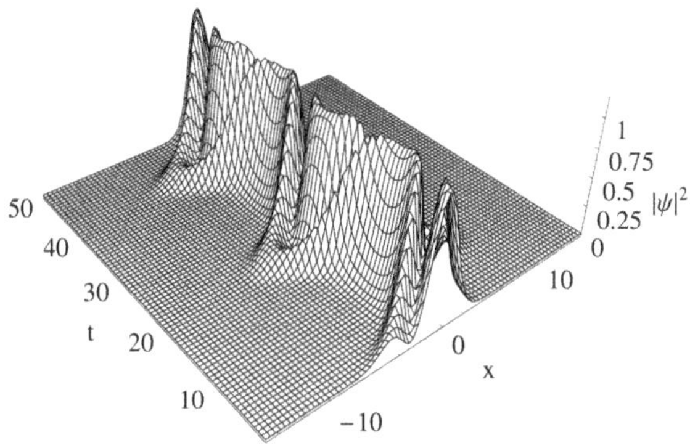

Figure 5.6. Two-soliton interaction in a confining trap ($M(t) = 0.09$) with $\gamma(t) = \gamma_0 \exp(0.3it)$, $\gamma_0 = 0.5$, $\alpha_{10} = 0.09$, $\beta_{10} = 0.71$, $\alpha_{20} = 0.031$, $\beta_{20} = 0.11$, $\phi_1 = 5.1$, $\delta_1 = 7.2$, $\phi_2 = 4.2$, $\delta_2 = 4.1$. Reprinted with permission from [17], Copyright (2008) by the American Physical Society.

5.3 Matter-wave interference pattern in the collision of solitons

This section has been reproduced with permission from [23].

Previous discussions have established that the dynamics of BECs in a time-dependent trap is governed by the quasi-one-dimensional GP equation (5.12), where the scattering length parameter is defined as $\gamma(t) = -2a_s(t)/a_B$, the trap frequency parameter is $M(t) = \omega_0^2(t)/\omega_\perp^2$, a_B is the Bohr radius, and $M(t)$ characterizes the harmonic trap, which can be confining ($M(t) > 0$) or expulsive ($M(t) < 0$). The solvability of the GP equation (5.12) depends on the compatibility of $\gamma(t)$ and $M(t)$, satisfying the condition $\gamma''(t)\gamma(t) - 2\gamma'(t)^2 - M(t)\gamma(t)^2 = 0$.

To explore the collisional dynamics of condensates in a time-dependent trap, one considers the two-bright-soliton solution, as detailed in section 5.2, given by

$$\psi^{(2)}(x, t) = \sqrt{\frac{1}{\gamma(t)}} \frac{A_1 + A_2 + A_3 + A_4}{B_1 + B_2} \mathrm{e}^{-\frac{\mathrm{i}}{2}c(t)x^2}, \tag{5.27}$$

where

$$\begin{aligned}
A_1 &= \left\{ -2\beta_2\left[(\alpha_2 - \alpha_1)^2 - (\beta_1^2 - \beta_2^2)\right] - 4\mathrm{i}\beta_1\beta_2(\alpha_2 - \alpha_1)\right\}\mathrm{e}^{\theta_1 + \mathrm{i}\xi_2}, \\
A_2 &= -2\beta_2\left[(\alpha_2 - \alpha_1)^2 + (\beta_1^2 + \beta_2^2)\right]\mathrm{e}^{-\theta_1 + \mathrm{i}\xi_2}, \\
A_3 &= \left\{ -2\beta_1\left[(\alpha_2 - \alpha_1)^2 + (\beta_1^2 - \beta_2^2)\right] + 4\mathrm{i}\beta_1\beta_2(\alpha_2 - \alpha_1)\right\}\mathrm{e}^{\mathrm{i}\xi_1 + \theta_2}, \\
A_4 &= -4\mathrm{i}\beta_1\beta_2[(\alpha_2 - \alpha_1) - \mathrm{i}(\beta_1 - \beta_2)]\mathrm{e}^{\mathrm{i}\xi_1 - \theta_2}, \\
B_1 &= -4\beta_1\beta_2[\sinh(\theta_1)\sinh(\theta_2) + \cos(\xi_1 - \xi_2)], \\
B_2 &= 2\cosh(\theta_1)\cosh(\theta_2)\left[(\alpha_2 - \alpha_1)^2 + (\beta_1^2 + \beta_2^2)\right],
\end{aligned} \tag{5.28}$$

and

$$\theta_j = 2\beta_j x - 4\int_0^t \alpha_j\beta_j\, dt' + 2\delta_j, \quad \xi_j = 2\alpha_j x - 2\int_0^t (\alpha_j^2 - \beta_j^2)\, dt' - 2\phi_j, \tag{5.29a}$$

$$\alpha_j = \alpha_{j0} e^{\int_0^t c(t')\, dt'}, \quad \beta_j = \beta_{j0} e^{\int_0^t c(t')\, dt'}, \quad \gamma(t) = \gamma_0 e^{\int_0^t c(t')\, dt'}, \tag{5.29b}$$

with α_{j0}, β_{j0}, δ_j, ϕ_j, and γ_0 as arbitrary parameters. To produce matter-wave interference patterns, the two bright solitons are allowed to collide in the presence of a trap, with appropriate choices of the scattering length $\gamma(t)$ and trap frequency $M(t)$ (or equivalently $c(t)$).

5.3.1 Case (i): expulsive trap

For $c(t) = -1$, the scattering length evolves as $\gamma(t) = \gamma_0 e^{-t}$ as shown in figure 5.7 (b), where γ_0 is a real constant, and the trap frequency $M(t) = -1$ is expulsive. The collision of two initially separated bright solitons (see figure 5.7(a)) in this expulsive harmonic trap is depicted in figure 5.7(c) showing the corresponding density evolution. The density pattern exhibits alternating bright and dark fringes, indicating high and low density regions, respectively. The phase difference between the condensates evolves continuously, as illustrated in figure 5.9(a).

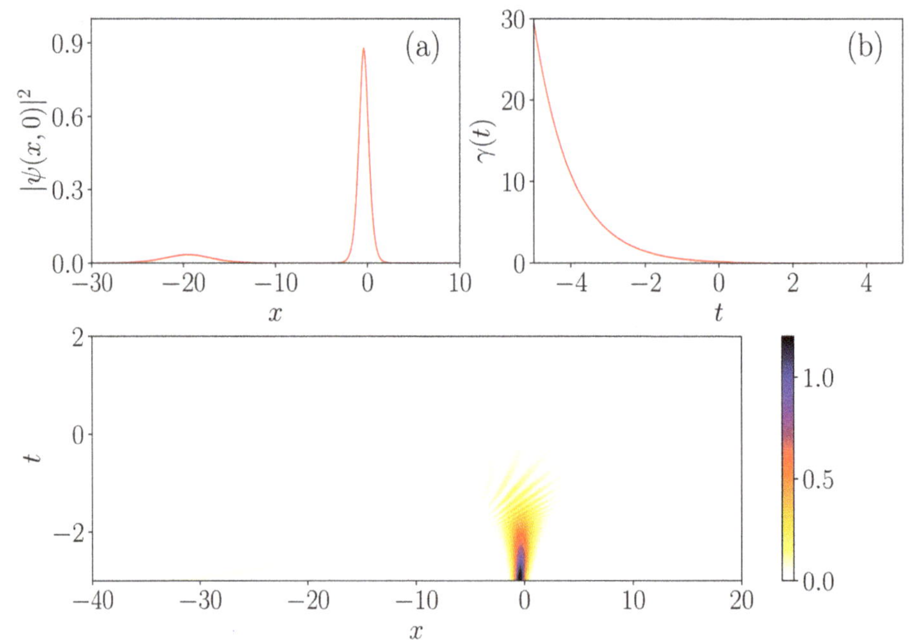

Figure 5.7. (a) Positions of two bright solitons at $t = -2.5$ with $c(t) = -1$, (b) time evolution of the scattering length $\gamma(t)$ for Case (i), and (c) collision of two bright solitons forming an interference pattern with $\gamma(t) = \gamma_0 e^{-t}$. The parameters used are $\alpha_{10} = 0.01$, $\alpha_{20} = 0.8$, $\beta_{10} = 0.06$, $\beta_{20} = 0.012$, $\gamma_0 = 0.2$, $\phi_1 = 0.01$, $\phi_2 = 0.1$, $\delta_1 = 0.1$, and $\delta_2 = 0.01$.

5.3.2 Case (ii): confining trap

For a trap frequency $M(t) = 0.3 - 0.09t^2$, shown in figure 5.8(a), and scattering length $\gamma(t) = \gamma_0 e^{0.15t^2}$, the soliton collision and resulting interference pattern in the confining regime are depicted in figures 5.8(b) and 5.8(c), respectively. As the trap transitions from expulsive to confining and back, the interference pattern emerges in the confining regime and vanishes when the trap becomes expulsive again, as evident in figures 5.8(a)–(c). These parameters are experimentally feasible. The phase difference between the condensates evolves over time, as shown in figure 5.9(b).

These results confirm that matter waves from colliding bright solitons in BECs produce interference patterns, akin to coherent laser beams, reflecting the long-range spatial coherence of the condensates. The interference patterns, generated through soliton collisions [23], resemble those observed experimentally by Andrews *et al* [24], where two condensates were separated by a light sheet and overlapped during

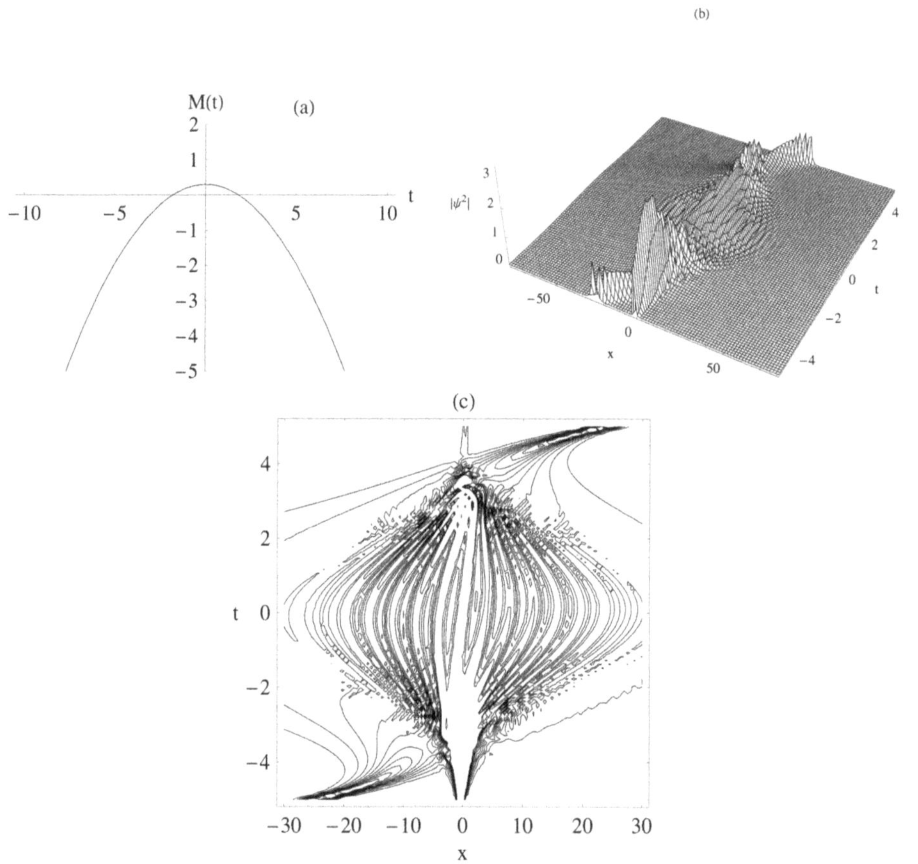

Figure 5.8. (a) Time evolution of the trap frequency $M(t) = 0.3 - 0.09t^2$ for Case (ii). (b) Two-soliton collision forming an interference pattern with $\gamma(t) = \gamma_0 e^{0.15t^2}$, using parameters $\alpha_{10} = 0.01$, $\alpha_{20} = 0.8$, $\beta_{10} = 0.06$, $\beta_{20} = 0.012$, $\gamma_0 = 0.02$, $\phi_1 = 0.2$, $\phi_2 = 0.1$, $\delta_1 = 0.1$, $\delta_2 = 0.2$. (c) Contour plot of the interference pattern in the confining regime. Reprinted from [23], Copyright (2009), with permission from Elsevier.

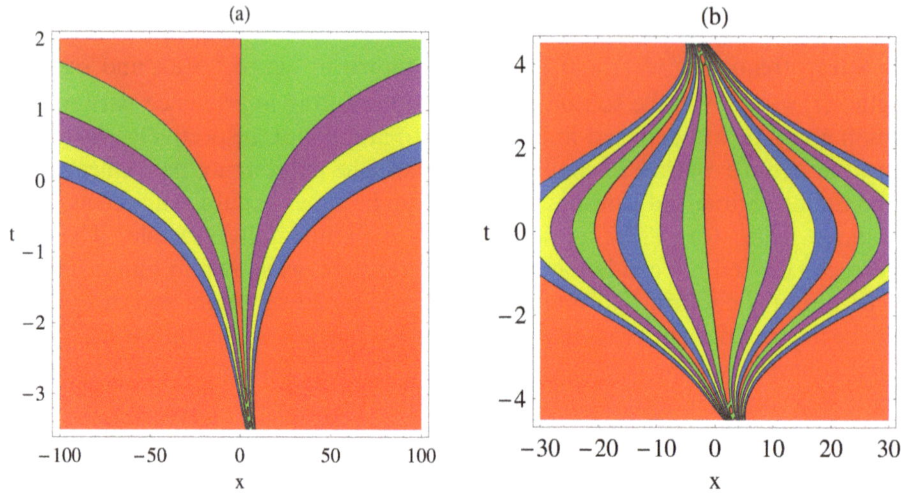

Figure 5.9. (a) Phase difference between condensates for Case (i). (b) Phase difference for Case (ii) with $\tau = 0$. Reprinted from [23], Copyright (2009), with permission from Elsevier.

ballistic expansion after switching off the trap. Here, the trap frequency $M(t)$ is tuned in coordination with the scattering length $\gamma(t)$, distinguishing this from the interference of independent condensates in separate traps, with or without phase coherence [25–27].

Optical traps enable diverse temporal variations of $M(t)$, while the scattering length $\gamma(t)$ can be controlled via Feshbach resonance or trap frequency adjustments. The phase evolution in figures 5.9(a) and 5.9(b) quantifies condensate coherence. Although coherence has been leveraged for atom lasers, the intra-trap soliton collisions further validate the coherent nature of BECs.

5.4 Bose–Einstein condensates with attractive and repulsive three-body interactions

This section has been reproduced with permission from [41].

5.4.1 Introduction

Studies of BEC dynamics with time-dependent two-body interactions, under either time-independent or time-dependent trapping potentials, reveal that bright solitons can achieve high local matter densities by increasing the absolute value of the scattering length via Feshbach resonance. A critical question emerges: how far can the scattering length be increased to produce high-density condensates? This question is technologically significant given the recent experiments realizing BECs on atomic chip surfaces [28] and in atomic waveguides [29], which involve strong matter-wave compression to enhance condensate density. However, condensate collapse occurs when density exceeds a critical threshold, and true one-dimensional systems are not expected to collapse, imposing constraints on density increases to maintain an effective one-dimensional BEC [4]. This necessitates exploring BEC dynamics within a safe parameter range. Consequently, alternative approaches are needed to achieve high-density condensates while preserving one-dimensionality without parametric restrictions. For large boson numbers, repulsive three-body interactions can counteract attractive two-body interactions, thereby enhancing condensate stability [30].

Two-body interactions alone are insufficient; combining them with three-body interactions extends the stability region while increasing condensate density in one-dimensional systems. Although BEC dynamics with both two- and three-body interactions have been studied [10, 31–34], the integrability of the associated model has not been thoroughly explored.

This section examines BEC dynamics with two- and three-body interactions in a harmonic trap. A modified GP equation is derived by applying a phase imprint to the order parameter of the cubic GP equation. Bright soliton solutions are generated, and the impact of three-body interactions on the condensates is analyzed.

5.4.2 Modified Gross–Pitaevskii equation with two- and three-body interactions

To develop an integrable model capturing the effects of two- and three-body interactions on condensates, one considers the effective quasi-one-dimensional GP equation in an expulsive harmonic trap [2–6]

$$\mathrm{i}\frac{\partial\psi(x,t)}{\partial t}+\frac{\partial^2\psi(x,t)}{\partial x^2}+2a(t)|\psi(x,t)|^2\psi(x,t)+\frac{1}{4}\lambda^2x^2\psi(x,t)=0, \tag{5.30}$$

where $\psi(x, t)$ is the condensate wavefunction, $a(t)$ is the time-dependent scattering length, and λ characterizes the harmonic trap. Introducing a phase imprint to transform the order parameter $\psi(x, t)$ into a new order parameter $q(x, t)$

$$q(x,t)=\psi(x,t)\mathrm{e}^{2\mathrm{i}\theta(x,t)}, \tag{5.31}$$

where $\theta(x, t)$ is the phase imprint, designed to satisfy the following partial differential equations

$$\frac{\partial\theta}{\partial x}=-\sqrt{\tau}|\psi|^2, \tag{5.32}$$

$$\frac{\partial\theta}{\partial t}=\mathrm{i}\sqrt{\tau}\left(\psi\frac{\partial\psi^*}{\partial x}-\psi^*\frac{\partial\psi}{\partial x}\right)+4\tau|\psi|^4, \tag{5.33}$$

ensuring that $q(x, t)$ obeys the evolution equation

$$\mathrm{i}\frac{\partial q}{\partial t}+\frac{\partial^2 q}{\partial x^2}+2a(t)|q|^2q+\frac{\lambda^2x^2}{4}q+4\tau|q|^4q+4\mathrm{i}\sqrt{\tau}\left(\frac{\partial|q|^2}{\partial x}\right)q=0, \tag{5.34}$$

where $\frac{\partial\theta}{\partial x}$ and $\frac{\partial\theta}{\partial t}$ conserve atom number and current density for the system described by equation (5.30). Here, $a(t)=\tilde{a}_0\mathrm{e}^{\lambda t}$ represents attractive ($\tilde{a}_0>0$) two-body interactions, and τ, a real parameter, denotes the strength of three-body interactions, assuming negligible three-atom collision losses [10, 31, 32, 35–37]. This phase-imprint engineering, which generates an integrable model for two- and three-body interactions, parallels the creation of dark solitons in sodium and rubidium BECs via phase engineering [38, 39].

When $\lambda = 0$ (no trapping potential), equation (5.34) reduces to the integrable cubic–quintic NLS equation in nonlinear optics, derived from the standard NLS equation via a U(1)-gauge transformation [40]

$$\mathrm{i}\frac{\partial q}{\partial t} + \frac{\partial^2 q}{\partial x^2} + 2\tilde{a}_0|q|^2 q + 4\tau|q|^4 q + 4\mathrm{i}\sqrt{\tau}\left(\frac{\partial |q|^2}{\partial x}\right)q = 0. \tag{5.35}$$

For $\tau = 0$, equation (5.34) reverts to the cubic GP equation with only attractive two-body interactions, justifying the term 'modified GP equation'. The quintic term $4\tau|q|^4 q$ accounts for attractive ($\tau > 0$) or repulsive ($\tau < 0$) three-body interactions, while the term $4\mathrm{i}\sqrt{\tau}\left(\frac{\partial |q|^2}{\partial x}\right)q$ mitigates modulation instability from three-body interactions, preserving integrability. Without this term, integrability is lost. The Lax pair for the modified GP equation can be derived from that of the cubic GP equation using transformations in equations (5.31)–(5.33).

5.4.3 Bright solitons and impact of three-body interactions

Starting with the trivial seed solution $q^{(0)} = 0$, the gauge transformation approach [7] yields the bright soliton solution for the modified GP equation [41]:

$$q^{(1)} = \frac{2}{\sqrt{\tilde{a}_0}}\beta_0 \exp\left(\frac{\lambda t}{2} - \frac{\mathrm{i}\lambda x^2}{4} + 2\mathrm{i}\theta\right)\mathrm{sech}(\chi_1)\mathrm{e}^{\mathrm{i}\xi_1}, \tag{5.36}$$

where

$$\chi_1 = 2\beta_1 x - 8\int \alpha_1\beta_1\,\mathrm{d}t + 2\delta_1, \tag{5.37a}$$

$$\xi_1 = 2\alpha_1 x - 4\int\left(\alpha_1^2 - \beta_1^2\right)\mathrm{d}t - 2\phi_1, \tag{5.37b}$$

$$\alpha_1 = \alpha_0\mathrm{e}^{\lambda t}, \quad \beta_1 = \beta_0\mathrm{e}^{\lambda t}, \tag{5.37c}$$

$$\theta = -\frac{2}{\tilde{a}_0}\tau\beta_0 \tanh\left[2\left(\frac{\mathrm{e}^{\lambda t}(x\lambda - 2\mathrm{e}^{\lambda t}\alpha_0)\beta_0}{\lambda} + \delta_1\right)\right], \tag{5.37d}$$

and α_0, β_0, δ_1, and ϕ_1 are arbitrary real constants. The phase imprint θ, given by equation (5.37d), depends on the density of the original order parameter ψ via equation (5.31), endowing the bright solitons with phase-dependent amplitude. The three-body interaction strength τ can be attractive ($\tau > 0$) or repulsive ($\tau < 0$), allowing analysis of both effects on the condensates.

5.4.3.1 Condensates with attractive three-body interactions

The bright soliton solution in equations (5.36)–(5.37d) includes a kink-like phase term, given by equation (5.37d), distinguishing it from cubic GP solitons ($\tau = 0$). Figure 5.10(a) illustrates the density evolution of condensates without three-body interactions, plotting the density $|q|^2$ of the new order parameter q. When attractive three-body interactions are included alongside attractive two-body interactions of equal strength, no significant change in matter-wave density is observed, as shown in figure 5.10(b). However, increasing the strength of attractive two-body interactions reduces condensate density, as depicted in figure 5.10(c), indicating instability that ejects atoms, lowering matter-wave density. To maintain long-term condensate stability, the attractive two-body interaction strength should be minimized.

The phase term in equation (5.37d) evolves spatially and temporally, affecting the real (or imaginary) part or derivatives of the order parameter. Figure 5.11(b) shows the impact of this phase on matter-wave solitons in the modified GP equation compared to the cubic GP equation in figure 5.11(a), revealing compression and rarefaction effects.

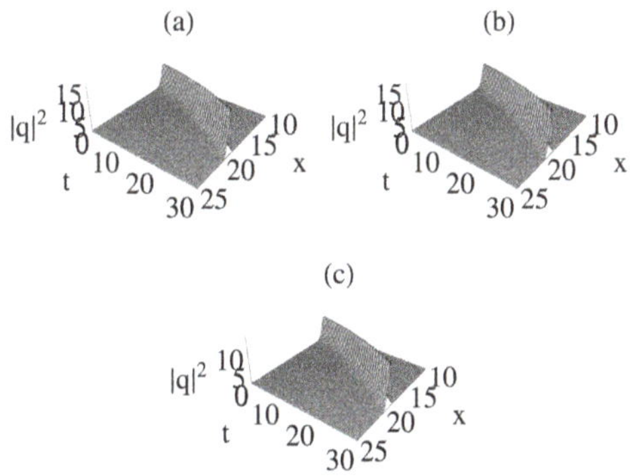

Figure 5.10. Density of condensates in the modified Gross–Pitaevskii equation with attractive two-body and three-body interactions, for parameters $\lambda = 0.02$, $\alpha_0 = 0.1$, $\beta_0 = 0.9$, $\delta_1 = -1.5$, $\phi_1 = 2.5$: (a) $\tilde{a}_0 = 0.4$, $\tau = 0$; (b) $\tilde{a}_0 = 0.4$, $\tau = 0.04$, and (d) $\tilde{a}_0 = 0.45$, $\tau = 0.04$. Reprinted from [41], copyright (2010) with the permission of The Physical Society of Japan.

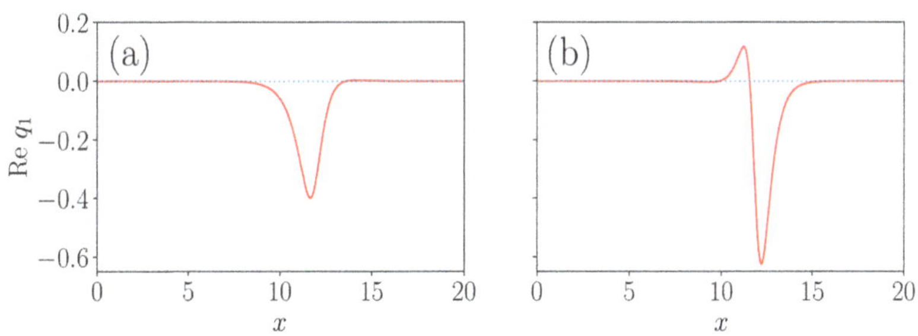

Figure 5.11. Real part of the order parameter for: (a) cubic GP equation; (b) modified GP equation with attractive three-body interactions.

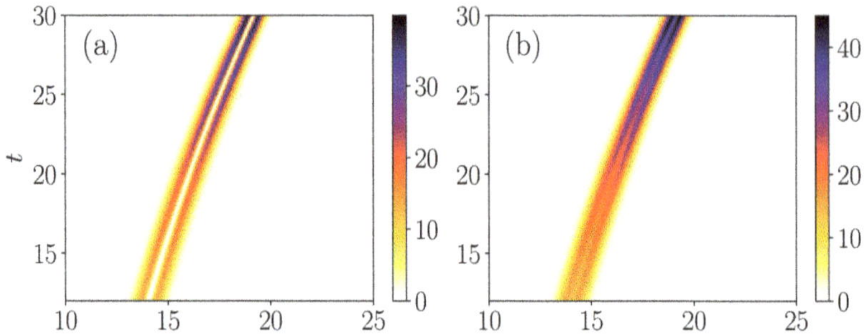

Figure 5.12. (a) Compression and splitting of $|q_x|$ in the cubic Gross–Pitaevskii equation. (b) Compression and suppression of splitting of $|q_x|$ in the modified GP equation with attractive three-body interactions.

The spatial derivative $|q_x|$, plotted in figures 5.12(a) and 5.12(b), indicates increased matter-wave density, leading to compression in the cubic GP equation. Additionally, figure 5.12(a) shows splitting of the matter wave, which is suppressed by attractive three-body interactions in figure 5.12(b).

5.4.3.2 Condensates with repulsive three-body interactions

For repulsive three-body interactions, the bright soliton solution becomes

$$\bar{q}^{(1)} = \frac{2}{\sqrt{\tilde{a}_0}}\beta_0 \operatorname{sech}(\chi_1)\mathrm{e}^{\mathrm{i}\xi_1} \exp\left(\frac{\lambda t}{2} - \frac{\mathrm{i}\lambda x^2}{4} + \bar{\theta}\right), \tag{5.38}$$

where

$$\bar{\theta} = \frac{4}{\tilde{a}_0}\sqrt{\bar{\tau}}\beta_0 \tanh\left[2\left(\frac{\mathrm{e}^{\lambda t}(x\lambda - 2\mathrm{e}^{\lambda t}\alpha_0)\beta_0}{\lambda} + \delta_1\right)\right], \tag{5.39}$$

and $\bar{\tau}$ is a real parameter representing repulsive interactions.

Figure 5.13(a) shows the density evolution for attractive two-body and repulsive three-body interactions, revealing an exponential increase in density [amplitude $\frac{2}{\sqrt{\tilde{a}_0}}\beta_0 \times \exp(\lambda t/2 + \bar{\theta})$] compared to figure 5.10(a) without three-body interactions. This suggests that a small repulsive three-body interaction allows condensates to hold more atoms, extending stability in one-dimensional systems. Increasing the attractive two-body interaction strength reduces density, as shown in figure 5.13(b), indicating instability. However, increasing the repulsive three-body interaction strength, as in figure 5.13(c), restores higher density, matching figure 5.13(a). Thus, even weak repulsive three-body interactions significantly increase condensed atom numbers compared to two-body interactions alone [31], reducing BEC instability relative to the cubic GP equation.

Extending the gauge transformation method, one can generate multisoliton solutions. The two matter-wave soliton assumes the following form

$$q^{(2)} = \frac{1}{\sqrt{\tilde{a}_0}}\frac{A_1 + A_2 + A_3 + A_4}{B_1 + B_2}\exp\left(-\frac{\lambda t}{2} - \frac{\mathrm{i}\lambda x^2}{4} + 2\mathrm{i}\theta\right), \tag{5.40}$$

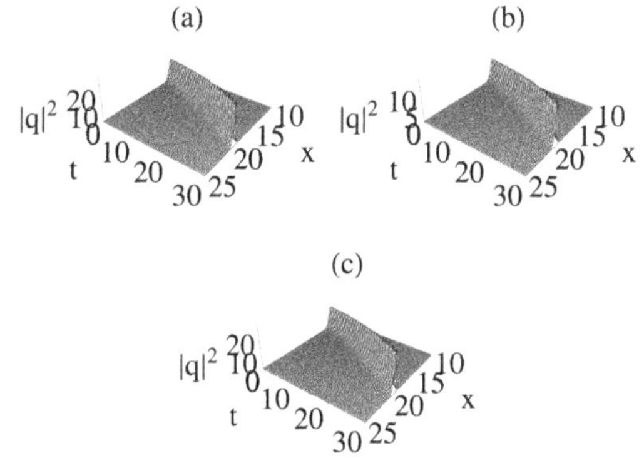

Figure 5.13. Density of condensates in the modified GP equation with attractive two-body and repulsive three-body interactions, with parameters as in figure 5.10: (a) $\tilde{a}_0 = 0.6$, $\bar{\tau} = 0.04$; (b) $\tilde{a}_0 = 0.8$, $\bar{\tau} = 0.04$; (c) $\tilde{a}_0 = 0.8$, $\bar{\tau} = 0.09$. Reprinted from [41], copyright (2010) with the permission of The Physical Society of Japan.

where

$$A_1 = \Big[-2\beta_2\big((\alpha_2 - \alpha_1)^2 - (\beta_1^2 - \beta_2^2)\big) - 4\mathrm{i}\beta_1\beta_2(\alpha_2 - \alpha_1)\Big]\mathrm{e}^{\chi_1 + \mathrm{i}\xi_2}, \tag{5.41a}$$

$$A_2 = -2\beta_2\Big[(\alpha_2 - \alpha_1)^2 + (\beta_1^2 + \beta_2^2)\Big]\mathrm{e}^{-\chi_1 + \mathrm{i}\xi_2}, \tag{5.41b}$$

$$A_3 = \Big[-2\beta_1\big((\alpha_2 - \alpha_1)^2 + (\beta_1^2 - \beta_2^2)\big) + 4\mathrm{i}\beta_1\beta_2(\alpha_2 - \alpha_1)\Big]\mathrm{e}^{\mathrm{i}\xi_1 + \chi_2}, \tag{5.41c}$$

$$A_4 = -4\mathrm{i}\beta_1\beta_2[(\alpha_2 - \alpha_1) - \mathrm{i}(\beta_1 - \beta_2)]\mathrm{e}^{\mathrm{i}\xi_1 - \chi_2}, \tag{5.41d}$$

$$B_1 = -4\beta_1\beta_2[\sinh(\chi_1)\sinh(\chi_2) + \cos(\xi_1 - \xi_2)], \tag{5.41e}$$

$$B_2 = 2\cosh(\chi_1)\cosh(\chi_2)\Big[(\alpha_2 - \alpha_1)^2 + (\beta_1^2 + \beta_2^2)\Big], \tag{5.41f}$$

and

$$\chi_1 = 2\beta_1 x - 8\int\alpha_1\beta_1\,\mathrm{d}t + 2\delta_1, \quad \xi_1 = 2\alpha_1 x - 4\int\left(\alpha_1^2 - \beta_1^2\right)\mathrm{d}t - 2\phi_1, \tag{5.41g}$$

$$\chi_2 = 2\beta_2 x - 8\int\alpha_2\beta_2\,\mathrm{d}t + 2\delta_2, \quad \xi_2 = 2\alpha_2 x - 4\int\left(\alpha_2^2 - \beta_2^2\right)\mathrm{d}t - 2\phi_2, \tag{5.41h}$$

$$\alpha_1 = \alpha_{10}\mathrm{e}^{\lambda t}, \quad \beta_1 = \beta_{10}\mathrm{e}^{\lambda t}, \quad \alpha_2 = \alpha_{20}\mathrm{e}^{\lambda t}, \quad \beta_2 = \beta_{20}\mathrm{e}^{\lambda t}. \tag{5.41i}$$

The soliton interactions in the modified GP equation mirror those in the cubic GP equation [6] for both attractive and repulsive three-body interactions, but repulsive interactions yield significantly higher condensate densities. To assess the impact of phase from attractive three-body interactions on soliton collisions, the asymptotic

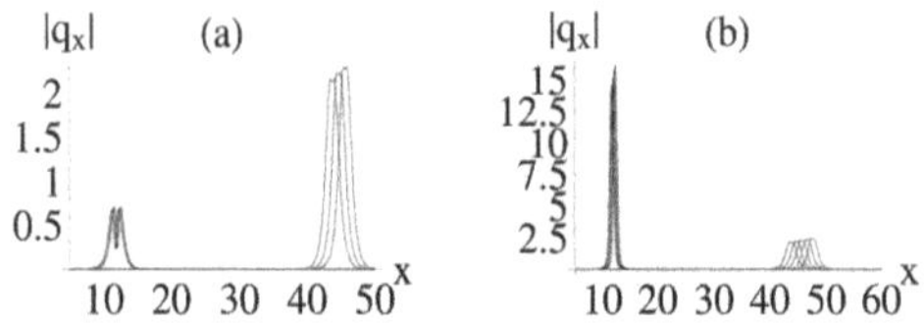

Figure 5.14. Asymptotic density profiles $|q_x|$ of the two-soliton solution for: (a) cubic GP equation; (b) modified GP equation with attractive three-body interactions. Reprinted from [41], copyright (2010) with the permission of The Physical Society of Japan.

density profiles $|q_x|$ of the two-soliton solution are plotted in figures 5.14(a) and (b), showing compression and suppression of splitting compared to the cubic GP equation.

5.5 Summary and future challenges

This chapter is devoted to the dynamics of quasi-one-dimensional scalar BECs with short-range binary and three-body interactions in time-independent and transient harmonic traps. Using gauge transformation approach, bright solitons were derived for time-independent traps, and their collisional dynamics was studied. The impact of transient traps on BEC dynamics was examined, revealing modified soliton behaviour. Matter-wave interference patterns were generated, highlighting the coherent nature of the condensates. A modified GP equation incorporating three-body interactions was developed, enabling the study of bright solitons under the combined impact of both binary and three-body interactions. These results elucidated the interplay of trapping potentials and nonlinear interactions in shaping quasi-1D BEC dynamics. In this connection, it would be interesting to explore the possibility of obtaining other nonlinear excitations like rogue waves and breathers in quasi-one-dimensional BECs. In addition, the nonlinear excitations admitted by the quasi-one-dimensional BECs in other trapping potentials like periodic optical lattice and anharmonic traps deserve to be explored analytically. The existence of completely localized solutions in two and three dimensions (2D and 3D solitons) in scalar BECs and their stability is yet to be explored. Additionally, the impact of higher-order interactions, such as quartic and quintic terms on quasi-1D BECs deserve to be unearthed to extend the understanding of nonlinear excitations beyond binary and three-body interactions.

Parts of this chapter have been reproduced with permission from [1].

References

[1] Radha R and Vinayagam P S 2015 An analytical window into the world of ultracold atoms *Rom. Rep. Phys.* **67** 89–142

[2] Pérez-García V M, Michinel H and Herrero H 1998 Bose-Einstein solitons in highly asymmetric traps *Phys. Rev.* A **57** 3837–42

[3] Salasnich L, Parola A and Reatto L 2002 Effective wave equations for the dynamics of cigar-shaped and disk-shaped Bose condensates *Phys. Rev.* A **65** 043614

[4] Liang Z X, Zhang Z D and Liu W M 2005 Dynamics of a bright soliton in Bose-Einstein condensates with time-dependent atomic scattering length in an expulsive parabolic potential *Phys. Rev. Lett.* **94** 050402
[5] Al Khawaja U 2006 Lax pairs of time-dependent Gross-Pitaevskii equation *J. Phys.* A **39** 9679–91
[6] Radha R and Ramesh Kumar V 2007 Bright matter wave solitons and their collision in Bose-Einstein condensates *Phys. Lett.* A **370** 46–50
[7] Chau L L, Shaw J C and Yen H C 1991 An alternative explicit construction of N-soliton solutions in 1.1 dimensions *J. Math. Phys.* **32** 1737–43
[8] Chong G and Hai W 2006 Dynamical evolutions of matter-wave bright solitons in an inverted parabolic potential *J. Phys. B: At. Mol. Opt. Phys.* **40** 211–20
[9] Hua-Mei L 2006 Dynamics of solitons in Bose-Einstein condensate with time-dependent atomic scattering length *Chin. Phys.* **15** 2216–22
[10] Kengne E and Talla P K 2006 Dynamics of bright matter wave solitons in Bose-Einstein condensates in an expulsive parabolic and complex potential *J. Phys. B: At. Mol. Opt. Phys.* **39** 3679–85
[11] Yuce C and Kilic A 2007 Dark and bright solitons in a Bose-Einstein condensate with a Feshbach resonance *Phys. Scr.* **75** 157–9
[12] Wu L, Zhang J F and Li L 2007 Modulational instability and bright solitary wave solution for Bose-Einstein condensates with time-dependent scattering length and harmonic potential *New J. Phys.* **9** 69
[13] Atre R, Panigrahi P K and Agarwal G S 2006 Class of solitary wave solutions of the one-dimensional Gross-Pitaevskii equation *Phys. Rev.* E **73** 056611
[14] Theocharis G, Rapti Z, Kevrekidis P G, Frantzeskakis D J and Konotop V V 2003 Modulational instability of Gross-Pitaevskii-type equations in 1.1 dimensions *Phys. Rev.* A **67** 063610
[15] Sulem C and Sulem P L 2004 *The Nonlinear Schrödinger Equation: Self-Focusing and Wave Collapse* (Springer)
[16] Khaykovich L, Schreck F, Ferrari G, Bourdel T, Cubizolles J, Carr L D, Castin T and Salomon C 2002 Formation of a matter-wave bright soliton *Science* **296** 1290–3
[17] Kumar V R, Radha R and Panigrahi P K 2008 Dynamics of Bose-Einstein condensates in a time-dependent trap *Phys. Rev.* A **77** 023611
[18] Xue J K 2005 Controllable compression of bright soliton matter waves *J. Phys. B: At. Mol. Opt. Phys.* **38** 3841–8
[19] Ciuryło R, Tiesinga E and Julienne P 2005 Optical tuning of the scattering length of cold alkaline-earth-metal atoms *Phys. Rev.* A **71** 030701
[20] Kevrekidis P G, Theocharis G, Frantzeskakis D J and Malomed B A 2003 Feshbach resonance management for Bose-Einstein condensates *Phys. Rev. Lett.* **90** 230401
[21] Pelinovsky D E, Kevrekidis P G and Frantzeskakis D J 2003 Averaging for solitons with nonlinearity management *Phys. Rev. Lett.* **91** 240201
[22] Engels P, Atherton C and Hoefer M A 2007 Observation of Faraday waves in a Bose-Einstein condensate *Phys. Rev. Lett.* **98** 095301
[23] Ramesh Kumar V, Radha R and Panigrahi P K 2009 Matter wave interference pattern in the collision of bright solitons *Phys. Lett.* A **373** 4381–5
[24] Andrews M R, Townsend C G, Miesner H J, Durfee D S, Kurn D M and Ketterle W 1997 Observation of interference between two Bose condensates *Science* **275** 637–41

[25] Javanainen J and Yoo S M 1996 Quantum phase of a Bose-Einstein condensate with an arbitrary number of atoms *Phys. Rev. Lett.* **76** 161–4
[26] Javanainen J 1986 Oscillatory exchange of atoms between traps containing Bose condensates *Phys. Rev. Lett.* **57** 3164–6
[27] Javanainen J 1991 Spontaneous symmetry breaking derived from a stochastic interpretation of quantum mechanics *Phys. Lett.* A **161** 207–11
[28] Zhang W, Wright E M, Pu H and Meystre P 2003 Fundamental limit for integrated atom optics with Bose-Einstein condensates *Phys. Rev.* A **68** 023605
[29] Liu X J, Li D J, Huang H, Li S Q and Wang Y Z 2001 Atom coherence propagation in a magnetic atomic waveguide *J. Opt. B: Quantum Semiclass. Opt.* **3** 171–7
[30] Josserand C and Rica S 1997 Coalescence and droplets in the subcritical nonlinear Schrödinger equation *Phys. Rev. Lett.* **78** 1215–8
[31] Gammal A, Frederico T, Tomio L and Chomaz P 2000 Atomic Bose-Einstein condensation with three-body interactions and collective excitations *J. Phys. B: At. Mol. Opt. Phys.* **33** 4053–67
[32] Abdullaev F K, Gammal A, Tomio L and Frederico T 2001 Stability of trapped Bose-Einstein condensates *Phys. Rev.* A **63** 043604
[33] Filho V S, Frederico T, Gammal A and Tomio L 2002 Stability of the trapped non-conservative Gross-Pitaevskii equation with attractive two-body interaction *Phys. Rev.* E **66** 036225
[34] Li Y and Hai W 2005 Three-body recombination in two coupled Bose-Einstein condensates *J. Phys.* A **38** 4105–14
[35] Gammal A, Frederico T and Tomio L 1999 Improved numerical approach for the time-independent Gross-Pitaevskii nonlinear Schrödinger equation *Phys. Rev.* E **60** 2421–4
[36] Gammal A, Frederico T, Tomio L and Abdullaev F 2000 Stability analysis of the D-dimensional nonlinear Schrödinger equation with trap and two- and three-body interactions *Phys. Lett.* A **267** 305–11
[37] Köhler T 2002 Three-body problem in a dilute Bose-Einstein condensate *Phys. Rev. Lett.* **89** 210404
[38] Denschlag J *et al* 2000 Generating solitons by phase engineering of a Bose-Einstein condensate *Science* **287** 97–101
[39] Burger S, Bongs K, Dettmer S, Ertmer W, Sengstock K, Sanpera A, Shlyapnikov G V and Lewenstein M 1999 Dark solitons in Bose-Einstein condensates *Phys. Rev. Lett.* **83** 5198–201
[40] Kundu A 1984 Landau-Lifshitz and higher-order nonlinear systems gauge generated from nonlinear Schrödinger-type equations *J. Math. Phys.* **25** 3433–8
[41] Ramesh Kumar V, Radha R and Wadati M 2010 Phase engineering and solitons of Bose-Einstein condensates with two-and three-body interactions *J. Phys. Soc. Japan* **79** 074005

IOP Publishing

An Introduction to Ultracold Atoms with Analytical and Numerical Methods

Paulsamy Muruganandam and Ramaswamy Radha

Chapter 6

Vectorial condensates

Scalar Bose–Einstein condensates (BECs), composed of a single type of bosonic atom in one internal state, exhibit dynamics governed by a single Gross–Pitaevskii (GP) equation, influenced by the trapping potential and s-wave scattering interactions. In contrast, two-component (vectorial) BECs, comprising two hyperfine states of the same atom or two distinct atomic species, display richer dynamics due to intra- and inter-species scattering interactions. Described by coupled GP equations, these systems pose significant theoretical and experimental challenges due to their complex nonlinear behaviour.

This chapter investigates the dynamics of two-component BECs in a time-dependent harmonic trap, deriving integrability conditions for vector bright solitons using Lax pair. It examines soliton collisions, demonstrating energy exchange between components and enhanced condensate stability. Feshbach resonance (FR) management is employed showing the long-lived nature of BECs in transient harmonic traps compared to time-independent traps. The dynamics of weakly coupled BECs either through temporal Rabi coupling or spatial coupling is analyzed. The chapter also explores stabilizing rogue waves, inherently unstable nonlinear excitations through FR or trap frequency modulation.

6.1 Vector Bose–Einstein condensates and Feshbach resonance management

This section has been reproduced with permission from [4].

A two-component BEC consisting of two hyperfine states of the same atom, confined in a transient harmonic trap with equal intra- and inter-species interaction strengths, is described by a pair of coupled GP equations in dimensionless, quasi-one-dimensional form [1–3]:

doi:10.1088/978-0-7503-5447-9ch6

$$i\frac{\partial\psi_1}{\partial t} + \frac{\partial^2\psi_1}{\partial x^2} + 2a(t)(|\psi_1|^2 + |\psi_2|^2)\psi_1 + \lambda(t)^2x^2\psi_1 = 0, \tag{6.1a}$$

$$i\frac{\partial\psi_2}{\partial t} + \frac{\partial^2\psi_2}{\partial x^2} + 2a(t)(|\psi_1|^2 + |\psi_2|^2)\psi_2 + \lambda(t)^2x^2\psi_2 = 0, \tag{6.1b}$$

where ψ_1 and ψ_2 represent the wavefunctions of the two components, $a(t)$ is the time-dependent scattering length, and $\lambda(t)$ denotes the trap frequency.

The system described by equations (6.1) is integrable, with associated Lax matrices $\mathcal{U}$ and $\mathcal{V}$ given by:

$$\mathcal{U} = \begin{pmatrix} \mathrm{i}\zeta(t) & Q_1 & Q_2 \\ -Q_1^* & -\mathrm{i}\zeta(t) & 0 \\ -Q_2^* & 0 & -\mathrm{i}\zeta(t) \end{pmatrix}, \quad \mathcal{V} = \begin{pmatrix} V_{11} & V_{12} & V_{13} \\ V_{21} & V_{22} & V_{23} \\ V_{31} & V_{32} & V_{33} \end{pmatrix}, \tag{6.2}$$

where:

$$\begin{aligned}
V_{11} &= -\mathrm{i}\zeta(t)^2 + \mathrm{i}\Gamma(t)x\zeta(t) + \frac{\mathrm{i}}{2}Q_1Q_1^* + \frac{\mathrm{i}}{2}Q_2Q_2^*, \quad V_{12} = (\Gamma(t)x - \zeta(t))Q_1 + \frac{\mathrm{i}}{2}\frac{\partial Q_1}{\partial x}, \\
V_{13} &= (\Gamma(t)x - \zeta(t))Q_2 + \frac{\mathrm{i}}{2}\frac{\partial Q_2}{\partial x}, \quad V_{21} = -(\Gamma(t)x - \zeta(t))Q_1^* + \frac{\mathrm{i}}{2}\frac{\partial Q_1^*}{\partial x}, \\
V_{22} &= \mathrm{i}\zeta(t)^2 - \mathrm{i}\Gamma(t)x\zeta(t) - \frac{\mathrm{i}}{2}Q_1Q_1^*, \quad V_{23} = -\frac{\mathrm{i}}{2}Q_2Q_1^*, \\
V_{31} &= -(\Gamma(t)x - \zeta(t))Q_2^* + \frac{i}{2}\frac{\partial Q_2^*}{\partial x}, \quad V_{32} = -\frac{\mathrm{i}}{2}Q_1Q_2^*, \quad V_{33} = \mathrm{i}\zeta(t)^2 - \mathrm{i}\Gamma(t)x\zeta(t) - \frac{i}{2}Q_2Q_2^*, \\
Q_1 &= \frac{1}{\sqrt{A(t)}}\psi_1(x,t)\mathrm{e}^{\mathrm{i}\Gamma(t)x^2/2}, \quad Q_2 = \frac{1}{\sqrt{A(t)}}\psi_2(x,t)\mathrm{e}^{\mathrm{i}\Gamma(t)x^2/2}.
\end{aligned}$$

The compatibility condition $\left(\frac{\partial\Phi}{\partial x}\right)_t = \left(\frac{\partial\Phi}{\partial t}\right)_x$ yields the zero-curvature condition $\mathcal{U}_t - \mathcal{V}_x + [\mathcal{U}, \mathcal{V}] = 0$, which ensures the integrability of equations (6.1) provided the spectral parameter $\zeta(t)$ satisfies the nonisospectral condition:

$$\zeta(t) = \mu \exp\left(-\int\Gamma(t)\,\mathrm{d}t\right), \tag{6.3}$$

where μ is a complex constant and $\Gamma(t)$ is an arbitrary time-dependent function related to the trap frequency:

$$\lambda^2(t) = \Gamma^2(t) - \frac{\mathrm{d}\Gamma(t)}{\mathrm{d}t}, \tag{6.4}$$

with:

$$\Gamma(t) = \frac{\mathrm{d}}{\mathrm{d}t}\ln a(t). \tag{6.5}$$

Substituting equation (6.5) into equation (6.4) relates the trap frequency $\lambda(t)$ to the scattering length $a(t)$, yielding the integrability condition:

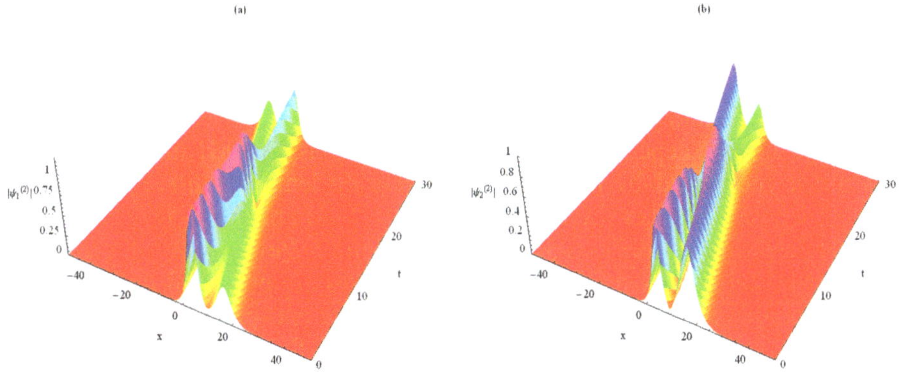

Figure 6.1. Energy switching between matter-wave (bright) solitons in a two-component BEC [4] with $a(t) = 0.01$ and $\lambda(t) = 0.01 + 0.0001t^2$. Reproduced with permission from [4].

$$-a(t)\frac{\mathrm{d}^2 a(t)}{\mathrm{d}t^2} + 2\left(\frac{\mathrm{d}a(t)}{\mathrm{d}t}\right)^2 - \lambda^2(t)a^2(t) = 0. \tag{6.6}$$

Equations (6.1) are completely integrable for appropriate choices of $\lambda(t)$ and $a(t)$ satisfying equation (6.6). For a constant trap frequency, $\lambda(t) = c_1$, equation (6.6) gives $a(t) = \exp(c_1 t)$.

Using the gauge transformation method, the bright soliton solutions of equations (6.1) are obtained as:

$$\psi_1^{(1)} = \sqrt{\frac{1}{a(t)}}\,\varepsilon_1^{(1)} 2\beta_1(t)\mathrm{sech}(\theta_1)\mathrm{e}^{\mathrm{i}\left(-\xi_1 + \Gamma(t)\frac{x^2}{2}\right)}, \tag{6.7a}$$

$$\psi_2^{(1)} = \sqrt{\frac{1}{a(t)}}\,\varepsilon_2^{(1)} 2\beta_1(t)\mathrm{sech}(\theta_1)\mathrm{e}^{\mathrm{i}\left(-\xi_1 + \Gamma(t)\frac{x^2}{2}\right)}, \tag{6.7b}$$

where:

$$\theta_1 = 4\int \alpha_1\beta_1\,\mathrm{d}t + 2x\beta_1 - 2\delta_1, \tag{6.8}$$

$$\xi_1 = 2\int \left(\alpha_1^2 - \beta_1^2\right)\mathrm{d}t + 2x\alpha_1 - 2\chi_1, \tag{6.9}$$

with $\alpha_1 = \alpha_{10}\mathrm{e}^{\int \Gamma(t)\,\mathrm{d}t}$, $\beta_1 = \beta_{10}\mathrm{e}^{\int \Gamma(t)\,\mathrm{d}t}$, and δ_1 and χ_1 are arbitrary parameters. The gauge transformation approach can be extended to generate multisoliton solutions [5]. Figure 6.1 illustrates the potential to transfer energy between modes in a vector BEC, enhancing the longevity of bright solitons or condensates, regardless of whether the trap is time-dependent or time-independent.

6.2 Enhancement of lifetime and collisional dynamics of vector BECs

This section has been reproduced with permission from [7].

Investigations of vector BECs reveal that their lifetime is prolonged by energy switching between the two components. FR management further demonstrates that vector BECs in a time-dependent harmonic trap exhibit greater longevity compared to those in a time-independent trap.

Initially, the time dependence of the harmonic trap is disabled and the condensate evolution is monitored by adjusting the scattering length to $a(t) = 0.5e^{-0.25t}$, as shown in figure 6.2(f), consistent with equations (6.4) and (6.6). This results in an expulsive trap with $\Gamma(t) = -0.25$, depicted in figure 6.2(e).

The analytical density profiles of the condensates are presented in figures 6.2(a) and (b). Corresponding numerical simulations, using real-time propagation via the split-step Crank–Nicolson method [6], are shown in figures 6.2(c) and (d). These figures demonstrate excellent agreement between analytical and numerical results. Doubling the trap frequency to $\Gamma(t) = -0.5$, maintaining the expulsive nature (figure 6.3(e)), and adjusting the scattering length to $a(t) = 0.5e^{-0.5t}$ (figure 6.3(f)), consistent with equations (6.4) and (6.6), induces condensate compression in both modes, as shown analytically in figures 6.3(a) and (b) and confirmed numerically in figures 6.3(c) and (d). Increasing the trap strength to 3.6 times the original with $\Gamma(t) = -0.9$, (figure 6.4(e)) and setting $a(t) = 0.5e^{-0.9t}$, (figure 6.4(f)) triggers the onset of condensate collapse, observed analytically in figures 6.4(a) and (b) and numerically in figures 6.4(c) and (d), with strong agreement.

To enhance condensate stability, a time-dependent expulsive trap is introduced with $\Gamma(t) = -0.25t$ (figure 6.5(e)) and a scattering length $a(t) = 0.5e^{-0.125t^2}$ (figure 6.5(f)), consistent with equations (6.4) and (6.6). The resulting density profiles are shown analytically in figures 6.5(a) and (b) and numerically in figures 6.5(c) and (d), confirming consistency. A 20-fold increase in the expulsive trap frequency ($\Gamma(t) = -5t$, figure 6.6(e)) with $a(t) = 0.5e^{-2.5t^2}$ (figure 6.6(f)), aligned with equations (6.4) and (6.6), yields density profiles in figures 6.6(a) and (b) that match numerical simulations in figures 6.6(c) and (d).

Further increasing the time-dependent expulsive trap frequency by 100 times ($\Gamma(t) = -25t$, figure 6.7(e)) and adjusting the scattering length to $a(t) = 0.5e^{-12.5t^2}$ (figure 6.7(f)), consistent with equations (6.4) and (6.6), does not lead to a sharp density increase, as shown analytically in figures 6.7(a) and (b) and numerically in figures 6.7(c) and 6.7(d). This indicates that further increases in trap frequency and scattering length manipulation via FR do not significantly elevate condensate density, despite rapidly increasing attractive interaction strength. Consequently, condensates in a time-dependent expulsive trap remain stable over extended periods, even with strong attractive interactions, with analytical and numerical results aligning closely [7]

The stability of vector BECs in a time-dependent expulsive trap is unaffected by added noise, as evidenced in figures 6.8, 6.9, and 6.10. Thus, two-component BECs with attractive interactions in a time-dependent expulsive trap, governed by an integrable system, exhibit prolonged lifetimes through FR management, outperforming those in time-independent traps. In contrast, scalar (single-component) condensates stabilized similarly via FR management decay rapidly during evolution.

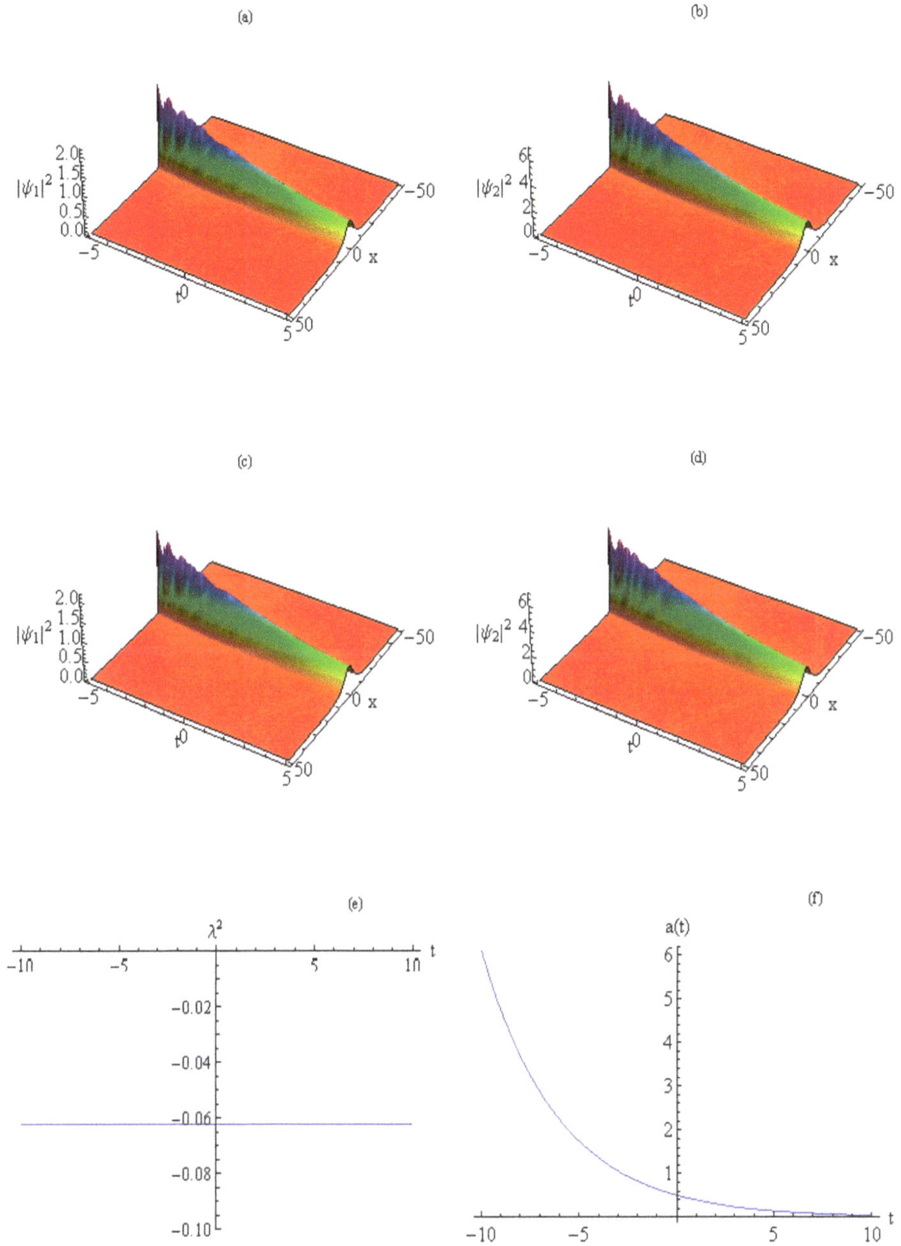

Figure 6.2. Upper panel (a)–(b): analytical density profile of condensates (bright solitons) in a time-independent expulsive trap with $\Gamma(t) = -0.25$, $a(t) = 0.5e^{-0.25t}$. Middle panel (c)–(d): numerically simulated density profile using the split-step Crank–Nicolson method [6] for the same parameters. Lower panel (e)–(f): trap strength and binary interaction strength. Reprinted from [7], Copyright (2015), with permission from Elsevier.

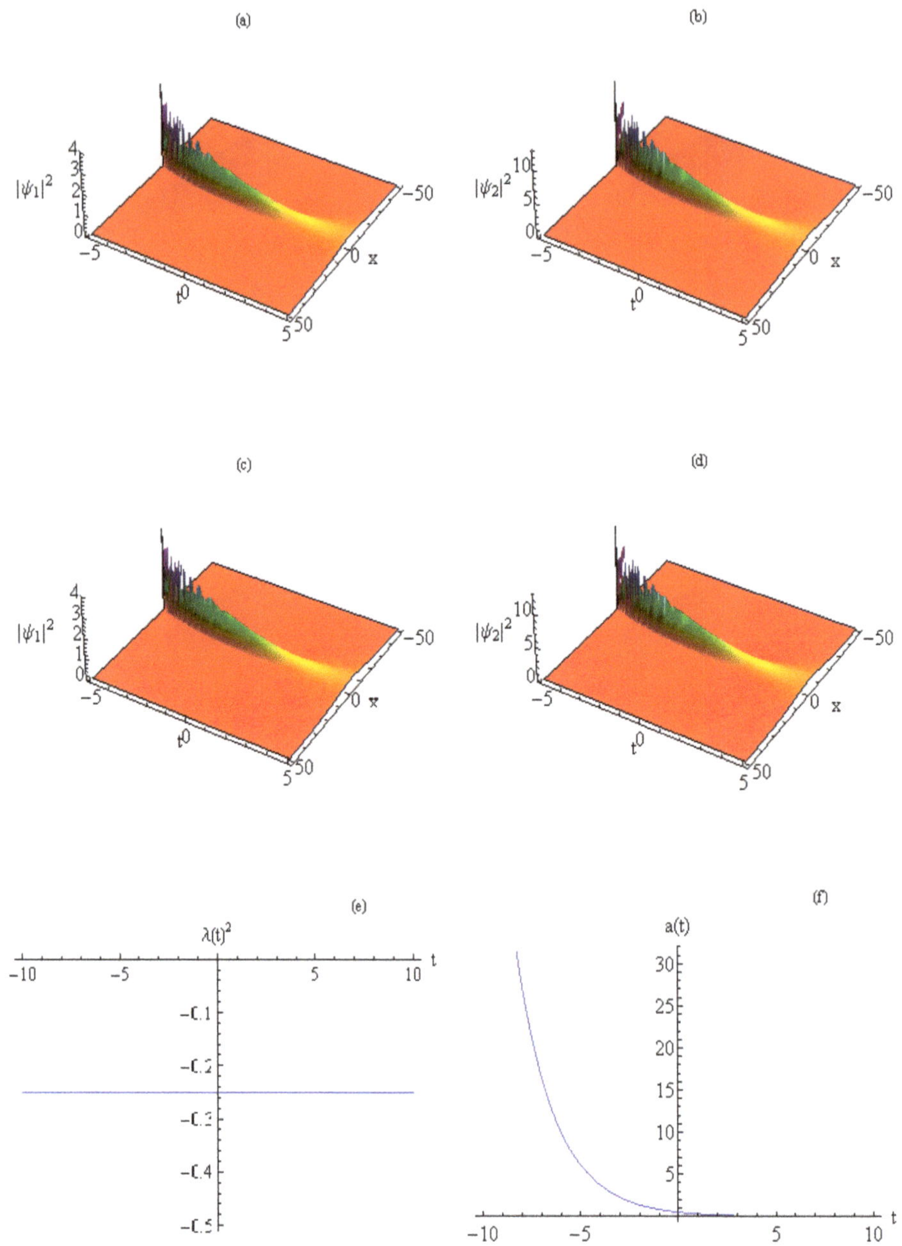

Figure 6.3. Upper panel (a)–(b): analytical compression of condensates in a time-independent expulsive trap with $\Gamma(t) = -0.5$, $a(t) = 0.5e^{-0.5t}$. Middle panel (c)–(d): numerically simulated density profile showing compression for the same parameters. Lower panel (e)–(f): trap strength and binary interaction strength. Reprinted from [7], Copyright (2015), with permission from Elsevier.

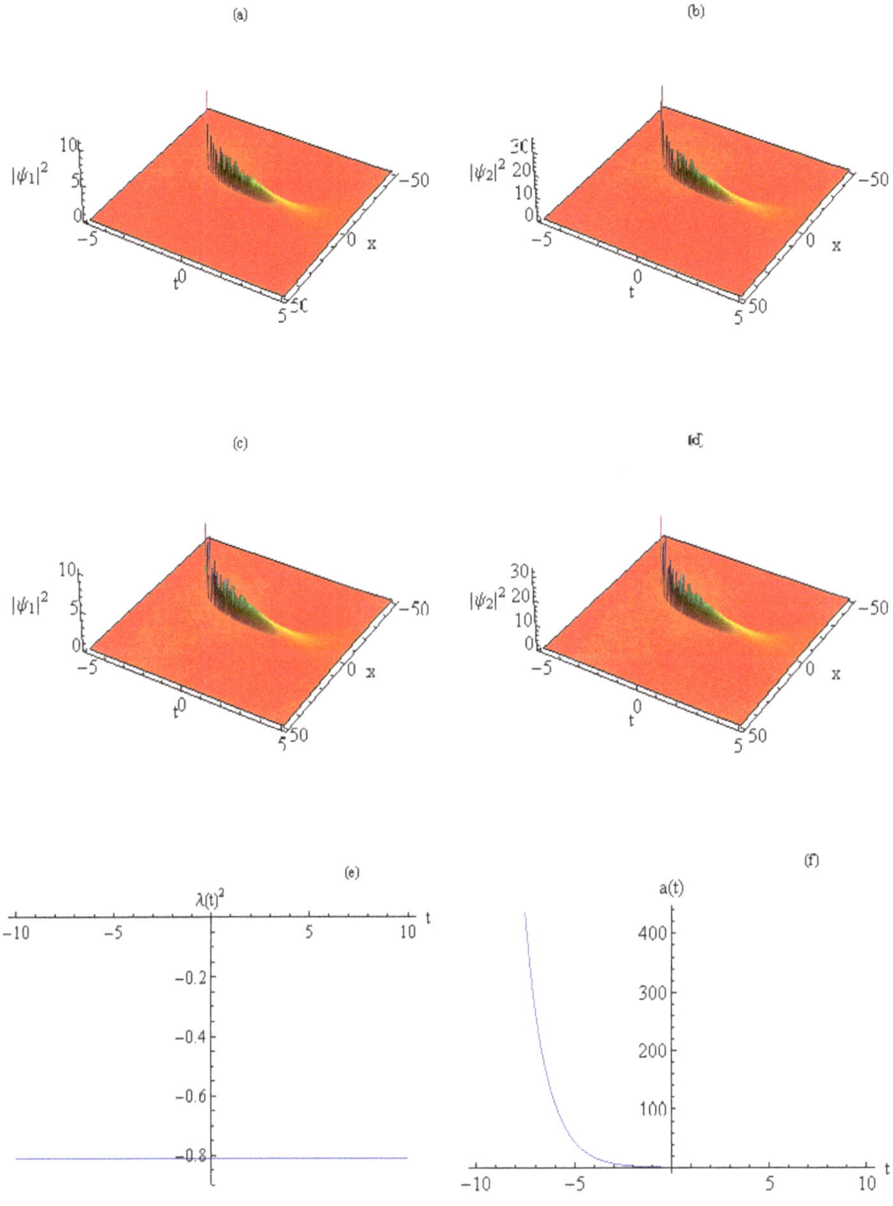

Figure 6.4. Upper panels (a)–(b): analytical onset of condensate collapse in a time-independent expulsive trap with $\Gamma(t) = -0.9$, $a(t) = 0.5e^{-0.9t}$. Middle panels (c)–(d): numerically simulated density profile showing collapse for the same parameters. Lower panels (e)–(f): trap strength and binary interaction strength. Reprinted from [7], Copyright (2015), with permission from Elsevier.

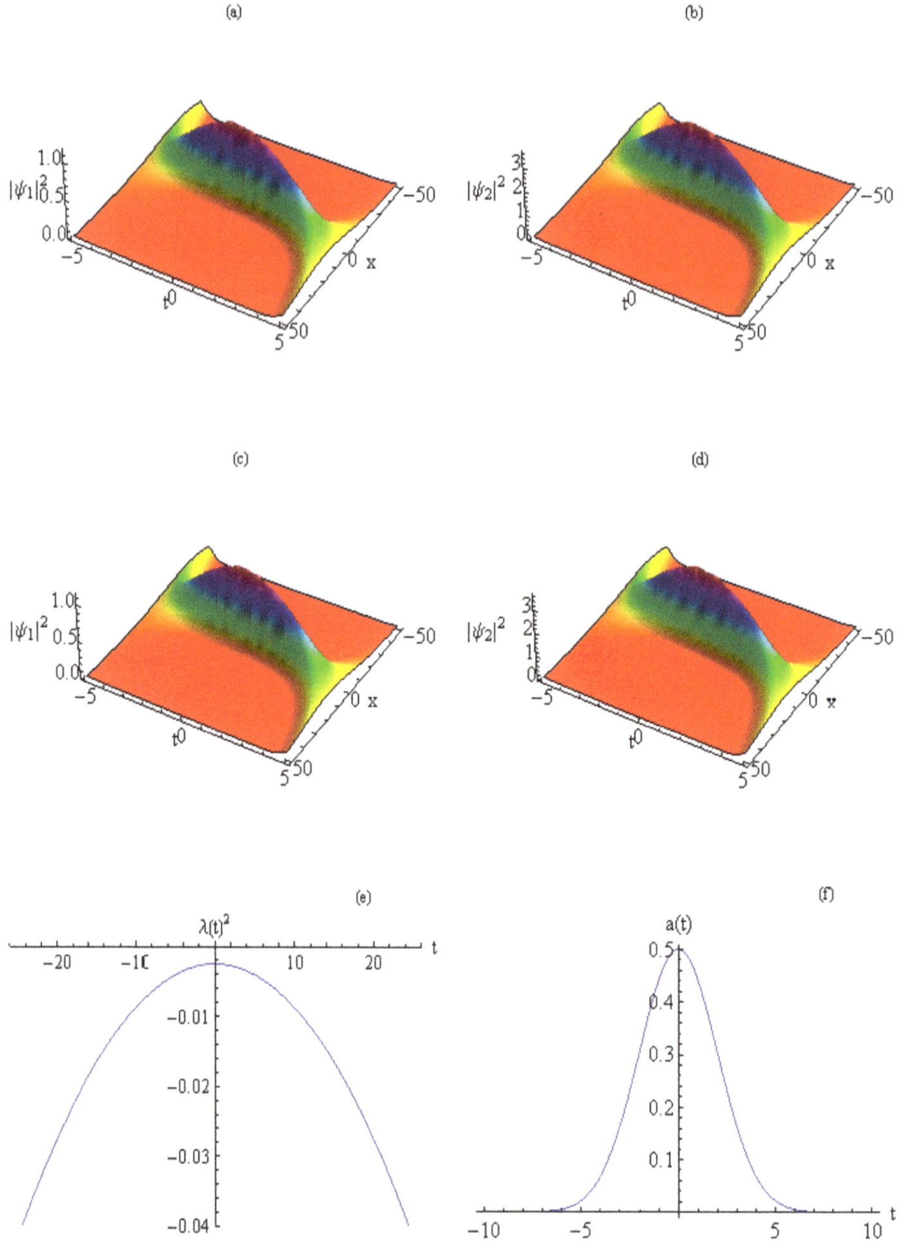

Figure 6.5. Upper panels (a)–(b): analytical density profile of condensates in a time-dependent expulsive trap with $\Gamma(t) = -0.25t$, $a(t) = 0.5\mathrm{e}^{-0.125t^2}$. Middle panels (c)–(d): numerically simulated density profile for the same parameters. Lower panels (e)–(f): transient trap strength and interaction strength. Reprinted from [7], Copyright (2015), with permission from Elsevier.

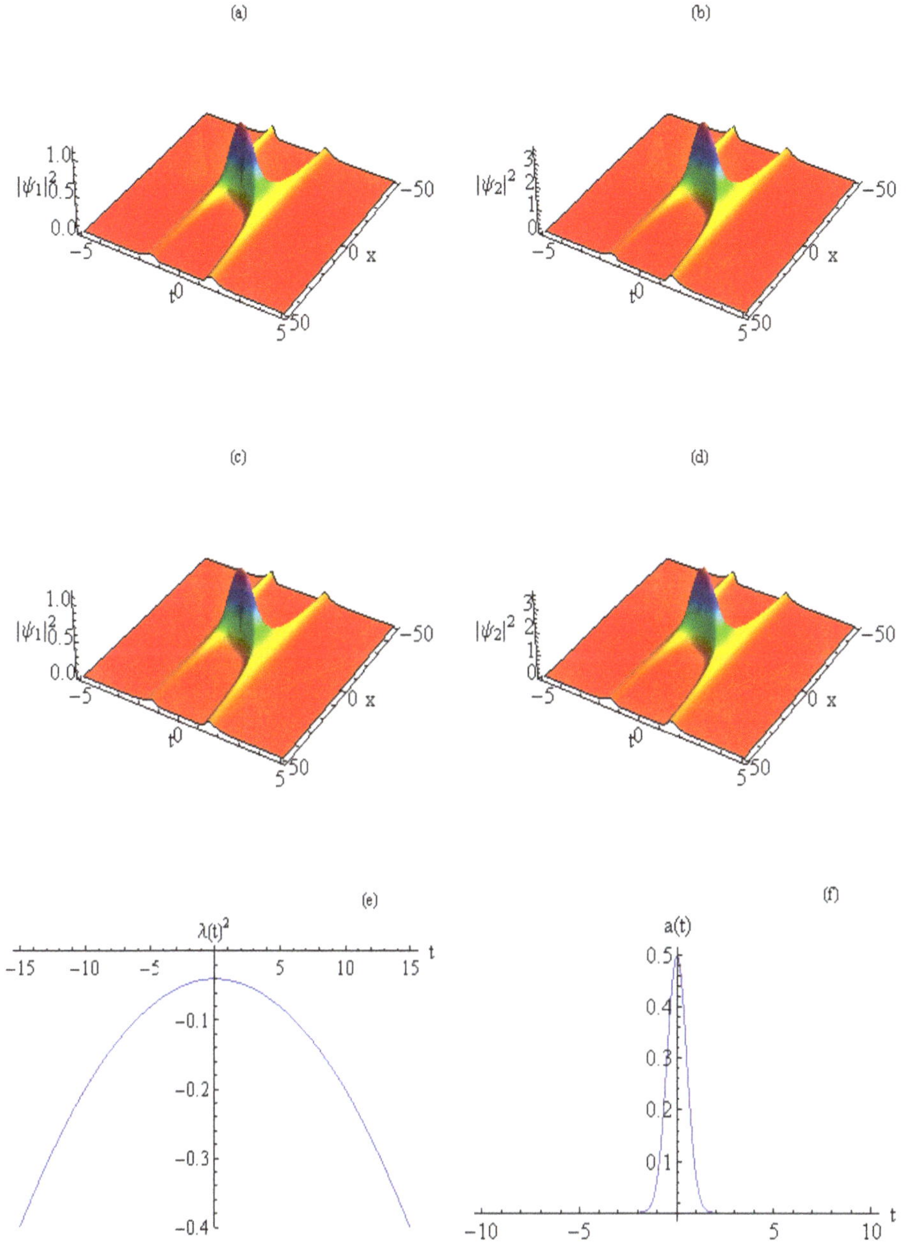

Figure 6.6. Upper panels (a)–(b): analytical density profile of condensates in a time-dependent expulsive trap with $\Gamma(t) = -5t$, $a(t) = 0.5\mathrm{e}^{-2.5t^2}$. Middle panels (c)–(d): numerically simulated density profile for the same parameters. Lower panels (e)–(f): transient trap strength and interaction strength. Reprinted from [7], Copyright (2015), with permission from Elsevier.

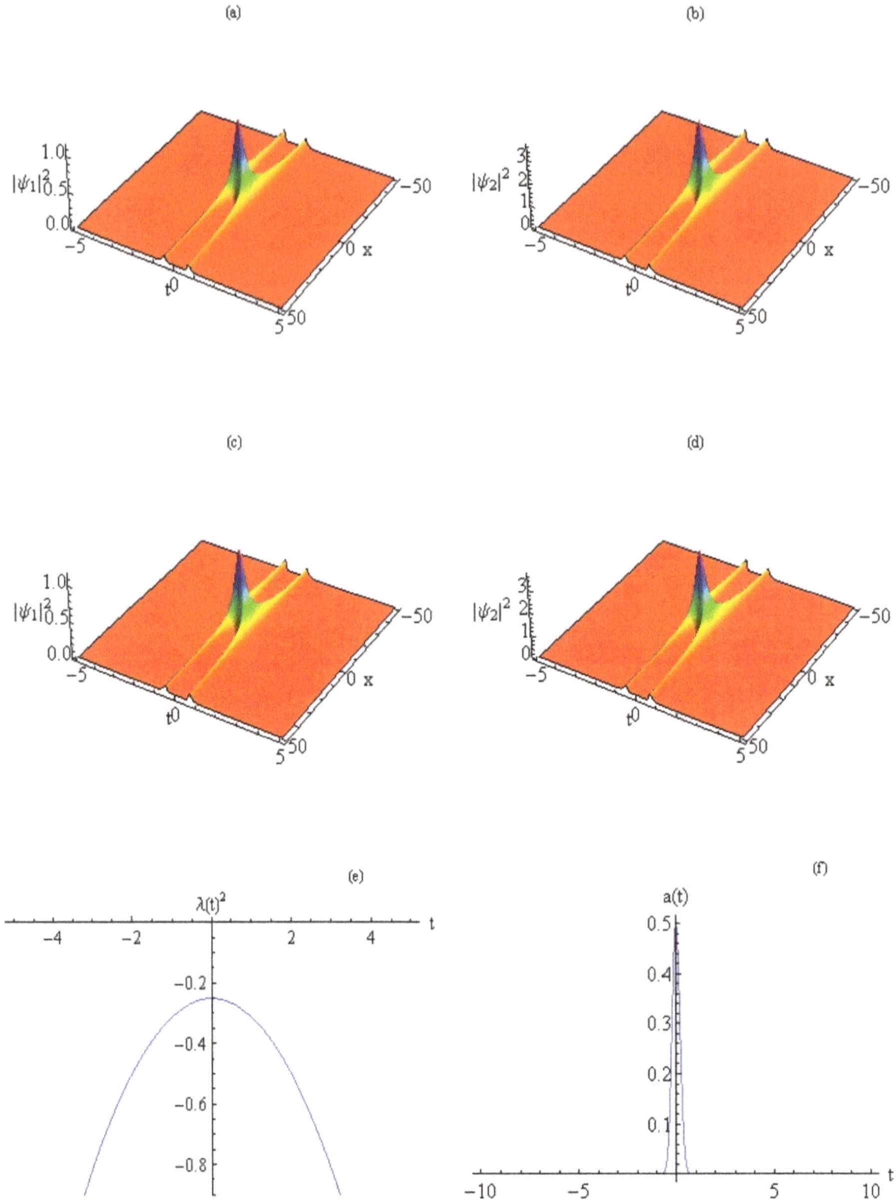

Figure 6.7. Upper panels (a)–(b): analytical density profile of condensates in a time-dependent expulsive trap with $\Gamma(t) = -25t$, $a(t) = 0.5e^{-12.5t^2}$. Middle panels (c)–(d): Numerically simulated density profile for the same parameters. Lower panels (e)–(f): Transient trap strength and interaction strength. Reprinted from [7], Copyright (2015), with permission from Elsevier.

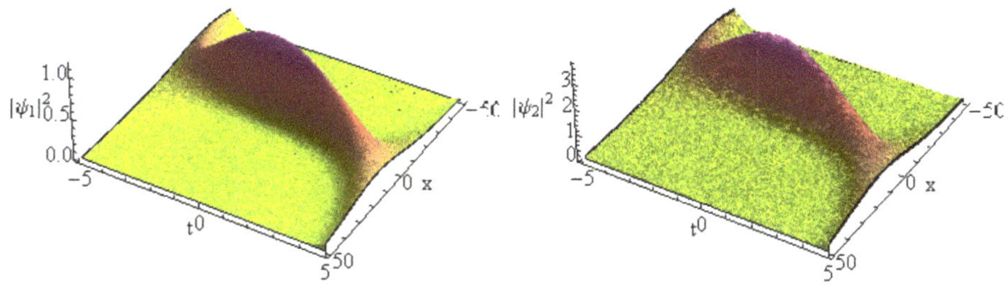

Figure 6.8. Numerically simulated density profile corresponding to figure 6.5, with added random noise. Reprinted from [7], Copyright (2015), with permission from Elsevier.

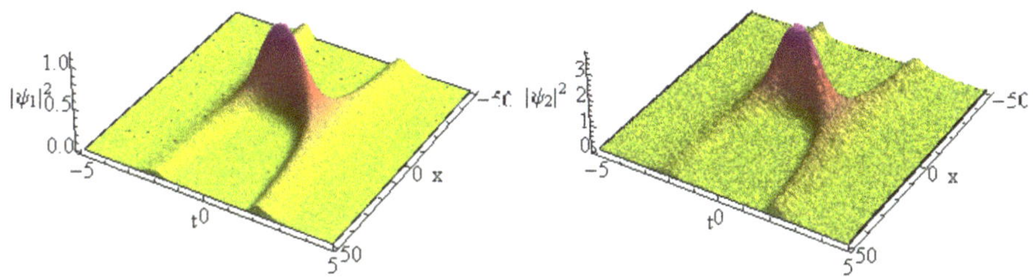

Figure 6.9. Numerically simulated density profile corresponding to figure 6.6, with added random noise. Reprinted from [7], Copyright (2015), with permission from Elsevier.

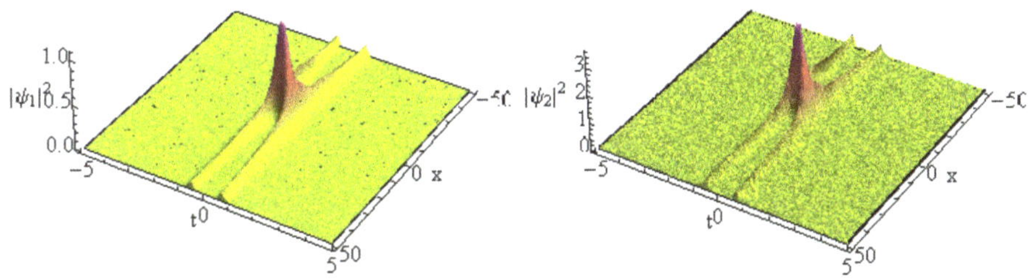

Figure 6.10. Numerically simulated density profile corresponding to figure 6.7, with added random noise. Reprinted from[7], Copyright (2015), with permission from Elsevier.

6.3 Impact of temporal Rabi coupling

This section has been reproduced with permission from [8].

The extended lifetime of two-component BECs results from intra-species and inter-species interactions. Additionally, incorporating weak time-dependent or spatially dependent coupling can further enhance the longevity of these condensates. Consider a two-component, quasi-one-dimensional, cigar-shaped BEC comprising two hyperfine states, such as $|F = 1, m_f = -1\rangle$ and $|F = 1, m_f = 1\rangle$ of ^{87}Rb atoms [9], confined at different vertical positions by parabolic traps and coupled via a time-dependent field. In the mean-field approximation, the dynamics is governed by the coupled GP equations

$$i\frac{\partial\psi_1}{\partial t} + \frac{\partial^2\psi_1}{\partial x^2} + 2a(t)(|\psi_1|^2 + |\psi_2|^2)\psi_1 + \lambda(t)^2x^2\psi_1 + iG(t)\psi_1 + \nu(t)\psi_2 = 0, \quad (6.10a)$$

$$i\frac{\partial\psi_2}{\partial t} + \frac{\partial^2\psi_2}{\partial x^2} + 2a(t)(|\psi_1|^2 + |\psi_2|^2)\psi_2 + \lambda(t)^2x^2\psi_2 + iG(t)\psi_2 + \nu(t)\psi_1 = 0, \quad (6.10b)$$

where ψ_1 and ψ_2 are the wavefunctions of the two components, $\nu(t)$ represents the coupling strength between the condensates, and $G(t)$ accounts for gain or loss from the thermal cloud. The addition of linear coupling to the Manakov model has been shown to preserve integrability [10]. The linearly coupled GP equations (6.10) have been studied, exploring phenomena such as domain walls [11] and symmetry breaking of solitons [12]. For $\nu(t) = 0$, the dynamics of vector BECs has been analyzed [3, 5]. The one-soliton solutions for equations (6.10) are obtained as [8]

$$\psi_1^{(1)} = \frac{2}{\sqrt{a(t)}}\varepsilon_1^{(1)}\beta_1(t)\mathrm{sech}(\theta_1)\mathrm{e}^{\mathrm{i}\left(-\xi_1 + \Gamma(t)\frac{x^2}{2}\right)}, \quad (6.11)$$

$$\psi_2^{(1)} = \frac{2}{\sqrt{a(t)}}\varepsilon_2^{(1)}\beta_1(t)\mathrm{sech}(\theta_1)\mathrm{e}^{\mathrm{i}\left(-\xi_1 + \Gamma(t)\frac{x^2}{2}\right)}, \quad (6.12)$$

where

$$\theta_1 = 8\int\alpha_1(t)\beta_1(t)\,\mathrm{d}t + 2x\beta_1(t) - 2\delta_1, \quad (6.13)$$

$$\xi_1 = 4\int(\alpha_1(t)^2 - \beta_1(t)^2)\mathrm{d}t + 2x\alpha_1(t) - 2\chi_1, \quad (6.14)$$

$$\Gamma(t) = f(t)\mathrm{e}^{\nu(t)}, \quad (6.15)$$

with

$$\alpha_1(t) = \alpha_{10}\mathrm{e}^{-2\int\Gamma(t)^2\,\mathrm{d}t}, \quad \beta_1(t) = \beta_{10}\mathrm{e}^{-2\int\Gamma(t)^2\,\mathrm{d}t}, \quad (6.16)$$

and δ_1, χ_1, and $f(t)$ are arbitrary parameters, while $\varepsilon_1^{(1)}$, $\varepsilon_2^{(1)}$ are coupling parameters. The soliton amplitude depends on the temporal scattering length $a(t)$ and trap frequency $\Gamma(t)$, with $\beta_1(t)$ varying exponentially with $\Gamma(t)$. Since $\Gamma(t)$ is exponentially dependent on the time-dependent coupling $\nu(t)$, the condensate density can increase rapidly, potentially driving the system into instability.

Figures 6.11(a) and (b) depict the condensate density profiles without Rabi coupling. Introducing coupling results in a sharp density increase, as shown in figures 6.12(a) and (b), indicating system instability. This instability can be mitigated by adjusting the temporal scattering length via FR, as illustrated in figures 6.13(a) and (b), or by fine-tuning the coupling coefficient $\nu(t)$, as shown in figures 6.14(a) and (b).

The gauge transformation approach [5] can be extended to generate multisoliton solutions. The collisional dynamics reveal intramodal inelastic interactions of bright

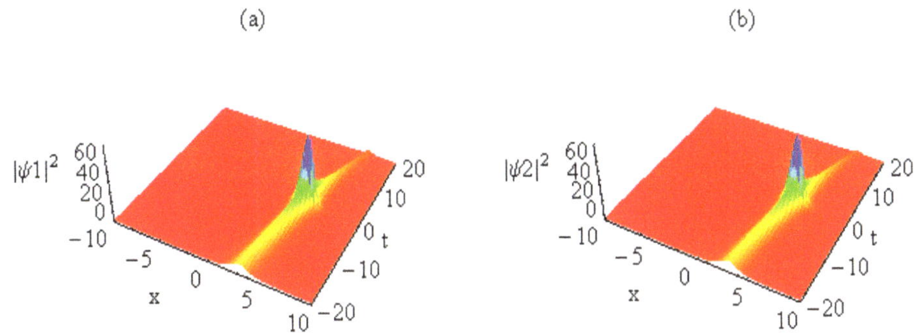

Figure 6.11. Density profiles of the condensates for $a(t) = 0.5t$, $\varepsilon_1^{(1)} = 0.3$, and $\Gamma(t) = 0.001t$, without coupling. Reprinted from [8], Copyright (2012), with permission from Elsevier.

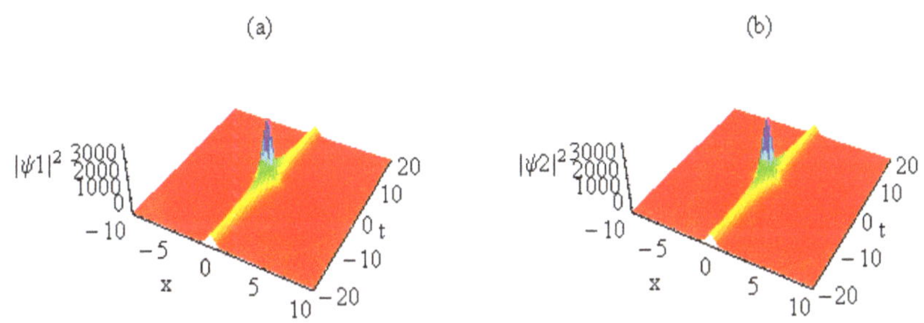

Figure 6.12. Effect of weak coupling on condensate density with $\nu(t) = 0.1t$, and other parameters as in figure 6.11. Reprinted from [8], Copyright (2012), with permission from Elsevier.

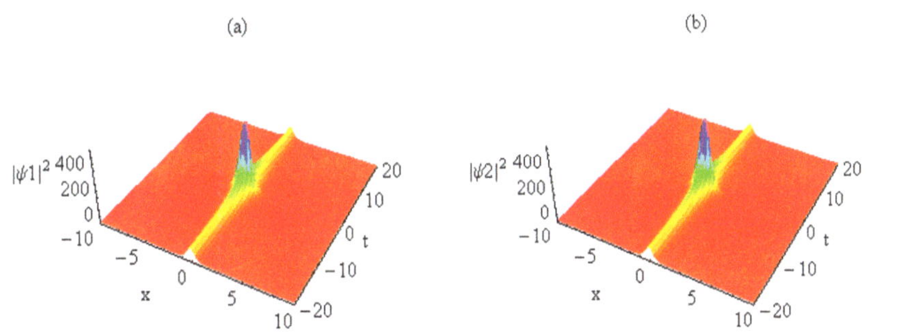

Figure 6.13. Stabilization of condensates by adjusting the scattering length to $a(t) = 0.25t$, with other parameters as in figure 6.12. Reprinted from [8], Copyright (2012), with permission from Elsevier.

solitons without coupling, as shown in figures 6.15(a) and (b). Introducing weak time-dependent coupling disrupts these interactions, as depicted in figures 6.16(a) and (b). However, fine-tuning the coupling coefficient $\nu(t)$ restores the intramodal inelastic collisions, as illustrated in figures 6.17(a) and (b).

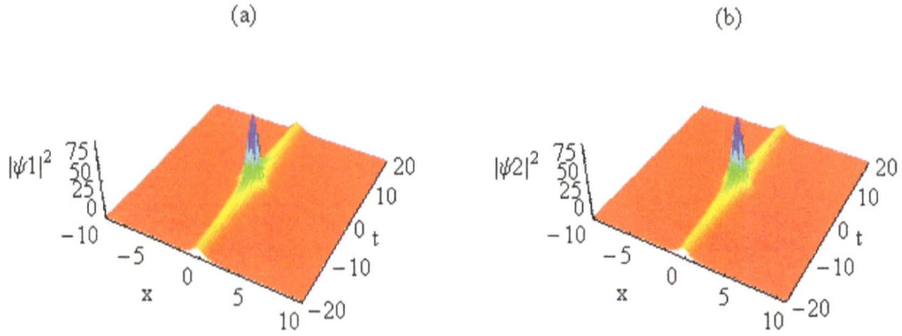

Figure 6.14. Stabilization of condensates by fine-tuning the coupling coefficient to $\nu(t) = 0.02t$, with other parameters as in figure 6.12. Reprinted from [8], Copyright (2012), with permission from Elsevier.

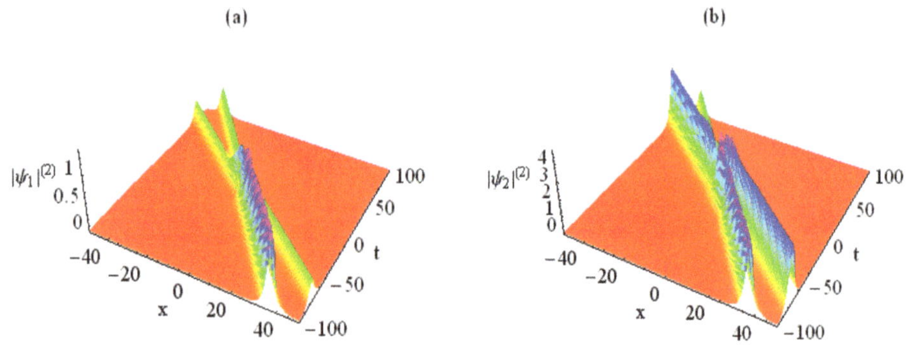

Figure 6.15. Intramodal inelastic collision of condensates without coupling, for $a(t) = 5$, $\Gamma(t) = 0.001t$, $\varepsilon_1^{(1)} = 0.89i$, and $\varepsilon_1^{(2)} = 0.6$. Reprinted from [8], Copyright (2012), with permission from Elsevier.

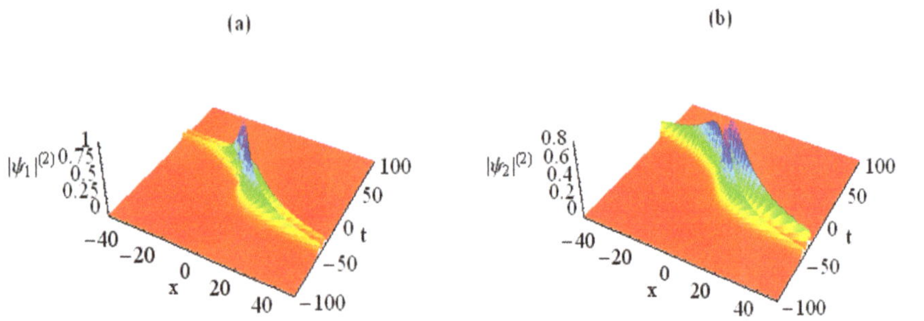

Figure 6.16. Effect of coupling on condensate collisions for $\nu(t) = 0.01t$, with other parameters as in figure 6.15. Reprinted from[8], Copyright (2012), with permission from Elsevier.

6.4 Spatially coupled Bose–Einstein condensates

This section has been reproduced with permission from [17].

To explore the effects of weak spatially dependent coupling, one considers cigar-shaped (quasi-one-dimensional) BECs comprising two hyperfine states of ^{87}Rb atoms [9], confined by a parabolic trapping potential and subjected to a spatially

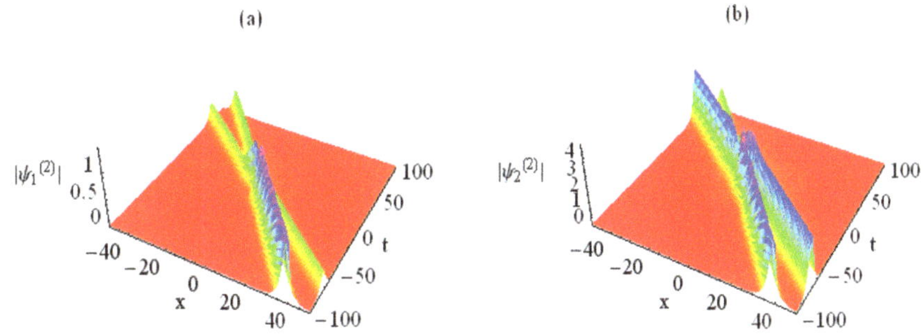

Figure 6.17. Recovery of intramodal inelastic collisions for $\nu(t) = 0.001t$, with other parameters as in figure 6.16. Reprinted from [8], Copyright (2012), with permission from Elsevier.

dependent coupling force $\nu(x)$. Within the mean-field framework, the system is governed by the coupled GP equations [13]:

$$i\frac{\partial\psi_1}{\partial t} + \frac{\partial^2\psi_1}{\partial x^2} + 2a(t)(|\psi_1|^2 + |\psi_2|^2)\psi_1 + \lambda(t)^2x^2\psi_1 + iG(x, t)\psi_1 + \nu(x)\psi_2 = 0, \quad (6.17)$$

$$i\frac{\partial\psi_2}{\partial t} + \frac{\partial^2\psi_2}{\partial x^2} + 2a(t)(|\psi_1|^2 + |\psi_2|^2)\psi_2 + \lambda(t)^2x^2\psi_2 + iG(x, t)\psi_2 + \nu(x)\psi_1 = 0, \quad (6.18)$$

where $\nu(x)$ represents the weak spatial coupling between the two components, and $G(x, t)$ accounts for feeding from a reservoir of condensate weakly coupled to the trap [14]. These equations have been analyzed for $\nu(x) = 0$ [3, 5], revealing the dynamics of vectorial BECs. When the trapping potential and coupling are absent ($\lambda(t) = \nu(x) = 0$), the scattering length $a(t)$ is constant, and gain/loss terms $G(x, t)$ are neglected, the system reduces to the integrable Manakov model [15, 16].

The bright soliton solutions of equations (6.17) and (6.18) are given by [17]:

$$\psi_1^{(1)} = \frac{2}{\sqrt{a(t)}}\varepsilon_1^{(1)}\beta_1(t)\text{sech}(\theta_1)e^{i(-\xi_1 + \Gamma(t)\int x\, dx)}, \quad (6.19)$$

$$\psi_2^{(1)} = \frac{2}{\sqrt{a(t)}}\varepsilon_2^{(1)}\beta_1(t)\text{sech}(\theta_1)e^{i(-\xi_1 + \Gamma(t)\int x\, dx)}, \quad (6.20)$$

where:

$$\theta_1 = 2\int\beta_1\, dx + \int(8\alpha_1\beta_1 - 4\alpha_1)dt - 2\delta_1, \quad (6.21)$$

$$\xi_1 = -2\int\alpha_1\, dx - 4\int\left(\alpha_1 + 4i\beta_1^2 - 4\beta_1\right)dt - 2\chi_1, \quad (6.22)$$

with $\alpha_1(t) = \alpha_{10}e^{-2\gamma'(t)}$, $\beta_1(t) = \beta_{10}e^{-2\gamma'(t)}$, and δ_1, χ_1, $\varepsilon_1^{(1)}$, $\varepsilon_2^{(1)}$ as arbitrary parameters, while $\varepsilon_{1,2}$ are coupling coefficients.

The soliton amplitude depends on the time-modulated scattering length $a(t)$ and $\beta_1(t)$ [17]. Density profiles in figures 6.18(a,b) and 6.19(a,b), and the time evolution in figure 6.20, show that without spatial coupling, condensate density increases over

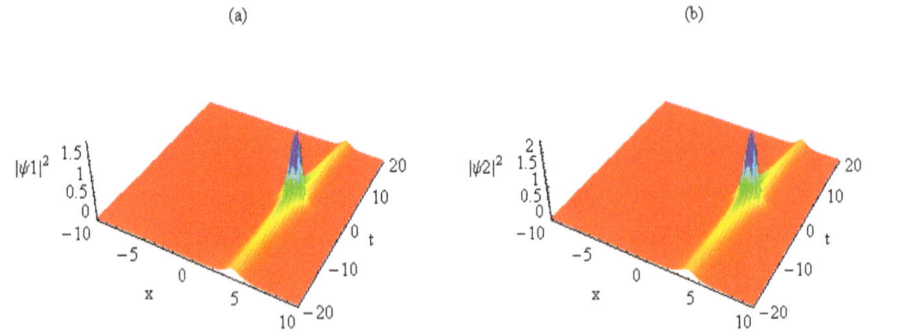

Figure 6.18. Density profiles of the condensates with parameters $a(t) = 0.5t$, $\varepsilon_1^{(1)} = 0.3$, and $\Gamma(t) = \int 0.005t \, \mathrm{d}t$, in the absence of coupling. Reprinted with permission from [17].

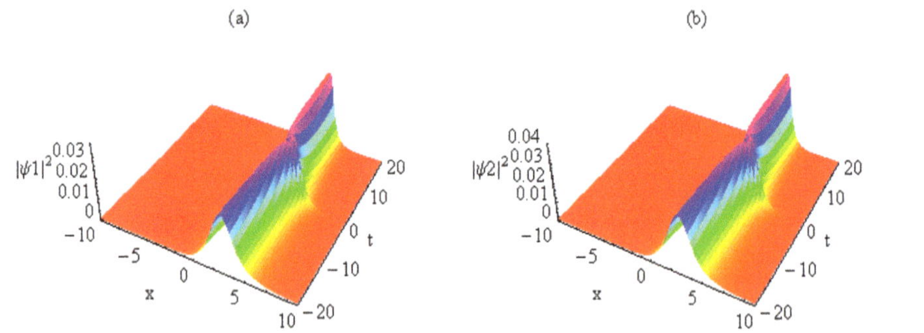

Figure 6.19. Effect of weak linear coupling $\nu(x) = 0.5x$ on condensate density, with other parameters as in figure 6.18. Reprinted with permission from [17].

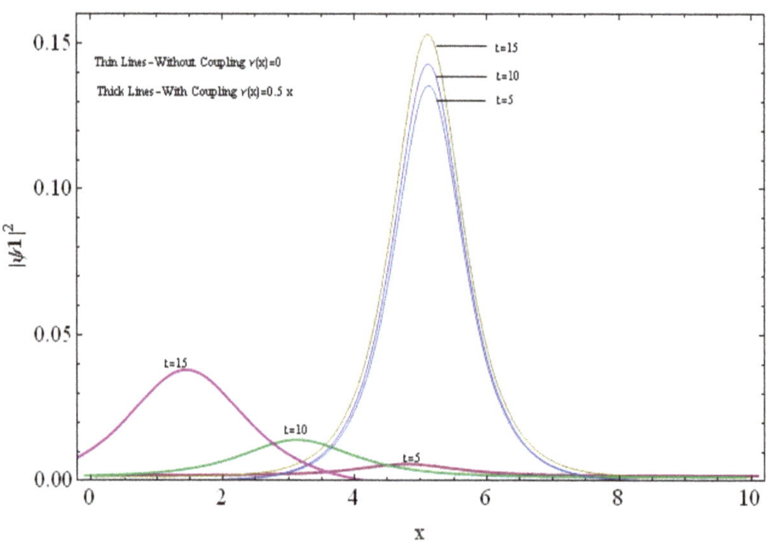

Figure 6.20. Time evolution of bright solitons (equation (6.19)) with and without linear coupling in a confining trap, for $\Gamma(t) = \int 0.005t \, \mathrm{d}t$. Reprinted with permission from [17].

time while remaining localized. Introducing linear coupling stretches the wave packet and shifts its localization centre, as depicted in figure 6.20.

Dark soliton solutions, derived from a nonzero seed, assume the following form:

$$\psi_1^{(1)} = \frac{2}{\sqrt{a(t)}}\varepsilon_1^{(1)}\beta_1(t)\tanh(\theta_1)e^{i(-\xi_1 + \Gamma(x,\, t)\int x\, dx)}, \tag{6.23}$$

$$\psi_2^{(1)} = \frac{2}{\sqrt{a(t)}}\varepsilon_2^{(1)}\beta_1(t)\tanh(\theta_1)e^{i(-\xi_1 + \Gamma(x,\, t)\int x\, dx)}. \tag{6.24}$$

The time evolution of dark solitons, shown in figure 6.21, exhibits similar effects of linear coupling, including wave packet stretching and localization centre shifts.

Spatially dependent linear coupling does not increase condensate density, unlike time-dependent coupling [8], but instead stretches the wave packet, facilitating stabilization of vectorial BECs. The gauge transformation approach can be extended to generate multiple dark solitons.

The shift in the matter-wave packet's centre due to spatial coupling can be leveraged to produce matter-wave interference patterns, as explored theoretically [18]. The density profiles for bright–bright soliton interactions in figures 6.22(a,b) show periodic peaks in matter-wave density, with a consistently bright central fringe indicating constructive interference and maximum intensity, accompanied by alternating bright and dark fringes. Contour plots in figure 6.23 confirm this pattern.

For dark–dark soliton interactions, contour plots in figure 6.24 reveal a dark central fringe, indicating destructive interference, with alternating dark and bright fringes.

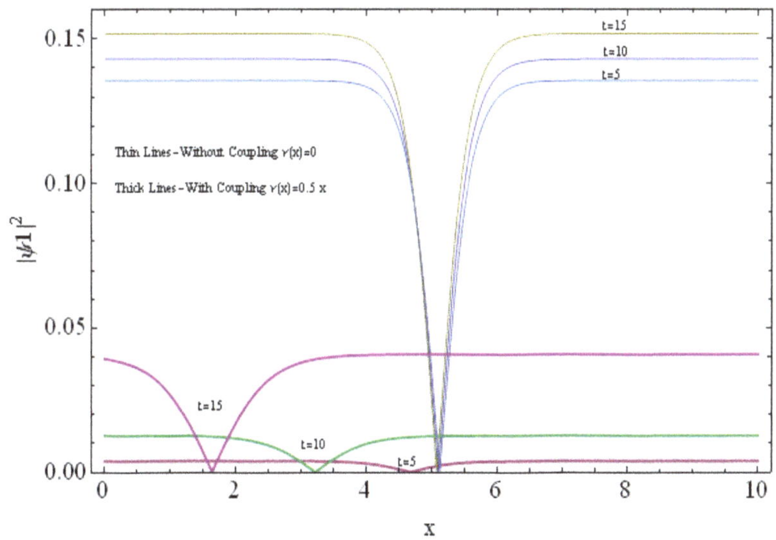

Figure 6.21. Time evolution of a dark soliton (equation (6.23)) with and without spatially dependent linear coupling in a confining trap, for $\Gamma(t) = \int 0.005t\ dt$. Reprinted with permission from [17].

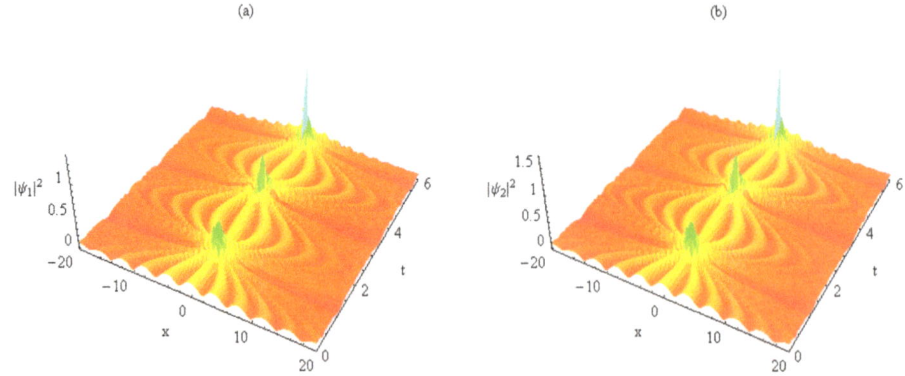

Figure 6.22. Density profiles of bright–bright soliton interactions in an expulsive trap, with parameters $a(t) = -\cos(\sqrt{2}\,t)$, $\Gamma(t) = \int 0.001t\ \mathrm{d}t$, $\nu(x) = [\log(-\cos(2x))]^2$, $\varepsilon_1^{(1)} = 0.89\mathrm{i}$, $\varepsilon_1^{(2)} = 0.6$. Reprinted with permission from [17].

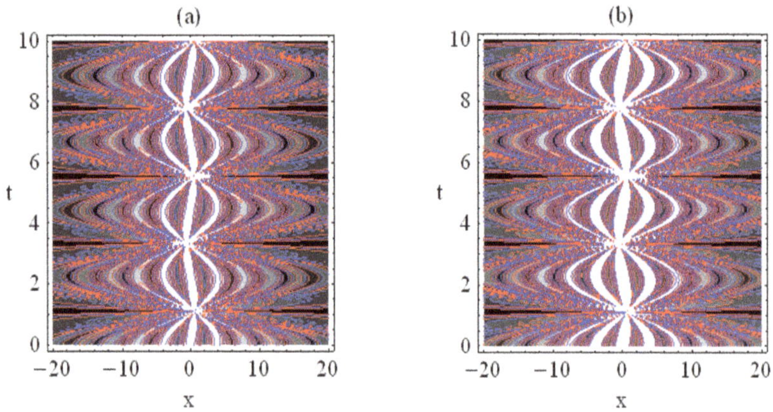

Figure 6.23. Contour plots of the bright–bright soliton interaction shown in figure 6.22. Reprinted with permission from [17].

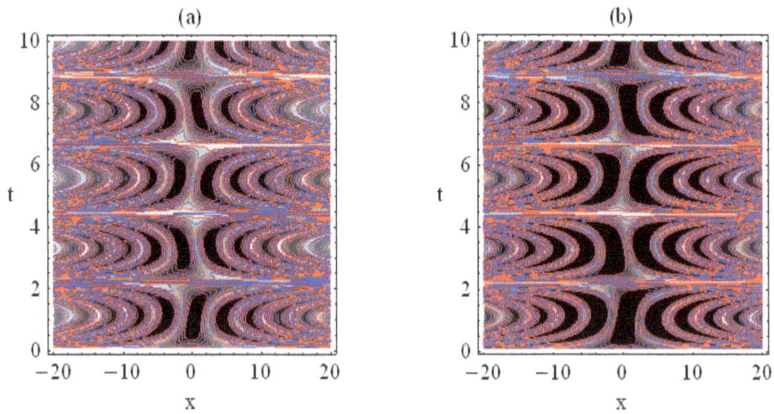

Figure 6.24. Interference patterns from dark–dark soliton interactions. Reprinted with permission from [17].

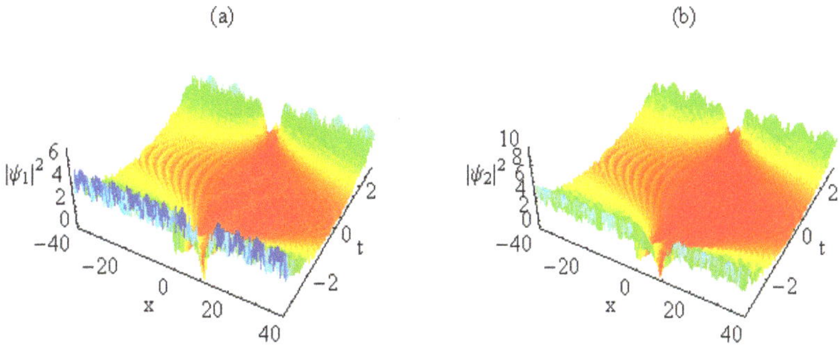

Figure 6.25. Density profiles of dark–bright soliton interactions in a transient trap, with parameters $\nu(x) = [\tan(0.03x)]^2 + 0.0075$, $a(t) = 0.001t$, $\Gamma(t) = -\int 0.05te^{-0.005t}\, dt$, $\varepsilon_1^{(1)} = 0.89i$, $\varepsilon_1^{(2)} = 0.6$. Reprinted with permission from [17].

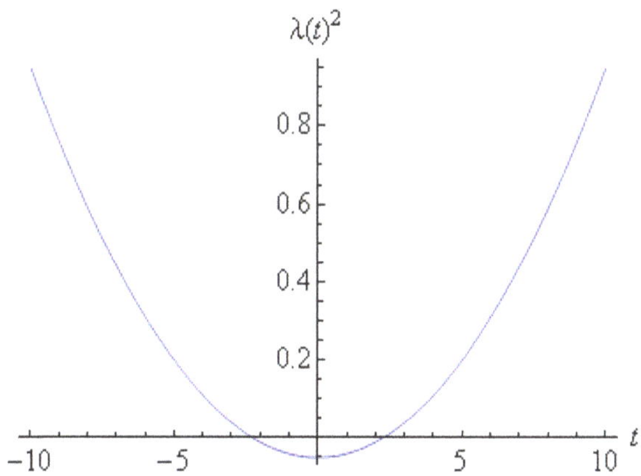

Figure 6.26. Transient trap profile for $\Gamma(t) = -\int 0.05te^{-0.005t}\, dt$. Reprinted with permission from [17].

In a transient trap, as shown in figure 6.26, the density profile of dark–bright soliton interactions in figure 6.25 indicates interference between dark and bright solitons in the confining region. As the trap reverts to the expulsive region, the interference pattern vanishes, and condensate density compresses, consistent with prior observations [19]. The interference pattern in figure 6.27 features a bright central fringe, reflecting the dominance of bright soliton amplitude, yielding maximum intensity. The fringe spacing in figures 6.23, 6.24, and 6.27 gives a measure of matter-wave coherence, a hallmark of the atomic system. The fringe spacing in the interference pattern shown in figure 6.27 narrows down due to the brief dwell time in the confining region.

Beyond bright and dark solitons, vector BECs support other localized excitations, such as algebraic rogue waves, whose dynamics warrant further exploration to fully understand the behaviour of vector BECs.

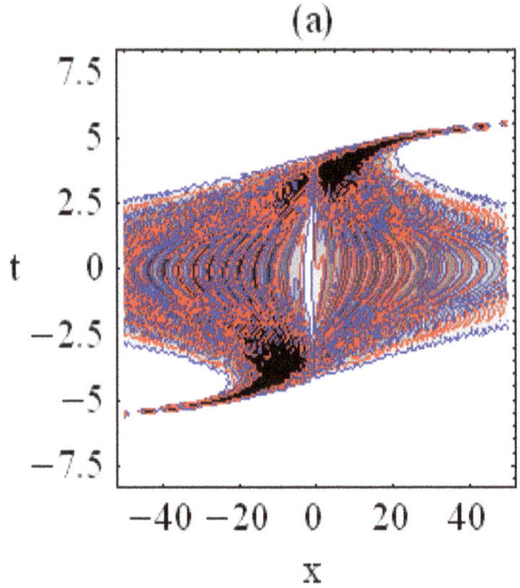

Figure 6.27. Contour plot of the dark–bright soliton interaction shown in figure 6.25. Reprinted with permission from [17].

6.5 Taming of rogue waves in vector BECs

Rogue waves are nonlinear oceanic waves with exceptionally large amplitudes, significantly exceeding surrounding wave crests, and are localized in both space and time. Like solitary waves, rogue waves, sometimes termed 'rogons', can re-emerge nearly unchanged in size or shape post-interaction [20]. These waves have caused significant maritime disasters [21, 22]. Unlike tsunamis, which are predictable hours or days in advance [23], rogue waves appear suddenly and vanish without trace, posing unique challenges [24].

Although their existence is confirmed by observations, their formation mechanisms remain incompletely understood. Recent studies suggest modulation instability as a primary cause [25, 26], with rogue wave phenomena reported in optics [27], plasma [28], and BECs [29]. Various nonlinear partial differential equations from physics, notably the nonlinear Schrödinger (NLS) equation, support rogue wave solutions. Recent work has proposed mechanisms for multi-rogue wave generation using the NLS equation [30–32], attributing their formation to collisions of multiple Akhmediev breathers arising from modulation instability [32]. Discrete integrable systems, such as the generalized Ablowitz–Ladik–Hirota lattice with variable coefficients, also support nonautonomous discrete rogue wave solutions [33]. The short lifespan of rogue waves complicates systematic study. Understanding their structure aids in elucidating their dynamics and controlling their amplitude and lifespan for applications in nonlinear optics and BECs.

Theoretical studies predict rogue wave phenomena in integrable multicomponent systems, including the Manakov model [34] and spinor $F = 1$ condensates [35], confirming novel bright–dark-rogue wave solutions. Mechanisms for rogue wave formation in two-dimensional coupled NLS equations, describing nonlinearly interacting waves in deep water, have been proposed [36, 37]. Numerical studies of two-component BECs with variable scattering lengths indicate that rogue wave solutions, generated through phase or density engineering, exist only for specific combinations of nonlinear coefficients [38]. These findings motivate efforts to identify simple mechanisms for generating and controlling rogue waves in vector BECs. While rogue waves have been manipulated in nonlinear optics [32, 39] and BECs [40], their dynamics in vector BECs governed by symmetric coupled GP equations remain underexplored. This section generates rogue waves for vector BECs using the coupled GP equation and controls their evolution by adjusting the scattering length via FR or the trapping frequency, a novel phenomenon in BECs.

6.5.1 Theoretical model and Lax pair

For a two-component BEC prepared in two hyperfine states, the dynamics at low temperatures are described by the coupled GP equations for cigar-shaped BECs [13]

$$\mathrm{i}\frac{\partial\psi_1}{\partial t}+\frac{\partial^2\psi_1}{\partial x^2}+2a(t)(|\psi_1|^2+|\psi_2|^2)\psi_1+\lambda(t)^2x^2\psi_1+\mathrm{i}G(t)\psi_1=0, \tag{6.25a}$$

$$\mathrm{i}\frac{\partial\psi_2}{\partial t}+\frac{\partial^2\psi_2}{\partial x^2}+2a(t)(|\psi_1|^2+|\psi_2|^2)\psi_2+\lambda(t)^2x^2\psi_2+\mathrm{i}G(t)\psi_2=0, \tag{6.25b}$$

where $\psi_j(j=1, 2)$ are the condensate order parameters, $a(t)$ is the time-dependent scattering length, $\lambda(t)$ is the trap frequency, and $G(t)$ accounts for atom gain or loss from the thermal cloud. These equations have previously been studied, and the dynamics of bright soliton collisions has been analyzed [3, 5]. Equations (6.25) admit the Lax pair

$$\frac{\partial\Phi}{\partial x}=\mathcal{U}\Phi, \tag{6.26a}$$

$$\frac{\partial\Phi}{\partial t}=\mathcal{V}\Phi, \tag{6.26b}$$

with matrices

$$\mathcal{U}=\begin{pmatrix}-i\zeta(t) & q_1 & q_2\\ -q_1^* & i\zeta(t) & 0\\ -q_2^* & 0 & i\zeta(t)\end{pmatrix}, \tag{6.27}$$

and

$$\mathcal{V}=\begin{pmatrix}\mathcal{V}_{11} & \mathcal{V}_{12} & \mathcal{V}_{13}\\ \mathcal{V}_{21} & \mathcal{V}_{22} & \mathcal{V}_{23}\\ \mathcal{V}_{31} & \mathcal{V}_{32} & \mathcal{V}_{33}\end{pmatrix}, \tag{6.28}$$

where

$$
\begin{aligned}
\mathcal{V}_{11} &= -2\mathrm{i}\zeta(t)^2 + 2\zeta(t)\mathrm{i}\Gamma(t)x + \mathrm{i}(|q_1|^2 + |q_2|^2),\\
\mathcal{V}_{12} &= 2\zeta(t)q_1 + \mathrm{i}\left(\frac{\partial q_1}{\partial x} + 2\mathrm{i}\Gamma(t)xq_1\right),\\
\mathcal{V}_{13} &= 2\zeta(t)q_2 + \mathrm{i}\left(\frac{\partial q_2}{\partial x} + 2\mathrm{i}\Gamma(t)xq_2\right),\\
\mathcal{V}_{21} &= -2\zeta(t)q_1^* + \mathrm{i}\left(\frac{\partial q_1^*}{\partial x} - 2\mathrm{i}\Gamma(t)xq_1^*\right),\\
\mathcal{V}_{22} &= 2\mathrm{i}\zeta(t)^2 - 2\zeta(t)\mathrm{i}\Gamma(t)x - \mathrm{i}|q_1|^2,\\
\mathcal{V}_{23} &= -\mathrm{i}q_2q_1^*,\\
\mathcal{V}_{31} &= -2\zeta(t)q_2^* + \mathrm{i}\left(\frac{\partial q_2^*}{\partial x} - 2\mathrm{i}\Gamma(t)xq_2^*\right),\\
\mathcal{V}_{32} &= -\mathrm{i}q_2^*q_1,\\
\mathcal{V}_{33} &= 2\mathrm{i}\zeta(t)^2 - 2\mathrm{i}\zeta(t)\Gamma(t)x - \mathrm{i}|q_2|^2,
\end{aligned}
$$

and

$$q_1(x, t) = \sqrt{a(t)}\,\mathrm{e}^{-\mathrm{i}\Gamma(t)x^2/2}\psi_1(x, t), \tag{6.29}$$

$$q_2(x, t) = \sqrt{a(t)}\,\mathrm{e}^{-\mathrm{i}\Gamma(t)x^2/2}\psi_2(x, t). \tag{6.30}$$

The nonisospectral parameter $\zeta(t)$ satisfies:

$$\zeta(t) = \mu \mathrm{e}^{-2\int \Gamma(t)\,\mathrm{d}t}, \tag{6.31}$$

where μ is a complex constant and $\Gamma(t)$ is an arbitrary time-dependent function. The compatibility condition $\frac{\partial \mathcal{U}}{\partial t} - \frac{\partial \mathcal{V}}{\partial x} + [\mathcal{U}, \mathcal{V}] = 0$ yields equations (6.25) with constraints

$$G(t) = \Gamma(t) + \frac{1}{2}\frac{a'(t)}{a(t)}, \tag{6.32}$$

$$\lambda(t)^2 = \Gamma(t)^2 + \frac{1}{2}\Gamma'(t), \tag{6.33}$$

subject to the integrability condition

$$\lambda(t)^2 = G(t)^2 + \frac{1}{2}\frac{a'(t)^2}{a(t)^2} - G(t)\frac{a'(t)}{a(t)} + \frac{1}{2}G'(t) - \frac{1}{4}\frac{a''(t)}{a(t)}. \tag{6.34}$$

To solve equations (6.25), one introduces the transformation

$$\psi_1(x, t) = \Lambda(x, t)U(X, T), \tag{6.35}$$

$$\psi_2(x, t) = \Lambda(x, t)V(X, T), \tag{6.36}$$

with coordinates

$$X = \sqrt{2}\, r_0 a(t)x - 2\sqrt{2}\, b r_0^3 \int a(t)^2 \, \mathrm{d}t, \tag{6.37}$$

$$T = r_0^2 \int a(t)^2 \, \mathrm{d}t, \tag{6.38}$$

and

$$\Lambda(x, t) = \sqrt{2r_0^2 a(t)} \exp\left[\mathrm{i}\left(-\frac{a'(t)}{2a(t)}x^2 + 2br_0^2 a(t)x - 2b^2 r_0^4 \int a(t)^2 \, \mathrm{d}t \right) \right], \tag{6.39}$$

where r_0 and b are arbitrary constants, reducing equations (6.25) to the Manakov model.

6.5.2 Construction of rogue waves

To construct rogue waves, one uses the nonzero plane wave seed solution

$$\psi_1[0] = c_1 \exp(\mathrm{i}\theta_1), \quad \psi_2[0] = c_2 \exp(\mathrm{i}\theta_2), \tag{6.40}$$

where

$$\theta_1 = g_1 x + \left(2c_1^2 + 2c_2^2 - g_1^2\right)t, \tag{6.41a}$$

$$\theta_2 = g_2 x + \left(2c_1^2 + 2c_2^2 - g_2^2\right)t. \tag{6.41b}$$

Substituting this seed into the Lax pair equations (6.26) yields

$$\frac{\partial \Phi_1}{\partial x} = \left(MUM^{-1} + \frac{\partial M}{\partial x} M^{-1} \right)\Phi_1 = \hat{\mathcal{U}}\Phi_1,$$
$$\frac{\partial \Phi_1}{\partial t} = \left(MVM^{-1} + \frac{\partial M}{\partial t} M^{-1} \right)\Phi_1 = \hat{\mathcal{V}}\Phi_1,$$

where $\Phi_1 = M\Phi$, and:

$$M = \mathrm{diag}\left[\exp\left(-\frac{\mathrm{i}}{3}(\theta_1 + \theta_2) \right), \exp\left(\frac{\mathrm{i}}{3}(2\theta_1 - \theta_2) \right), \exp\left(\frac{\mathrm{i}}{3}(\theta_1 + \theta_2) \right) \right], \tag{6.42}$$

with

$$\hat{\mathcal{U}} = \begin{pmatrix} \chi_{11} & c_1 & c_2 \\ -c_1 & \chi_{22} & 0 \\ -c_2 & 0 & \chi_{33} \end{pmatrix}, \tag{6.43}$$

$$\hat{\mathcal{V}} = \mathrm{i}\hat{\mathcal{U}}^2 - \left[\frac{2}{3}(g_1 + g_2) - 2\zeta_1\right]\hat{\mathcal{U}} + mI, \tag{6.44}$$

where

$$\begin{aligned}
\chi_{11} &= -2\mathrm{i}\zeta_1 - \frac{\mathrm{i}}{3}(g_1 + g_2),\\
\chi_{22} &= \mathrm{i}\zeta_1 - \frac{\mathrm{i}}{3}(2g_1 - g_2),\\
\chi_{33} &= \mathrm{i}\zeta_1 + \frac{\mathrm{i}}{3}(2g_2 - g_1),\\
m &= 2\mathrm{i}\zeta_1^2 + \frac{2\mathrm{i}}{3}\left(c_1^2 + c_2^2 + \frac{2\mathrm{i}}{9}(g_1^2 - g_1g_2 + g_2^2) + \frac{2\mathrm{i}\zeta_1}{3}(g_1 + g_2)\right).
\end{aligned}$$

For rational solutions, one chooses

$$\sigma = g_2 + 3\zeta_{1R}, \quad g_1 = g_2 - 2\sigma, \quad c_1 = c_2 = 2\sigma, \tag{6.45}$$

where g_2 and ζ_{1R} are real. The fundamental solution matrix for the Lax pair at $\zeta(t) = \zeta_1$ and $\psi_j = \psi_j[0]$ is $\Phi = M^{-1}\Theta$, where

$$\Theta = \begin{pmatrix} \phi_{11} & 4\sigma^2\nu + 2\sqrt{3}\sigma & 4\sigma \\ \phi_{21} & -2(\sqrt{3} - \mathrm{i})\nu - 2\sigma & -2\sigma^2(\sqrt{3} - \mathrm{i}) \\ \phi_{31} & [-2(\sqrt{3} - \mathrm{i})\nu - 2\sigma]^* & [-2\sigma^2(\sqrt{3} - \mathrm{i})]^* \end{pmatrix}, \tag{6.46}$$

with

$$\begin{aligned}
\phi_{11} &= 4\sigma^2(\nu + 2\mathrm{i}t) + 4\sqrt{3}\sigma\nu + 2,\\
\phi_{21} &= -2(\sqrt{3} - \mathrm{i})\sigma^2(\nu^2 + 2\mathrm{i}t) - 4\sigma\nu,\\
\phi_{31} &= -2(\sqrt{3} + \mathrm{i})\sigma^2(\nu^2 + 2\mathrm{i}t) - 4\sigma\nu,\\
\nu &= x + 2\sqrt{3}(\sigma - \mathrm{i}\sqrt{3}\zeta_{1R})\mathrm{i}t.
\end{aligned}$$

Using the gauge transformation approach [41], one obtains

$$\psi_1[1] = \psi_1[0] - 2\mathrm{i}(\zeta_1 - \zeta_1^*)\frac{\phi_1\phi_2^*}{|\phi_1|^2 + |\phi_2|^2 + |\phi_3|^2}, \tag{6.47a}$$

$$\psi_2[1] = \psi_2[0] - 2\mathrm{i}(\zeta_1 - \zeta_1^*)\frac{\phi_1\phi_3^*}{|\phi_1|^2 + |\phi_2|^2 + |\phi_3|^2}. \tag{6.47b}$$

The first-order rogue wave solution can be written as [42]

$$\psi_1 = \sqrt{\frac{2}{a(t)}}\varepsilon_1^{(1)}\beta(t)[-1 - \mathrm{i}\sqrt{3} + a]\exp\left[\mathrm{i}\theta_1 - \xi_1 + \frac{\Gamma(t)x^2}{2}\right], \tag{6.48a}$$

$$\psi_2 = \sqrt{\frac{2}{a(t)}}\varepsilon_1^{(2)}\beta(t)[-1 - \mathrm{i}\sqrt{3} + a]\exp\left[\mathrm{i}\theta_2 - \xi_1 + \frac{\Gamma(t)x^2}{2}\right], \tag{6.48b}$$

where $a = f_1/f_2$, and

$$\begin{aligned} f_1 &= -6\delta\sigma\sqrt{3} - 36t\sigma^2\sqrt{3} - 3 + \mathrm{i}(36t\sigma^2 + 6\delta\sigma + 5\sqrt{3}), \\ f_2 &= 12\sigma^2\delta^2 + 8\delta\sigma\sqrt{3} + 144t^2\sigma^4 + 5, \\ \delta &= x + 6\zeta_{1R}t. \end{aligned}$$

The gauge transformation approach [5] can be extended to generate multi-rogue wave solutions. The second-order rogue wave solution is written as

$$\psi_1 = \sqrt{\frac{2}{a(t)}}\varepsilon_1^{(1)}\beta(t)[-1 - \mathrm{i}\sqrt{3} + a_1]\exp\left[\mathrm{i}\theta_1 - \xi_1 + \frac{\Gamma(t)x^2}{2}\right], \tag{6.49a}$$

$$\psi_2 = \sqrt{\frac{2}{a(t)}}\varepsilon_1^{(2)}\beta(t)[-1 - \mathrm{i}\sqrt{3} + a_2]\exp\left[\mathrm{i}\theta_2 - \xi_1 + \frac{\Gamma(t)x^2}{2}\right], \tag{6.49b}$$

where

$$\begin{aligned} a_1 &= \frac{J_1 + \mathrm{i}K_1}{D}, \quad a_2 = \frac{J_2 + \mathrm{i}K_2}{D}, \\ J_1 &= -864\sqrt{3}\sigma^6t^3 - 144\sqrt{3}\sigma^5\delta t^2 - 72\sqrt{3}\sigma^4\delta^2 t - 216\sigma^4t^2 - 12\sqrt{3}\sigma^3\delta^3 - 144\sigma^3\delta t \\ &\quad - 18\sigma^2\delta^2 - 12\sqrt{3}\sigma^2 t + 3, \\ J_2 &= 864\sqrt{3}\sigma^6t^3 - 144\sqrt{3}\sigma^5\delta t^2 + 72\sqrt{3}\sigma^4\delta^2 t - 216\sigma^4t^2 - 12\sqrt{3}\sigma^3\delta^3 + 144\sigma^3\delta t \\ &\quad - 18\sigma^2\delta^2 + 12\sqrt{3}\sigma^2 t + 3, \\ K_1 &= 864\sigma^6t^3 + 144\sigma^5\delta t^2 + 72\sigma^4\delta^2 t + 312\sqrt{3}\sigma^4t^2 + 12\sigma^3\delta^3 + 96\sqrt{3}\sigma^3\delta t + 18\sqrt{3}\sigma^2\delta^2 \\ &\quad + 108\sigma^2 t + 12\sigma\delta + \sqrt{3}, \\ K_2 &= 864\sigma^6t^3 - 144\sigma^5\delta t^2 + 72\sigma^4\delta^2 t - 312\sqrt{3}\sigma^4t^2 - 12\sigma^3\delta^3 + 96\sqrt{3}\sigma^3\delta t - 18\sqrt{3}\sigma^2\delta^2 \\ &\quad + 108\sigma^2 t - 12\sigma\delta - \sqrt{3}, \\ D &= 1728\sigma^8t^4 + 384\sqrt{3}\sigma^5\delta t^2 + 12\sigma^4\delta^4 + 432\sigma^4t^2 + 16\sqrt{3}\sigma^3\delta^3 + 24\sigma^2\delta^2 + 4\sqrt{3}\sigma\delta + 1, \end{aligned}$$

and

$$a(t) = 2\sigma f(t), \quad \alpha(t) = \alpha_0\sigma\exp\left(-2\int\Gamma(t)\,\mathrm{d}t\right), \quad \beta(t) = \beta_0\sigma\exp\left(-2\int\Gamma(t)\,\mathrm{d}t\right). \tag{6.50}$$

6.5.3 Stabilization of rogue waves

Figure 6.28 displays the density profiles of first-order rogue waves with $a(t) = 0.0006$, $f(t) = 0.001$, showing high densities that suggest rapid collapse or disappearance. Since densities $|\psi_j|^2 (j = 1, 2)$ are inversely proportional to the scattering length $a(t)$, which scales with $f(t)$, increasing $a(t)$ through the FR stabilizes rogue waves by reducing density, as shown in figure 6.29 for $a(t) = 0.006$, $f(t) = 0.01$. Further stabilization occurs at $a(t) = 0.06$, $f(t) = 0.1$,

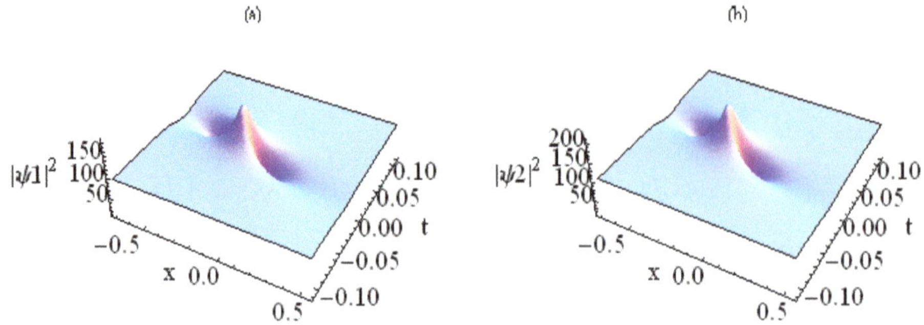

Figure 6.28. Density profiles of first-order rogue waves for $a(t) = 0.0006$, $g_2 = 0$, $e_1 = 0.1$, $\Gamma(t) = 0.01t$, $f(t) = 0.001$. Adapted with permission from [42], Copyright (2013) by the American Physical Society.

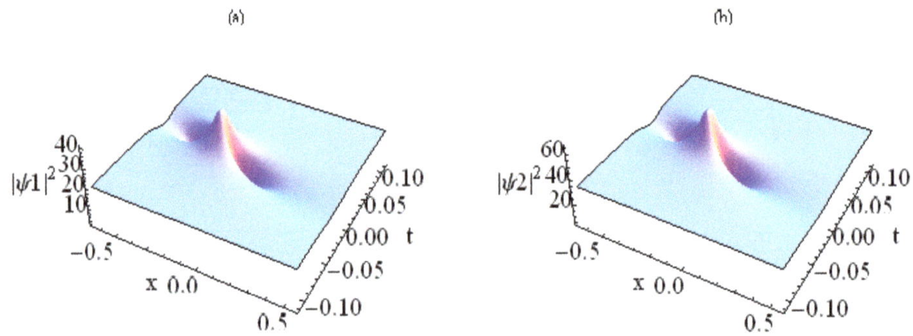

Figure 6.29. Stabilization of first-order rogue waves by adjusting the scattering length to $a(t) = 0.006$, $f(t) = 0.01$, with other parameters as in figure 6.28. Adapted with permission from [42], Copyright (2013) by the American Physical Society.

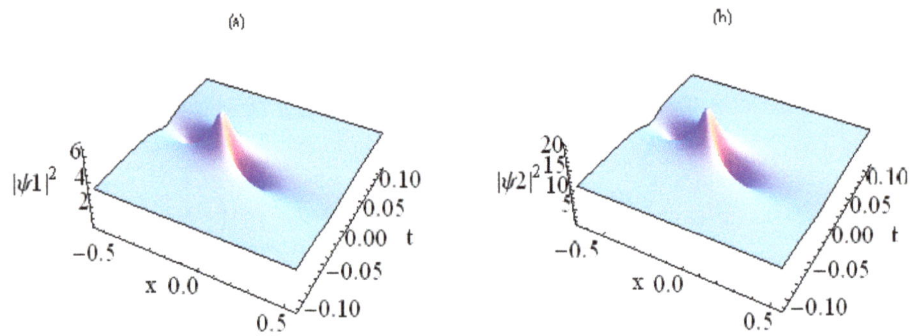

Figure 6.30. Further stabilization of first-order rogue waves for $a(t) = 0.06$, $f(t) = 0.1$, with other parameters as in figure 6.28. Adapted with permission from [42], Copyright (2013) by the American Physical Society.

as depicted in figure 6.30. This process, termed 'taming', extends the rogue wave lifespan.

Figure 6.31 shows second-order rogue wave density profiles for time-dependent scattering lengths $a(t) = 0.12t$, $f(t) = 0.05t$. Fine-tuning $a(t) = 0.168t$, $f(t) = 0.07t$

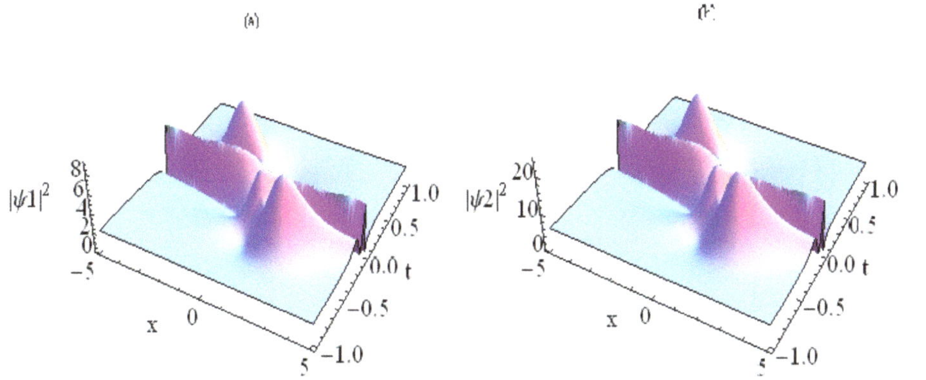

Figure 6.31. Density profiles of second-order rogue waves for $a(t) = 0.12t$, $f(t) = 0.05t$, $g_2 = 0.9$, $e_1 = 0.1$, $\Gamma(t) = 0.1t$. Adapted with permission from [42], Copyright (2013) by the American Physical Society.

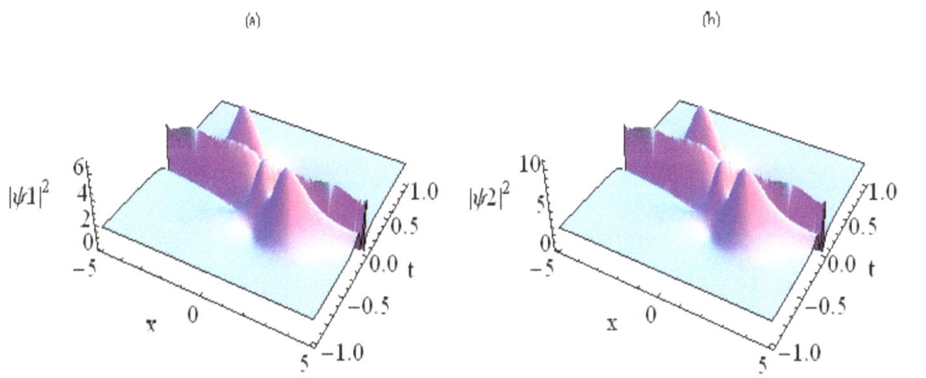

Figure 6.32. Stabilization of second-order rogue waves by adjusting the time-dependent scattering length to $a(t) = 0.168t$, $f(t) = 0.07t$, with other parameters as in figure 6.31. Adapted with permission from [42], Copyright (2013) by the American Physical Society.

(figure 6.32) and $a(t) = 0.36t$, $f(t) = 0.15t$ (figure 6.33) further reduces density, delaying collapse and extending lifespan compared to figures 6.28–6.30.

The trapping frequency $\lambda(t)$, related to $\Gamma(t)$ via equation (6.33), also influences stabilization. Figure 6.34 shows second-order rogue wave density profiles for periodically varying scattering lengths $a(t) = 2\cos(0.15t)$, $f(t) = \cos(0.15t)$, with corresponding contour plots in figure 6.35. Adjusting $\Gamma(t) = 0.1t$ enhances lifespan, as shown in the contour plot of figure 6.36. Further tuning $\Gamma(t) = 0.03t$ extends lifespan, as depicted in figure 6.37.

Thus, rogue waves can be stabilized by adjusting the scattering length (constant or time-dependent) via FR or the trapping frequency, extending their lifespan, a novel phenomenon with potential applications in BECs and nonlinear optics.

The two-component BECs discussed involve only temporal variations in intra- and inter-species scattering lengths. Recent interest in collisionally inhomogeneous BECs suggests exploring vector BECs with spatiotemporal interactions.

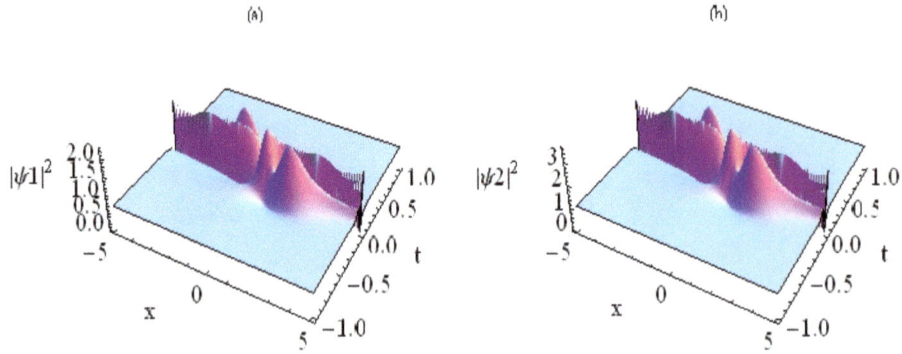

Figure 6.33. Further stabilization of second-order rogue waves by fine-tuning the time-dependent scattering length to $a(t) = 0.36t$, $f(t) = 0.15t$, with other parameters as in figure 6.31. Adapted with permission from [42], Copyright (2013) by the American Physical Society.

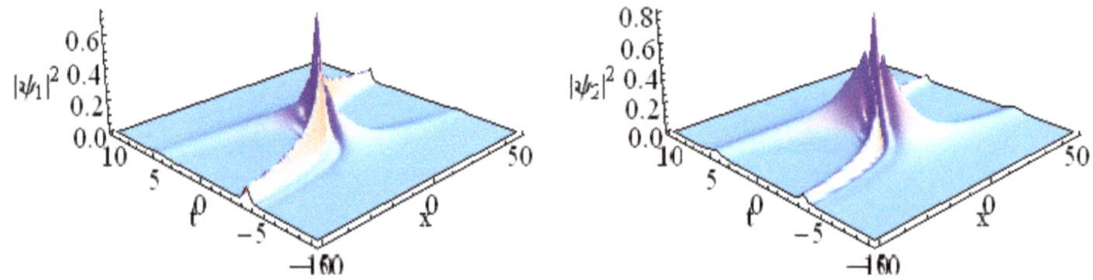

Figure 6.34. Density profiles of second-order rogue waves for $a(t) = 2\cos(0.15t)$, $g_2 = 0.5$, $e_1 = 0.5/3$, $\Gamma(t) = 0.15t$, $f(t) = \cos(0.15t)$. Adapted with permission from [42], Copyright (2013) by the American Physical Society.

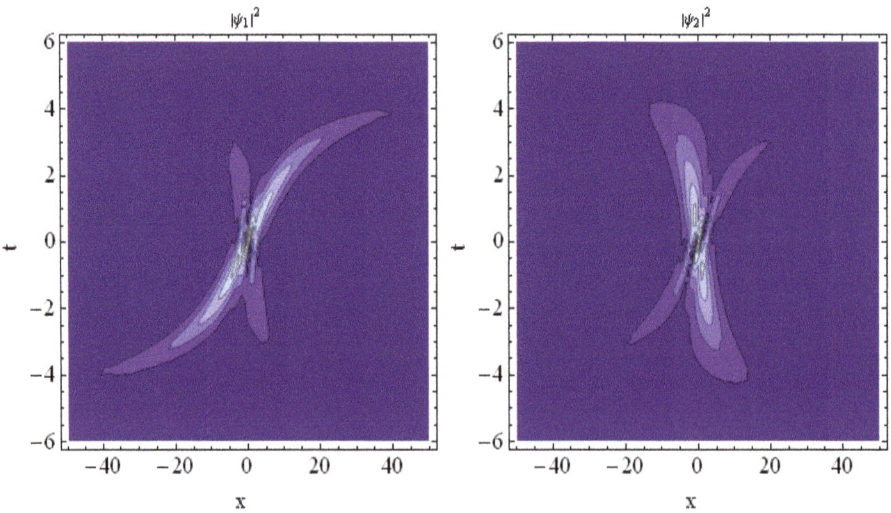

Figure 6.35. Contour plots of second-order rogue waves in figure 6.34. Adapted with permission from [42], Copyright (2013) by the American Physical Society.

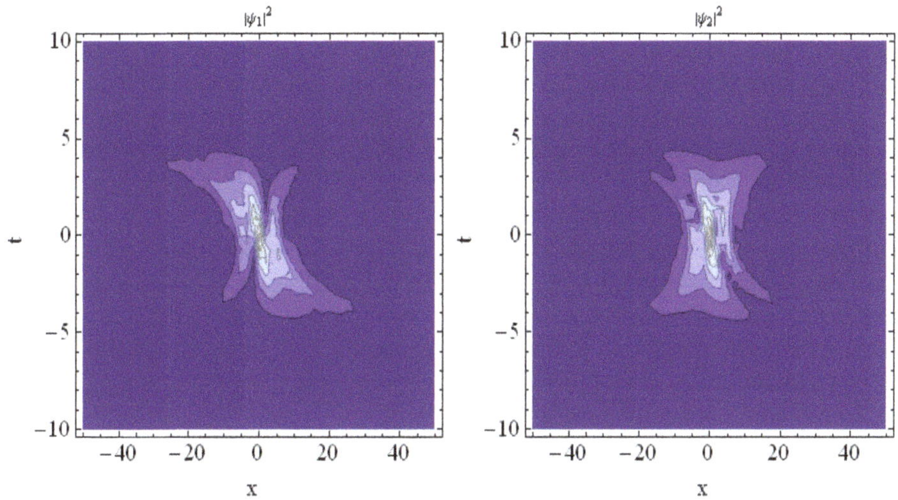

Figure 6.36. Time evolution of second-order rogue waves with extended lifespan by adjusting $\Gamma(t) = 0.1t$, with other parameters as in figure 6.34. Adapted with permission from [42], Copyright (2013) by the American Physical Society.

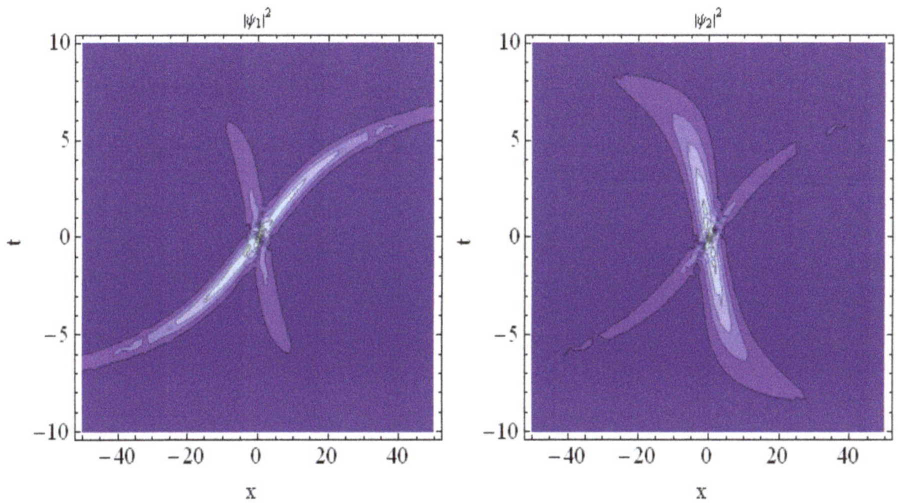

Figure 6.37. Second-order rogue wave profiles with extended lifespan by fine-tuning $\Gamma(t) = 0.03t$, with other parameters as in figure 6.34. Adapted with permission from [42], Copyright (2013) by the American Physical Society.

6.6 Summary and future challenges

This chapter is devoted to the investigation of the dynamics of two-component (vectorial) BECs, described by coupled GP equations under various conditions. It was shown that vector BECs in transient harmonic traps are found to be long-lived compared to their counterparts in the time-independent traps. The dynamics of both temporally and spatially coupled vector BECs were then analyzed and the nature of bright solitons admitted by them was brought out. The stability of the bright solitons admitted under these circumstances was also addressed. It was then demonstrated that rogue waves, which are inherently unstable in nature can be tamed by manipulating the scattering lengths through FR.

It should be pointed out at this juncture that there are several challenges which need to be overcome to penetrate deep into the dynamics of vector BECs. The identification of an integrable model for vectorial BECs with asymmetric interaction strengths and the associated nonlinear excitations/collisional dynamics is still open and promises to unravel several unexplored phenomena. The identification of such an integrable model may have wider ramifications in spin–orbit coupled BECs. The question of extending the stabilization of rogue waves to diverse trapping geometries is still open and challenging even today.

References

[1] Liu X, Pu H, Xiong B, Liu W M and Gong J 2009 Formation and transformation of vector solitons in two-species Bose-Einstein condensates with a tunable interaction *Phys. Rev.* A **79** 013423

[2] Zhang X F, Hu X H, Liu X X and Liu W M 2009 Vector solitons in two-component Bose-Einstein condensates with tunable interactions and harmonic potential *Phys. Rev.* A **79** 033630

[3] Rajendran P M S and Lakshmanan M 2009 Interaction of dark-bright solitons in two-component Bose-Einstein condensates *J. Phys. B: At. Mol. Opt. Phys.* **42** 145307

[4] Radha R and Vinayagam P S 2015 An analytical window into the world of ultracold atoms *Rom. Rep. Phys.* **67** 89–142

[5] Ramesh Kumar V, Radha R and Wadati M 2010 Collision of bright vector solitons in two-component Bose-Einstein condensates *Phys. Lett.* A **374** 3685–94

[6] Muruganandam P and Adhikari S 2009 Fortran programs for the time-dependent Gross-Pitaevskii equation in a fully anisotropic trap *Comput. Phys. Commun.* **180** 1888–912

[7] Radha R, Vinayagam P, Sudharsan J, Liu W M and Malomed B A 2015 Engineering bright solitons to enhance the stability of two-component Bose-Einstein condensates *Phys. Lett.* A **379** 2977–83

[8] Radha R and Vinayagam P 2012 Stabilization of matter wave solitons in weakly coupled atomic condensates *Phys. Lett.* A **376** 944–9

[9] Matthews M R, Anderson B P, Haljan P C, Hall D S, Holland M J, Williams J E, Wieman C E and Cornell E A 1999 Watching a superfluid untwist itself: recurrence of Rabi oscillations in a Bose-Einstein condensate *Phys. Rev. Lett.* **83** 3358–61

[10] Tratnik M V and Sipe J E 1988 Bound solitary waves in a birefringent optical fiber *Phys. Rev.* A **38** 2011–7

[11] Dror N, Malomed B A and Zeng J 2011 Domain walls and vortices in linearly coupled systems *Phys. Rev.* E **84** 046602

[12] Sakaguchi H and Malomed B A 2011 Symmetry breaking of solitons in two-component Gross-Pitaevskii equations *Phys. Rev.* E **83** 036608

[13] Pethick C and Smith H 2008 *Bose–Einstein Condensation in Dilute Gases* (Cambridge University Press) (includes bibliographical references and index)

[14] Chen P Y P and Malomed B A 2005 A model of a dual-core matter-wave soliton laser *J. Phys. B: At. Mol. Opt. Phys.* **38** 4221–34
[15] Manakov S V 1974 On the theory of two-dimensional stationary self-focusing of electromagnetic waves *Sov. Phys. -JETP* **38** 248–53
[16] Kaup D J and Malomed B A 1993 Soliton trapping and daughter waves in the Manakov model *Phys. Rev.* A **48** 599–604
[17] Radha R, Vinayagam P S and Porsezian K 2014 Soliton dynamics of spatially coupled vector becs *Rom. Rep. Phys.* **66** 427
[18] Adhikari S K and Malomed B A 2009 Two-component gap solitons with linear interconversion *Phys. Rev.* A **79** 015602
[19] Ramesh Kumar V, Radha R and Panigrahi P K 2009 Matter wave interference pattern in the collision of bright solitons *Phys. Lett.* A **373** 4381–5
[20] Yan Z 2010 Nonautonomous 'rogons' in the inhomogeneous nonlinear Schrödinger equation with variable coefficients *Phys. Lett.* A **374** 672–9
[21] Smith R 1976 Giant waves *J. Fluid Mech.* **77** 417–31
[22] Lavrenov I V 1998 The wave energy concentration at the Agulhas current off South Africa *Nat. Hazards* **17** 117–27
[23] Pelinovsky E N and Kharif C 2008 *Extreme Ocean Waves* (Springer) (includes bibliographical references and index)
[24] Akhmediev N, Ankiewicz A and Taki M 2009 Waves that appear from nowhere and disappear without a trace *Phys. Lett.* A **373** 675–8
[25] Peregrine D H 1983 Water waves, nonlinear Schrödinger equations and their solutions *J. Aust. Math. Soc. Ser. B Appl. math* **25** 16–43
[26] Benjamin T B and Feir J E 1967 The disintegration of wave trains on deep water Part 1. Theory *J. Fluid Mech.* **27** 417–30
[27] Solli D R, Ropers C, Koonath P and Jalali B 2007 Optical rogue waves *Nature* **450** 1054–7
[28] Moslem W M, Shukla P K and Eliasson B 2011 Surface plasma rogue waves *Europhys. Lett.* **96** 25002
[29] Bludov Y V, Konotop V V and Akhmediev N 2009 Matter rogue waves *Phys. Rev.* A **80** 033610
[30] Wang L H, Porsezian K and He J S 2013 Breather and rogue wave solutions of a generalized nonlinear Schrödinger equation *Phys. Rev.* E **87** 053202
[31] Guo B, Ling L and Liu Q P 2012 Nonlinear Schrödinger equation: generalized Darboux transformation and rogue wave solutions *Phys. Rev.* E **85** 026607
[32] Akhmediev N, Soto-Crespo J M and Ankiewicz A 2009 How to excite a rogue wave *Phys. Rev.* A **80** 043818
[33] Yan Z and Jiang D 2012 Nonautonomous discrete rogue wave solutions and interactions in an inhomogeneous lattice with varying coefficients *J. Math. Anal. Appl.* **395** 542–9
[34] Guo B L and Ling L M 2011 Rogue wave, breathers and bright-dark-rogue solutions for the coupled Schrödinger equations *Chin. Phys. Lett.* **28** 110202
[35] Qin Z and Mu G 2012 Matter rogue waves in an $F = 1$ spinor Bose-Einstein condensate *Phys. Rev.* E **86** 036601
[36] Onorato M, Osborne A R and Serio M 2006 Modulational instability in crossing sea states: a possible mechanism for the formation of freak waves *Phys. Rev. Lett.* **96** 014503
[37] Shukla P K, Kourakis I, Eliasson B, Marklund M and Stenflo L 2006 Instability and evolution of nonlinearly interacting water waves *Phys. Rev. Lett.* **97** 094501

[38] Bludov Y, Konotop V and Akhmediev N 2010 Vector rogue waves in binary mixtures of Bose-Einstein condensates *Eur. Phys. J. Spec. Top.* **185** 169–80

[39] Bludov Y V, Driben R, Konotop V V and Malomed B A 2013 Instabilities, solitons and rogue waves in pt-coupled nonlinear waveguides *J. Opt.* **15** 064010

[40] Wen L, Li L, Li Z D, Song S W, Zhang X F and Liu W M 2011 Matter rogue wave in Bose-Einstein condensates with attractive atomic interaction *Eur. Phys. J.* D **64** 473–8

[41] Chau L L, Shaw J C and Yen H C 1991 An alternative explicit construction of N-soliton solutions in 1.1 dimensions *J. Math. Phys.* **32** 1737–43

[42] Vinayagam P S, Radha R and Porsezian K 2013 Taming rogue waves in vector Bose-Einstein condensates *Phys. Rev.* E **88** 042906

IOP Publishing

An Introduction to Ultracold Atoms with Analytical and Numerical Methods

Paulsamy Muruganandam and Ramaswamy Radha

Chapter 7

Spin–orbit-coupled Bose–Einstein condensates

This chapter explores spin–orbit (SO)- and Rabi-coupled Bose–Einstein condensates, focusing on synthetic SO and Rabi coupling, including Rashba and Dresselhaus interactions. The theoretical framework is established through the coupled GP type equation with SO and Rabi coupling terms. Then, we focus on the investigation of the associated dynamics employing Darboux transformation based on a suitable correlation between the transient trap and interatomic interaction. Nonlinear excitations like rogue waves, breathers, bright/dark solitons, and mixed bound states, both with and without transient trapping potentials are generated analytically and the impact of SO and Rabi coupling studied. Stabilization mechanism for rogue waves in SO-Rabi-coupled BECs is also examined. The focus then shifts to the numerical investigation of the dynamics of SO-Rabi-coupled BECs for a general scenario including real-time and imaginary-time propagation. In particular, split-step Crank–Nicolson (SSCN), split-step Fourier transform (SSFT) and Newton conjugate gradient techniques, are exploited to study the stationary and dynamical properties of SO and Rabi-coupled BECs. The chapter concludes with a summary of findings, future research challenges, and proposed problems, providing a comprehensive overview of the rich physics endowed with SO-coupled quantum systems.

7.1 Synthetic spin–orbit and Rabi coupling in Bose–Einstein condensates

The spin is an intrinsic quantum property of particles such as electrons, protons, and neutrons. Unlike other properties, spin has no direct classical counterpart. However, the velocity or momentum of a quantum particle is closely related to its classical equivalent. When the spin of an electron interacts with its orbital angular momentum, an interesting phenomenon known as SO coupling (SOC) occurs.

doi:10.1088/978-0-7503-5447-9ch7

Quantum mechanics provides a precise explanation for this mechanism. The benefits of SO interaction are significant, enabling us to explore various applications in condensed matter physics, including spintronics, topological insulators, and the spin-Hall effect.

Consider an electron revolving around a nucleus with velocity $\mathbf{v}$ in an electric field $\mathbf{E} = -E_0\hat{z}$. This electron will experience an effective magnetic field $\mathbf{B}$ given by

$$\mathbf{B} = -\frac{1}{c^2}(\mathbf{v} \times \mathbf{E}), \tag{7.1}$$

The above magnetic field $\mathbf{B}$ interacts with the electron's spin and orbital angular momentum. This interaction is known as SOC. The Hamiltonian that describes this interaction is the Zeeman interaction Hamiltonian, represented by

$$H_{\mathrm{SO}} = \boldsymbol{\mu}_s \cdot \mathbf{B} \equiv \frac{g_s \mu_{\mathrm{B}} E_0}{2mc^2}(\mathbf{k} \times \boldsymbol{\sigma}) \cdot \hat{z}, \tag{7.2}$$

or simply represented as

$$H_{\mathrm{SO}} = \alpha_{\mathrm{SO}}(\mathbf{k} \times \boldsymbol{\sigma}) \cdot \hat{z}, \tag{7.3}$$

In the above, g_s is the Landé g-factor and μ_{B} is the Bohr magneton, and $\alpha_{\mathrm{SO}} = g_s \mu_B E_0/2mc^2$ is the SOC coefficient. The magnetic moment $\boldsymbol{\mu}_s$ is parallel to the spin $\boldsymbol{\sigma}$ (Pauli matrices). A magnetic field $\mathbf{B}$ contributes to a SO Hamiltonian as described above. The resulting momentum can produce three distinct types of SOC: Rashba, Dresselhaus, or a combination of both.

7.1.1 Rashba and Dresselhaus spin–orbit coupling

In materials, SOC directly links the momentum-dependent Zeeman energy. For example, the Lorentz-invariant Maxwell's equations establish that a static electric field, $\mathbf{E} = E_0\hat{z}$, in the laboratory frame (at rest) generates an SO magnetic field B_{SO} in a moving frame. An object with momentum $\hbar\mathbf{k} = \hbar(-k_y, k_x, 0)$ experiences this SO magnetic field given by

$$\mathbf{B}_{\mathrm{SO}} = \frac{\hbar E_0}{mc^2} \times (k_x\hat{y} - k_y\hat{x}). \tag{7.4}$$

One can derive spin-momentum-dependent Zeeman interaction from equations (7.3) and (7.4) as

$$-\boldsymbol{\mu}_s \cdot \mathbf{B}_{\mathrm{SO}} = \alpha_{\mathrm{R}} \times (k_y\sigma_x - k_x\sigma_y). \tag{7.5}$$

The above equation is well known as the Rashba SOC [1]. In all condensed matter systems, due to the motion of an electron on a crystal surface, there is a potential gradient that gives rise to effective SO interactions. These interactions typically originate from the lack of mirror symmetry in the two-dimensional system, leading to the Rashba SOC whose graphical representation is depicted in figure 7.1(a). On the other hand, in a bulk crystal, a lack of inversion symmetry results in another form of SOC, such as the linear Dresselhaus SOC [2]. Its schematic representation is

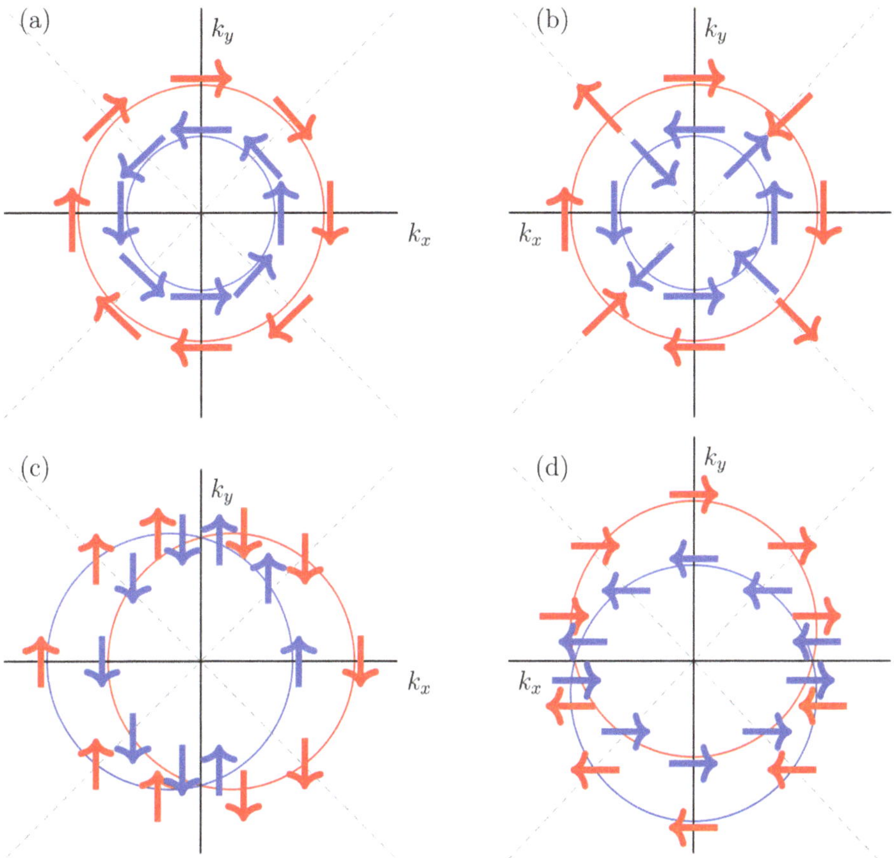

Figure 7.1. Spin orientation of the various SOC schemes with respect to 2D (a, b) and resultant 1D (c, d): (a) Rashba SOC $\alpha_R \times (k_y\sigma_x - k_x\sigma_y)$ depicts the lack of mirror symmetry; (b) Dresselhaus coupling $\alpha_D \times (-k_y\sigma_x - k_x\sigma_y)$ shows bulk inversion symmetry. And the equal combination of the Rashba and Dresselhaus couplings give two different unidirectional couplings as (c) $-\alpha_{RD}k_x\sigma_y$ and (d) $\alpha_{RD}k_y\sigma_x$. Reproduced with permission from [3] CC-BY-NC 4.0.

illustrated in figure 7.1(b), and the Dresselhaus interaction Hamiltonian can be described as follows:

$$-\boldsymbol{\mu}_s \cdot \mathbf{B}_{SO} = \alpha_D \times (-k_y\sigma_x - k_x\sigma_y), \tag{7.6}$$

The object in the moving frame has a momentum of $\hbar\mathbf{k} = \hbar(k_y, k_x, 0)$. In solids, the symmetric combination of the above-mentioned SOC gives rise to new SOC interactions [4, 5], as schematically illustrated in figures 7.1(c) and (d). These SOCs couple the atoms in a one-dimensional direction only. In the standard terminology, such coupling is referred to as a 'persistent spin-helix symmetry point'. The Dresselhaus interaction Hamiltonian can be described as follows

$$-\boldsymbol{\mu}_s \cdot \mathbf{B} = -\alpha_{RD} \times k_x\sigma_y, \tag{7.7a}$$

$$-\boldsymbol{\mu}_s \cdot \mathbf{B} = \alpha_{RD} \times k_y\sigma_x. \tag{7.7b}$$

Generally, there are two primary types of SOC: the Rashba effect, caused by inversion symmetry breaking due to an induced electric field, and the Dresselhaus SO splitting, arising from the inherent lack of inversion symmetry in materials. Both effects are fundamental to various physical phenomena, including spin-Hall effects, spintronics, topological insulators, quantum simulations, and more. Studying these phenomena in a cold environment can reveal richer physics. Additionally, dilute quantum gases, particularly multi-component BECs (spinor condensates), are ideal for manipulating and studying condensed matter phenomena. Before delving into the details of synthetic SO-coupled BEC experiments, one must review the basics of fine and hyperfine splitting.

From atomic physics, the SO interaction is the coupling between the total orbital angular momentum, $\mathbf{L} = \hbar\sqrt{l(l+1)}$, and the total spin angular momentum, $\mathbf{S} = \hbar\sqrt{s(s+1)}$, where l is the orbital quantum number and s is the electron spin quantum number. If $l = 0$ (ground state), there is no total angular momentum, resulting in no splitting. However, if $l = 1$, there is a splitting known as fine-structure splitting. The eigenstates, $j = 1/2$ and $3/2$, are calculated using the total angular momentum quantum number $j = |l \pm s|$. The respective energies of these eigenstates are calculated using the operator $\mathbf{J} = \mathbf{L} + \mathbf{S}$, where $\mathbf{J} = \sqrt{j(j+1)}$.

The SO interaction is a weaker force than the Coulomb interaction and is responsible for the fine structure of atomic lines. For example, in an atom with a ground state ($l = 0$) electronic manifold, the transition to an excited state ($l = 1$) electronic manifold can result in two distinct lines: the D_1 line (where $j = 1/2$ for both states) and the D_2 line (where $j = 1/2$ for the ground state and $j = 3/2$ for the excited state).

Furthermore, the fine structure can be further split due to the coupling between the nuclear spin quantum number (i) and the total angular momentum quantum number (j). Heavier nuclei, such as ^{87}Rb, ^{39}K, and ^{23}Na, have a nuclear spin quantum number of $i = 3/2$. In the lowest ground electronic state ($j = 1/2, l = 0$), the corresponding total angular momentum of the nucleus can be either $F = 1$ or $F = 2$, leading to a doubly degenerate hyperfine splitting. When a magnetic field is applied, this degeneracy is lifted through Zeeman splitting, resulting in $(2F + 1)$ distinct energy levels for each value of F. The eigenstates of the nuclear total angular momentum are given by $F = \hbar\sqrt{F(F+1)}$, and their energies can be calculated using the operator $F = J + I$, where $I = \hbar\sqrt{i(i+1)}$.

Spinor BECs of dilute, ultracold atomic gases serve as a powerful platform for investigating a wide array of quantum and condensed matter phenomena. These systems leverage the hyperfine electronic manifold of the atomic ground state, where atoms occupy distinct energy levels, or sublevels, determined by the hyperfine interaction between the nucleus and electrons. The number of occupied sublevels defines the spinor type: a binary system corresponds to two levels, a spin-1 system to three, a spin-2 system to five, and so forth. Spinor BECs, particularly those with SOC, have been extensively studied theoretically through coupled GP equations, which describe the dynamics of the multi-component wavefunction. Experimentally, spinor BECs are realized using internal atomic states, such as Zeeman levels within the same hyperfine manifold of atoms like sodium, rubidium, or potassium, or mixtures, such as rubidium–potassium. For spin-1 BECs, comprehensive reviews provide detailed insights into their theoretical and experimental properties, highlighting their role in exploring magnetic and topological phases [6].

A key phenomenon in spinor BECs is SOC, which links the spin of an atom to its momentum and is ubiquitous in condensed matter systems, from single atoms (e.g., hydrogen) to bulk materials like semiconductors. In electronic systems, SOC underlies novel quantum phenomena, including topological insulators, topological superconductors, topological semimetals, and the anomalous Hall effect. However, naturally occurring SOC is challenging to control, limiting experimental flexibility. Synthetic SOC in BECs overcomes this limitation by offering highly tunable coupling strengths achieved through laser-induced techniques. For instance, the Spielman group at NIST demonstrated synthetic SOC in a two-component BEC of rubidium-87, using two hyperfine states to form a pseudo-spin-1/2 system [7]. These states, labelled as 'spin-up' and 'spin-down', were coupled using two Raman laser beams, which imparted momentum to the atoms, creating SOC. Unlike conventional BECs, where the zero-momentum state is the unique ground state, SOC BECs exhibit a degenerate ground state with multiple low-energy states, enabling the emergence of exotic quantum phases [7, 8].

The tunability of synthetic SOC in BECs provides a versatile platform for simulating the behaviour of charged particles in electromagnetic fields and exploring strong correlations in quantum many-body systems. This flexibility has led to the discovery of symmetry-broken ground-state phases, such as the half-quantum vortex and stripe phases, the latter exhibiting supersolid-like properties characterized by simultaneous superfluid and crystalline order [9]. The stripe phase, a linear combination of plane waves with opposite momenta, results in periodic density modulations, offering insights into supersolidity and spin-density waves. These findings have advanced the use of ultracold atoms as quantum simulators, capable of modelling complex condensed matter systems with unprecedented control [10, 11]. By enabling precise manipulation of SOC, spinor BECs not only deepen the understanding of fundamental quantum phenomena but also pave the way for applications in quantum technologies, such as topological quantum computing.

In experiments, researchers use a ^{87}Rb Bose–Einstein condensate with its electronic ground state in the $5S_{1/2}$ manifold. By applying a bias magnetic field along the y-direction, Zeeman splitting is induced, resulting in three hyperfine states: $|F=1, m_F=+1\rangle$, $|F=1, m_F=0\rangle$, and $|F=1, m_F=-1\rangle$. This process creates a spinor or multi-component BEC. The coupling between spinor components can be controlled using lasers, enabling the generation of synthetic SOC through Raman transitions.

The schematic representation of the Raman process is displayed in figure 7.2. In this method, two laser beams, oriented at a 90° angle and propagating along the $y \pm x$ directions, respectively, have a wavelength λ_L with a small frequency difference ω_L and $\omega_L + \Delta\omega_L$. The wave vector difference is given by $\boldsymbol{k}_0 = k_0\hat{x}$, where $\hat{x}$ is the unit vector along the x direction. The transition between the states, induced by

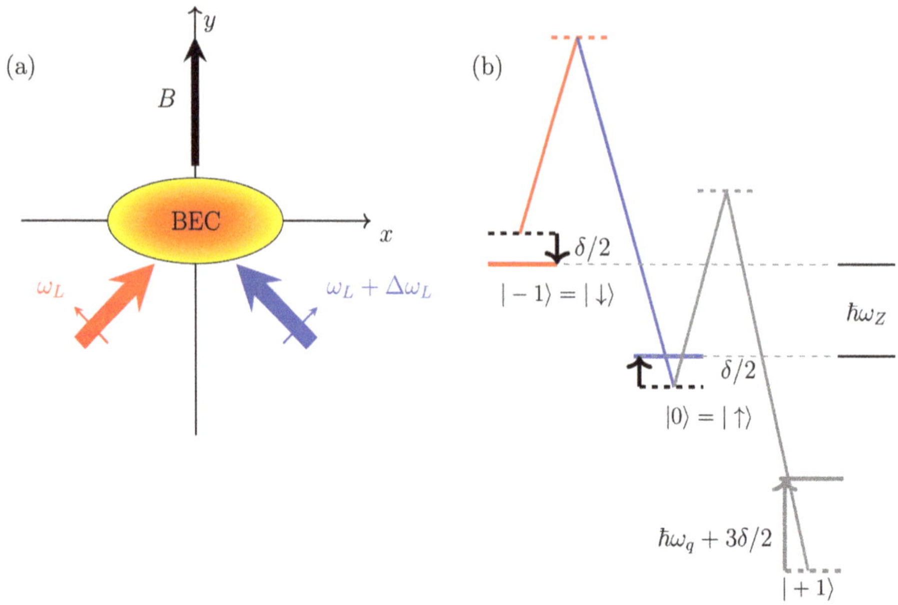

Figure 7.2. Level diagram: two-photon Raman transition. Two-photon Raman transition couples the states $|F = 1, m_F = -1\rangle$ and $|F = 1, m_F = 0\rangle$ of the $F = 1$ hyperfine manifold in ^{87}Rb. The energy difference between these states is given by the Zeeman splitting $\hbar\omega_Z$. The lasers have a frequency shift $\Delta\omega_L = \omega_Z + \delta$, where δ is a small detuning. Due to the large quadratic Zeeman shift ω_a, the state $|F = 1, m_F = +1\rangle$ is decoupled from $|F = 1, m_F = 0\rangle$ and can be omitted from the dynamics. Reproduced with permission from [3] CC-BY-NC 4.0.

the laser field, is characterized by the Rabi frequency Ω, which is determined by the laser's intensity.

Consider a spin-1 BEC with hyperfine states $|F = 1, m_F = -1\rangle$, $|F = 1, m_F = 0\rangle$, and $|F = 1, m_F = +1\rangle$. Two Raman lasers couple the states $|F = 1, m_F = -1\rangle$ and $|F = 1, m_F = 0\rangle$, with an energy difference of $\hbar\omega_Z$, which matches the frequency difference $\Delta\omega_L$ between the lasers. The energy splitting between $|F = 1, m_F = 0\rangle$ and $|F = 1, m_F = +1\rangle$ is $\hbar(\omega_Z - \omega_q)$, where ω_q accounts for the quadratic Zeeman effect. This splitting is significantly larger than $\hbar\omega_Z$ and includes an additional shift from the Raman resonance, making the $|F = 1, m_F = +1\rangle$ state a high-energy state that can be neglected. By excluding this state, the system reduces to an effective two-level system, with $|F = 1, m_F = 0\rangle$ labelled as pseudo-spin-up $|\uparrow\rangle$ and $|F = 1, m_F = -1\rangle$ as pseudo-spin-down $|\downarrow\rangle$. The resulting single-particle Hamiltonian for this pseudo-spin-1/2 system is given by

$$H_{sp} = \frac{p_x^2}{2m} + \frac{\hbar k_L}{m} p_x \sigma_z + \frac{\hbar\Omega}{2}\sigma_x + \frac{\delta}{2}\sigma_z, \tag{7.8}$$

where $p_x = -i\hbar\partial_x$ is the momentum operator along the x-axis, m is the atomic mass, k_L is the Raman laser recoil wavenumber, Ω is the Raman coupling strength, δ is the detuning, and σ_x, σ_z are the 2×2 Pauli matrices. This Hamiltonian, derived from a three-level system, captures the dynamics of SOC in BECs, with detailed derivations available in the literature [7, 12].

The term $\frac{\delta}{2}\sigma_z$ in equation (7.8) represents a detuning between the two pseudo-spin states. In many experimental and theoretical treatments, this term is neglected ($\delta \approx 0$) by fine-tuning the Raman lasers to resonance, ensuring that the two-photon transition between $|\uparrow\rangle$ and $|\downarrow\rangle$ is on-resonance. This simplifies the Hamiltonian while retaining the essential physics of SOC.

SO-coupled BECs confined in a strong transverse trapping potential can be described by the single-particle Hamiltonian

$$H_{\mathrm{sp}} = \frac{p_x^2}{2m} + \frac{\hbar k_{\mathrm{L}}}{m} p_x \sigma_z + \frac{\hbar \Omega}{2} \sigma_x + V(x), \tag{7.9}$$

where $V(x)$ represents a harmonic trapping potential, k_{L} is the recoil wavenumber which quantifies the momentum transfer from the Raman lasers coupling the two hyperfine states, and Ω governs the Rabi mixing between them. The derivation of this effective Hamiltonian for a spin-1 BEC is provided in appendix B. The dynamics and ground state properties of SO-coupled BECs are governed by this Hamiltonian. By solving the associated coupled GP equations, one can obtain the condensate wavefunction and analyze the system's dispersion relations. Depending on the strength of the Raman coupling and external parameters, the dispersion can exhibit a single minimum at zero momentum (zero-momentum phase), two minima at finite momentum (plane-wave phase), or a double-well structure leading to spontaneous breaking of translational symmetry (stripe phase). The presence of SOC fundamentally modifies the superfluid properties and ground state phases of the condensate, leading to the emergence of exotic phenomena such as supersolidity, topological phases, and skyrmion textures [13–16].

7.2 Model and Lax pair

Considering a SO-coupled quasi-one-dimensional BEC in a harmonic trap with longitudinal and transverse frequencies satisfying the relation $\omega_x \ll \omega_\perp$, it can be described at sufficiently low temperatures by the two-coupled GP equation of the following form

$$\mathrm{i}\partial_t \psi_1 = \left[-\partial_x^2 + V_{\mathrm{trap}}(x, t) - 2\gamma(t)(|\psi_1|^2 + |\psi_2|^2) + \mathrm{i}k_{\mathrm{L}}\partial_x\right]\psi_1 - \Omega\psi_2, \tag{7.10a}$$

$$\mathrm{i}\partial_t \psi_2 = \left[-\partial_x^2 + V_{\mathrm{trap}}(x, t) - 2\gamma(t)(|\psi_1|^2 + |\psi_2|^2) - \mathrm{i}k_{\mathrm{L}}\partial_x\right]\psi_2 - \Omega\psi_1. \tag{7.10b}$$

In the above equation, ψ_i, $i = 1{,}2$ are field variables, $\pm ik_{\mathrm{L}}\partial_x$ represents the momentum transfer between the laser beams and the atoms due to SOC, $\gamma(t) = \gamma_0 \exp \int \sigma(t)\mathrm{d}t$ represents binary interaction(scattering length), which depends on the choice of $\sigma(t)$ defined by equation (7.16) as $\frac{\mathrm{d}}{\mathrm{d}t}\log(\eta(t))$, where $\eta(t)$

is an arbitrary time-dependent parameter [17]. Ω denotes Rabi coupling and $V_{\text{trap}}(x, t) = (\lambda(t))^2 x^2/2$, where $\lambda(t) = \omega_x/\omega_\perp$. To construct soliton solutions, one begins with the celebrated Manakov model. It should be mentioned that the model equations (7.10a) and (7.10b) reduce to the celebrated Manakov model by eliminating SOC and Rabi coupling one by one in the following manner. By considering $V_{\text{trap}}(x, t)$=0 and employing the following transformation

$$\psi_1(x, t) = \hat{\psi}_1(x, t) \exp\left[\frac{\mathrm{i}}{2} k_{\mathrm{L}}(k_L t - 2x)\right] \tag{7.11a}$$

$$\psi_2(x, t) = \hat{\psi}_2(x, t) \exp\left[\frac{\mathrm{i}}{2} k_{\mathrm{L}}(k_L t + 2x)\right] \tag{7.11b}$$

one converts the two-coupled GP equation (7.10) into constant a coefficient NLS equation after elimination of SOC as

$$\mathrm{i}\partial_t \hat{\psi}_1 = [-\partial_x^2 - 2(|\hat{\psi}_1|^2 + |\hat{\psi}_2|^2)]\hat{\psi}_1 - \Omega\hat{\psi}_2 \tag{7.12a}$$

$$\mathrm{i}\partial_t \hat{\psi}_2 = [-\partial_x^2 - 2(|\hat{\psi}_1|^2 + |\hat{\psi}_2|^2)]\hat{\psi}_2 - \Omega\hat{\psi}_1 \tag{7.12b}$$

By invoking the following transformation,

$$\begin{pmatrix} \hat{\psi}_1(x, t) \\ \hat{\psi}_2(x, t) \end{pmatrix} = \begin{pmatrix} a\cos(\Omega t) & b\sin(\Omega t) \\ b\sin(\Omega t) & a\cos(\Omega t) \end{pmatrix} \cdot \begin{pmatrix} \tilde{\psi}_1(x, t) \\ \tilde{\psi}_2(x, t) \end{pmatrix} . \tag{7.13}$$

where the constants are $a = 1$ and $b = -\mathrm{i}$, one can remove Rabi coupling to convert equation (7.13) to the celebrated Manakov model (after dropping the bar)

$$\mathrm{i}\partial_t \psi_1 = \left[-\partial_x^2 - 2(|\psi_1|^2 + |\psi_2|^2)\right]\psi_1 \tag{7.14a}$$

$$\mathrm{i}\partial_t \psi_2 = \left[-\partial_x^2 - 2(|\psi_1|^2 + |\psi_2|^2)\right]\psi_2 \tag{7.14b}$$

The above Manakov model admits the following Lax pair [18, 19]

$$\mathbf{\Phi}_x = \mathbf{U\Phi} = \mathbf{U}_0\, \mathbf{\Phi} + \mathbf{U}_1\, \mathbf{\Phi}\, \mathbf{\Lambda}, \tag{7.15a}$$

$$\mathbf{\Phi}_t = \mathbf{V\Phi} = \mathbf{V}_0\, \mathbf{\Phi} + \mathbf{V}_1\, \mathbf{\Phi}\, \mathbf{\Lambda} + \mathbf{V}_2\, \mathbf{\Phi}\, \mathbf{\Lambda}^2, \tag{7.15b}$$

where,

$$\mathbf{U}_0 = \begin{pmatrix} 0 & \psi_1(x, t) & \psi_2(x, t) \\ -\psi_1^*(x, t) & 0 & 0 \\ -\psi_2^*(x, t) & 0 & 0 \end{pmatrix}, \quad \mathbf{U}_1 = \begin{pmatrix} 1 & 0 & 0 \\ 0 & -1 & 0 \\ 0 & 0 & -1 \end{pmatrix}, \quad \mathbf{\Lambda} = \begin{pmatrix} \zeta_1 & 0 & 0 \\ 0 & \zeta_2 & 0 \\ 0 & 0 & \zeta_3 \end{pmatrix},$$

$$\mathbf{V}_0 = \frac{\mathrm{i}}{2}\begin{pmatrix} \psi_1(x,t)\psi_1^*(x,t) + \psi_2(x,t)\psi_2^*(x,t) & \psi_{1x}(x,t) & \psi_{2x}(x,t) \\ \psi_{1x}^*(x,t) & -\psi_1(x,t)\psi_1^*(x,t) & -\psi_2(x,t)\psi_1^*(x,t) \\ \psi_{2x}^*(x,t) & -\psi_1(x,t)\psi_2^*(x,t) & -\psi_2(x,t)\psi_2^*(x,t) \end{pmatrix},$$

$$\mathbf{V}_1 = \begin{pmatrix} 0 & -\psi_1(x,t) & -\psi_2(x,t) \\ \psi_1^*(x,t) & 0 & 0 \\ \psi_2^*(x,t) & 0 & 0 \end{pmatrix}, \quad \mathbf{V}_2 = \mathrm{i}\begin{pmatrix} 1 & 0 & 0 \\ 0 & -1 & 0 \\ 0 & 0 & -1 \end{pmatrix},$$

where $\zeta_{1,2,3}$ are the spectral parameters. $\Phi = (\phi_1, \phi_2, \phi_3)^T$ is a three-component Jost function, **U** and **V**, known as the Lax pair, are functionals of the solutions of the model equations. The consistency condition $\mathbf{\Phi}_{xt} = \mathbf{\Phi}_{tx}$ leads to $\mathbf{U}_t - \mathbf{V}_x + [\mathbf{U}, \mathbf{V}] = \mathbf{0}$, which is equivalent to the coupled NLS equations (CNLSE) termed as Manakov model. Mapping of Manakov to the coupled GP equation can be done by employing similarity transformation given in [20]. While invoking the similarity transformation, the following constraints will have to be complied with for the successful mapping of Manakov to the coupled GP equation. They are of the following form

$$\frac{\mathrm{d}}{\mathrm{d}t}\eta(t) = \sigma(t)\eta(t) \tag{7.16}$$

with

$$\lambda(t)^2 = \frac{\mathrm{d}}{\mathrm{d}t}\sigma(t) - \sigma(t)^2 \tag{7.17}$$

and

$$\sigma(t) = \frac{\mathrm{d}}{\mathrm{d}t}\ln\gamma(t) \tag{7.18}$$

From equation (7.16), one can choose $\sigma(t)$ satisfying the condition $\sigma(t) = \frac{\mathrm{d}}{\mathrm{d}t}\log(\eta(t))$, where, $\eta(t)$ is an arbitrary time dependent parameter. Accordingly, one can fix the nature of the trapping potential $\lambda(t)$ given by (7.17) and scattering length $\gamma(t)$ by equation (7.18). Equation (7.17) represents the parabolic (harmonic) trapping potential $\lambda(t)^2$, related to the interaction strength $\gamma(t)$ of the coupled GP equation through the integrability condition given in [17, 21].

7.3 Darboux transformation

It is well known that the Darboux transformation for equation (7.10) is given by [22]

$$T = \zeta I - [\zeta_1^* - (\zeta_1^* - \zeta_1)P_1], \quad P_1 = \frac{\phi_1\phi_1^\dagger}{\phi_1^\dagger\phi_1} \tag{7.19}$$

where $\phi_1 = \phi(x, t, \zeta_1)(m_1, m_2, m_3)^{\mathrm{Tr}}$, m_1, m_2, m_3 are constants, P_1 is a projection matrix, I is 3×3 identity matrix and $\phi(x, t, \zeta_1)$ is the fundamental solution of the Lax pair equation at $\zeta = \zeta_1$, $q_i = q_i[0]$ $(i = 1{,}2)$ which leads to the following transformation between the fields

$$\psi_1[1] = \psi_1[0] + 2I \frac{(\zeta_1 - \zeta_1^*)\phi_1\phi_2^*}{|\phi_1|^2 + |\phi_2|^2 + |\phi_3|^2}, \tag{7.20a}$$

$$\psi_2[1] = \psi_2[0] + 2I \frac{(\zeta_1 - \zeta_1^*)\phi_1\phi_3^*}{|\phi_1|^2 + |\phi_2|^2 + |\phi_3|^2}. \tag{7.20b}$$

In the above equation, $\psi_i[j]$ represents the field variables where $j = 0, 1, 2$ indicates its zero-order, first-order and second-order iterations, respectively. In other words, $\psi_i[0]$ represents the seed (vacuum) solution and $\psi_i[1]$ the first iterated solution and so on.

7.4 Rogue waves, breathers, bright and dark solitons

It should be emphasized that seed solutions play a significant role by helping us obtain the required form of localized solutions such as rogue waves, breathers and classical solitons like bright or dark solitons. Hence, one can obtain the required form of localized excitations in Darboux transformation by a careful selection of the seed solutions. Rogue waves are inherently transient in nature and localized in both space and time while breathers exhibit periodicity and stable within their oscillatory domains. This distinction stems from the underlying symmetry and the chosen eigenvalues during the process of Darboux transformation. Since rogue waves, breathers and dark solitons are intrinsically related to modulated wave backgrounds, one has to choose a nonzero seed solution. The physics of these phenomena necessitates the existence of a complex interplay between their background wave and localized structures. Hence, for rogue waves, breathers and dark solitons, one has to begin with nonzero plane wave as the seed solution. Since bright solitons are self contained solutions of the models arising due to the delicate balance between the dispersion and nonlinearity, they are exempted from nonzero seed. They do not rely on background field. Instead, they are stand alone structures where all the energy is localized within the soliton itself. The selection of nonzero plane wave as the seed solution for both field variables leads to rogue waves, whereas selection of nonzero plane wave for one field and zero seed for another field variable leads to breathers.

7.4.1 Rogue waves

In order to derive the rational solutions, one employs Darboux transformation method and chooses the seed solution of the following form

$$\psi_1[0] = c_1 e^{i\theta_1(x,\, t)}, \tag{7.21}$$

$$\psi_2[0] = c_2 e^{i\theta_2(x,\, t)} \tag{7.22}$$

where $\theta_i(x, t) = d_i x + (2c_1^2 + 2c_2^2 - d_i^2)t$ with c_i and d_i being arbitrary constants. One now inserts these seed solutions into the Lax pair equations (7.15) and transforms it into the following form

$$\Psi_x = (MUM^{-1} + M_x M^{-1})\Psi = U_1\Psi, \tag{7.23}$$

$$\Psi_t = (MUM^{-1} + M_x M^{-1})\Psi = V_1\Psi \tag{7.24}$$

where $\Psi = M\phi$, $M = \mathrm{diag}\{\exp(-(\theta_1(x, t) + \theta_2(x, t))), \exp(2\theta_1(x, t) - \theta_2(x, t)), \exp(-(2\theta_2(x, t) - \theta_1(x, t)))\}$. Accordingly, U_1 and V_1 can be computed as

$$\mathbf{U}_1 = \begin{pmatrix} -2\mathrm{i}\zeta_1 - \mathrm{i}(d_1 + d_2) & c_1 & c_2 \\ -c_1 & \mathrm{i}\zeta_1 + \mathrm{i}(2d_1 - d_2) & 0 \\ -c_2 & 0 & \mathrm{i}\zeta_1 + \mathrm{i}(2d_2 - d_1) \end{pmatrix},$$

and $V_1 = \mathrm{i}U_1^2 - \frac{2}{3}(d_1 + d_2) - (2\zeta_1)U_1 + mI$ where, $m = 2\mathrm{i}\zeta_1^2 + \frac{2}{3}\mathrm{i}(c_1^2 + c_2^2) + \frac{2}{9}(d_1^2 - d_1 d_2 + d_2^2) + \frac{2}{3}\mathrm{i}\zeta_1(d_1 + d_2)$. One can choose the parameters to satisfy the relation $c_1 = c_2 = \pm 2\alpha$, $d_1 = d2 - 2\alpha$, $\alpha = d_2 + 3e_1$, $e_1 = \mathrm{Re}(\zeta_1)$, $\mathrm{Im}(\zeta_1) = \pm\sqrt{3\alpha}$ where d_1 and $\mathrm{Re}(\zeta_1)$ are arbitrary real numbers and after tedious calculation, one obtains the fundamental solution for the Lax pair matrix at $\zeta = \zeta_1$, $\psi_i = \psi_i[0]$. By choosing the parameters $m_1 = 0$, $m_2 = 1$, $m_3 = 0$, one arrives at the rogue wave solution given by [19]

$$\psi_1(x, t) = \alpha\, \mathrm{e}^{\mathrm{i}\theta_1(x, t)}\left[-1 - \mathrm{i}\sqrt{3} + \frac{-6\sqrt{3}\alpha\delta + \mathrm{i}(6\alpha\delta + 5\sqrt{3}) - 3}{12\alpha^2\delta^2 + 8\sqrt{3}\alpha\delta + 5}\right] \tag{7.25a}$$

$$\psi_2(x, t) = \alpha\, \mathrm{e}^{\mathrm{i}\theta_2(x, t)}\left[-1 + \mathrm{i}\sqrt{3} + \frac{-6\sqrt{3}\alpha\delta + \mathrm{i}(-6\alpha\delta - 5\sqrt{3}) - 3}{12\alpha^2\delta^2 + 8\sqrt{3}\alpha\delta + 5}\right] \tag{7.25b}$$

where $\theta_i(x, t) = d_i x + (2c_1^2 + 2c_2^2 - d_i^2)t$ and $\delta = x + 6e_1 t$ with c_i and d_i being arbitrary constants. For a suitable choice of parameters, the first-order rogue wave is displayed in figure 7.3.

7.4.2 Rogue wave without trap

The first-order rogue waves described by equation (7.25), for the model given by equation (7.10) without a trapping potential are depicted in figure 7.3. The panels in the first row, (a) and (b), illustrate rogue waves in the absence of Rabi coupling ($\Omega = 0$) while the panels (c) and (d) show the same rogue waves with the inclusion of Rabi coupling ($\Omega = 2$). By comparing panels (c, d) with (a, b), it can be observed that the inclusion of Rabi coupling results in the appearance of stripe-like structures along the temporal axis.

In the middle row, panels (e), (f), (g) and (h) represent the corresponding contour plots of the figures shown in the first row represented by the panels (a), (b), (c), and (d) respectively. Panels (e) and (f) show petal like structures of rogue waves while the panels shown in (g) and (h) reconfirm the appearance of stripes along the temporal

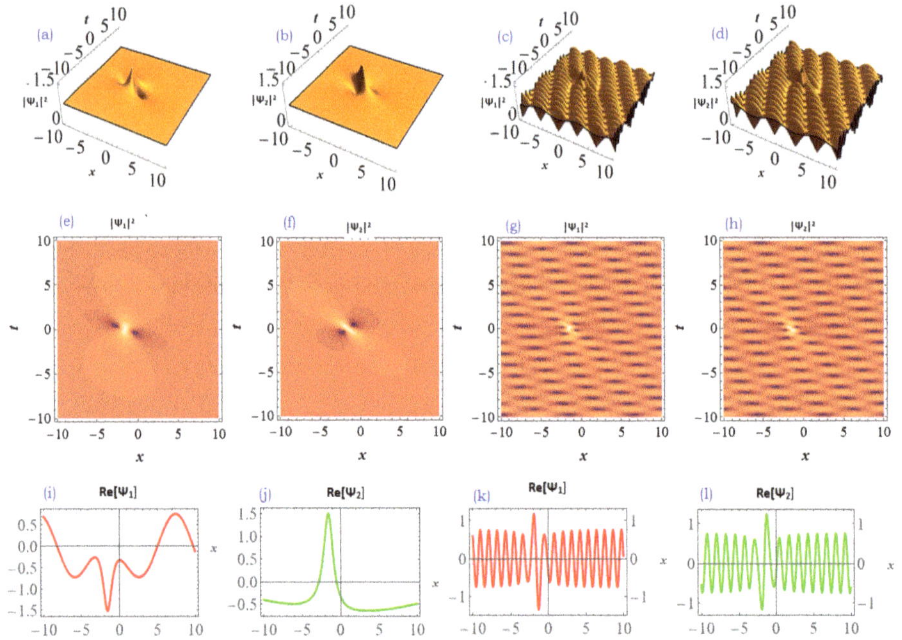

Figure 7.3. First-order rogue wave without trap: for the parametric choice $\sigma(t) = 0$, $\alpha = d_2 + 3e_1$, $c_1 = c_2 = 2\alpha$, $d_1 = d_2 = 2\alpha$, and $d_2 = 0.08$, $e_1 = 0.1$. First row: (a, b) rogue waves without Rabi coupling ($\Omega = 0$). (c, d) Rogue waves with Rabi coupling ($\Omega = 2$). Middle row: (e, f) and (g, h) represent the corresponding contour plots of the 3D plots in the first row. Last row: (i, j) Real part of the field variables ψ_1 and ψ_2 without SOC ($k_L = 0$). (k, l) Real part of the field variables ψ_1 and ψ_2 with SOC ($k_L = 8$). Reprinted from [19], Copyright (2025), with permission from Elsevier.

axis. Finally, in the last row, panels (i) and (j) display the real part of the field variables ψ_1 and ψ_2 without SOC ($k_L = 0$) while panels (k, l) show the real part of the same field variables with SOC for $k_L = 8$. A simple comparison of the panels shown in the last row indicates that the inclusion of SOC introduces rapid oscillations in the real parts of the field variables.

7.4.3 Rogue wave with transient trap

When the transient trap shown in figure 7.4 (a), is switched on, the corresponding first-order rogue wave profiles are depicted as in figure 7.5. Panels (a, b) display the rogue waves without the Rabi coupling. Introduction of Rabi coupling ($\Omega = 2$) in the presence of the transient trap results in striped bands along the temporal direction. Notably, these bands appear stable within the confining region and begin to overlap exponentially with the band gaps shrinking in the expulsive region, as shown in panels (c) and (d).

In the middle row, panels (e), (f), (g) and (h) show the corresponding contour plots of the rogue waves without and with Rabi coupling shown by the panels (a), (b), (c) and (d), respectively. Panels (e) and (f) indicate that the petal structure gets disrupted in the transient trap. Panels (i) and (j) shown in the last row in figure 7.5

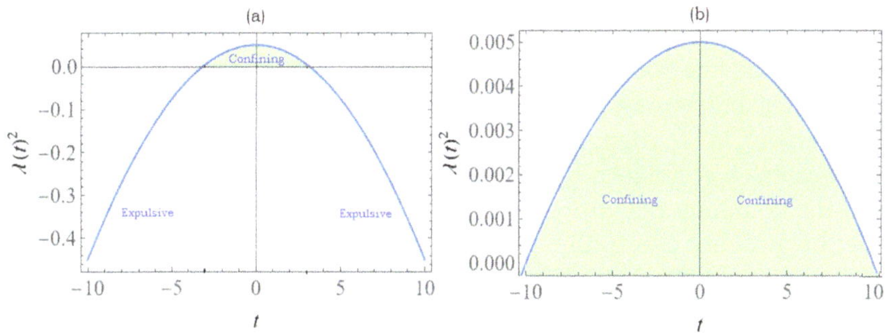

Figure 7.4. Evolution of time dependent trapping potential for the choice of parameters (a) $\sigma(t) = 0.05t$ and (b) $\sigma(t) = 0.025t$. Reprinted from [19], Copyright (2025), with permission from Elsevier.

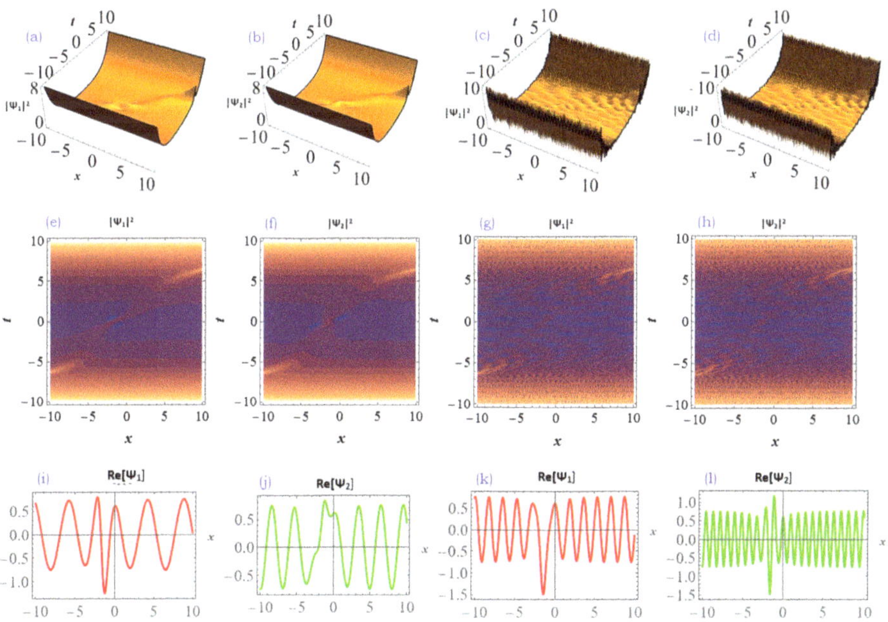

Figure 7.5. First-order rogue wave with transient trap: the rogue waves for the parametric choice $\sigma(t) = 0.05t$, with all other parameters being the same as in figure 7.3. First row: (a, b) rogue waves without Rabi coupling ($\Omega = 0$). (c, d) Rogue waves with Rabi frequency ($\Omega = 2$). Middle row: (e, f) and (g, h) are the corresponding contour plots of the 3D figures in the first row. Last row: (i, j) and (k, l) show the real part of the field variables ψ_1 and ψ_2, without SOC ($k_L = 0$) and with SOC ($k_L = 8$), respectively. Reprinted from [19], Copyright (2025), with permission from Elsevier.

depict the real part of the field variables ψ_1 and ψ_2 without SOC while panels (k) and (l) display the real part of the same field variables with SOC. It is again evident that SOC introduces rapid oscillations in the field variables. In addition, the introduction of the transient trap itself induces more oscillations, as evident by comparing panels (i) and (j) of figure 7.5 with that of figure 7.3.

7.4.4 Stabilization of rogue waves

It is obvious that the moment the transient trap is switched on, the stripe bands overlap in the expulsive region leading to instability in the system. To overcome this instability arising in the rogue wave structures, one manipulates the scattering length through Feshbach resonance by changing $\exp(0.05t)$ to $\exp(0.025t)$ to extend the range of the confining region from -10 to 10, as shown in figure 7.4(b) consistent with the integrability constraints given by equation (7.16), (7.17), and (7.18). This extended domain of the confining region is in stark contrast to the domain shown earlier in figure 7.4(a) spanning over -3 to 3 along the temporal axis. It should be emphasized that the stabilization of the rogue waves [23] which are inherently unstable has become possible as one allows the condensates to occupy the confining region.

The results are presented in figure 7.6. The first row displays the rogue waves without Rabi coupling (shown by panels (a) and (b)) and with Rabi coupling (shown by panels (c) and (d)) while the corresponding contour plots for the first row are shown in the middle row by panels (e),(f), (g) and (h), respectively. It is obvious from the panels shown in panels (e) and (f) that one can retrieve the petal structure shown earlier in panels (e) and (f) of figure 7.3. In addition, one can also recover the striped

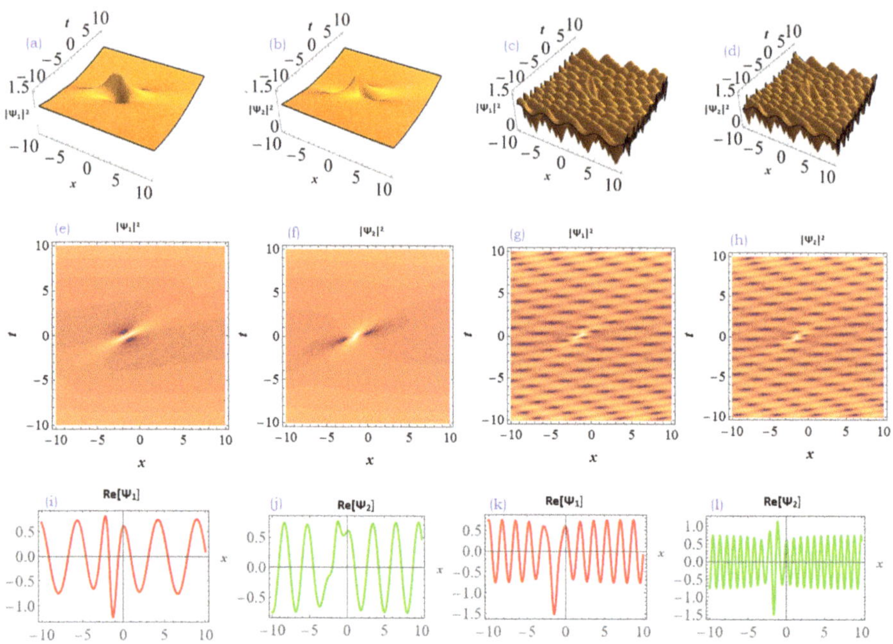

Figure 7.6. Stabilized first-order rogue wave with transient trap: The rogue waves are analyzed for the parametric choice $\sigma(t) = 0.025t$, with all other parameters being the same as in figure 7.3. First row: (a, b) rogue waves without Rabi coupling ($\Omega = 0$). (c, d) Rogue waves with Rabi coupling ($\Omega = 2$). Middle row: (e, f) and (g, h) show the corresponding contour plots of the 3D plots in the first row. Last row: (i, j) and (k, l) depict the real parts of the field variables ψ_1 and ψ_2 without SOC ($k_{\rm L} = 0$) and with SOC ($k_{\rm L} = 8$), respectively. Reprinted from [19], Copyright (2025), with permission from Elsevier.

pattern shown in panels (g) and (h) of figure 7.3. The last row shown by panels (i) and (j) show the real part of the field variables ψ_1 and ψ_2 without SOC ($k_1 = 0$) while the panels (k) and (l) with SOC ($k_1 = 8$).

7.4.5 Breathers

The breathers solutions can also be generated by choosing the plane wave and zero as seed solution of the following form

$$\psi_1[0] = c\mathrm{e}^{\mathrm{i}\theta(x,\,t)}, \tag{7.26}$$

$$\psi_2[0] = 0 \tag{7.27}$$

where $\theta(x, t) = \mathrm{d}x + (2c^2 - d^2)t$ and $M = \mathrm{diag}\{1, (\exp(\mathrm{i}\theta(x, t)), 1)\}$ and derive the respective U_1 and V_1. Then, by analysing the characteristic equation of U_1 and V_1 matrices and choosing $m_1 = 1$, $m_2 = \mathrm{e}^{\tau_1 + ik_1}$, $m_3 = \mathrm{e}^{\tau_2 + ik_2}$, where τ_i, k_i are real numbers, one obtains breather solution of the following form

$$\psi_1(x, t) = c_1\mathrm{e}^{\mathrm{i}\theta_i(x,\,t)} + 6(\zeta_1\mathrm{i})\frac{\phi_1\phi_2^*}{\phi_1\phi_1^* + \phi_2\phi_2^* + \phi_3\phi_3^*}, \tag{7.28a}$$

$$\psi_2(x, t) = c_2\mathrm{e}^{\mathrm{i}\theta_i(x,\,t)} + 6(\zeta_1\mathrm{i})\frac{\phi_1\phi_2^*}{\phi_1\phi_1^* + \phi_2\phi_2^* + \phi_3\phi_3^*}, \tag{7.28b}$$

where,

$$\begin{aligned}
\phi_1 &= c\mathrm{e}^{\frac{1}{3}\mathrm{i}(\theta_1(x,\,t)+\theta_2(x,\,t))}k_2m_2 + \mathrm{e}^{\frac{1}{3}\mathrm{i}(\theta_1(x,\,t)+\theta_2(x,\,t))}k_3m_3(\zeta_1 + 2\mathrm{i}\zeta_1),\\
\phi_2 &= c\mathrm{e}^{-\frac{1}{3}\mathrm{i}(2\theta_1(x,\,t)-\theta_2(x,\,t))}k_3m_3 + \mathrm{e}^{-\frac{1}{3}\mathrm{i}(2\theta_1(x,\,t)-\theta_2(x,\,t))}k_2m_2(\zeta_1 + 2\mathrm{i}\zeta_1),\\
\phi_3 &= \mathrm{e}^{-\frac{1}{3}\mathrm{i}(2\theta_2(x,\,t)-\theta_1(x,\,t))}k_1m_1, \quad k_1 = \exp\left(3\mathrm{i}\zeta_1^2t + \mathrm{i}\zeta_1x\right),\\
k_2 &= \exp\left[\zeta_1x + t\left(2\mathrm{i}\left(c^2 + \zeta_1^2\right) + 2\zeta_1\zeta_1 + \mathrm{i}\zeta_1^2\right)\right],\\
k_3 &= \exp\left[\zeta_2x + t\left(2\mathrm{i}\left(c^2 + \zeta_1^2\right) + 2\zeta_1\zeta_2 + \mathrm{i}\zeta_2^2\right)\right],
\end{aligned}$$

where, $\theta_i(x, t) = t(2c_{i+1}^2 + 2c_i^2 - d_i^2) + xd_i$, $\zeta_2 = -\zeta_1 + \mathrm{i}(d - \zeta_1)$, $m_2 = \mathrm{e}^{a_1 + ib_1}$, $m_3 = e^{a_2 + ib_2}$ with ζ_1 and ζ_2 being arbitrary complex constants and a_i, b_i, c_i, d_i and m_1 are arbitrary real parameters. For a suitable choice of parameters, the breather solutions without and with trapping potential are displayed in figures. 7.7 and 7.8, respectively.

7.4.6 Breathers without trap

The breather solution given by equation (7.28) is illustrated in figure 7.7. Panels (a) and (b) show the behaviour of breathers without Rabi coupling while panels (c) and (d) display breathers with Rabi coupling. The inclusion of Rabi coupling introduces temporal stripes with double- and single-mode peaks centred around $t = 0$ for ψ_1 and ψ_2, as seen in panels (c) and (d). The middle row represents the corresponding

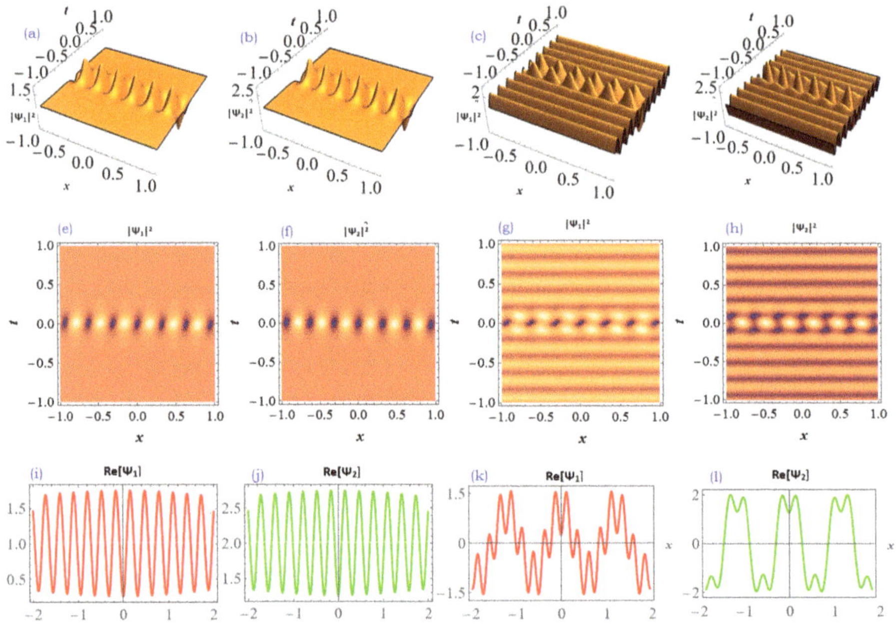

Figure 7.7. Breathers without trap: the parameters used are $\sigma(t) = 0$, $a_i = b_i = 1$, $c_1 = 1$, $c_2 = 2$, $m_1 = 1$, $\zeta_1 = 10i$, and $\zeta_2 = -0.25i$. First row: (a, b) breathers without Rabi coupling ($\Omega = 0$); (c, d) breathers with Rabi coupling ($\Omega = 2$). Middle row: (e, f) and (g, h) represent the corresponding contour plots corresponding of the first row shown by panels (a, b) and (c, d). Last row: (i, j) real part of the field variables ψ_1 and ψ_2 without SOC ($k_L = 0$); (k, l) real part of the field variables ψ_1 and ψ_2 with SOC ($k_L = 8$). Reprinted from [19], Copyright (2025), with permission from Elsevier.

contour plots of the first row showing the periodical variation in the amplitude of breathers and the stripes arising by virtue of Rabi coupling. The impact of SOC is illustrated in the last row with the panels (i, j) showing the results for $k_L = 0$ and panels (k, l) for $k_L = 8$. Thus, it is obvious that the amplitude of the breathers stays constant between a maxima and minima in a given region of space despite oscillating with time in the absence of SOC while the addition of SOC induces rapid fluctuations in the same spatial domain.

7.4.7 Breathers with transient trap

When the trapping potential (as shown in figure 7.4) is switched on, the positioning of the breathers is compressed and tilted, as depicted in panels (a) and (b) of figure 7.8. The inclusion of Rabi coupling again introduces striped bands along the temporal axis but does not significantly impact the double- and single-mode peaks around $t = 0$, as shown in panels (c) and (d) of figure 7.8. The contour plots shown in panels (e), (f), (g) and (h) confirm the above observation. Comparison of panels (k) and (l) with (i) and (j) demonstrates that one witnesses increased fluctuations arising due to the influence of SOC in the amplitudes of the field variables ψ_1 and ψ_2.

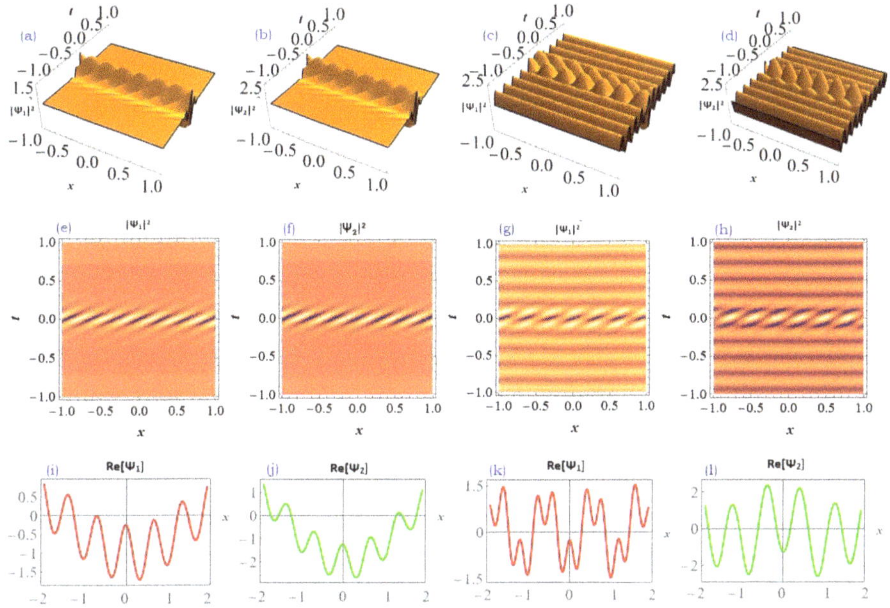

Figure 7.8. Breathers with transient trap: The parameter $\sigma(t) = 0.05t$ while all other parameters are the same as in figure 7.7. First row: (a, b) breathers without Rabi coupling ($\Omega = 0$); (c, d) breathers with Rabi coupling ($\Omega = 2$). Middle row: (e, f) and (g, h) represent the corresponding contour plots of the first row shown by panels (a, b) and (c, d). Last row: (i, j) real part of the field variables ψ_1 and ψ_2 for $k_L = 0$; (k, l) Real part of the field variables ψ_1 and ψ_2 for $k_L = 8$. Reprinted from [19], Copyright (2025), with permission from Elsevier.

7.4.8 Rogue–dark–bright

The matrices $\mathbf{U}_1$ and $\mathbf{V}_1$ are obtained by constructing rogue wave solutions combined with dark and bright soliton solutions, using the same seed solution as in breather construction and applying successive Darboux transformations. By analyzing the characteristic equation of the matrices and its roots and choosing multiple root options instead of a single root along with the the parametric choice $m_1 = a_1 + b_1 i$, $m_2 = 1$, $m_3 = a_3 + b_3 i$, one can construct the amalgamated rogue-dark–bright solution which are termed as 'mixed bound states' of the following form:

$$\psi_1(x, t) = c e^{i\theta(x, t)} \left[1 - \frac{4 e^{-\eta_{1R}} (\alpha^2 - \alpha + \beta^2 + i\beta)}{e^{-\eta_{1R}} [2(\alpha^2 + \beta^2) - 2\alpha + 1] + e^{2\eta_{1R}} \left[a_3^2 + b_3^2 \right]} \right], \tag{7.29a}$$

$$\psi_2(x, t) = -\frac{4c[(\alpha - 1) a_3 + i(a_3\beta + (\alpha - 1) b_3) - \beta b_3] e^{\frac{\eta_{1R}}{2} + i[t(3c^2 - d^2) + dx]}}{e^{-\eta_{1R}} (2(\alpha^2 + \beta^2) - 2\alpha + 1) + e^{2\eta_{1R}} \left(a_3^2 + b_3^2 \right)}, \tag{7.29b}$$

where $\theta(x, t) = dx + (2c^2 - d^2)t$, $\mu = \frac{1}{2} \log\left(\frac{a_3^2 + b_3^2}{2c^2} \right)$, $\eta_{1R} = \frac{2}{3} c(x - 2dt)$, $\alpha = a_1 c + c(x - 2dt)$ and $\beta = b_1 c - 2c^2 t$ with a_i, b_i, c and d being arbitrary real parameters.

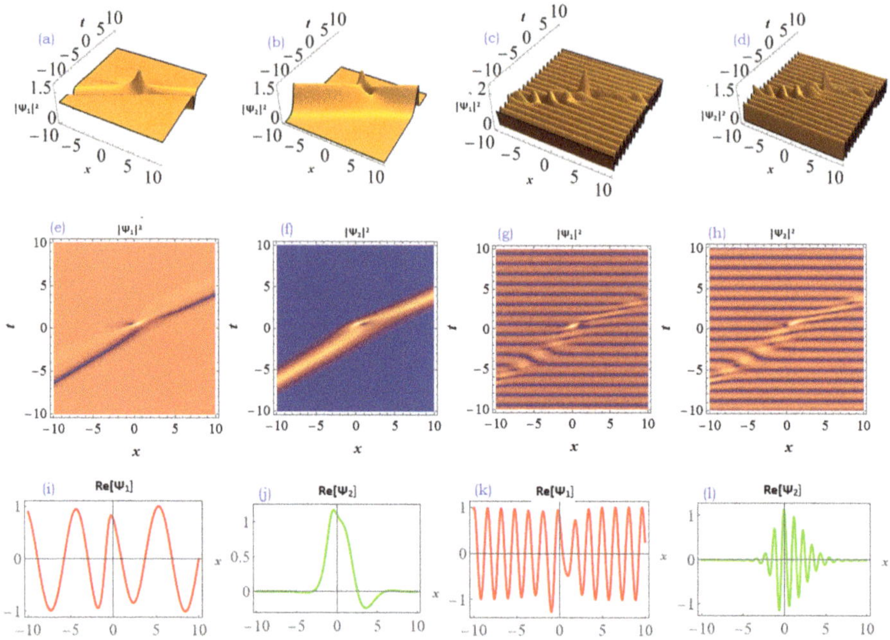

Figure 7.9. Mixed bound states without trap: The parameters $\sigma(t) = 0$, $a_i = b_i = c = d = 1$. First row: (a, b) dark–bright–rogue waves without Rabi coupling ($\Omega = 0$), (c, d) dark–bright–rogue waves with Rabi coupling ($\Omega = 2$). Middle row: (e, f) and (g, h) represent the corresponding contour plots of the panels (a, b) and (c, d), respectively shown in the first row. Last row: (i, j) real part of the field variables ψ_1 and ψ_2 without SOC ($k_L = 0$); (k, l) real part of the field variables ψ_1 and ψ_2 with SOC ($k_L = 8$). Reprinted from [19], Copyright (2025), with permission from Elsevier.

Again, for a suitable choice of parameters, the mixed bound states with rogue waves are displayed in figures 7.9 and 7.10 without and with transient trap, respectively.

7.4.9 Mixed bound states without transient trap

The collision of rogue waves with dark and bright solitons, as given by equation (7.29) in the absence of harmonic trap without Rabi coupling is shown in panels (a) and (b) of figure 7.9. Introduction of Rabi coupling in panels results in striped bands along the temporal axis, as seen in panels (c, d). The contour plots in the middle row shown by panels (g) and (h) reconfirm the impact of Rabi coupling resulting in the formation of striped bands in the temporal direction while one does not observe any such stripe bands in the panels (e) and (f). Comparison of panels (i)and (j) with (k) and (l) highlights the impact of SOC which introduces rapid fluctuations in the amplitude of the real part of the rogue-dark–bright solitons.

7.4.10 Mixed bound states with transient trap

Introduction of the time dependent trap compresses the amplitude of mixed bound state shown in panels (a) and (b) of figure 7.10 which is further confirmed by the

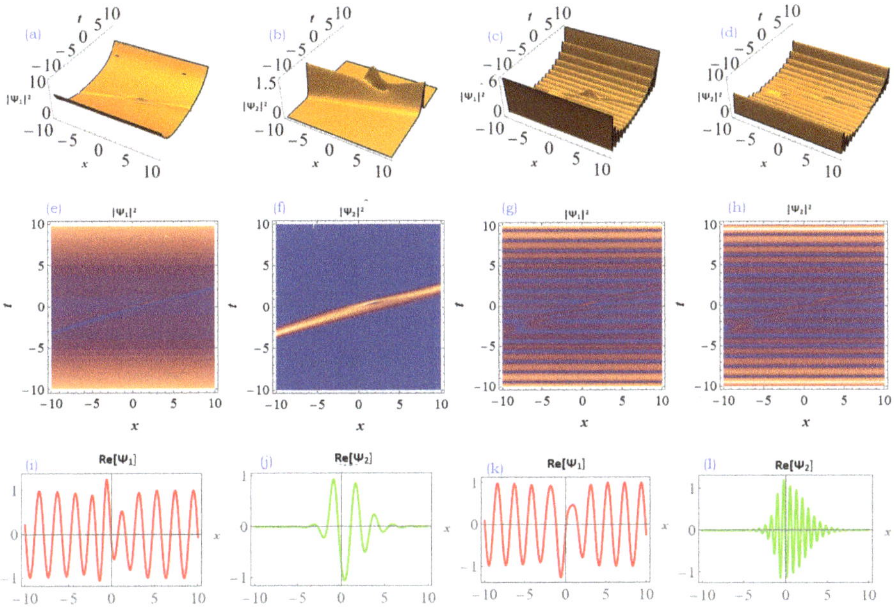

Figure 7.10. Mixed bound states with transient trap: The parameter $\sigma(t) = 0.05t$ is used, while all other parameters are the same as in figure 7.9. First row: (a, b) dark–bright–rogue waves without Rabi coupling ($\Omega = 0$); (c, d) dark–bright–rogue waves with Rabi coupling ($\Omega = 2$). Middle row: (e, f) and (g, h) represent the corresponding contour plots of the panels (a, b) and (c, d), respectively shown in the first row. **Last row:** (e, f) real part of the field variables ψ_1 and ψ_2 without SOC ($k_L = 0$) and with SOC ($k_L = 8$). Reprinted from [19], Copyright (2025), with permission from Elsevier.

contour plots shown in panels (e) and (f). The inclusion of Rabi coupling creates striped bands again shown by panels (g) and (h). In addition, when the mixed bound states cross over to the expulsive domain, their amplitude overshoots the upper bound as shown in panels (c) and (d) of figure 7.10. As observed earlier, the rapid oscillations due to SOC are more pronounced in panels (k) and (l) as compared to panels (i) and (j).

7.4.11 Dark–bright–solitons without and with transient trap

The classical dark–bright soliton solution is generated for the following choice of parameters $m_1 = 1$, $m_2 = 0$, $m_3 = a_3 + b_3 i$ to obtain

$$\psi_1(x, t) = c e^{i\theta(x, t)} \tanh\left[\frac{3\eta_{1R}}{2} + \mu\right], \tag{7.30a}$$

$$\psi_2(x, t) = -4c^2(a_3 + ib_3)\,\mathrm{sech}\left[\frac{3\eta_{1R}}{2} + \mu\right] e^{i[t(3c^2 - d^2) + dx]}, \tag{7.30b}$$

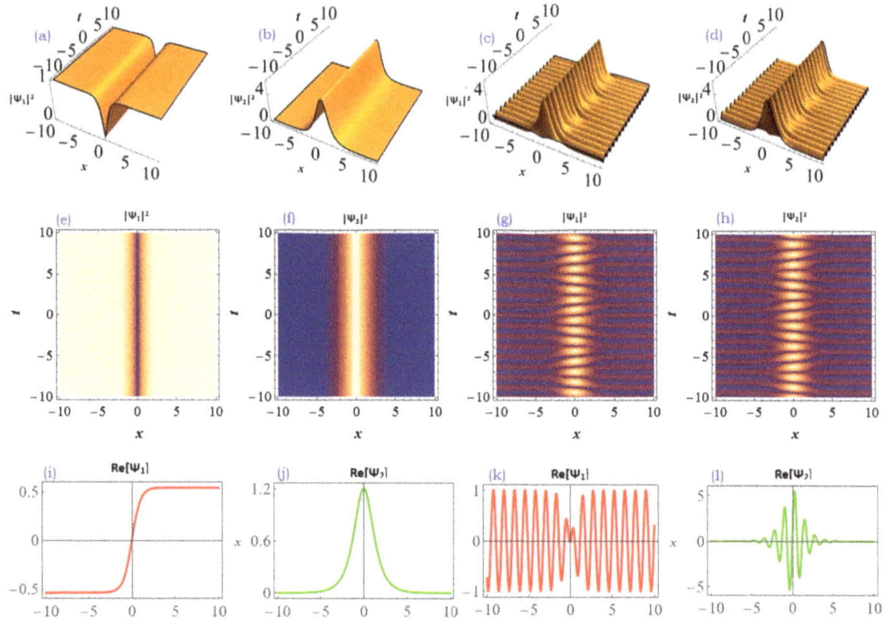

Figure 7.11. Dark–bright solitons without trap for $\sigma(t) = 0$, $a_3 = b_3 = c = 1$, and $d = 0$. First row: (a,b) dark–bright–without Rabi coupling ($\Omega = 0$) and c,d dark–bright–rogue waves with Rabi coupling $\Omega = 2$. Middle row: (e,f) represent the contour plots of (a,b) and (g,h) contour plots (c,d), respectively. Last row: (i,j) and (k,l) show the real part of the field variables ψ_1 and ψ_2 without and with SOC for $k_L = 8$, respectively. Reprinted from [19], Copyright (2025), with permission from Elsevier.

where the parameters $\theta(x, t)$, μ and η_{1R} are the same as in the previous section. The profiles of dark and bright solitons without and with trapping potential are displayed in figures 7.11 and 7.12, respectively.

The conventional trapless dark and bright solitons without Rabi coupling, are shown in panels (a) and (b) of figure 7.11 and this is complimented by the contour plots shown by panels (e) and (f). One observes that the inclusion of Rabi coupling introduces stripes along the temporal axis in the density profile of both dark and bright solitons, as witnessed in panels (c) and (d), which is again confirmed by the contour plots shown by panels (g) and (h). In addition, one also notices the flipping of dark solitons (shown by panel (a)) to attain positive density profile as shown in panel (c) quite similar to a bright soliton and this occurs by virtue of Rabi coupling. From the panels (i), (j), (k) and (l) shown in the last row, one notices that the addition of SOC introduces rapid fluctuations in the dark solitons while the fluctuations are centred around the origin in the bright solitons.

When the transient trap is switched on (shown in figure 7.4), one witnesses a 45° shift in the trajectory of both dark and bright solitons shown in panels (a) and (b) which are further endorsed by the contour plots shown in panels (e) and (f) in addition to the stripes due to Rabi coupling shown in panels (c) and (d) which are in confirmity with panels (g) and (h). In addition, one also observes that the width of

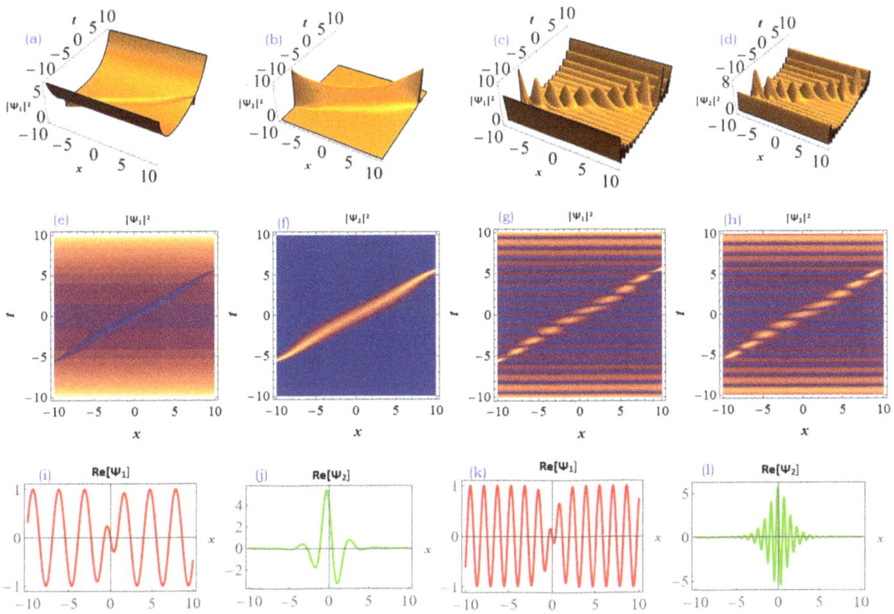

Figure 7.12. Dark–bright solitons With transient trap for $\sigma(t) = 0.05t$ with all other parameters being the same as in figure 7.11. First row: (a,b) and (c,d) dark–bright solitons without and with Rabi coupling for ($\Omega = 0$) and ($\Omega = 2$) respectively. Middle row: (e,f) and (g,h) represent the corresponding contour plots of the first row. Last row: (i,j) and (k,l) illustrate the real part of the field variables ψ_1 and ψ_2 without and with SOC for $k_L = 8$. Reprinted from [19], Copyright (2025), with permission from Elsevier.

the solitons widen in the confining trap while they shrink in the expulsive domain. Besides, one also notices that when the transient trap is switched on, it introduces oscillations in the real part of the order parameters ψ_1 and ψ_2 (shown in panels (i) and (j)) which are further enhanced by the addition of SOC shown in panels (k) and (l). It can also be observed that the amplitude of real part of ψ_1 around the origin almost becomes zero for dark solitons while that of ψ_2 becomes maximum around the origin.

7.4.12 Second-order rogue waves

Darboux transformation approach can be extended to generate multiple rogue waves. For example, the second-order rogue waves are of the following form:

$$\psi_1(x, t) = c_1 \, e^{2it}\left[1 + \frac{G_2(x, t) + iH_2(x, t)t}{d_2(x, t)}\right], \tag{7.31a}$$

$$\psi_2(x, t) = c_2 e^{2it}\left[1 + \frac{G_2(x, t) + iH_2(x, t)t}{d_2(x, t)}\right], \tag{7.31b}$$

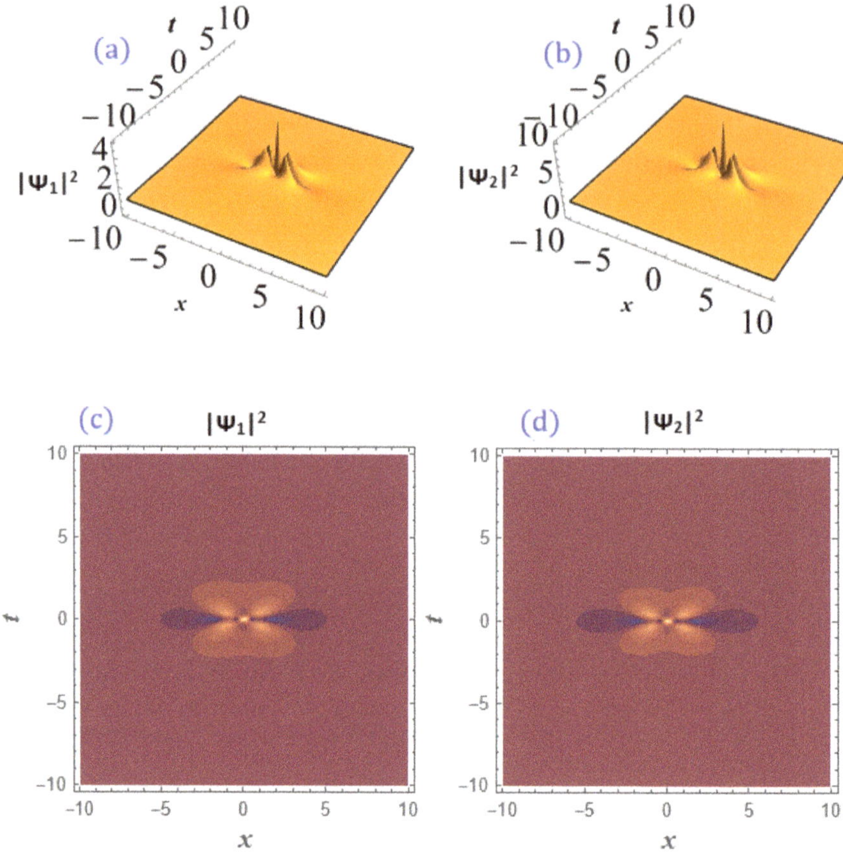

Figure 7.13. Panels (a) and (b) represent a second-order rogue wave for $\sigma(t) = 0$, $c_1 = 1$, $c_2 = 2$, $m_1 = 0.1$, $m_2 = 0.5$ with a primary crest at the centre surrounded by four secondary crests (c), (d) contour plots of (a,b) with a four petal structure. Reprinted from [19], Copyright (2025), with permission from Elsevier.

$$
\begin{aligned}
G_2(x, t) &= m_2^2(-7680t^4 - 2304t^2x^2 - 1728t^2 - 96x^4 - 144x^2 + 72x + 18) - 384m_1m_2x \\
H_2(x, t) &= m_2^2(-6144t^4 - 3072t^2x^2 - 768t^2 - 384x^4 - 576x^2 + 288x + 360) - 1536m_1m_2x \\
d_2(x, t) &= 128m_1^2 + m_1m_2(1536t^2x - 128x^3 + 96x - 48) + m_2^2(2048t^6 + 1536t^4x^2 + 3456t^4 \\
&\quad + 384t^2x^4 - 576t^2x^2 - 288t^2x + 792t^2 + 32x^6 + 24x^4 + 24x^3 + 54x^2 - 18x + 9)
\end{aligned}
$$

For a suitable choice of parameters, second-order rogue waves with a four petal structure are shown in figures 7.13(c) and (d) while one observes a primary crest with a peak intensity at the centre surrounded by four secondary crests, shown in figures 7.13(a) and (b).

7.5 Numerical methods

Analytical solutions for SO and Rabi-coupled BECs in a transient harmonic trap, as outlined in preceding sections, are achievable only when a functional correlation exists between the trap frequency and the scattering length. In more general scenarios, where these parameters are independent or an arbitrary trapping potential is employed, the system of equations does not yield closed-form solutions, necessitating numerical simulations. To address such scenarios, the SSCN and split-step Fourier methods (introduced in chapter 3) are employed to solve the time-dependent GP equations. Alternatively, conjugate gradient methods, described in section 3.4, can be adapted to tackle the time-independent case. The implementation of these numerical methods for solving the coupled GP equations is elaborated below. The one-dimensional GP equations for a SO-coupled BEC, incorporating equal contributions of Rashba and Dresselhaus SOC in dimensionless form, are expressed as [24]:

$$\mathrm{i}\frac{\partial \psi_\uparrow}{\partial t} = \left[-\frac{1}{2}\frac{\partial^2}{\partial x^2} - \mathrm{i}k_\mathrm{L}\frac{\partial}{\partial x} + V(x) + g_\uparrow|\psi_\uparrow|^2 + g_{\uparrow\downarrow}|\psi_\downarrow|^2\right]\psi_\uparrow + \Omega\psi_\downarrow, \tag{7.32a}$$

$$\mathrm{i}\frac{\partial \psi_\downarrow}{\partial t} = \left[-\frac{1}{2}\frac{\partial^2}{\partial x^2} + \mathrm{i}k_\mathrm{L}\frac{\partial}{\partial x} + V(x) + g_{\downarrow\uparrow}|\psi_\uparrow|^2 + g_\downarrow|\psi_\downarrow|^2\right]\psi_\downarrow + \Omega\psi_\uparrow, \tag{7.32b}$$

where $g_{\downarrow\uparrow} = g_{\uparrow\downarrow}$. The associated chemical potential is given by [25]:

$$\begin{aligned}\mu = \int \Bigg[&\sum_{j\in\uparrow,\downarrow}\left(\frac{1}{2}\left|\frac{\partial \psi_j}{\partial x}\right|^2 + V_j|\psi_j|^2\right) + 2\Omega\,\mathrm{Re}\left(\psi_\uparrow\psi_\downarrow^*\right) \\ &- \mathrm{i}k_\mathrm{L}\left(\psi_\uparrow^*\frac{\partial \psi_\uparrow}{\partial x} - \psi_\downarrow^*\frac{\partial \psi_\downarrow}{\partial x}\right) + g_\uparrow|\psi_\uparrow|^4 + 2g_{\uparrow\downarrow}|\psi_\uparrow|^2|\psi_\downarrow|^2 + g_\downarrow|\psi_\downarrow|^4\Bigg]\mathrm{d}x.\end{aligned} \tag{7.33}$$

To investigate the dynamics of SO-coupled BECs in a quasi-one-dimensional setting, researchers utilize a SSCN-based numerical scheme [26–28] to solve the coupled GP equations (equation (7.32)). The numerical approach splits equation (7.32a) into two parts:

$$\mathrm{i}\frac{\partial \psi_\uparrow}{\partial t} = [V(x) + g_\uparrow|\psi_\uparrow|^2 + g_{\uparrow\downarrow}|\psi_\downarrow|^2]\psi_\uparrow + \Omega\psi_\downarrow, \tag{7.34a}$$

$$\mathrm{i}\frac{\partial \psi_\uparrow}{\partial t} = -\frac{1}{2}\frac{\partial^2 \psi_\uparrow}{\partial x^2} - \mathrm{i}k_\mathrm{L}\frac{\partial \psi_\uparrow}{\partial x} \equiv H_\uparrow, \tag{7.34b}$$

and similarly splits equation (7.32b) into:

$$\mathrm{i}\frac{\partial \psi_\downarrow}{\partial t} = [V(x) + g_{\downarrow\uparrow}|\psi_\uparrow|^2 + g_\downarrow|\psi_\downarrow|^2]\psi_\downarrow + \Omega\psi_\uparrow, \tag{7.35a}$$

$$\mathrm{i}\frac{\partial \psi_\downarrow}{\partial t} = -\frac{1}{2}\frac{\partial^2 \psi_\downarrow}{\partial x^2} + \mathrm{i}k_\mathrm{L}\frac{\partial \psi_\downarrow}{\partial x} \equiv H_\downarrow, \tag{7.35b}$$

The numerical solution process begins by solving equations (7.34a) and (7.35a) using trial solutions $\psi_\uparrow(x;\, t_0)$ and $\psi_\downarrow(x;\, t_0)$ at time $t = t_0$, yielding an intermediate solution at $t = t_0 + \Delta$, where Δ is the time step. This intermediate solution serves as the initial input for solving equations (7.34b) and (7.35b), producing the final solutions $\psi_\uparrow(x;\, t_0 + \Delta)$ and $\psi_\downarrow(x;\, t_0 + \Delta)$. This procedure is iterated n times, with the time variable discretized as $t_n = n\Delta$, to obtain the converged solution at the desired final time $t_{\text{final}} = t_0 + n\Delta$.

7.5.1 Real time propagation

To study the dynamics of SO-coupled spinor BECs, real-time propagation must be employed. The solution at time t_n, denoted as $(\psi_\uparrow^n, \psi_\downarrow^n)$, is first advanced over a time step Δ by solving the equations (7.34a) and (7.35a), yielding an intermediate solution $(\psi_\uparrow^{n+1/2}, \psi_\downarrow^{n+1/2})$. Here, $\psi_\uparrow^n$ and $\psi_\downarrow^n$ represent the discretized wave functions at time t_n. Since equations (7.34a) and (7.35a) lack spatial derivative terms, this propagation is performed with high accuracy for small Δ using the following method:

$$\mathrm{i}\frac{\partial}{\partial t}\Psi^{n+1/2} = \mathcal{O}_{nd}(\mathbf{S})\Psi^n, \tag{7.36}$$

where

$$\mathbf{S} = \begin{pmatrix} V(x) + g_{\uparrow\uparrow}|\psi_\uparrow|^2 + g_{\uparrow\downarrow}|\psi_\downarrow|^2 & \Omega \\ \Omega & V(x) + g_{\downarrow\uparrow}|\psi_\uparrow|^2 + g_{\downarrow\downarrow}|\psi_\downarrow|^2 \end{pmatrix}, \quad \Psi = \begin{pmatrix} \psi_\uparrow \\ \psi_\downarrow \end{pmatrix}, \tag{7.37}$$

with $\Psi = (\psi_\uparrow, \psi_\downarrow)^T$. Here, $\mathcal{O}_{\text{nd}}(\mathbf{S})$ denotes the time-evolution operation governed by the matrix $\mathbf{S}$, where the subscript 'nd' indicates the absence of derivative terms

Subsequently, the time propagation corresponding to the operators H_j is executed numerically using the semi-implicit Crank–Nicolson (CN) method as:

$$\frac{\psi_j^{n+1} - \psi_j^{n+1/2}}{-\mathrm{i}\Delta} = \frac{1}{2}H_j\left(\psi_j^{n+1} + \psi_j^{n+1/2}\right). \tag{7.38}$$

The formal solution to equation (7.38) is given by

$$\psi_j^{n+1} = \mathcal{O}_{\text{CN}}(H_j)\psi_j^{n+1/2} \equiv \frac{1 - \mathrm{i}\Delta H_j/2}{1 + \mathrm{i}\Delta H_j/2}\psi_j^{n+1/2}, \tag{7.39}$$

which, when combined with equation (7.36), yields

$$\Psi^{n+1} = \mathcal{O}_{\text{CN}}(H_j)\mathcal{O}_{\text{nd}}(\mathbf{S})\Psi^n. \tag{7.40}$$

Here, $j = \uparrow, \downarrow$, $\mathcal{O}_{\text{CN}}$ represents the time-evolution operation involving H_j, and the subscript 'CN' refers to the Crank–Nicolson algorithm. The operation $\mathcal{O}_{\text{CN}}$ propagates the intermediate solution $\psi_j^{n+1/2}$ by the time step Δ, generating the solution ψ_j^{n+1} at the next time step $t_{n+1} = (n + 1)\Delta$.

The solution of the equation (7.36) is not an explicit one as in reference [27]. So one has to solve the equation in the following manner. For that, one first rewrites equations (7.34a) and (7.35a) as

$$\mathrm{i}\frac{\partial \psi_\uparrow}{\partial t} = a\psi_\uparrow, + b\psi_\downarrow, \quad \mathrm{i}\frac{\partial \psi_\downarrow}{\partial t} = c\psi_\uparrow + d\psi_\downarrow, \tag{7.41}$$

where

$$a = V(x) + g_{\uparrow\uparrow}\left|\psi_\uparrow\right|^2 + \beta\left|\psi_\downarrow\right|^2, \quad d = V(x) + \beta\left|\psi_\uparrow\right|^2 + g_{\downarrow\downarrow}\left|\psi_\downarrow\right|^2, \quad b = c = \Omega.$$

The formal solution for the above equations can be written as (in terms of the discretized variables):

$$\begin{pmatrix} \psi_\uparrow^{n+1/2} \\ \psi_\downarrow^{n+1/2} \end{pmatrix} = \frac{1}{2R}\mathrm{e}^{-\frac{\mathrm{i}}{2}\Delta(R+S)}\begin{pmatrix} A_\mathrm{R} & B_\mathrm{R} \\ C_\mathrm{R} & D_\mathrm{R} \end{pmatrix}\begin{pmatrix} \psi_\uparrow^n \\ \psi_\downarrow^n \end{pmatrix}, \tag{7.42}$$

where $S = a + d$ and $T = a - d$, $A_\mathrm{R} = R[1 + \mathrm{e}^{\mathrm{i}R\Delta}] + T[1 - \mathrm{e}^{\mathrm{i}A\Delta}]$, $B_\mathrm{R} = 2b[1 - \mathrm{e}^{\mathrm{i}R\Delta}]$, $C_\mathrm{R} = 2c[1 - \mathrm{e}^{\mathrm{i}R\Delta}]$, $D_\mathrm{R} = R[1 + \mathrm{e}^{\mathrm{i}R\Delta}] - T[1 - \mathrm{e}^{\mathrm{i}A\Delta}]$ and $R = \sqrt{(a-d)^2 + 4bc}$.

The time propagation of the remaining equations (7.34b) and (7.35b) are performed numerically by the semi-implicit CN scheme [29]. For example, the CN scheme is employed by mapping the equations (7.34b) and (7.35b) onto one-dimensional spatial grid points in x. The discretization of equations (7.34b) and (7.35b) along the spatial grid point x is carried out as follows [29–31]:

$$\begin{aligned}\frac{\mathrm{i}\left(\psi_{\uparrow,i}^{n+1} - \psi_{\uparrow,i}^{n+1/2}\right)}{\Delta} = &\frac{-1}{4h^2}\left[\left(\psi_{\uparrow,i+1}^{n+1} - 2\psi_{\uparrow,i}^{n+1} + \psi_{\uparrow,i-1}^{n+1}\right) + \left(\psi_{\uparrow,i+1}^{n+1/2} - 2\psi_{\uparrow,i}^{n+1/2} + \psi_{\uparrow,i-1}^{n+1/2}\right)\right] \\ &- \frac{\mathrm{i}k_\mathrm{L}}{4h}\left[\left(\psi_{\uparrow,i+1}^{n+1} - \psi_{\uparrow,i-1}^{n+1}\right) + \left(\psi_{\uparrow,i+1}^{n+1/2} - \psi_{\uparrow,i-1}^{n+1/2}\right)\right],\end{aligned} \tag{7.43}$$

$$\begin{aligned}\frac{\mathrm{i}\left(\psi_{\downarrow,i}^{n+1} - \psi_{\uparrow,i}^{n+1/2}\right)}{\Delta} = &\frac{-1}{4h^2}\left[\left(\psi_{\downarrow,i+1}^{n+1} - 2\psi_{\downarrow,i}^{n+1} + \psi_{\downarrow,i-1}^{n+1}\right) + \left(\psi_{\uparrow,i+1}^{n+1/2} - 2\psi_{\downarrow,i}^{n+1/2} + \psi_{\downarrow,i-1}^{n+1/2}\right)\right] \\ &+ \frac{\mathrm{i}k_\mathrm{L}}{4h}\left[\left(\psi_{\downarrow,i+1}^{n+1} - \psi_{\downarrow,i-1}^{n+1}\right) + \left(\psi_{\downarrow,i+1}^{n+1/2} - \psi_{\downarrow,i-1}^{n+1/2}\right)\right],\end{aligned} \tag{7.44}$$

which subsequently become

$$\begin{aligned}\psi_{\uparrow,i}^{n+1} - \psi_{\uparrow,i}^{n+1/2} = &\frac{\mathrm{i}\Delta}{4h^2}\left[\left(\psi_{\uparrow,i+1}^{n+1} - 2\psi_{\uparrow,i}^{n+1} + \psi_{\uparrow,i-1}^{n+1}\right) + \left(\psi_{\uparrow,i+1}^{n+1/2} - 2\psi_{\uparrow,i}^{n+1/2} + \psi_{\uparrow,i-1}^{n+1/2}\right)\right] \\ &- \frac{\Delta k_\mathrm{L}}{4h}\left[\left(\psi_{\uparrow,i+1}^{n+1} - \psi_{\uparrow,i-1}^{n+1}\right) + \left(\psi_{\uparrow,i+1}^{n+1/2} - \psi_{\uparrow,i-1}^{n+1/2}\right)\right],\end{aligned} \tag{7.45}$$

$$\begin{aligned}\psi_{\downarrow,i}^{n+1} - \psi_{\uparrow,i}^{n+1/2} = &\frac{\mathrm{i}\Delta}{4h^2}\left[\left(\psi_{\downarrow,i+1}^{n+1} - 2\psi_{\downarrow,i}^{n+1} + \psi_{\downarrow,i-1}^{n+1}\right) + \left(\psi_{\uparrow,i+1}^{n+1/2} - 2\psi_{\downarrow,i}^{n+1/2} + \psi_{\downarrow,i-1}^{n+1/2}\right)\right] \\ &+ \frac{\Delta k_\mathrm{L}}{4h}\left[\left(\psi_{\downarrow,i+1}^{n+1} - \psi_{\downarrow,i-1}^{n+1}\right) + \left(\psi_{\downarrow,i+1}^{n+1/2} - \psi_{\downarrow,i-1}^{n+1/2}\right)\right],\end{aligned} \tag{7.46}$$

where $\psi^n_{\uparrow,i} = \psi_\uparrow(x_i; t_n)$ and $\psi^n_{\downarrow,i} = \psi_\downarrow(x_i; t_n)$ refer to $x \equiv x_i = ih$, $i = 0, 1, 2, \ldots, N_x$ and h is the space step. For the discretization along x, one chooses $x \equiv x_i = -N_x h/2 + ih$, $i = 0, 1, 2, \ldots, N_x$. The above equations are essentially a set of algebraic equations in $\psi^{n+1}_{\uparrow,i+1}$, $\psi^{n+1}_{\uparrow,i}$, $\psi^{n+1}_{\uparrow,i-1}$, $\psi^{n+1}_{\downarrow,i+1}$, $\psi^{n+1}_{\downarrow,i}$ and $\psi^{n+1}_{\downarrow,i-1}$ at time t_{n+1}, which are solved using the proper boundary conditions. The algebraic equations emerging from (7.45) are written explicitly as [29]

$$A_i^-\psi^{n+1}_{\uparrow,i-1} + A_i^0\psi^{n+1}_{\uparrow,i} + A_i^+\psi^{n+1}_{\uparrow,i+1} = b_i, \tag{7.47a}$$

$$D_i^-\psi^{n+1}_{\downarrow,i-1} + D_i^0\psi^{n+1}_{\downarrow,i} + D_i^+\psi^{n+1}_{\downarrow,i+1} = d_i, \tag{7.47b}$$

where $A_i^0 = D_i^0 = 1 + \mathrm{i}\Delta/(2h^2)$, $A_i^+ = D_k^- = \Delta k_\mathrm{L}/(4h) - \mathrm{i}\Delta/(4h^2)$, and $A_i^- = D_i^+ = -\Delta k_\mathrm{L}/(4h) - \mathrm{i}\Delta/(4h^2)$ and

$$b_i = \frac{\mathrm{i}\Delta}{4h^2}\left(\psi^{n+1/2}_{\uparrow,i+1} - 2\psi^{n+1/2}_{\uparrow,i} + \psi^{n+1/2}_{\uparrow,i-1}\right) - \frac{\Delta K_\mathrm{L}}{4h}\left(\psi^{n+1/2}_{\uparrow,i+1} - \psi^{n+1/2}_{\uparrow,i-1}\right) + \psi^{n+1/2}_{\uparrow,i}, \tag{7.48a}$$

$$d_i = \frac{\mathrm{i}\Delta}{4h^2}\left(\psi^{n+1/2}_{\downarrow,i+1} - 2\psi^{n+1/2}_{\downarrow,i} + \psi^{n+1/2}_{\downarrow,i-1}\right) + \frac{\Delta K_\mathrm{L}}{4h}\left(\psi^{n+1/2}_{\downarrow,i+1} - \psi^{n+1/2}_{\downarrow,i-1}\right) + \psi^{n+1/2}_{\downarrow,i}, \tag{7.48b}$$

All quantities in b_i and d_i refer to time step $t_{n+1/2}$ and are considered known. The only unknowns in (7.47) are the wave forms $\psi^{n+1}_{\uparrow,i\pm1}$, $\psi^{n+1}_{\uparrow,i}$, $\psi^{n+1}_{\downarrow,i\pm1}$ and $\psi^{n+1}_{\downarrow,i}$ at time step t_{n+1} which can be obtained by the term forward recursion relation,

$$\psi^{n+1}_{j,i+1} = \alpha_{j,i}\psi^{n+1}_{j,i} + \beta_{j,i} \tag{7.49}$$

where, $j = \uparrow, \downarrow$, $\alpha_{j,i}$ and $\beta_{j,i}$ are coefficients to be calculated. Substituting equation (7.49) in equation (7.47) provides the solution,

$$\psi^{n+1}_{\uparrow,i} = \gamma_{\uparrow,i}(A_i^-\psi^{n+1}_{\uparrow,i-1} + A_i^+\beta_{\uparrow,i} - b_i) \tag{7.50a}$$

$$\psi^{n+1}_{\downarrow,i} = \gamma_{\downarrow,i}(D_i^-\psi^{n+1}_{\downarrow,i-1} + D_i^+\beta_{\downarrow,i} - d_i) \tag{7.50b}$$

with

$$\gamma_{\uparrow,i-1} = -1/(A_i^0 + A_i^+\alpha_{\uparrow,i}), \tag{7.51a}$$

$$\gamma_{\downarrow,i-1} = -1/(D_i^0 + D_i^+\alpha_{\downarrow,i}), \tag{7.51b}$$

$$\alpha_{\uparrow,i-1} = \gamma_{\uparrow,i}A_i^-, \quad \beta_{\uparrow,i-1} = \gamma_{\uparrow,i}(A_i^+\beta_{\uparrow,i} - b_k) \tag{7.52a}$$

$$\alpha_{\downarrow,i-1} = \gamma_{\downarrow,i}A_i^-, \quad \beta_{\downarrow,i-1} = \gamma_{\downarrow,i}(D_i^+\beta_{\downarrow,i} - d_k). \tag{7.52b}$$

By using the recursion relations (7.50), (7.51) and (7.52), one can find $\alpha_{j,i}$ and $\beta_{j,i}$, where i is starting from $N_x - 2$ to 0. The initial conditions are set as $\alpha_{N_x-1} = 0$ and $\beta_{j,N_x-1} = \psi^{n+1}_{j,N_x}$. Once the coefficients $\alpha_{j,i}$, $\beta_{j,i}$, and $\gamma_{j,i}$ are determined, the solution can be computed across the spatial grid from $j = 0$ to $j = N_x - 1$ using the recursive relation equation (7.49), starting from the initial value $\psi_0^{n+1} = 0$ subject to boundary conditions, i.e., $\psi(x) = 0$ at $x = \pm\infty$ [27]. For real-time propagation, the initial solution at $t = 0$ is typically the Gaussian ground state of a harmonic potential with zero nonlinearity ($\alpha = \beta = 0$). During time evolution, the nonlinearity is gradually introduced until the desired value is reached, yielding the final solution of the GP equations.

7.5.2 Imaginary-time propagation

Although real-time propagation offers several advantages, as previously described, this method involves complex wavefunctions when dealing with non-stationary properties. For obtaining stationary ground state solutions, one must solve equations (7.32a) and (7.32b) using imaginary-time propagation. In this approach, the wavefunction becomes fundamentally real-valued, making the imaginary-time propagation method particularly convenient as it deals with real variables. This transformation is achieved by substituting time with an imaginary quantity through the transformation $t \to -it$. In this case, equations (7.41) can be rewritten as

$$-\frac{\partial\psi_\uparrow}{\partial t} = a\psi_\uparrow + b\psi_\downarrow, \quad -\frac{\partial\psi_\downarrow}{\partial t} = c\psi_\uparrow + d\psi_\downarrow. \tag{7.53}$$

Now, the formal solution for the above equations can be written as (in terms of the discretized variables):

$$\begin{pmatrix} \psi_\uparrow^{n+1/2} \\ \psi_\downarrow^{n+1/2} \end{pmatrix} = \frac{1}{2R} e^{-\frac{1}{2}\Delta(R+S)} \begin{pmatrix} A_I & B_I \\ C_I & D_I \end{pmatrix} \begin{pmatrix} \psi_\uparrow^n \\ \psi_\downarrow^n \end{pmatrix}, \tag{7.54}$$

where $R = \sqrt{(a-d)^2 + 4bc}$, $S = a + d$, $T = a - d$, $A_I = R[1 + e^{R\Delta}] + T[1 - e^{A\Delta}]$, $B_I = 2b[1 - e^{R\Delta}]$, $C_I = 2c[1 - e^{R\Delta}]$ and $D_I = R[1 + e^{R\Delta}] - T[1 - e^{A\Delta}]$. Similarly, equations (7.34) and (7.35) can be written as

$$-\frac{\partial\psi_\uparrow}{\partial t} = [V(x) + g_{\uparrow\uparrow}|\psi_\uparrow|^2 + \beta|\psi_\downarrow|^2]\psi_\uparrow + \Omega\psi_\downarrow, \tag{7.55a}$$

$$-\frac{\partial\psi_\uparrow}{\partial t} = -\frac{1}{2}\frac{\partial^2\psi_\uparrow}{\partial x^2} - ik_L\frac{\partial\psi_\uparrow}{\partial x}, \tag{7.55b}$$

and

$$-\frac{\partial\psi_\downarrow}{\partial t} = [V(x) + \beta|\psi_\uparrow|^2 + g_{\downarrow\downarrow}|\psi_\downarrow|^2]\psi_\downarrow + \Omega\psi_\uparrow, \tag{7.56a}$$

$$-\frac{\partial\psi_\downarrow}{\partial t} = -\frac{1}{2}\frac{\partial^2\psi_\downarrow}{\partial x^2} + ik_L\frac{\partial}{\partial x}\psi_\downarrow, \tag{7.56b}$$

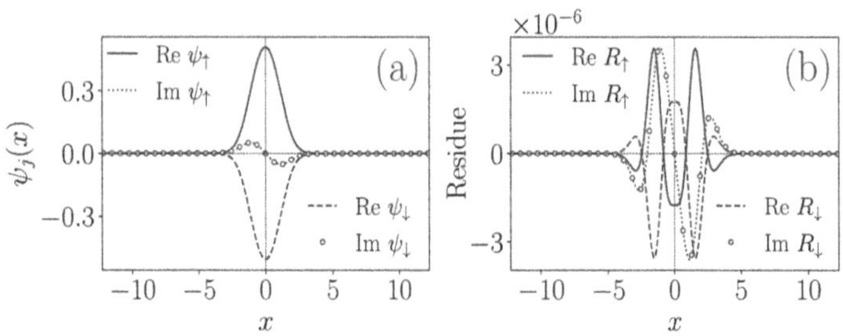

Figure 7.14. Plots illustrating the performance of the SSCN numerical method: the real and imaginary parts of (a) the pseudo-spin-component wavefunctions and (b) their corresponding residuals, obtained by solving the coupled GP equations (7.32a) with a harmonic trap potential, $V(x) = \frac{1}{2}x^2$, using the SSCN method with imaginary-time propagation, $dx = 0.01$, and $dt = 0.0001$. The parameters are $g_\uparrow = g_\downarrow = g_{\uparrow\downarrow} = 1$, $k_L = 0.5$, and $\Omega = 1$. Residuals are computed by substituting the numerically obtained wavefunctions and chemical potential into the time-independent GP equations (see equation (7.57a) below).

respectively. Subsequently, equations (7.55b) and (7.56b) can be discretized using the CN scheme similar to real-time propagation.

Figure 7.14 presents numerical results demonstrating the effectiveness of the SSCN method in solving the coupled GP equations for a pseudo-spin-1/2 BEC confined in a one-dimensional harmonic trap with $g_{\uparrow\uparrow} = g_{\downarrow\downarrow} = g_{\uparrow\downarrow} = 1$, $\Omega = 1$, $k_L = 0.5$, $dx = 0.01$, and $dt = 0.0001$. Figure 7.14(a) displays the real and imaginary parts of the converged spinor wavefunctions $\psi_\uparrow(x)$ and $\psi_\downarrow(x)$, obtained via imaginary-time propagation and 7.14(b) shows the corresponding residuals, which quantify the deviation from the stationary (time-independent) GP equations when the numerical solutions and chemical potential are substituted into equation (7.57). The results validate the suitability of the SSCN method for computing stationary states of SO-coupled BECs.

Appendix A provides a comprehensive overview of numerical codes for solving the time-dependent GP equations, including spinor, and SO-coupled BECs. These codes employ the SSCN scheme.

7.5.3 Split-step Fourier transform (SSFT) method

The SSFT method can also be employed to solve the coupled GP equations (7.32). This approach follows a procedure similar to the SSCN method, with the key difference lying in the treatment of the kinetic energy terms. Rather than using the CN scheme, the SSFT method utilizes a Fourier transform procedure to handle the spatial derivatives in equations (7.34b) and (7.35b) (or equations (7.55b) and (7.56b) for imaginary-time propagation). The implementation details of the SSFT method remain consistent with those described in section 3.3.

Figure 7.15 presents numerical results demonstrating the performance of the SSFT method in solving the coupled GP equation with parameters chosen to match those used in figure 7.14, specifically $g_{\uparrow\uparrow} = g_{\downarrow\downarrow} = g_{\uparrow\downarrow} = 1$, $\Omega = 1$, $k_L = 0.5$, with spatial and temporal discretization given by $dx = 0.01$ and $dt = 0.0001$, respectively. Figure 7.15(a) displays the real and imaginary parts of the numerically

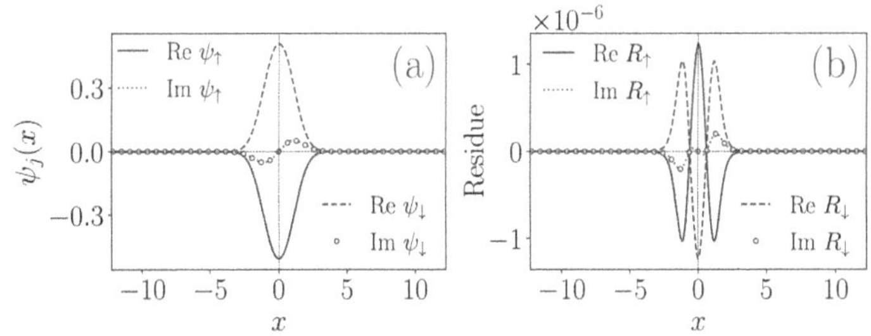

Figure 7.15. Plots depicting the real and imaginary parts of (a) the pseudo-spin-component wavefunctions and (b) their corresponding residuals, obtained by solving the coupled GP equations (7.32a) with a harmonic trap potential, $V(x) = \frac{1}{2}x^2$, using the SSFT method with imaginary-time propagation. The spatial and temporal step sizes are $dx = 0.01$ and $dt = 0.0001$, respectively, with other parameters matching those in figure 7.14.

converged spinor wavefunctions $\psi_\uparrow(x)$ and $\psi_\downarrow(x)$, obtained using imaginary-time propagation, and figure 7.15(b) shows the corresponding residuals.

7.5.4 Newton conjugate gradient method

In addition to the SSCN and SSFT methods discussed previously, the conjugate gradient methods presented in section 3.4 can be employed to solve the time-independent version of the coupled GP equations (7.32) of the form

$$\mu\psi_\uparrow = \left[-\frac{1}{2}\frac{\partial^2}{\partial x^2} - \mathrm{i}k_\mathrm{L}\frac{\partial}{\partial x} + V(x) + g_\uparrow|\psi_\uparrow|^2 + g_{\uparrow\downarrow}|\psi_\downarrow|^2\right]\psi_\uparrow + \Omega\psi_\downarrow, \tag{7.57a}$$

$$\mu\psi_\downarrow = \left[-\frac{1}{2}\frac{\partial^2}{\partial x^2} + \mathrm{i}k_\mathrm{L}\frac{\partial}{\partial x} + V(x) + g_{\downarrow\uparrow}|\psi_\uparrow|^2 + g_\downarrow|\psi_\downarrow|^2\right]\psi_\downarrow + \Omega\psi_\uparrow, \tag{7.57b}$$

where μ represents the chemical potential, as defined in equation (7.33).

The time independent equations may be rewritten in matrix form for convenience as

$$\begin{pmatrix} \partial_x^2 + \mathrm{i}k_\mathrm{L}\partial_x & 0 \\ 0 & \partial_x^2 - \mathrm{i}k_\mathrm{L}\partial_x \end{pmatrix}\begin{pmatrix} \psi_\uparrow \\ \psi_\downarrow \end{pmatrix} + \begin{pmatrix} F_\uparrow & \Omega \\ \Omega & F_\downarrow \end{pmatrix}\begin{pmatrix} \psi_\uparrow \\ \psi_\downarrow \end{pmatrix} = 0. \tag{7.58a}$$

where

$$F_\uparrow = \mu - V(x) - g_\uparrow|\psi_\uparrow|^2 - g_{\uparrow\downarrow}|\psi_\downarrow|^2, \tag{7.58b}$$

$$F_\downarrow = \mu - V(x) - g_{\downarrow\uparrow}|\psi_\uparrow|^2 - g_\downarrow|\psi_\downarrow|^2. \tag{7.58c}$$

For numerical computation, the equation is reformulated by separating the real and imaginary components, as $\psi_\uparrow = u_\uparrow + \mathrm{i}v_\uparrow$ and $\psi_\downarrow = u_\downarrow + \mathrm{i}v_\downarrow$, where $u_\uparrow, v_\uparrow, u_\downarrow, v_\downarrow \in \mathbb{R}$. This decomposition enables the matrix equation (7.58) to be rewritten in terms of real-valued components, in matrix form, as

$$\begin{pmatrix} -\frac{1}{2}\partial_x^2 + F_\uparrow & k_\mathrm{L}\partial_x & \Omega & 0 \\ -k_\mathrm{L}\partial_x & -\frac{1}{2}\partial_x^2 + F_\uparrow & 0 & \Omega \\ \Omega & 0 & -\frac{1}{2}\partial_x^2 + F_\downarrow & -k_\mathrm{L}\partial_x \\ 0 & \Omega & k_\mathrm{L}\partial_x & -\frac{1}{2}\partial_x^2 + F_\downarrow \end{pmatrix} \begin{pmatrix} u_\uparrow \\ v_\uparrow \\ u_\downarrow \\ v_\downarrow \end{pmatrix} = 0, \tag{7.59}$$

which takes the form $\mathbf{L}_0\mathbf{u} = 0$ (see section 3.4). This representation separates the coupled equations into their real and imaginary components while preserving the structure of the original system.

To employ Newton conjugate gradient method, an easily invertible positive definite self-adjoint operator may be defined as [32]

$$\mathbf{M} = \begin{pmatrix} c_1 - \frac{1}{2}\frac{\partial^2}{\partial x^2} & 0 & 0 & 0 \\ 0 & c_2 - \frac{1}{2}\frac{\partial^2}{\partial x^2} & 0 & 0 \\ 0 & 0 & c_3 - \frac{1}{2}\frac{\partial^2}{\partial x^2} & \\ 0 & 0 & 0 & c_2 - \frac{1}{2}\frac{\partial^2}{\partial x^2} \end{pmatrix}, \tag{7.60}$$

where c_k's, $k = 1, 2, 3, 4$ are positive constants. In this case, the linearization operator $\mathbf{L}_1$ becomes

$$\mathbf{L}_1 = \begin{pmatrix} \frac{1}{2}\frac{\partial^2}{\partial x^2} + F_\uparrow & k_\mathrm{L}\frac{\partial}{\partial x} & \Omega & 0 \\ -k_\mathrm{L}\frac{\partial}{\partial x} & \frac{1}{2}\frac{\partial^2}{\partial x^2} + F_\uparrow & 0 & \Omega \\ \Omega & 0 & \frac{\partial^2}{\partial x^2} + F_\downarrow & -k_\mathrm{L}\frac{\partial}{\partial x} \\ 0 & \Omega & k_\mathrm{L}\frac{\partial}{\partial x} & \frac{\partial^2}{\partial x^2} + F_\downarrow \end{pmatrix} - 2\begin{pmatrix} g_\uparrow u_\uparrow^2 & g_\uparrow u_\uparrow v_\uparrow & g_{\uparrow\downarrow} u_\uparrow u_\downarrow & g_{\uparrow\downarrow} u_\uparrow v_\downarrow \\ g_\uparrow u_\uparrow v_\uparrow & g_\uparrow v_\uparrow^2 & g_{\uparrow\downarrow} u_\downarrow v_\uparrow & g_{\uparrow\downarrow} v_\uparrow v_\downarrow \\ g_{\uparrow\downarrow} u_\uparrow u_\downarrow & g_{\uparrow\downarrow} u_\downarrow v_\uparrow & g_\downarrow u_\downarrow^2 & g_\downarrow u_\downarrow v_\downarrow \\ g_{\uparrow\downarrow} u_\uparrow v_\downarrow & g_{\uparrow\downarrow} v_\uparrow v_\downarrow & g_\downarrow u_\downarrow v_\downarrow & g_\downarrow v_\downarrow^2 \end{pmatrix}. \tag{7.61}$$

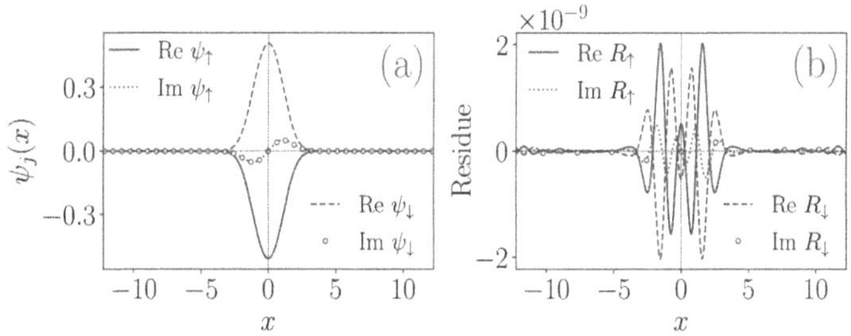

Figure 7.16. Plots depicting the real and imaginary components of (a) the pseudo-spin-component wavefunctions and (b) their corresponding residuals, obtained by solving the time-independent GP equations (7.57a) using the Newton conjugate gradient method with a spatial step size of $dx = 0.01$. Other parameters are consistent with those in figure 7.14.

For a given set of intra- and inter-component interaction strengths ($g_\uparrow$, $g\downarrow$, $g\uparrow\downarrow$), trap potential $V(x)$, SO and Rabi coupling strengths (k_L, Ω), chemical potential μ, and appropriate initial conditions, the conjugate gradient method yields highly accurate solutions for the spin-component wavefunctions $\psi_\uparrow$ and $\psi_\downarrow$.

Figure 7.16 presents numerical results demonstrating the performance of the Newton conjugate gradient method in solving the time-independent GP equations for a spin-1/2 BEC with SOC. The simulation uses the same physical parameters as in figure 7.14. Figure 7.16(a) shows the real and imaginary parts of the converged component wavefunctions $\psi_\uparrow(x)$ and $\psi_\downarrow(x)$, while figure 7.16(b) displays the corresponding residuals. Residuals of $\sim 10^{-9}$ demonstrate high accuracy and convergence of the Newton conjugate gradient method, confirming its effectiveness for obtaining stationary solutions in SO-coupled BEC systems.

A natural extension of these algorithms can be made to spin-1 spinor BEC systems with SOC, where the additional spin-component introduces new computational considerations. Furthermore, the framework is readily adaptable to higher-dimensional systems, including two- and three-dimensional binary BECs with Rashba-type SOC. The technical aspects of such extensions, particularly focusing on implementations using the CN approach, are comprehensively addressed in reference [33].

The SSFT and conjugate gradient methods can be similarly modified to accommodate these more complex systems. The SSFT method proves particularly effective for multidimensional configurations due to its efficient handling of spatial derivatives through spectral methods, while the conjugate gradient approach offers advantages in solving the resulting linear systems. This adaptability significantly expands the potential applications of these numerical schemes, making them valuable tools for investigating a wide range of multi-component BEC systems across different spatial dimensions.

7.6 Plane wave and stripe patterns

An interesting and unique feature of SO-coupled BECs is their ability to exhibit exotic quantum phases, notably the plane-wave phase and the stripe phase. These phases arise due to the interplay between SOC, interactions, and external potentials, and they are characterized by distinct spin textures and excitation spectra.

In the plane-wave phase, the condensate acquires a nonzero momentum in a preferred direction, spontaneously breaking time-reversal and parity symmetries. The condensate predominantly occupies a single momentum state associated with a particular spin component, resulting in a uniform density profile but a nontrivial spin polarization. This phase can be thought of as a macroscopic quantum state with a finite superfluid velocity [7, 34]. On the other hand, the stripe phase is characterized by the simultaneous occupation of two momentum states with opposite momenta, leading to an interference pattern in real space. As a result, the condensate exhibits periodic density modulations, forming stripes, a hallmark of supersolid-like behaviour where both superfluidity and crystalline order coexist. The stripe phase thus represents a self-organized density structure intertwined with spin textures [13, 35].

The excitation spectra of these two phases are also markedly different. In the plane-wave phase, the system exhibits a roton-like minimum in the excitation spectrum, suggesting a tendency towards the formation of periodic structures, although the density remains uniform. In contrast, the stripe phase features a Goldstone mode associated with the spontaneous breaking of translational symmetry, in addition to the usual phonon mode of superfluids These phenomena have been theoretically predicted and experimentally realized, most notably in the pioneering works on SO-coupled BECs achieved via Raman coupling schemes in ultracold atomic gases [7, 36]. The study of these phases provides important insights into new forms of quantum matter and stimulates exploration of supersolidity and quantum phase transitions in highly controllable settings. One can derive the dispersion relation for both phases, starting from the Hamiltonian of a SO-coupled BEC and using the Bogoliubov approximation for excitations.

7.7 Dynamics of SO-coupled BECs

This section explores the behaviour of bright and stripe solitons in SO-coupled BECs by numerically solving the coupled GP equations (7.32), which describe the evolution of the two-component wavefunction of the condensate. The numerical simulations employ the SSCN method, as detailed above. The spatial domain is discretized with a step size $\mathrm{d}x = 0.025$, spanning $x \in [-51.2, 51.2]$, with periodic boundary conditions to mimic an infinite system, while the external confinement is a harmonic potential $V(x) = \frac{1}{2}\lambda^2 x^2$. A time step of $\mathrm{d}t = 6.25 \times 10^{-4}$ is used for both imaginary-time and real-time propagation.

To analyze the dynamics of SO-coupled BECs, it is useful to determine the chemical potentials associated with the individual spin components, denoted as $\mu_\uparrow$ and $\mu_\downarrow$ for the spin-up ($\psi_\uparrow$) and spin-down ($\psi_\downarrow$) components, respectively. These chemical potentials are derived by assuming stationary solutions to the coupled GP equations (7.32a) and (7.32b), of the form:

$$\psi_\uparrow(x, t) = \mathrm{e}^{-\mathrm{i}\mu_\uparrow t}[\psi_{\uparrow R}(x) + \mathrm{i}\psi_{\uparrow I}(x)], \quad \psi_\downarrow(x, t) = \mathrm{e}^{-\mathrm{i}\mu_\downarrow t}[\psi_{\downarrow R}(x) + \mathrm{i}\psi_{\downarrow I}(x)], \tag{7.62}$$

where $\psi_{\uparrow R}$ and $\psi_{\uparrow I}$ are the real and imaginary parts of the time-independent spin-up wavefunction, and similarly for the spin-down component. The negative sign in the exponential reflects the conventional time evolution in quantum mechanics, where the phase factor is $e^{-iEt/\hbar}$, and $\mu_\uparrow, \mu_\downarrow$ are the chemical potentials scaled appropriately.

Substituting these expressions into the coupled GP equations and separating the real and imaginary parts, one obtains a set of equations that can be used to derive the chemical potentials. After performing the substitution and simplifying, the chemical potentials $\mu_\uparrow$ and $\mu_\downarrow$ are expressed as:

$$\mu_\uparrow = \frac{1}{N_\uparrow}\int \{\frac{1}{2}\left[\left(\frac{\partial\psi_{\uparrow R}}{\partial x}\right)^2 + (V(x) + g_{\uparrow\uparrow}|\psi_\uparrow|^2 + g_{\uparrow\downarrow}|\psi_\downarrow|^2)\psi_{\uparrow R}^2\right] + \left[\Omega\psi_{\downarrow R} + k_{\mathrm{L}}\frac{\partial\psi_{\uparrow I}}{\partial x}\right]\psi_{\uparrow R}\}\mathrm{d}x, \tag{7.63a}$$

$$\mu_\downarrow = \frac{1}{N_\downarrow}\int \{\frac{1}{2}\left[\left(\frac{\partial\psi_{\downarrow R}}{\partial x}\right)^2 + (V(x) + g_{\uparrow\downarrow}|\psi_\uparrow|^2 + g_{\downarrow\downarrow}|\psi_\downarrow|^2)\psi_{\downarrow R}^2\right] + \left[\Omega\psi_{\uparrow R} - k_{\mathrm{L}}\frac{\partial\psi_{\downarrow I}}{\partial x}\right]\psi_{\downarrow R}\}\mathrm{d}x, \tag{7.63b}$$

where $N_\uparrow = \int\psi_{\uparrow R}^2\,\mathrm{d}x$ and $N_\downarrow = \int\psi_{\downarrow R}^2\,\mathrm{d}x$ represent the normalization integrals for the real parts of the spin-up and spin-down wavefunctions, respectively. Here, $V(x)$ is the external potential, α and β are the intra- and inter-component interaction strengths, Ω is the Rabi coupling strength, and k_{L} is the SOC strength.

These expressions for $\mu_\uparrow$ and $\mu_\downarrow$ provide a functional relationship between the chemical potentials and the wavefunction profiles, enabling the study of stationary states and their stability in SO-coupled BECs. The integrals encapsulate the kinetic energy, potential energy, nonlinear interaction energy, and coupling terms, weighted by the wavefunction's real part, ensuring that the chemical potentials reflect the energy per particle in each component.

To obtain stationary states, the GP equations are solved using imaginary-time propagation, which converges to the ground state by evolving the wavefunction in imaginary time. The dynamics of solitons are then studied via real-time propagation, allowing the wavefunction to evolve under the influence of the SOC and nonlinear interactions. The initial wavefunctions are chosen as Gaussian profiles, reflecting the localized nature of solitons, with the form

$$\begin{bmatrix}\psi_\uparrow(x,0)\\ \psi_\downarrow(x,0)\end{bmatrix} = \begin{cases}\dfrac{1}{\sqrt{2\pi}}\begin{bmatrix}1\\ \pm1\end{bmatrix}\exp\left(-\dfrac{x^2}{2}\right), & \text{if}\lambda = 0,\\[2ex] \dfrac{\sqrt{\lambda}}{\sqrt{2\pi}}\begin{bmatrix}1\\ \pm1\end{bmatrix}\exp\left(-\dfrac{x^2}{2}\right), & \text{if}\lambda \neq 0,\end{cases} \tag{7.64}$$

where $\Psi = (\psi_\uparrow, \psi_\downarrow)^T$ represents the two-component wavefunction, λ is a normalization parameter, and the ± 1 choice determines the relative phase between the spin components.

For running the simulations, one can fix the interaction parameters as $g_{\uparrow\uparrow} = g_{\downarrow\downarrow} = g_{\uparrow\downarrow} = -0.8$, the SOC strength $k_L = 0.5$, and the Rabi coupling $\Omega = 0.5$. Imaginary-time propagation generates initial wavefunction profiles, which then evolve in real time to study their dynamical stability and motion.

For the chosen parameters, two distinct stationary states emerge, characterized by chemical potentials $\mu = -0.623$ and $\mu = -0.537$. The symmetry of the initial wavefunctions significantly influences the soliton dynamics. Consider first the antisymmetric initial condition, defined as $\psi_\uparrow(x) = -\psi_\downarrow(-x)$. Figures 7.17(a) and (b) illustrate the real (solid red line) and imaginary (dashed blue line) parts of the antisymmetric wavefunctions for the spin-up ($\psi_\uparrow$) and spin-down ($\psi_\downarrow$) components, respectively. The corresponding density profiles, $|\psi_\uparrow|^2$ and $|\psi_\downarrow|^2$, shown in figure 7.17(c) and 7.17(d), reveal single-soliton structures. During real-time evolution, these solitons remain stationary, as depicted in figure 7.17(e) and 7.17(f), indicating dynamic stability for the antisymmetric configuration with $\mu_\uparrow = \mu_\downarrow = -0.623$.

In contrast, the cross-symmetric initial condition, $\psi_\uparrow(x) = \psi_\downarrow(-x)$, leads to different behaviour. Figures 7.18(a) and (b) display the real and imaginary parts of the cross-symmetric wavefunctions, with density profiles in figure 7.18(c) and (d) showing two-soliton complexes. These solitons are dynamically unstable, as evidenced by their motion along the $\pm x$ directions during time evolution, illustrated in figure 7.18(e) and (f). This instability arises from the SOC, which induces effective interactions between the spin components, as noted in reference [38]. The chemical potentials for this state are $\mu_\uparrow = \mu_\downarrow \approx -0.537$, indicating a higher energy compared to the antisymmetric case.

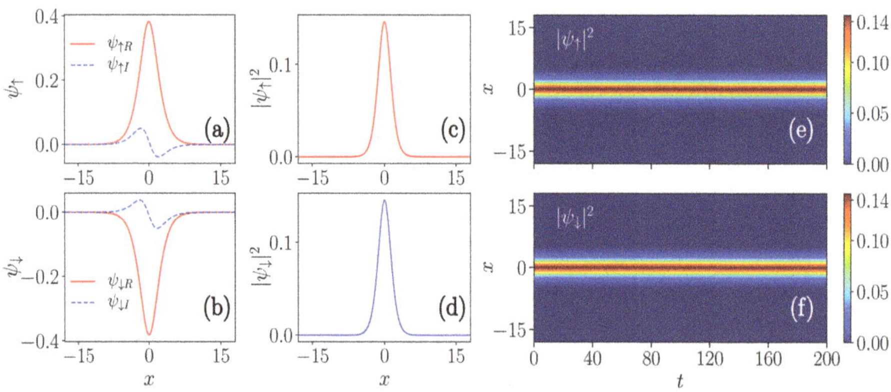

Figure 7.17. Plots of the real (solid red line) and imaginary (dashed blue line) parts of the antisymmetric initial wavefunctions of the spin components (a) $\psi_\uparrow$ and (b) $\psi_\downarrow$, and the corresponding single-soliton density profiles, (c) $|\psi_\uparrow|^2$ and (d) $|\psi_\downarrow|^2$, for $\lambda = 0$, $g_{\uparrow\uparrow} = g_{\downarrow\downarrow} = g_{\uparrow\downarrow} = -0.8$, $k_L = 0.5$, and $\Omega = 0.5$. Time evolution of the densities, (e) $|\psi_\uparrow|^2$ and (f) $|\psi_\downarrow|^2$, obtained from the numerical solution of the coupled GP equations (7.32a) and (7.32b), confirming the stationary nature of the solitons. The chemical potentials are $\mu_\uparrow = \mu_\downarrow \approx -0.623$. Reprinted from [37]. Copyright 2020 IOP Publishing Ltd.

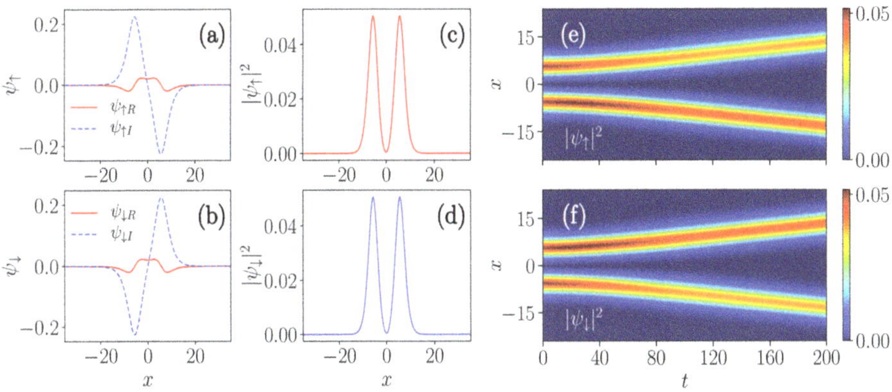

Figure 7.18. Plots of the real (solid red line) and imaginary (dashed blue line) parts of the cross-symmetric initial wavefunctions, (a) $\psi_\uparrow$ and (b) $\psi_\downarrow$, and the corresponding two-soliton density profiles, (c) $|\psi_\uparrow|^2$ and (d) $|\psi_\downarrow|^2$, for $g_{\uparrow\uparrow} = g_{\downarrow\downarrow} = g_{\uparrow\downarrow} = -0.8$, $k_{\rm L} = 0.5$, and $\Omega = 0.5$. Time evolution of the density, (e) $|\psi_\uparrow|^2$ and (f) $|\psi_\downarrow|^2$, showing the unstable nature as the solitons move in the $\pm x$ directions. The chemical potentials are $\mu_\uparrow = \mu_\downarrow \approx -0.537$. Reprinted from [37]. Copyright 2020 IOP Publishing Ltd.

The stability differences can be understood through the phase relationship between solitons. States with opposite phases (out-of-phase) tend to repel each other, while they attract when they are in phase, as discussed in reference [39]. In the antisymmetric case, the opposite phases stabilize the single-soliton structure, whereas the cross-symmetric configuration results in repulsive dynamics, leading to the separation of the two-soliton complex.

The dynamics of stripe solitons, characterized by periodic density modulations, are investigated for larger SOC and Rabi coupling parameters: $\lambda = 0$, $g_{\uparrow\uparrow} = g_{\downarrow\downarrow} = g_{\uparrow\downarrow} = -0.8$, $k_{\rm L} = 4$, and $\Omega = 2$. These parameters favour the formation of stripe phases due to the modified dispersion relation induced by SOC. The antisymmetric initial wavefunctions, shown in figures 7.19(a) and (b), produce stable stripe solitons with density profiles in figures 7.19(c) and (d). The chemical potentials are $\mu_\uparrow = \mu_\downarrow = -8.207$, indicating a lower-energy state. Time evolution, depicted in figure 7.19(e) and (f), confirms the stationary nature of these stripe solitons.

Conversely, the cross-symmetric initial wavefunctions, illustrated in figures 7.20 (a) and (b), yield stripe solitons with density profiles in figures 7.20(c) and (d). These solitons are unstable, propagating in the $\pm x$ directions during time evolution, as shown in figures 7.20(e) and (f). The chemical potentials are $\mu_\uparrow = \mu_\downarrow \approx -8.148$, higher than the antisymmetric case, reflecting the energetic instability of the cross-symmetric configuration.

The observed soliton phases are governed by the single-particle dispersion relation, derived from the linearization of the GP equations about homogeneous state

$$\omega_\pm(k_x) = \frac{k_x^2}{2} \pm \sqrt{k_{\rm L}^2 k_x^2 + \Omega^2}. \tag{7.65}$$

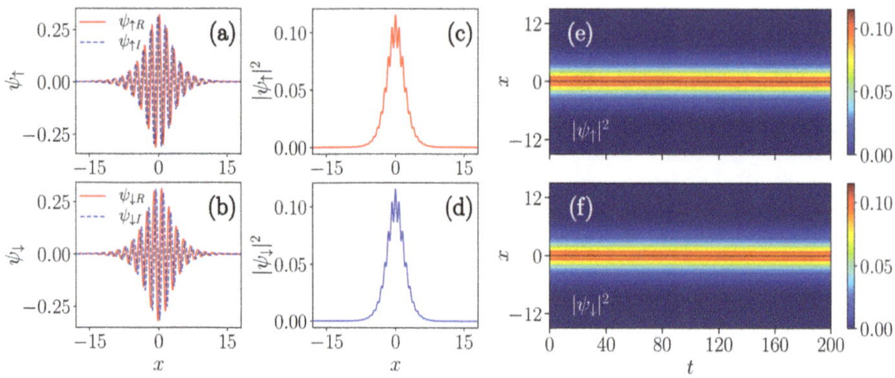

Figure 7.19. Plots of the real (solid red line) and imaginary (dashed blue line) parts of the antisymmetric initial wavefunctions, (a) $\psi_\uparrow$ and (b) $\psi_\downarrow$, and the density profiles of the stripe solitons, (c) $|\psi_\uparrow|^2$ and (d) $|\psi_\downarrow|^2$, for $\lambda = 0$, $g_{\uparrow\uparrow} = g_{\downarrow\downarrow} = g_{\uparrow\downarrow} = -0.8$, $k_{\rm L} = 4$, and $\Omega = 2$. The chemical potentials are $\mu = -8.207$. Time evolution of the density, (e) $|\psi_\uparrow|^2$ and (f) $|\psi_\downarrow|^2$, confirming the stationary nature of the stripe solitons. Reprinted from [37]. Copyright 2020 IOP Publishing Ltd.

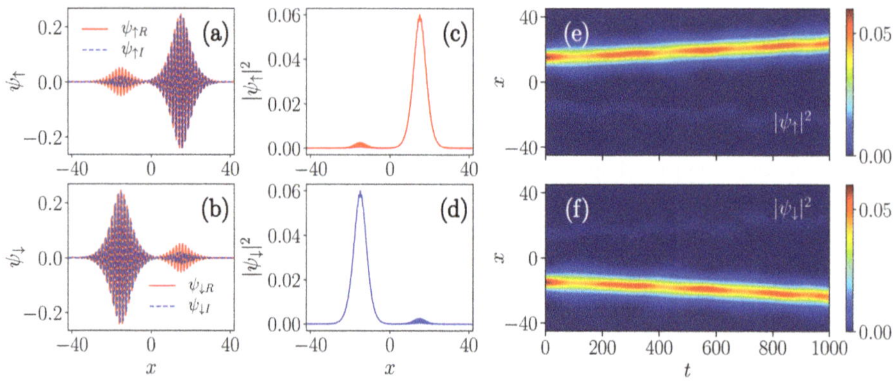

Figure 7.20. Plots of the real (solid red line) and imaginary (dashed blue line) parts of the cross-symmetric initial wavefunctions, (a) $\psi_\uparrow$ and (b) $\psi_\downarrow$, and the density profiles of the stripe solitons, (c) $|\psi_\uparrow|^2$ and (d) $|\psi_\downarrow|^2$, for $\lambda = 0$, $\alpha = \beta = -0.8$, $k_{\rm L} = 4$, and $\Omega = 2$. The chemical potentials are $\mu_\uparrow = \mu_\downarrow \approx -8.148$. Time evolution of the density, (e) $|\psi_\uparrow|^2$ and (f) $|\psi_\downarrow|^2$, showing the propagation of the spin components in the $\pm x$ directions. Reprinted from [37]. Copyright 2020 IOP Publishing Ltd.

This relation is obtained by assuming plane-wave solutions $\psi_{\uparrow,\downarrow} = \phi_{\uparrow,\downarrow} \exp[{\rm i}(k_x x - \omega t)]$, where $\phi_{\uparrow,\downarrow}$ are small amplitudes, and substituting into the coupled GP equations (7.32a) and (7.32b). The dispersion relation exhibits two branches, ω_+ and ω_-, which reflect the interplay between SOC ($k_{\rm L}$) and Rabi coupling (Ω). The bright solitons, observed in figures 7.17 and 7.18, exist when $\Omega > k_{\rm L}^2$, corresponding to a regime where the dispersion favours localized states. Stripe solitons, seen in figures 7.19 and 7.20, emerge when $\Omega < k_{\rm L}^2$, leading to periodic density modulations due to the modified dispersion, as discussed in references [24, 33]. A detailed phase diagram based on this dispersion relation can be found in reference [33].

The dynamics of bright and stripe solitons in SO-coupled BECs reveal the critical role of wavefunction symmetry and SOC parameters. Antisymmetric initial conditions lead to stable, stationary solitons, while cross-symmetric conditions result in unstable, propagating solitons. The dispersion relation provides a theoretical framework for understanding these phases, highlighting the transition between bright and stripe solitons as a function of Ω and k_{L}. These findings underscore the rich physics of SO-coupled BECs, offering a platform to explore quantum solitons and their applications in quantum simulation.

7.8 Summary and future challenges

This chapter examined the dynamics of SO-coupled (SOC) and Rabi-coupled BECs with hyperfine spin-$F = 1/2$, focusing on nonlinear excitations and their stability. Analytical solutions were derived using a Lax pair and Darboux transformation, assuming a correlation between the time-dependent trap frequency and scattering length, which enabled the study of bright solitons, dark solitons, breathers, rogue waves, rogue-dark–bright states, and mixed bound states. Collisional dynamics and stability of these excitations, both with and without transient traps were also analyzed thoroughly. A comparison between Rashba and Dresselhaus SOC, highlighting their role in spin dynamics is also brought out. For general system parameters, numerical methods, including the split-step Fourier and Newton conjugate gradient techniques, were employed to explore the dynamics. These findings have brought out the rich nonlinear behaviour of SOC and Rabi-coupled BECs, driven by tunable spin interactions.

It should be pointed out that there are several grey areas in the domain of SO-coupled BECs which are yet to be explored giving scope for further research. The dynamics of spin-$F = 1$ and spin-$F = 3/2$ SOC and Rabi-coupled BECs, with specific reference to the identification of the associated nonlinear excitations, warranted analytical and numerical exploration to build on the progress made in the spin-$F = 1/2$ BECs. The role of free parameters in SOC systems, such as coupling strengths and trap frequencies, invited deeper study to uncover novel excitations. The stability and interactions of higher-order rogue waves and mixed states in multi-component BECs posed unresolved challenges. Additionally, the impact of other external trapping potentials, such as optical lattices, on soliton and breather dynamics remained unexplored, offering opportunities to probe SOC-induced phenomena in diverse experimental settings.

This chapter has been reproduced with permission from [19].

7.9 Problems

Exercise 7.1

1. Find the ground state wavefunctions using the SSCN method with imaginary-time propagation corresponding to the one-dimensional SO-coupled pseudo-spin-1/2 BEC described by the coupled GP equations (7.32a) with $V(x) = 0$, $g_{\uparrow\uparrow} = g_{\downarrow\downarrow} = 1$, $g_{\uparrow\downarrow} = g_{\downarrow\uparrow} = 0.5$, $\Omega = 0.5$, $k_{\mathrm{L}} = 0.5$, and normalization $\int(|\psi_{\uparrow}|^2 + |\psi_{\downarrow}|^2)\mathrm{d}x = 1$. Find the ground state wavefunctions using the

SSCN method with imaginary-time propagation and evaluate the chemical potential.

2. Investigate the collision dynamics of two solitons in a SO-coupled BEC described by the GP equations (7.32a) with the initial wavefunctions

$$\psi_\uparrow(x, 0) = \frac{1}{\sqrt{2\pi}}(\mathrm{e}^{-(x+5)^2/2} + \mathrm{e}^{-(x-5)^2/2}),$$
$$\psi_\downarrow(x, 0) = \frac{1}{\sqrt{2\pi}}(\mathrm{e}^{-(x+5)^2/2} - \mathrm{e}^{-(x-5)^2/2}),$$

with parameters $\alpha = 1$, $\beta = 0.5$, $\Omega = 0.2$, and $k_\mathrm{L} = 1$.

Using the SSCN method with spatial domain $x \in [-100, 100]$ ($\mathrm{d}x = 0.1$) and time step $\mathrm{d}t = 0.001$, simulate the system for $t \in [0, 20]$. Implement real-time propagation by separately solving the kinetic and potential steps as in previous problems. Track the density profiles and compute the centre-of-mass motion throughout the evolution.

Analyze the results, focusing on how the SOC affects the soliton interaction, and verify whether the solitons exhibit predicted repulsive dynamics leading to separation along $\pm x$ directions.

3. Consider a spin-1 spinor BEC described by the coupled GP equations for the components ψ_{+1}, ψ_0, and ψ_{-1}:

$$\mu\psi_m = \left(-\frac{\hbar^2}{2m}\frac{\partial^2}{\partial x^2} + V(x) + c_0\sum_{m'}|\psi_{m'}|^2\right)\psi_m + c_2\sum_{m',m''}\langle m, m'|\mathbf{F}\cdot\mathbf{F}|m'', m\rangle\,\psi_{m'}\psi^*_{m''}, \tag{7.66}$$

where $V(x) = \frac{1}{2}m\omega^2x^2$ is the trapping potential with $\omega = 0.1$, $c_0 = 1$ and $c_2 = 0.1$ are the spin-independent and spin-dependent interaction strengths, respectively, and the system is normalized such that $\sum_m \int |\psi_m|^2\mathrm{d}x = 1$. $\mathbf{F} = (F_x, F_y, F_z)$ is spin-1 angular momentum matrices in the basis $|m\rangle = |+1\rangle, |0\rangle, |-1\rangle$:

$$F_x = \frac{\hbar}{\sqrt{2}}\begin{pmatrix}0 & 1 & 0\\ 1 & 0 & 1\\ 0 & 1 & 0\end{pmatrix},\ F_y = \frac{\hbar}{\sqrt{2}}\begin{pmatrix}0 & -i & 0\\ i & 0 & -i\\ 0 & i & 0\end{pmatrix},\ F_z = \hbar\begin{pmatrix}1 & 0 & 0\\ 0 & 0 & 0\\ 0 & 0 & -1\end{pmatrix}.$$

Using the SSCN method, compute the ground state of this system via imaginary-time propagation.

4. Starting from the spinor GP equations in problem 3 (7.66), introduce a linear Zeeman term $-pF_z\psi_m$ with strength $p = 0.2$. Using the SSFT method, numerically evolve the ground state obtained in problem 3 for time $t \in [0, 10]$ and compute the kinetic energy operator in Fourier space at each time step. Track and plot the time evolution of the magnetization $\langle F_z\rangle = \int(|\psi_{+1}|^2 - |\psi_{-1}|^2)\mathrm{d}x$ throughout the simulation and discuss the nature of the oscillatory dynamics.

5. Find the ground state using the CG method of the spin-1 BEC including SOC with

$$H_{\mathrm{SO}} = \mathrm{i}k_{\mathrm{L}}\sum_{m}\langle m|F_x|m'\rangle\,\psi_m^*\frac{\partial\psi_{m'}}{\partial x}.$$

with $k_{\mathrm{L}} = 0.5$, $c_0 = 1$, $c_2 = -0.1$, $V(x) = 0$.

6. Investigate phase separation in a pseudo-spin-1/2 BEC governed by the coupled GP equations (7.32a) using the SSFT method over the time interval $t \in [0, 20]$. Fix the interaction parameters as $g_{\uparrow\uparrow} = g_{\downarrow\downarrow} = 1$, $g_{\uparrow\downarrow} = g_{\downarrow\uparrow} = 1.5$, and set the Raman coupling $\Omega = 0.1$, the SOC strength $k_{\mathrm{L}} = 0.5$, and the external potential $V(x) = 0$. Use the following initial wavefunctions:

$$\psi_\uparrow(x, 0) = \frac{1}{\sqrt{\pi}}\mathrm{e}^{-(x-5)^2/2}, \quad \psi_\downarrow(x, 0) = \frac{1}{\sqrt{\pi}}\mathrm{e}^{-(x+5)^2/2}.$$

Simulate the time evolution and plot the resulting density profiles $|\psi_\uparrow(x, t)|^2$ and $|\psi_\downarrow(x, t)|^2$ to analyze the emergence of phase separation.

References

[1] Bychkov Y A and Rashba E I 1984 Oscillatory effects and the magnetic susceptibility of carriers in inversion layers *J. Phys. C Solid State Phys.* **17** 6039–45

[2] Dresselhaus G 1955 Spin-orbit coupling effects in zinc blende structures *Phys. Rev.* **100** 580–6

[3] Ravisankar R 2020 Spin-orbit coupled spinor Bose-Einstein condensates: ground states, dynamics and excitations *PhD Thesis* Bharathidasan University, Tiruchirappalli, India

[4] Ganichev S D and Golub L E 2014 Interplay of Rashba/Dresselhaus spin splittings probed by photogalvanic spectroscopy - a review *Phys. Status Solidi* B **251** 1801–23

[5] Tao L L and Tsymbal E Y 2018 Persistent spin texture enforced by symmetry *Nat. Commun.* **9** 2763

[6] Stamper-Kurn D M and Ueda M 2013 Spinor Bose gases: symmetries, magnetism, and quantum dynamics *Rev. Mod. Phys.* **85** 1191–244

[7] Lin Y J, Jiménez-García K and Spielman I B 2011 Spin-orbit-coupled Bose-Einstein condensates *Nature* **471** 83–6

[8] Jiménez-García K, LeBlanc L, Williams R A, Beeler M C, Qu C, Gong M, Zhang C and Spielman I B 2015 Tunable spin-orbit coupling via strong driving in ultracold-atom systems *Phys. Rev. Lett.* **114** 125301

[9] Li J R, Lee J, Huang W, Burchesky S, Shteynas B, Top F C, Jamison A O and Ketterle W 2017 A stripe phase with supersolid properties in spin-orbit-coupled Bose-Einstein condensates *Nature* **543** 91–4

[10] Galitski V and Spielman I B 2013 Spin-orbit coupling in quantum gases *Nature* **494** 49–54S

[11] Goldman N and Dalibard J 2014 Periodically driven quantum systems: effective Hamiltonians and engineered gauge fields *Phys. Rev.* X **4** 031027

[12] Radić J, Sedrakyan T A, Spielman I B and Galitski V 2011 Vortices in spin-orbit-coupled Bose-Einstein condensates *Phys. Rev.* A **84** 063604

[13] Wang C, Gao C, Jian C -M and Zhai H 2010 Spin-orbit coupled spinor Bose-Einstein condensates *Phys. Rev. Lett.* **105** 160403

[14] Zhai H 2015 Degenerate quantum gases with spin-orbit coupling: a review *Rep. Progr. Phys.* **78** 026001

[15] Cao S, Shan C -J, Zhang D -W, Qin X and Xu J 2015 Dynamical generation of dark solitons in spin-orbit-coupled Bose-Einstein condensates *J. Opt. Soc. Am.* B **32** 201

[16] Tononi A, Wang Y and Salasnich L 2019 Quantum solitons in spin-orbit-coupled Bose-Bose mixtures *Phys. Rev.* A **99** 063618

[17] Serkin V and Hasegawa A 2002 Exactly integrable nonlinear Schrödinger equation models with varying dispersion, nonlinearity and gain: application for soliton dispersion *IEEE J. Sel. Top. Quantum Electron.* **8** 418–31

[18] Radha R, Vinayagam P S and Porsezian K 2013 Rotation of the trajectories of bright solitons and realignment of intensity distribution in the coupled nonlinear Schrödinger equation *Phys. Rev.* E **88** 032903

[19] Vinayagam P and Radha R 2025 Robust dynamics of rogue waves, breathers and mixed bound state solutions in spin-orbit and Rabi coupled condensates *Phys. Lett.* A **531** 130169

[20] Belkroukra H, Sameut H C and Benarous M 2022 New families of breathers in trapped two-component condensates *Phys. Wave Phenom.* **30** 67–72

[21] Vinayagam P, Radha R, Bhuvaneswari S, Ravisankar R and Muruganandam P 2017 Bright soliton dynamics in spin orbit-Rabi coupled Bose-Einstein condensates *Commun. Nonlinear Sci.* **50** 68–76

[22] Matveev V B and Salle M 1991 *Darboux Transformations and Solitons* (Springer)

[23] Vinayagam P S, Radha R and Porsezian K 2013 Taming rogue waves in vector Bose-Einstein condensates *Phys. Rev.* E **88** 042906

[24] Achilleos V, Frantzeskakis D J, Kevrekidis P G and Pelinovsky D E 2013 Matter-wave bright solitons in spin-orbit coupled Bose-Einstein condensates *Phys. Rev. Lett.* **110** 264101

[25] Bao W and Cai Y 2015 Ground states and dynamics of spin-orbit-coupled Bose-Einstein condensates *SIAM J. Appl. Math.* **75** 492–517

[26] Adhikari S K and Muruganandam P 2002 Bose-Einstein condensation dynamics from the numerical solution of the Gross-Pitaevskii equation *J. Phys. B: At. Mol. Opt. Phys.* **35** 2831–43

[27] Muruganandam P and Adhikari S 2009 Fortran programs for the time-dependent Gross-Pitaevskii equation in a fully anisotropic trap *Comput. Phys. Commun.* **180** 1888–912

[28] Kumar R K, Young S L E, Vudragović D, Balaž A, Muruganandam P and Adhikari K S 2015 Fortran and C programs for the time-dependent dipolar Gross-Pitaevskii equation in an anisotropic trap *Comput. Phys. Commun.* **195** 117–28

[29] Koonin S E 2018 *Computational Physics: Fortran Version* (CRC Press)

[30] Ames W F 2014 *Numerical Methods for Partial Differential Equations* 3rd edn (Academic) (description based upon print version of record)

[31] Dautray R and Lions J L 2012 *Mathematical Analysis and Numerical Methods for Science and Technology: Vol 3 Spectral Theory and Applications* (Springer)

[32] Yang J 2010 *Nonlinear Waves in Integrable and Nonintegrable Systems* (Society for Industrial and Applied Mathematics)

[33] Ravisankar R, Vudragović D, Muruganandam P, Balaž A and Adhikari S K 2021 Spin-1 spin-orbit- and Rabi-coupled Bose-Einstein condensate solver *Comput. Phys. Commun.* **259** 107657

[34] Li Y, Pitaevskii L P and Stringari S 2012 Quantum tricriticality and phase transitions in spin-orbit coupled Bose-Einstein condensates *Phys. Rev. Lett.* **108** 225301

[35] Ho T L and Zhang S 2011 Bose-Einstein condensates with spin-orbit interaction *Phys. Rev. Lett.* **107** 150403
[36] Zhang J Y *et al* 2012 Collective dipole oscillations of a spin-orbit coupled Bose-Einstein condensate *Phys. Rev. Lett.* **109** 115301
[37] Ravisankar R, Sriraman T, Salasnich L and Muruganandam P 2020 Quenching dynamics of the bright solitons and other localized states in spin-orbit coupled Bose-Einstein condensates *J. Phys. B: At. Mol. Opt. Phys.* **53** 195301
[38] Kartashov Y V, Konotop V V and Zezyulin D A 2014 Bose-Einstein condensates with localized spin-orbit coupling: soliton complexes and spinor dynamics *Phys. Rev.* A **90** 063621
[39] Ostrovskaya E A, Kivshar Y S, Chen Z and Segev M 1999 Interaction between vector solitons and solitonic gluons *Opt. Lett.* **24** 327

Paulsamy Muruganandam and Ramaswamy Radha

Chapter 8

Bose–Einstein condensates with long-range interactions

Atoms with large magnetic dipole moments can form Bose–Einstein condensates (BECs) known as dipolar BECs. The theoretical and experimental progress in the study of ultracold atomic gases, particularly involving ^{52}Cr, ^{168}Er, and ^{164}Dy, has generated significant interest in exploring the unique properties of these dipolar systems. Unlike typical alkali atoms used in early BEC experiments, such as rubidium or sodium, certain atomic species like chromium (^{52}Cr), dysprosium (^{164}Dy), and erbium (^{168}Er) possess significant permanent magnetic dipole moments. These strong dipolar interactions introduce long-range and anisotropic character to the interatomic forces, which contrasts sharply with the short-range, isotropic contact interactions that dominate in conventional BECs. Dipolar quantum gases have attracted extensive attention from both theorists and experimentalists, primarily because the long-range and anisotropic dipole–dipole interactions (DDIs) can be precisely controlled and tuned.

Since the realization of BECs in these highly magnetic species, particularly ^{52}Cr and ^{164}Dy, there has been growing interest in exploring the rich many-body physics enabled by DDIs. These interactions give rise to a variety of novel quantum phenomena, including the formation of quantum droplets, roton excitations, structured ground states, and the emergence of dipolar supersolids. While quantum degenerate gases of other species continue to yield fascinating insights, the unique interaction landscape of dipolar BECs opens the door to previously inaccessible states of matter and complex quantum phase transitions [1].

8.1 Dipolar Bose–Einstein condensates

BECs with dipolar atoms and molecules offer unparalleled advantages due to their long-range, anisotropic interactions, tunability, and ability to access exotic quantum

doi:10.1088/978-0-7503-5447-9ch8

phases. These systems serve as powerful platforms for quantum simulation, quantum information processing, precision metrology, ultracold chemistry, and fundamental physics tests, while also providing analogues to condensed matter and astrophysical phenomena. Experimental advances, such as the realization of Dy and Er BECs or K–Rb molecular BECs, have already demonstrated transformative applications, and ongoing research promises further breakthroughs in controlling and harnessing these quantum systems.

The first experimental realization of a dipolar BEC was reported in 2005 by the Pfau group in Stuttgart, using ^{52}Cr atoms, which possess a magnetic dipole moment of 6 μ_B [2]. Bose–Einstein condensation was achieved at a critical temperature of approximately $T_c \approx 700$ nK. Chromium is particularly suitable for realizing a dipolar BEC due to its electronic configuration [Ar]3d^{5}4s^1. According to Hund's rules, this configuration leads to a large total spin, resulting in a magnetic dipole moment of 6 μ_B (Bohr magnetons).

Dipolar BECs exhibit unique characteristics due to the anisotropic nature of long-range dipolar interactions. For instance, after release from the magnetic trap, the condensate aspect ratio behaves differently due to the anisotropy of the interaction depending on whether the dipoles are aligned along or perpendicular to the trap symmetry axis [2]. Furthermore, the stability of dipolar BECs depends strongly on trap geometry, as explored by the Pfau group [3]. In an elongated (cigar-shaped) trap with trap aspect ratio $\lambda < 1$, the dipolar interactions are predominantly attractive, as shown in figure 8.1(A). Conversely, in a flattened (pancake-shaped) trap with $\lambda > 1$, the interactions are primarily repulsive, as depicted in figure 8.1(B). As a result,

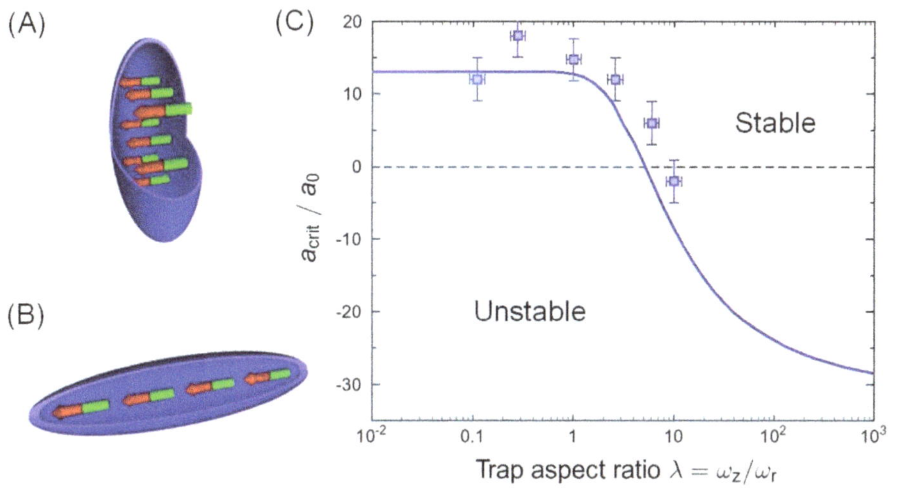

Figure 8.1. (A) In a pancake-shaped trap, dipoles predominantly repel each other. (B) In a cigar-shaped trap, dipolar interactions are primarily attractive. (A) and (B) Adapted from reference [3], with permission from Springer Nature. (C) Stability diagram of a dipolar BEC in the scattering length (a) versus trap aspect ratio (λ) plane. Experimental data (squares) and theoretical predictions (solid line) for the critical scattering length a_{crit}. Reproduced with permission from [4]. CC-BY-NC 4.0.

dipolar BECs exhibit greater stability in pancake-shaped traps. The stability diagram, shown in figure 8.1(C), maps the stability of the dipolar BEC as a function of scattering length and trap aspect ratio.

The theoretical foundation for describing interatomic interactions in ultracold gases in the presence of dipolar forces was laid by You and collaborators [5–8]. They analyzed the effective interaction potential (pseudopotential) for atoms at ultralow temperatures in the presence of a static (dc) electric field. When such a field is applied along the z-axis, the effective interaction potential takes the form:

$$U(\mathbf{R}) = g\delta(\mathbf{R}) + U_{\mathrm{dd}}(\mathbf{R}), \tag{8.1}$$

where the first term represents the usual short-range contact interaction with interaction strength $g = \frac{4\pi\hbar^2 a}{m}$, and the second term $U_{\mathrm{dd}}(\mathbf{R})$ describes the long-range DDI between atoms in the condensate.

8.1.1 Short range or contact interaction

The contact interaction arises when the wavefunctions of two atoms overlap in space. This interparticle interaction in condensate is analogous to the interaction between hard spheres, where the interaction is short-range and isotropic, meaning it has no preferred direction since the spheres only interact upon direct contact.

At ultra low temperatures, the contact interaction is effectively described by low-energy scattering between atoms. In quantum scattering theory, it is well established that at low energies, the details of the interaction potential become irrelevant, and the scattering behaviour depends solely on a single parameter: the s-wave scattering length a [9, 10]. Consequently, atomic collisions in a BEC are dominated by two-body s-wave interactions. The condensate wavefunction under such interactions is well described by a scalar field.

In the two-body interaction potential, two distinct spatial regions can be identified based on the interatomic separation $\mathbf{R} = \mathbf{r} - \mathbf{r}'$. For distances $r = |\mathbf{R}| > r_0$, where r_0 is the effective range of the potential, the interaction is dominated by the Van der Waals force, which decays as r^{-6} and can be neglected in most cases. In contrast, for $r < r_0$, the potential has a more complex shape and cannot be ignored.

To calculate the low-energy properties of a dilute Bose gas, the whole interaction potential can be replaced with an effective Hamiltonian in which only the scattering length appears explicitly. This approach, introduced by Fermi and formalized by Huang, is known as the pseudopotential [11]. The simplest pseudopotential for describing two-body short-range interactions is given by:

$$V^c(\mathbf{R}) = g\delta(\mathbf{R}), \tag{8.2}$$

where $g = 4\pi\hbar^2 a/m$ is the contact interaction strength for indistinguishable bosons of mass m.

8.1.2 Long range or dipole–dipole interaction

In addition to the contact interaction discussed above, atoms can also interact through other interactions. When atoms possess a permanent magnetic moment, they also exhibit a magnetic dipole moment, allowing them to interact via DDI.

Atoms can exhibit two kinds of dipole moments: the electric dipole moment (μ_e) and the magnetic dipole moment (μ_m). The electric dipole moment arises from the spatial separation of positive and negative charges. For example, when an external electric field is applied to a polar molecule, the field can induce a separation between the negatively charged electrons and the positively charged nucleus, causing the molecule to develop a net electric dipole moment, as illustrated in figure 8.2.

Consequently, in addition to the usual short-range contact interaction described above, the atoms also experience a long-range magnetic DDI, given by:

$$U_{\mathrm{dd}}(\mathbf{R}) = -\frac{\mu_0}{4\pi}\frac{(\boldsymbol{\mu}_1 \cdot \boldsymbol{\mu}_2)\mathbf{R}^2 - 3(\boldsymbol{\mu}_1 \cdot \mathbf{R})(\boldsymbol{\mu}_2 \cdot \mathbf{R})}{|\mathbf{R}|^5}, \tag{8.3}$$

where μ_0 is the vacuum permeability, and $\mathbf{R}$ is the relative separation vector between two dipoles.

Assuming that the dipoles are identical and polarized along the z-axis, such that $\boldsymbol{\mu}_1 = \boldsymbol{\mu}_2 = \bar{\boldsymbol{\mu}}$, the DDI simplifies to:

$$U_{\mathrm{dd}}(\mathbf{R}) = d^2\frac{1 - 3\cos^2\theta}{|\mathbf{R}|^3}, \tag{8.4}$$

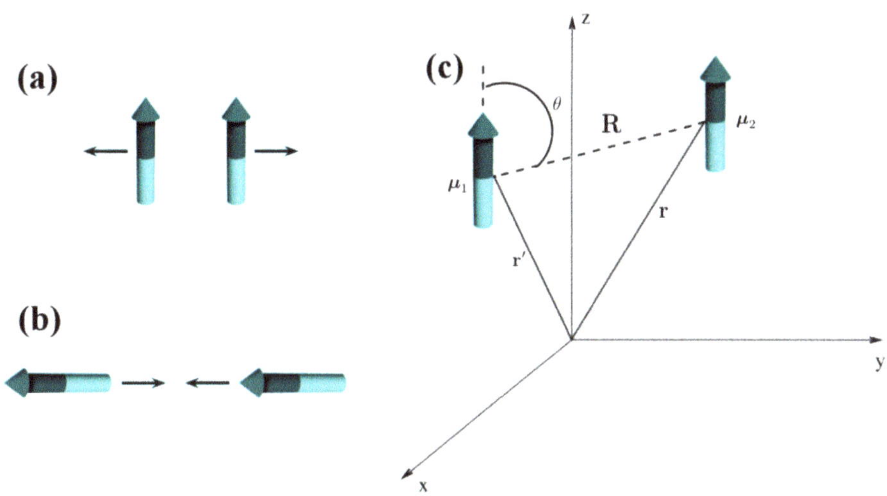

Figure 8.2. (a) Two polarized dipoles aligned side by side repel each other as indicated in black arrows; (b) two polarized dipoles in a head-to-tail configuration attract each other as indicated in black arrows; (c) geometry for the interaction of two aligned dipoles. Reproduced with permission from [12]. CC-BY-NC 4.0.

where $d^2 = \mu_0\bar{\mu}^2/(4\pi)$ is the dipolar coupling constant, θ is the angle between $\mathbf{R}$ and the polarization direction (z-axis), and $\bar{\mu} = |\bar{\boldsymbol{\mu}}|$ is the magnitude of the dipole moment.

The anisotropic nature of the interaction influences various properties of the condensate. For instance, attractive dipolar interactions can increase the critical temperature for Bose–Einstein condensation, while repulsive interactions lower it. Another key difference between contact and dipolar interactions lies in their spatial range: the contact interaction is short-ranged, modelled as a delta function, whereas the DDI decays as $1/r^3$, making it effectively long-range.

The relative strength between dipolar and contact interactions can be quantified using a dimensionless parameter:

$$a_{\mathrm{dd}} = \frac{\mu_0\bar{\mu}^2 m}{12\pi\hbar^2 a}, \tag{8.5}$$

where a_{dd} is known as the dipolar length, a is the s-wave scattering length, and m is the atomic mass. The ratio a_{dd}/a plays a crucial role in determining the stability and collective behaviour of dipolar condensates.

8.2 Mean-field dipolar Gross–Pitaevskii (GP) equation

Solving the N-body Schrödinger equation for the dipolar BEC is numerically very challenging. However, for many dilute Bose gases, the system can be accurately described using a mean-field approximation. In this approach, the complex N-body system is reduced to a one-particle picture, where each particle experiences an effective potential known as the mean-field potential arising from interactions with all other particles.

The time-dependent mean-field GP equation which governs the dynamics of a dipolar BEC with N atoms describes the evolution of the normalized wavefunction $\Psi(\mathbf{r}, t)$ under the condition $\int |\Psi(\mathbf{r}, t)|^2 d\mathbf{r} = 1$ and is expressed in three dimensions as [8, 13–15]

$$\begin{aligned} i\hbar\frac{\partial\Psi(\mathbf{r}, t)}{\partial t} = & \Bigg[-\frac{\hbar^2}{2M}\nabla^2 + V_{\mathrm{trap}}(\mathbf{r}) + 4\pi\hbar^2\frac{a}{M}N|\Psi(\mathbf{r}, t)|^2 \\ & + N\int V_{\mathrm{dd}}(\mathbf{r} - \mathbf{r}')|\Psi(\mathbf{r}', t)|^2 d\mathbf{r}'\Bigg]\Psi(\mathbf{r}, t), \end{aligned} \tag{8.6}$$

where $\mathbf{r} = (x, y, z)$ is the position vector, M is the atomic mass, $\hbar$ is the reduced Planck constant, and ∇^2 is the Laplacian operator. The harmonic trapping potential is $V_{\mathrm{trap}}(\mathbf{r}) = \frac{M}{2}(\omega_x^2 x^2 + \omega_y^2 y^2 + \omega_z^2 z^2)$, with ω_x, ω_y, ω_z being the trap frequencies. The interaction potential consists of two terms: the contact interaction, characterized by the s-wave scattering length a, and the dipolar interaction, represented by the potential $V_{\mathrm{dd}}(\mathbf{r} - \mathbf{r}') = \frac{\mu_0\mu^2}{4\pi|\mathbf{r} - \mathbf{r}'|^3}(1 - 3\cos^2\theta)$, where μ_0 is the vacuum permeability, μ is the dipole moment, and θ is the angle between the dipole orientation and $\mathbf{r} - \mathbf{r}'$. The Hamiltonian encapsulates the kinetic energy, trapping potential, and both short-range and long-range interactions, providing a comprehensive description of the dipolar BEC's mean-field dynamics.

The external potential $V_{\text{trap}}(\mathbf{r})$ is assumed to be a confining, axially symmetric harmonic trap of the form:

$$V_{\text{trap}}(\mathbf{r}) = \frac{1}{2}m\omega^2\left(\Omega_\rho^2\rho^2 + \Omega_z^2 z^2\right), \tag{8.7}$$

where $\Omega_z\omega$ and $\Omega_\rho\omega$ are the axial and radial trapping frequencies, respectively. Here, $\mathbf{r} \equiv (\rho, z)$, with $\rho = \sqrt{x^2 + y^2}$ denoting the radial coordinate and z the axial coordinate.

8.2.1 Dipolar GP equation in dimensionless variables

Convenient dimensionless parameters can be defined using the characteristic trap frequency ω and the corresponding harmonic oscillator length $l = \sqrt{\hbar/(m\omega)}$. Introducing dimensionless variables: $\bar{\mathbf{r}} = \mathbf{r}/l$, $\bar{t} = \omega t$, $\bar{\rho} = \rho/l$, $\bar{z} = z/l$, and $\bar{\psi} = l^{3/2}\psi$, equation (8.6) can be rewritten in the dimensionless form as

$$\begin{aligned} i\frac{\partial\bar{\psi}(\bar{\mathbf{r}}, \bar{t})}{\partial\bar{t}} = \Bigg[&-\frac{1}{2}\nabla^2 + \frac{1}{2}\left(\Omega_\rho^2\bar{\rho}^2 + \Omega_z^2\bar{z}^2\right) + 4\pi\bar{a}N|\bar{\psi}|^2 \\ &+ 3N\bar{a}_{\text{dd}}\int \bar{V}_{\text{dd}}(\bar{\mathbf{R}})|\bar{\psi}(\bar{\mathbf{r}}', \bar{t})|^2\, \mathrm{d}\bar{\mathbf{r}}'\Bigg]\bar{\psi}(\bar{\mathbf{r}}, \bar{t}), \end{aligned} \tag{8.8}$$

where the dimensionless DDI potential is given by

$$\bar{V}_{\text{dd}}(\bar{\mathbf{R}}) = \frac{1 - 3\cos^2\theta}{|\bar{\mathbf{R}}|^3}, \tag{8.9}$$

with $\bar{a} = a/l$ and $\bar{a}_{\text{dd}} = a_{\text{dd}}/l$.

For notational simplicity, the bars on all variables are omitted. The dimensionless GP equation then takes the following form

$$\begin{aligned} i\frac{\partial\psi(\mathbf{r}, t)}{\partial t} = \Bigg[&-\frac{1}{2}\nabla^2 + \frac{1}{2}\left(\Omega_\rho^2\rho^2 + \Omega_z^2 z^2\right) \\ &+ 4\pi aN|\psi(\mathbf{r}, t)|^2 + 3Na_{\text{dd}}\int V_{\text{dd}}(\mathbf{R})\, |\psi(\mathbf{r}', t)|^2\, \mathrm{d}\mathbf{r}'\Bigg]\psi(\mathbf{r}, t). \end{aligned} \tag{8.10}$$

8.2.2 One-dimensional GP equation for a cigar-shaped dipolar BEC

For a cigar-shaped dipolar BEC with strong radial confinement ($\Omega_\rho > \Omega_z$), it is assumed that the condensate remains in the ground state along the radial direction [16]

$$\psi(\boldsymbol{\rho}) = \frac{1}{d_\rho\sqrt{\pi}}\exp\left(-\frac{\rho^2}{2d_\rho^2}\right), \tag{8.11}$$

of the transverse trap and the wavefunction $\psi(\mathbf{r}) = \psi_{1D}(z) \times \psi(\boldsymbol{\rho})$ can be written as [17, 18]

$$\psi(\mathbf{r}) = \frac{1}{\sqrt{\pi d_\rho^2}} \exp\left(-\frac{\rho^2}{2d_\rho^2}\right)\psi_{1D}(z); \quad \Omega_\rho d_\rho^2 = 1, \tag{8.12}$$

where d_ρ is the radial harmonic oscillator length. The contribution of the dipole potential to the energy is

$$H_{\mathrm{dd}} = \frac{N}{2}\int \mathrm{d}^3r \int \mathrm{d}^3r' n(\mathbf{r}) U_{\mathrm{dd}}(\mathbf{r} - \mathbf{r}') n(\mathbf{r}') = \frac{1}{2}\frac{N}{(2\pi)^3}\int \mathrm{d}^3k \tilde{n}(\mathbf{k}) \tilde{U}_{\mathrm{dd}}(\mathbf{k}) \tilde{n}(-\mathbf{k}), \tag{8.13}$$

where the density is defined as $n(\mathbf{r}) \equiv |\psi(\mathbf{r})|^2$. In equation (8.13), a convolution of the corresponding variables is performed in Fourier space, where the tilde denotes Fourier transformations [8, 19–21]:

$$\tilde{U}_{\mathrm{dd}}(\mathbf{k}) = \frac{4\pi}{3} 3a_{\mathrm{dd}}\left[\frac{3k_z^2}{k^2} - 1\right], \tag{8.14}$$

$$\tilde{n}(\mathbf{k}) = \exp\left[-\frac{k_\rho^2 d_\rho^2}{4}\right]\tilde{n}_{1D}(k_z). \tag{8.15}$$

The details of the Fourier transformation of DDI potential, $\tilde{U}_{\mathrm{dd}}(\mathbf{k})$ is given in appendix C. The k_x, k_y integrals in (8.13) can now be done and

$$\begin{aligned} H_{\mathrm{dd}} &= \frac{4\pi N}{3}\frac{3a_{\mathrm{dd}}}{2}\frac{1}{2\pi}\int_{-\infty}^{\infty} \mathrm{d}k_z \tilde{n}_{1D}(k_z)\tilde{n}_{1D}(-k_z) \\ &\times \frac{1}{(2\pi)^2}\int_{-\infty}^{\infty}\int_{-\infty}^{\infty} \mathrm{d}k_x \mathrm{d}k_y \left(\frac{3k_z^2}{k_\rho^2 + k_z^2} - 1\right)\exp\left(-\frac{k_\rho^2 d_\rho^2}{2}\right), \\ &= \frac{N}{2}\frac{1}{2\pi}\int_{-\infty}^{\infty} \mathrm{d}k_z \tilde{n}_{1D}(k_z)\tilde{n}_{1D}(-k_z) V_{1D}(k_z), \end{aligned} \tag{8.16}$$

where the 1D potential in Fourier space can be written as

$$V_{1\mathrm{D}}(k_z) = 2a_{\mathrm{dd}}\int_0^{\infty} \mathrm{d}k_\rho k_\rho \left(\frac{3k_z^2}{k_\rho^2 + k_z^2} - 1\right)\exp\left(-\frac{k_\rho^2 d_\rho^2}{2}\right), \tag{8.17}$$

or

$$V_{1\mathrm{D}}(k_z) = \frac{2a_{\mathrm{dd}}}{d_\rho^2} s_{1\mathrm{D}}\left(\frac{k_z d_\rho}{\sqrt{2}}\right), \quad s_{1D}(\zeta) = \int_0^{\infty} du\left[\frac{3\zeta^2}{u + \zeta^2} - 1\right]e^{-u}. \tag{8.18}$$

The one-dimensional form of the dipolar potential in configuration space is given by

$$U_{\mathrm{dd}}^{1\mathrm{D}}(Z) = \frac{1}{2\pi}\int_{-\infty}^{\infty} \mathrm{d}k_z \mathrm{d}^{ik_z z} V_{1\,\mathrm{D}}(k_z) = \frac{6a_{\mathrm{dd}}}{(\sqrt{2}d_\rho)^3}\left[\frac{4}{3}\delta(\sqrt{t}) + 2\sqrt{t} - \sqrt{\pi}(1 + 2t)e^t \operatorname{erfc}(\sqrt{t})\right]. \tag{8.19}$$

where $t = [Z/(\sqrt{2}\, d_\rho)]^2$, $Z = |z - z'|$. Similar, but not identical, 1D reduced potential was derived in [17, 18], where the δ-function term was absent. To derive the effective one-dimensional equation for a cigar-shaped dipolar BEC, the ansatz (8.12) is substituted into equation (8.10). The resulting expression is then multiplied by the ground-state wavefunction $\psi(\rho)$ and integrated over the radial coordinate ρ to obtain the 1D equation

$$\mathrm{i}\frac{\partial \psi_{1D}(z, t)}{\partial t} = \left[-\frac{1}{2}\frac{\partial^2}{z^2} + \frac{1}{2}\Omega_z^2 z^2 + \frac{2aN|\psi_{1D}|^2}{d_\rho^2} + N\int_{-\infty}^{\infty} U_{\mathrm{dd}}^{1\mathrm{D}}(Z)|\psi_{1\,\mathrm{D}}(z', t)|^2 \mathrm{d}z' \right] \psi_{1\,\mathrm{D}}(z, t). \quad (8.20)$$

8.2.3 Two-dimensional GP equation for a disk-shaped dipolar BEC

In the disk-shaped geometry, where strong axial confinement is applied ($\Omega_z > \Omega_\rho$), the dipolar BEC is assumed to occupy the ground state of the axial harmonic oscillator given by $\psi(z) = \exp(-z^2/2d_z^2)/(\pi d_z^2)^{1/4}$. Consequently, the full 3D wavefunction $\psi(\mathbf{r})$ can be expressed as a product state [16, 20, 21]:

$$\psi(\mathbf{r}) = \frac{1}{(\pi d_z^2)^{1/4}} \exp\left(-\frac{z^2}{2d_z^2} \right) \psi_{2D}(x, y), \quad (8.21)$$

where $\psi_{2D}(x, y)$ is the effective 2D wavefunction in the radial plane and $d_z = \sqrt{1/\Omega_z}$ characterizes the axial confinement length. Substituting the ansatz in equation (8.21) into the 3D GP equation (8.10), and integrating out the z-dependence, one obtains an effective 2D equation [20, 21]. Fischer further proposed that the axial width d_z is not necessarily fixed but should be determined self-consistently through the relation [20]:

$$\Omega_z^2 d_z^4 = 1 + \left(g + \frac{8\pi}{3} g_d \right) \frac{n d_z}{\sqrt{2\pi}}, \quad (8.22)$$

where g and g_d denote the contact and dipolar interaction strengths, respectively, and n is the 2D density. The contribution of the dipole potential to energy is

$$H_{\mathrm{dd}} = \frac{N}{2} \int \mathrm{d}^3 r \int \mathrm{d}^3 r' n(\mathbf{r}) U_{\mathrm{dd}}(\mathbf{r} - \mathbf{r}') n(\mathbf{r}') = \frac{1}{2}\frac{N}{(2\pi)^3} \int \mathrm{d}^3 k \tilde{n}(\mathbf{k}) \tilde{U}_{\mathrm{dd}}(\mathbf{k}) \tilde{n}(-\mathbf{k}), \quad (8.23)$$

where the quantity $n(\mathbf{r}) \equiv |\psi(\mathbf{r})|^2$ represents the density. In equation (8.23), a convolution of the respective variables is performed in Fourier space, where the tilde symbol denotes Fourier transforms:

$$\tilde{n}(\mathbf{k}) = \exp(-k_z^2 d_z^2/4) \tilde{n}_{2D}(k_x, k_y). \quad (8.24)$$

The k_z integral in equation (8.23) can now be done and

$$H_{\rm dd} = \frac{4\pi}{3}\frac{N}{2}\frac{3a_{\rm dd}}{(2\pi)^2}\int {\rm d}^2k\; \tilde{n}_{\rm 2D}(k_x, k_y)\tilde{n}_{\rm 2D}(-k_x, -k_y) \\ \times \left[\frac{2}{d_z\sqrt{2\pi}} - \frac{3k_\rho}{2}\exp\left(\frac{k_\rho^2 d_z^2}{2}\right){\rm erfc}\left(\frac{k_\rho d_z}{\sqrt{2}}\right)\right], \tag{8.25}$$

where ${\rm erfc}(z) = 1 - {\rm erf}(z)$ and $k_\rho = \sqrt{k_x^2 + k_y^2}$, with erf being the error function. Then, the 2D potential in Fourier space is given by

$$V_{\rm 2D}(\mathbf{k}_\rho) = 4\pi a_{\rm dd}\left[\frac{2}{d_z\sqrt{2\pi}} - \frac{3k_\rho}{2}\exp\left(\frac{k_\rho^2 d_z^2}{2}\right){\rm erfc}\left(\frac{k_\rho d_z}{\sqrt{2}}\right)\right] = \frac{4\pi a_{\rm dd}}{d_z\sqrt{2\pi}}h_{\rm 2D}\left(\frac{k_\rho d_z}{\sqrt{2}}\right), \tag{8.26}$$

where $h_{\rm 2D}(\xi) = 2 - 3\sqrt{\pi}\xi e^{\xi^2}{\rm erfc}(\xi)$ [21].

With this introduction, the effective 2D equation for the disk-shaped dipolar BEC can now be derived as

$${\rm i}\frac{\partial\psi_{\rm 2D}(\vec{\rho}, t)}{\partial t} = \left[-\frac{\nabla_\rho^2}{2} + \frac{\Omega_\rho^2\rho^2}{2} + \frac{4\pi a N}{\sqrt{2\pi}\,d_z}|\psi_{\rm 2D}|^2 + \frac{4\pi a_{\rm dd}N}{\sqrt{2\pi}\,d_z} \right. \\ \left. \times \int\frac{{\rm d}^2k_\rho}{(2\pi)^2}\exp({\rm i}\mathbf{k}_\rho\cdot\vec{\rho})\tilde{n}(\mathbf{k}_\rho)h_{\rm 2D}\left(\frac{k_\rho d_z}{\sqrt{2}}\right)\right]\psi_{\rm 2D}(\vec{\rho}, t), \tag{8.27}$$

where $\tilde{n}(\mathbf{k}_\rho) = \int \exp(i\mathbf{k}_\rho\cdot\vec{\rho})|\psi_{\rm 2D}(\vec{\rho})|^2 d\vec{\rho}$, $h_{\rm 2D}(\xi) = 2 - 3\sqrt{\pi}\xi e^{\xi^2}{\rm erfc}(\xi)$ [16, 21], $\mathbf{k}_\rho \equiv (k_x, k_y)$, and the dipolar term is written in Fourier space.

8.3 Split-step Crank–Nicolson method for dipolar Gross–Pitaevskii equations

8.3.1 One-dimensional dipolar Gross–Pitaevskii equation

The one-dimensional GP equation, as presented in equation (8.28), describes the dynamics of a dipolar BEC tightly confined in the transverse directions and subject to a harmonic trap along the z-axis. For notational clarity, the wavefunction is denoted as $\psi(z, t)$ replacing the alternative notation $\psi_{\rm 1D}$ and the equation is expressed as

$${\rm i}\frac{\partial\psi(z, t)}{\partial t} = \left[-\frac{1}{2}\frac{\partial^2}{\partial z^2} + \frac{1}{2}\Omega_z^2 z^2 + \frac{2aN}{d_\rho^2}|\psi(z, t)|^2 \right. \\ \left. + N\int_{-\infty}^{\infty} U_{\rm dd}^{\rm 1D}(z - z')|\psi(z', t)|^2{\rm d}z'\right]\psi(z, t) \equiv H\psi(z, t). \tag{8.28}$$

In the above, the wavefunction $\psi(z, t)$ is subjected to the normalization condition $\int|\psi(z, t)|^2{\rm d}z = 1$.

The Hamiltonian encapsulates four distinct contributions: the kinetic energy, associated with the second derivative term; the harmonic potential energy, arising from the trapping potential; the contact interaction, proportional to the local density $|\psi|^2$; and the nonlocal dipolar interaction, represented by the integral term. The nonlocal character of the dipolar interaction, which couples the wavefunction at position z with its density at all other positions z', poses a computational challenge. To address this, numerical solutions employ fast Fourier transforms (FFTs) to evaluate the convolution in Fourier space. This approach significantly enhances computational efficiency, making the numerical treatment of the 1D dipolar GP equation feasible for studying ground states and dynamics of dipolar BECs.

The split-step Crank–Nicolson (SSCN) method addresses this equation by discretizing the spatial domain and time splitting the Hamiltonian into two parts: $H = H_1 + H_2$, where

$$H_1 = \frac{1}{2}\Omega^2 z^2 + \frac{2aN}{d_\rho^2}|\psi(z, t)|^2 + N\int_{-\infty}^{\infty} U_{\rm dd}^{\rm 1D}(z - z')|\psi(z', t)|^2 {\rm d}z', \quad H_2 = -\frac{1}{2}\frac{\partial^2}{\partial z^2}. \tag{8.29}$$

The time evolution over a small time step Δ approximates the full propagator as

$$\psi(z, t + \Delta) \approx {\rm e}^{-{\rm i}H\Delta}\psi(z, t) \approx {\rm e}^{-{\rm i}H_1\Delta/2}{\rm e}^{-{\rm i}H_2\Delta}{\rm e}^{-{\rm i}H_1\Delta/2}\psi(z, t).$$

The propagation proceeds in two stages. First, the non-derivative step solves

$${\rm i}\frac{\partial\psi}{\partial t} = H_1\psi,$$

starting from $\psi(z, t_n)$, yielding an intermediate solution

$$\psi^{n+1/2}(z) = {\rm e}^{-{\rm i}\Delta H_1}\psi^n(z),$$

where $\psi^n(z) = \psi(z, t_n)$. Since H_1 lacks spatial derivatives, this step is computed directly, with the dipolar integral evaluated via FFTs. In other words, the density $|\psi(z, t)|^2$ is transformed to Fourier space, multiplied by the Fourier transform of $U_{\rm dd}^{1D}$, and transformed back to real space.

Second, the derivative step solves

$${\rm i}\frac{\partial\psi}{\partial t} = H_2\psi,$$

using the Crank–Nicolson scheme on the intermediate solution $\psi^{n+1/2}$. The scheme is

$$\frac{\psi^{n+1} - \psi^{n+1/2}}{-{\rm i}\Delta} = \frac{1}{2}H_2(\psi^{n+1} + \psi^{n+1/2}),$$

resulting in

$$\psi^{n+1} = \frac{1 - i\Delta H_2/2}{1 + {\rm i}\Delta H_2/2}\psi^{n+1/2}.$$

The spatial domain is discretized over the interval $[-L/2, L/2]$ with grid points defined as $z_i = -L/2 + ih$, for $i = 0, 1, \ldots, N_z$, resulting in $N_z + 1$ points, where $h = L/N_z$ is the spatial step size. The second derivative in the governing equation is approximated using a three-point finite-difference formula within the Crank–Nicolson time-stepping scheme, which is semi-implicit and ensures second-order accuracy in both space and time. This discretization yields a tridiagonal system of linear equations, which is solved by a forward-backward recursion algorithm, as detailed in chapter 3, with a computational complexity of $O(N_z)$. Boundary conditions are typically imposed such that the wavefunction $\psi(z_i, t)$ vanishes at the domain edges ($\psi(z_0, t) = \psi(z_{N_z}, t) = 0$), consistent with the physical behaviour of a BEC in a finite domain.

For numerical implementation, the initial state at $t = 0$ is typically chosen as the ground state of the harmonic oscillator for the non-interacting case ($a = a_{\rm dd} = 0$)

$$\psi(z, 0) = \left(\frac{\Omega_z}{\pi}\right)^{1/4} \mathrm{e}^{-\Omega_z z^2/2}.$$

The nonlinear interaction terms are introduced gradually during time iteration to reach the desired values of a and $a_{\rm dd}$, facilitating convergence to the physical solution.

To compute the ground state, imaginary-time propagation replaces real time with $t = -it$, transforming the GP equation into the following form

$$\begin{aligned} -\frac{\partial\psi(z, t)}{\partial t} = \Bigg[&-\frac{1}{2}\frac{\partial^2}{\partial z^2} + \frac{1}{2}\Omega_z^2 z^2 + \frac{2aN}{d_\rho^2}|\psi(z, t)|^2 \\ &+ N\int_{-\infty}^{\infty} U_{\rm dd}^{\rm 1D}(z - z')|\psi(z', t)|^2 \mathrm{d}z' \Bigg]\psi(z, t) \equiv H\psi(z, t). \end{aligned} \tag{8.30}$$

The SSCN method for imaginary time employs the same Hamiltonian splitting. The non-derivative step is

$$\psi^{n+1/2} = \mathrm{e}^{-\Delta H_1}\psi^n,$$

and the Crank–Nicolson step for imaginary time reads as

$$\frac{\psi^{n+1} - \psi^{n+1/2}}{-\Delta} = \frac{1}{2}H_2(\psi^{n+1} + \psi^{n+1/2}).$$

The resulting tridiagonal system is solved similarly, with normalization enforced after each iteration to ensure a stable ground state solution.

For a stationary state, the wavefunction takes the form $\psi(z, t) = \hat{\psi}(z)\mathrm{e}^{-\mathrm{i}\mu t}$, where μ is the chemical potential. Substituting the wavefunction into the GP equation yields

$$\mu\hat{\psi}(z) = \left[-\frac{1}{2}\frac{d^2}{dz^2} + \frac{1}{2}\Omega_z z^2 + \frac{2aN}{d_\rho^2}|\hat{\psi}(z)|^2 + N\int_{-\infty}^{\infty} U_{\rm dd}^{\rm 1D}(z - z')|\hat{\psi}(z')|^2 dz'\right]\hat{\psi}(z). \tag{8.31}$$

Multiplying by $\hat{\psi}^*(z)$ and integrating over all space, with normalization $\int|\hat{\psi}|^2 \mathrm{d}z = 1$, the chemical potential becomes

$$\mu = \int_{-\infty}^{\infty}\left[\frac{1}{2}\left|\frac{d\hat{\psi}}{dz}\right|^2 + \frac{1}{2}\Omega_z^2 z^2|\hat{\psi}|^2 + \frac{2aN}{d_\rho^2}|\hat{\psi}|^4 + N|\hat{\psi}(z)|^2\int_{-\infty}^{\infty}U_{\rm dd}^{\rm 1D}(z-z')|\hat{\psi}(z')|^2 dz'\right]dz. \tag{8.32}$$

The energy per particle, as derived in Dalfovo *et al* [22], is given by

$$E = \int_{-\infty}^{\infty}\left[\frac{1}{2}\left|\frac{d\hat{\psi}}{dz}\right|^2 + \frac{1}{2}\Omega_z^2 z^2|\hat{\psi}|^2 + \frac{aN}{d_\rho^2}|\hat{\psi}|^4 + \frac{N}{2}|\hat{\psi}(z)|^2\int_{-\infty}^{\infty}U_{\rm dd}^{\rm 1D}(z-z')|\hat{\psi}(z')|^2 dz'\right]dz, \tag{8.33}$$

where the nonlinear terms are halved relative to the chemical potential, reflecting the mean-field energy contribution.

8.3.2 Two-dimensional dipolar Gross–Pitaevskii equation

The two-dimensional GP equation (8.8) governs the dynamics of a BEC confined in the x–y plane with strong confinement along the z-axis. For notational consistency, the wavefunction is denoted as $\psi(\boldsymbol{\rho}, t)$, replacing the alternative notation $\psi_{\rm 2D}$, and the equation is expressed as:

$$\begin{aligned} i\frac{\partial\psi(\boldsymbol{\rho}, t)}{\partial t} =& \left[-\frac{1}{2}\left(\frac{\partial^2}{\partial x^2} + \frac{\partial^2}{\partial y^2}\right) + \frac{\Omega_\rho^2}{2}(x^2+y^2) + \frac{4\pi aN}{\sqrt{2\pi}\,d_z}|\psi|^2\right. \\ &\left. + \frac{4\pi a_{\rm dd}N}{\sqrt{2\pi}\,d_z}\int\frac{d^2k_\rho}{(2\pi)^2}e^{i\mathbf{k}_\rho\cdot\boldsymbol{\rho}}\tilde{n}(\mathbf{k}_\rho)h_{\rm 2D}\left(\frac{k_\rho d_z}{\sqrt{2}}\right)\right]\psi(\boldsymbol{\rho}, t) \equiv H\psi(\boldsymbol{\rho}, t), \end{aligned} \tag{8.34}$$

where $\boldsymbol{\rho} = (x, y)$, $\mathbf{k}_\rho = (k_x, k_y)$, $k_\rho = |\mathbf{k}_\rho|$, $\tilde{n}(\mathbf{k}_\rho) = \mathcal{F}\{|\psi|^2\}$ is the Fourier transform of the density, $h_{\rm 2D}$ is the 2D dipolar interaction kernel, $a_{\rm dd}$ is the dipolar scattering length, and d_z is the axial confinement length. The Hamiltonian is decomposed into three parts:

$$\begin{aligned} H_1 &= \frac{\Omega_\rho^2}{2}(x^2+y^2) + \frac{4\pi aN}{\sqrt{2\pi}\,d_z}|\psi|^2 + \frac{4\pi a_{\rm dd}N}{\sqrt{2\pi}\,d_z}\int\frac{d^2k_\rho}{(2\pi)^2}e^{i\mathbf{k}_\rho\cdot\boldsymbol{\rho}}\tilde{n}(\mathbf{k}_\rho)h_{\rm 2D}\left(\frac{k_\rho d_z}{\sqrt{2}}\right), \\ H_2 &= -\frac{1}{2}\frac{\partial^2}{\partial x^2}, \quad H_3 = -\frac{1}{2}\frac{\partial^2}{\partial y^2}. \end{aligned} \tag{8.35}$$

The SSCN method extends the 1D approach by applying three sub-steps over a time step Δ. The non-derivative step (H_1) is computed using FFTs to evaluate the dipolar term, while the derivative steps (H_2 and H_3) employ the Crank–Nicolson scheme, solving tridiagonal systems for the x- and y-directions sequentially.

For a stationary state $\psi(\boldsymbol{\rho}, t) = \hat{\psi}(\boldsymbol{\rho})e^{-i\mu t}$, the chemical potential is given by

$$\begin{aligned} \mu = \int_{-\infty}^{\infty}\int_{-\infty}^{\infty}&\left[\frac{1}{2}\left(\left|\frac{\partial\hat{\psi}}{\partial x}\right|^2 + \left|\frac{\partial\hat{\psi}}{\partial y}\right|^2\right) + \frac{\Omega_\rho^2}{2}(x^2+y^2)|\hat{\psi}|^2 + \frac{4\pi aN}{\sqrt{2\pi}\,d_z}|\hat{\psi}|^4\right. \\ &\left. + \frac{4\pi a_{\rm dd}N}{\sqrt{2\pi}\,d_z}|\hat{\psi}|^2\int\frac{d^2k_\rho}{(2\pi)^2}e^{i\mathbf{k}_\rho\cdot\boldsymbol{\rho}}\tilde{n}(\mathbf{k}_\rho)h_{2D}\left(\frac{k_\rho d_z}{\sqrt{2}}\right)\right]dx\,dy. \end{aligned} \tag{8.36}$$

This expression is derived using integration by parts to express the kinetic energy in terms of first derivatives, ensuring computational efficiency.

8.3.3 Numerical solution to three-dimensional dipolar Gross–Pitaevskii equation

To implement the SSCN scheme, the three-dimensional GP equation (8.10) is decomposed into four parts as

$$\mathrm{i}\frac{\partial \psi(\mathbf{r}, t)}{\partial t} = (H_1 + H_2 + H_3 + H_4)\psi(\mathbf{r}, t),$$

where

$$\begin{aligned} H_1 &= \frac{\Omega_\rho^2}{2}(x^2 + y^2 + \lambda^2 z^2) + 4\pi a N|\psi|^2 + 3Na_{\mathrm{dd}}\int V_{\mathrm{dd}}(\mathbf{r} - \mathbf{r}')|\psi(\mathbf{r}', t)|^2 \mathrm{d}\mathbf{r}', \\ H_2 &= -\frac{1}{2}\frac{\partial^2}{\partial x^2}, \quad H_3 = -\frac{1}{2}\frac{\partial^2}{\partial y^2}, \quad H_4 = -\frac{1}{2}\frac{\partial^2}{\partial z^2}. \end{aligned}$$

The SSCN method applies four sub-steps over a time step Δ, with the non-derivative step (H_1) computed using 3D FFTs to handle the dipolar interaction, and the derivative steps (H_2, H_3, H_4) solved via the Crank–Nicolson scheme, addressing each spatial dimension independently.

For a stationary state $\psi(\mathbf{r}, t) = \hat{\psi}(\mathbf{r})\mathrm{e}^{-\mathrm{i}\mu t} \equiv \hat{\psi}\mathrm{e}^{-\mathrm{i}\mu t}$, the chemical potential is given by

$$\begin{aligned} \mu = \int_{-\infty}^{\infty}\int_{-\infty}^{\infty}\int_{-\infty}^{\infty}\Bigg[\frac{1}{2}\left(\left|\frac{\partial\hat{\psi}}{\partial x}\right|^2 + \left|\frac{\partial\hat{\psi}}{\partial y}\right|^2 + \left|\frac{\partial\hat{\psi}}{\partial z}\right|^2\right) + \frac{\Omega_\rho^2}{2}(x^2 + y^2 + \lambda^2 z^2)|\hat{\psi}|^2 \\ + 4\pi a N|\hat{\psi}|^4 + 3Na_{\mathrm{dd}}|\hat{\psi}|^2\int V_{\mathrm{dd}}(\mathbf{r} - \mathbf{r}')|\hat{\psi}(\mathbf{r}')|^2 d\mathbf{r}'\Bigg]\mathrm{d}x\,\mathrm{d}y\,\mathrm{d}z. \end{aligned} \tag{8.37}$$

Integration by parts simplifies the kinetic term, aligning the expression with numerical implementations.

8.4 Ground state properties of dipolar BECs

8.4.1 Gaussian variational approximation

The properties of a dipolar BEC are investigated using an approximate method. Specifically, a time-dependent variational procedure is employed to solve the dipolar GP equations in three, two, and one dimensions. This approach reduces the GP equation to a system of second-order nonlinear ordinary differential equations governing the variational parameters.

8.4.1.1 Variational approximation for the 3D dipolar GP equation

The Lagrangian density corresponding to the three-dimensional dipolar GP equation (8.10) is given by

$$\mathcal{L} = \frac{\mathrm{i}}{2}(\psi\partial_t\psi^* - \psi^*\partial_t\psi) + \frac{|\nabla\psi|^2}{2} + V(\mathbf{r})|\psi|^2 + 2\pi aN|\psi|^4 + \frac{N}{2}|\psi|^2 \int U_{\mathrm{dd}}(\mathbf{r} - \mathbf{r}')|\psi(\mathbf{r}')|^2\,\mathrm{d}^3\mathbf{r}'. \tag{8.38}$$

The variational approximation employs a Gaussian ansatz, expressed as

$$\psi(\mathbf{r}, t) = \frac{\pi^{-3/4}}{w_\rho\sqrt{w_z}} \exp\left(-\frac{\rho^2}{2w_\rho^2} - \frac{z^2}{2w_z^2} + \mathrm{i}\alpha\rho^2 + \mathrm{i}\beta z^2\right), \tag{8.39}$$

where $w_\rho(t)$ and $w_z(t)$ denote the time-dependent radial and axial widths, respectively, and $\alpha(t)$ and $\beta(t)$ represent the time-dependent phase parameters.

Substituting the trial wavefunction (8.39) into the Lagrangian density (8.38), the effective Lagrangian per particle, $L = \int\mathcal{L}\,\mathrm{d}^3\mathbf{r}$, is derived as

$$L = \left(w_\rho^2\dot{\alpha} + \frac{w_z^2\dot{\beta}}{2}\right) + \frac{\Omega_\rho^2 w_\rho^2}{2} + \frac{\Omega_z^2 w_z^2}{4} + \frac{1}{2w_\rho^2} + \frac{1}{4w_z^2} + 2w_\rho^2\alpha^2 + w_z^2\beta^2 + \frac{N}{\sqrt{2\pi}}\frac{1}{w_\rho^2 w_z}[a - a_{\mathrm{dd}}f(\kappa)], \tag{8.40}$$

where Ω_ρ and Ω_z are the radial and axial trap frequencies, a_{dd} is the DDI strength, and the function $f(\kappa)$ is defined as:

$$f(\kappa) = \frac{1 + 2\kappa^2 - 3\kappa^2 d(\kappa)}{1 - \kappa^2}, \tag{8.41}$$

$$d(\kappa) = \frac{\operatorname{arctanh}\sqrt{1 - \kappa^2}}{\sqrt{1 - \kappa^2}}, \qquad \kappa = \frac{w_\rho}{w_z}. \tag{8.42}$$

The dynamics of the variational parameters are governed by the Euler–Lagrange equations:

$$\frac{\mathrm{d}}{\mathrm{d}t}\left(\frac{\partial L}{\partial \dot{q}_j}\right) - \frac{\partial L}{\partial q_j} = 0, \quad q_j \in \{w_\rho, w_z, \alpha, \beta\}. \tag{8.43}$$

Applying the above Euler–Lagrange equation to w_ρ, w_z, α, and β, the equations for the widths w_ρ and w_z are obtained as

$$\ddot{w}_\rho + \Omega_\rho^2 w_\rho = \frac{1}{w_\rho^3} + \frac{N}{\sqrt{2\pi}}\frac{[2a - a_{\mathrm{dd}}g(\kappa)]}{w_\rho^3 w_z}, \tag{8.44}$$

$$\ddot{w}_z + \Omega_z^2 w_z = \frac{1}{w_z^3} + \frac{2N}{\sqrt{2\pi}}\frac{[a - a_{\mathrm{dd}}c(\kappa)]}{w_\rho^2 w_z^2}, \tag{8.45}$$

where the functions $g(\kappa)$ and $c(\kappa)$ are given by:

$$g(\kappa) = \frac{2 - 7\kappa^2 - 4\kappa^4 + 9\kappa^4 d(\kappa)}{(1 - \kappa^2)^2}, \tag{8.46}$$

$$c(\kappa) = \frac{1 + 10\kappa^2 - 2\kappa^4 - 9\kappa^2 d(\kappa)}{(1 - \kappa^2)^2}. \tag{8.47}$$

Equations (8.44) and (8.45) describe the time evolution of the radial and axial widths, respectively. From these, expressions for the frequencies of the lowest-lying modes can be derived [8]. The chemical potential μ for a stationary state is calculated as

$$\mu = \frac{1}{2w_\rho^2} + \frac{1}{4w_z^2} + \frac{2N[a - a_{\mathrm{dd}} f(\kappa)]}{\sqrt{2\pi}\, w_z w_\rho^2} + \frac{\Omega_\rho^2 w_\rho^2}{2} + \frac{\Omega_z^2 w_z^2}{4}. \tag{8.48}$$

8.4.1.2 Variational approximation for the 1D dipolar GP equation

To solve equation (8.20), a Gaussian variational ansatz is employed, given by

$$\psi_{\mathrm{1D}}(z) = \frac{\pi^{-1/4}}{\sqrt{w_z}} \exp\left(-\frac{z^2}{2w_z^2} + i\beta z^2\right). \tag{8.49}$$

From equation (8.12), it follows that the variational 1D ansatz (8.49) corresponds to the following 3D wavefunction:

$$\psi(\mathbf{r}, t) = \frac{\pi^{-3/4}}{d_\rho \sqrt{w_z}} \exp\left(-\frac{\rho^2}{2d_\rho^2} - \frac{z^2}{2w_z^2} + i\beta z^2\right). \tag{8.50}$$

This variational wavefunction (8.50) is a special case of the 3D variational wavefunction (8.39), with $w_\rho = d_\rho$ and $\alpha = 0$. Consequently, the 1D variational Lagrangian can be derived from the 3D Lagrangian (8.40) by setting $w_\rho = d_\rho$ and $\alpha = 0$, yielding:

$$L_{\mathrm{1D}} = \frac{w_z^2 \dot{\beta}}{2} + \frac{1}{4w_z^2} + w_z^2 \beta^2 + \frac{\Omega_z^2 w_z^2}{4} + \frac{N}{\sqrt{2\pi}\, d_\rho^2 w_z}[a - a_{\mathrm{dd}} f(\kappa_0)], \quad \kappa_0 = \frac{d_\rho}{w_z}, \tag{8.51}$$

where constant terms have been omitted for simplicity. This derivation of the 1D Lagrangian (8.51) avoids the need to construct a Lagrangian density involving error functions, as would be required for the 1D potential (8.19), and its subsequent integration.

The Euler–Lagrange equation for the variational parameter w_z of the Lagrangian (8.51) is given by

$$\ddot{w}_z + \Omega_z^2 w_z = \frac{1}{w_z^3} + \frac{2N[a - a_{\mathrm{dd}} c(\kappa_0)]}{\sqrt{2\pi}\, w_z^2 d_\rho^2}. \tag{8.52}$$

The variational chemical potential is expressed as

$$\mu = \frac{1}{4w_z^2} + \frac{2N[a - a_{\rm dd}f(\kappa_0)]}{\sqrt{2\pi}\, w_z d_\rho^2} + \frac{\Omega_z^2 w_z^2}{4}. \tag{8.53}$$

These variational results provide a simple yet effective approximation to the 1D GP equation (8.20).

8.4.1.3 Variational approximation for the 2D dipolar GP equation

To solve equation (8.27), a Gaussian ansatz is employed for the 2D wavefunction, given by:

$$\psi_{\rm 2D}(\rho) = \frac{1}{w_\rho\sqrt{\pi}} \exp\left(-\frac{\rho^2}{2w_\rho^2} + i\alpha\rho^2\right). \tag{8.54}$$

From equation (8.21), it follows that the 2D wavefunction (8.54) corresponds to the following 3D wavefunction:

$$\psi(\mathbf{r}, t) = \frac{\pi^{-3/4}}{w_\rho\sqrt{d_z}} \exp\left(-\frac{\rho^2}{2w_\rho^2} - \frac{z^2}{2d_z^2} + i\alpha\rho^2\right). \tag{8.55}$$

This variational wavefunction (8.55) represents a special case of the 3D variational wavefunction (8.39) with $w_z = d_z$ and $\beta = 0$. Consequently, the 2D variational Lagrangian is derived from the 3D Lagrangian (8.40) as

$$L_{\rm 2D} = w_\rho^2\dot{\alpha} + \frac{w_\rho^2\Omega_\rho^2}{2} + \frac{1}{2w_\rho^2} + 2w_\rho^2\alpha^2 + \frac{N}{\sqrt{2\pi}\, w_\rho^2 d_z}[a - a_{\rm dd}f(\bar{\kappa})], \quad \bar{\kappa} = \frac{w_\rho}{d_z}, \tag{8.56}$$

where constant terms have been omitted for simplicity. The Euler–Lagrange variational equation for the radial width w_ρ is then obtained as

$$\ddot{w}_\rho + w_\rho\Omega_\rho^2 = \frac{1}{w_\rho^3} + \frac{N}{\sqrt{2\pi}\, w_\rho^3 d_z}[2a - a_{\rm dd}g(\bar{\kappa})]. \tag{8.57}$$

The chemical potential μ for a stationary state is expressed as

$$\mu = \frac{1}{2w_\rho^2} + \frac{2N[a - a_{\rm dd}f(\bar{\kappa})]}{\sqrt{2\pi}\, d_z w_\rho^2} + \frac{w_\rho^2\Omega_\rho^2}{2}. \tag{8.58}$$

In a quasi-2D system, the radial width is significantly larger than the axial oscillator length, i.e., $w_\rho \gg d_z$. As a result, $\bar{\kappa} \to \infty$, and the function $f(\bar{\kappa}) \to -2$. In this limit, the interaction term in equation (8.56) simplifies to $N(a + 2a_{\rm dd})/(\sqrt{2\pi}\, w_\rho^2 d_z)$. Thus, the variational approximation indicates that the effect of the dipolar interaction in equation (8.27) is to modify the contact interaction by replacing a with $a + 2a_{\rm dd}$.

8.4.2 Thomas–Fermi (TF) approximation

In the time-dependent, axially symmetric GP equation, when the atomic interaction energy significantly exceeds the kinetic energy gradient term, the kinetic energy can be neglected, leading to the TF approximation. For time-dependent harmonic traps, the condensate density is assumed to take the form [22–25]:

$$n(\mathbf{r}, t) = n_0(t)\left[1 - \frac{\rho^2}{R_\rho^2(t)} - \frac{z^2}{R_z^2(t)}\right], \tag{8.59}$$

where $n_0(t) = \frac{15N}{8\pi R_\rho^2 R_z}$ is the normalization constant, and R_ρ and R_z represent the radial and axial sizes of the condensate, respectively.

Using the parabolic density profile given by equation (8.59), the TF energy functional E_{TF} can be expressed as [24]:

$$E_{\mathrm{TF}} = \frac{2\gamma^2 R_\rho^2 + \lambda^2 R_z^2}{14} + \frac{15}{28\pi}\frac{4\pi a_s N}{R_\rho^2 R_z}\left[1 - \frac{a_{\mathrm{dd}}}{a_s} f(\kappa)\right], \tag{8.60}$$

where $\kappa = R_\rho / R_z$ denotes the aspect ratio of the condensate, and $f(\kappa)$ is defined in equation (8.41). In the TF regime, the evolution of the condensate sizes is governed by the following set of coupled ordinary differential equations [23, 26]

$$\ddot{R}_\rho = -\gamma^2 R_\rho + \frac{15 a_s N}{R_\rho R_z}\left[\frac{1}{R_\rho^2} - \frac{a_{\mathrm{dd}}}{a_s}\left(\frac{1}{R_\rho^2} + \frac{3}{2}\frac{f(\kappa)}{R_\rho^2 - R_z^2}\right)\right], \tag{8.61a}$$

$$\ddot{R}_z = -\lambda^2 R_z + \frac{15 a_s N}{R_\rho^2}\left[\frac{1}{R_z^2} + \frac{2a_{\mathrm{dd}}}{a_s}\left(\frac{1}{R_z^2} + \frac{3}{2}\frac{f(\kappa)}{R_\rho^2 - R_z^2}\right)\right]. \tag{8.61b}$$

The equilibrium sizes of the condensate can be determined by setting the time derivatives $\ddot{R}_\rho$ and $\ddot{R}_z$ to zero in equations (8.61a) and (8.61b), yielding the transcendental equation for κ

$$3\kappa^2 \frac{a_{\mathrm{dd}}}{a_s}\left[\left(1 + \frac{\eta^2}{2}\right)\frac{f(\kappa)}{1 - \kappa^2} - 1\right] + \left(\frac{a_{\mathrm{dd}}}{a_s} - 1\right)(\kappa^2 - \eta^2) = 0, \tag{8.62}$$

and the radial size is given by

$$R_\rho = \left[15 a_s N \kappa \gamma^{-2}\left\{1 + \frac{a_{\mathrm{dd}}}{a_s}\left(\frac{3}{2}\frac{\kappa^2 f(\kappa)}{1 - \kappa^2} - 1\right)\right\}\right]^{1/5}, \tag{8.63}$$

with $R_z = R_\rho / \kappa$. The chemical potential in the TF approximation is given by

$$\mu_{\mathrm{TF}} = 4\pi a_s n_0\left[1 - \frac{a_{\mathrm{dd}}}{a_s} f(\kappa)\right]. \tag{8.64}$$

8.4.3 Analysis of dipolar condensates in various trap geometries

The properties of dipolar condensates are primarily governed by three key parameters: the dipolar interaction strength, trap aspect ratio (λ), and scattering length (a_s). Figure 8.3 presents a phase diagram in the $\lambda - a_{\rm crit}/a_0$ plane, showing stable and unstable regions for $\omega = 2\pi \times 700$ Hz, with experimental data points indicated by error bars. The diagram compares critical scattering lengths derived from variational calculations for different atom numbers, alongside numerical and experimental stability curves for $N = 20\ 000$ atoms. Both theoretical approaches show reasonable agreement with experimental observations.

The stability analysis reveals that for small atom numbers (e.g., $N = 250$), the condensate remains strongly stable in both cigar ($\lambda \ll 1$) and pancake ($\lambda \gg 1$) geometries. However, increasing the atom number reduces stability in cigar-shaped traps, while pancake configurations maintain stability regardless of atom number. These findings are consistent with previous work by Wilson *et al* [27].

Further comparison of stability curves obtained through variational methods, TF approximation, and direct numerical integration of the dipolar GP equation is shown in figure 8.4 for $N = 10\ 000$ atoms. The TF and variational results agree well for moderate aspect ratios ($1 \lessapprox \lambda \lessapprox 10$) but diverge in extreme pancake or cigar regimes. Notably, the numerical critical scattering lengths align closely with variational predictions, while the TF results remain relatively insensitive to large variations in atom number.

The stability threshold with respect to atom number and ground state structure reveals that dipolar BECs remain stable only below a critical atom number [28]. Beyond this threshold, collapse occurs regardless of trap parameters. This behaviour can be investigated through imaginary-time propagation of the 3D GP equation

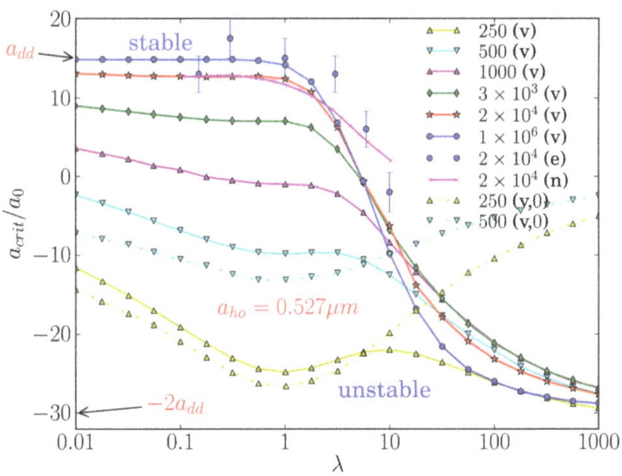

Figure 8.3. Phase diagram depicting the stable and unstable regions in the $\lambda - a_{\rm crit}/a_0$ plane for $\omega = 2\pi \times 700$ Hz [4]. The points with error bars represent experimental data. Reproduced with permission from [4].

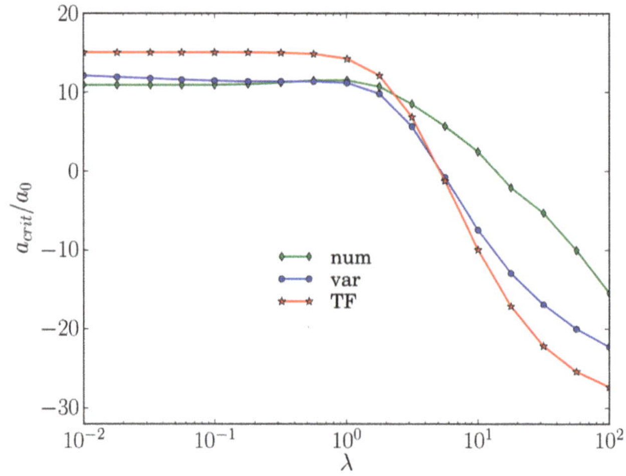

Figure 8.4. Phase diagram illustrating the stable and unstable regions in the λ–a_{crit}/a_0 plane for $N = 10,000$ atoms. Reproduced with permission from [4]. CC-BY-NC 4.0.

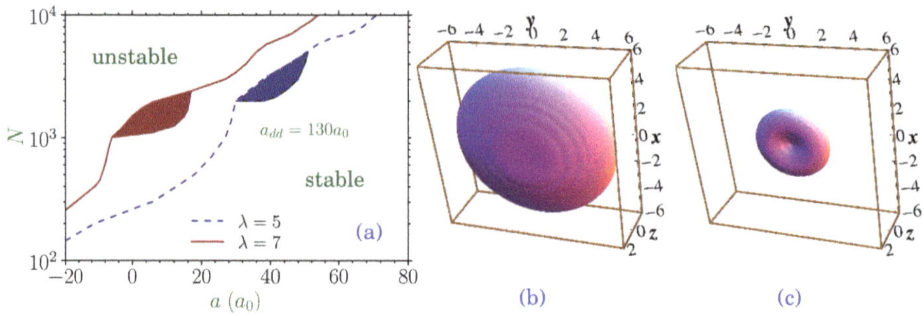

Figure 8.5. (a) The N–a stability phase diagram for a ^{164}Dy BEC with a dipole–dipole scattering length $a_{\text{dd}} = 130a_0$ in a disk-shaped trap with aspect ratios $\lambda = 5$ and $\lambda = 7$, and a harmonic oscillator length $l = 1$ μm. (b, c) Three-dimensional contour plots of the density $|\psi(x, y, z)|^2$ for a disk-shaped ^{164}Dy BEC with $a_{\text{dd}} = 130a_0$, $\nu = \gamma = 1$, $\lambda = 5$, $l = 1$ μm, $N = 3000$, and $a = 40a_0$, showing contour levels at (b) $|\psi|^2 = 0.001$ and (c) $|\psi|^2 = 0.027$. Reprinted from [29] with permission from Elsevier.

with gradually increased nonlinearities. Figure 8.5(a) displays the $N - a$ stability phase diagram for ^{164}Dy BECs ($a_{\text{dd}} = 130a_0$) in disk-shaped traps ($\lambda = 5, 7$) with $l = 1$μm. The shaded region indicates metastability, characterized by biconcave density structures resulting from a local (rather than global) energy minimum. This metastable state arises from roton instability [28], where dipolar repulsion in the disk plane redistributes atoms from the centre to the periphery.

The biconcave structure, a hallmark of dipolar interactions, is visualized in figures 8.5(b) and (c) through 3D density contours ($|\psi(x, y, z)|^2$) for $\lambda = 5$. While the lower-density contour (0.001) in (b) shows minimal structure, the higher-density contour (0.027) in (c) clearly reveals the characteristic biconcave profile, particularly prominent in the central region of metastable dipolar BECs.

8.4.4 Energy, chemical potential and root-mean-square sizes of the dipolar BEC

The energy, chemical potential, and root-mean-square (rms) sizes of dipolar BECs can be determined by solving the 1D (8.20), 2D (equation (8.27)), and 3D (equation (8.10)) dipolar GP equations using numerical and approximate methods. The energy (E), chemical potential (μ), and rms size ($\langle z \rangle$) are calculated through imaginary-time propagation of the 1D dipolar GP equation (8.20), with results presented in table 8.1. The calculations consider ^{52}Cr atoms with a scattering length $a = 6$ nm (approximately $113a_0$, where a_0 is the Bohr radius) and a dipolar length $a_{\rm dd} = 16a_0$, for parameters $\lambda = 1$, $d_\rho = 1$, $l = 1$ μm, and varying numbers of atoms N. The computations account for different spatial and temporal step sizes, dz and dt. To ensure convergence for a given spatial and temporal step, a sufficient number of spatial discretization points and time iterations are employed. In table 8.1, Gaussian variational results are included for comparison. These results offer a more accurate approximation to the numerical solution for smaller numbers of atoms. Table 8.2 presents the results for the energy E, chemical potential μ, and rms size ρ obtained from the two-dimensional GP equation (8.27) with parameters $\gamma = 1$, $d_z = 1$, and $l = 1$ μm. The numerical results were computed using various spatial and temporal step sizes, dx and dt, and different numbers of ^{52}Cr atoms, N, with a dipole–dipole scattering length $a_{\rm dd} = 16a_0$ and an s-wave scattering length $a = 6$ nm. For comparison, results from an axially symmetric Gaussian variational approach are also included.

Next, the three-dimensional dipolar GP equation (8.10) is analyzed with a scattering length $a = 0$ and varying dipolar interaction strengths $g_{\rm dd} = 3a_{\rm dd}N$, where $N = 1, 2, 3, 4$, within an axially symmetric trap characterized by an aspect ratio $\lambda = 1/2$. Numerical results obtained for different numbers of spatial and temporal steps are compared with those derived using a Gaussian variational approach and are summarized in table 8.3.

The solution of the 3D GP equation is considered for a model BEC of ^{52}Cr atoms in a cigar-shaped, axially symmetric trap with an aspect ratio $\lambda = 1/2$, as first studied by Bao *et al* [30]. The nonlinearities adopted in that study, specifically $-4\pi a = 0.207\,16$ and $4\pi a_{\rm dd} = 0.033\,146$, correspond to the approximate values of the scattering lengths and characteristic length: $a \approx 100a_0$, $a_{\rm dd} \approx 16a_0$, and $l = 0.321$ μm. Results for the energy E and chemical potential μ are presented in table 8.4, while the rms sizes $\langle z \rangle$ and $\langle x \rangle$ are provided in table 8.5. The variational and TF results are included alongside the numerical calculations of Bao *et al* [30].

For weak nonlinearities or small atom numbers, the Gaussian variational results agree well with numerical calculations, as the wavefunction maintains a quasi-Gaussian profile in these regimes. However, for strong nonlinearities or large atom numbers, where the wavefunction approaches the TF shape described by equation (8.59), the TF approximation provides better agreement with numerical results, as demonstrated in tables 8.4 and 8.5.

Appendix A provides a comprehensive overview of numerical algorithms and codes for solving the time-dependent GP equation with dipolar interactions in 1D, 2D, and 3D geometries, using the SSCN method.

Table 8.1. Ground-state properties of the 1D GP equation (8.20) for a ^{52}Cr BEC, showing energy (E), chemical potential (μ), and rms size ($\langle z \rangle$). System parameters are $\lambda = 1$, $d_\rho = 1$ μm (radial confinement), $a = 6$ nm (s-wave scattering length), $a_{dd} = 16a_0$ (dipolar length), and varying atom number N. Length scales in equations (8.10) and (8.20) are expressed in harmonic oscillator units ($l = 1$ μm). Numerical results compare three discretization schemes: (A) $dz = 0.01$, $dt = 0.0001$; (B) $dz = 0.05$, $dt = 0.0005$; and (C) $dz = 0.2$, $dt = 0.001$, with corresponding variational results. Reproduced with permission from [4].

N	$\langle z \rangle$ var	$\langle z \rangle$ (C)	$\langle z \rangle$ (B)	$\langle z \rangle$ (A)	E var	E (C)	E (B)	E (A)	μ var	μ (C)	μ (B)	μ (A)
100	0.7939	0.7929	0.7937	0.7937	0.7239	0.7222	0.7222	0.7222	0.9344	0.9299	0.9297	0.9297
500	1.0425	1.0378	1.0381	1.0380	1.4371	1.4166	1.4166	1.4166	2.2157	2.1693	2.1691	2.1691
1000	1.2477	1.2374	1.2375	1.2375	2.1376	2.0920	2.0920	2.0920	3.4165	3.3235	3.3235	3.3235
5000	2.0249	1.9938	1.9939	1.9939	5.8739	5.6912	5.6912	5.6911	9.6671	9.3490	9.3490	9.3490
10 000	2.5233	2.4815	2.4815	2.4815	9.2129	8.9131	8.9131	8.9131	15.223	14.715	14.715	14.715
50 000	4.2451	4.1720	4.1720	4.1720	26.505	25.623	25.623	25.623	43.993	42.529	42.529	42.528

Table 8.2. Ground-state properties of the 2D GP equation solution for a ^{52}Cr BEC showing energy (E), chemical potential (μ), and rms size ($\langle\rho\rangle$). Numerical results are presented for two discretization schemes: (A) $dx = dy = 0.05$, $dt = 0.000\,25$ and (B) $dx = dy = 0.2$, $dt = 0.002$, with comparison to variational results. Parameters include $\gamma = 1$, $d_z = 1\ \mu$m (axial confinement), $a = 6$ nm (s-wave scattering length), $a_{dd} = 16a_0$ (dipolar length), and varying atom number N. Length scales in equations (8.10) and (8.27) are in units of the harmonic oscillator length $l = 1\ \mu$m. Reproduced with permission from [4]. CC-BY-NC 4.0.

N	$\langle\rho\rangle$ var	$\langle\rho\rangle$ (B)	$\langle\rho\rangle$ (A)	E var	E (B)	E (A)	μ var	μ (B)	μ (A)
100	1.0985	1.0966	1.0974	1.2182	1.2155	1.2156	1.4187	1.4120	1.4118
500	1.3514	1.3418	1.3420	1.8653	1.8382	1.8383	2.5437	2.4840	2.4839
1000	1.5482	1.5303	1.5304	2.4571	2.3988	2.3988	3.5070	3.3900	3.3900
5000	2.2549	2.2076	2.2075	5.2206	4.9987	4.9988	7.8005	7.4245	7.4247
10 000	2.6824	2.6191	2.6191	7.3787	7.0289	7.0290	11.090	10.522	10.523
50 000	4.0420	3.9343	3.9343	16.680	15.792	15.792	25.161	23.786	23.787

Table 8.3. Ground-state energy (E) and chemical potential (μ) from numerical solutions of the 3D GP equation (8.10) for $\lambda^2 = 0.25$, $a = 0$, and varying dipolar interaction strengths $g_{dd} = 3a_{dd}N$. Results are compared with Gaussian variational calculations. Three discretization schemes were employed using the Crank–Nicolson method: (A) $dx = dy = dz = 0.05$, $dt = 0.0004$ with 256^3 mesh; (B) $dx = dy = dz = 0.1$, $dt = 0.001$ with 128^3 mesh; and (C) $dx = dy = dz = 0.2$, $dt = 0.003$ with 128^3 mesh. Reproduced with permission from [4]. CC-BY-NC 4.0.

g_{dd}	E var	E (C)	E (B)	E (A)	μ var	μ (C)	μ (B)	μ (A)
0	1.2500	1.2498	1.2500	1.2500	1.2500	1.2498	1.2500	1.2500
1	1.2230	1.2220	1.2222	1.2222	1.1934	1.1907	1.1910	1.1911
2	1.1907	1.1872	1.1874	1.1874	1.1203	1.1089	1.1097	1.1099
3	1.1521	1.1433	1.1436	1.1437	1.0253	0.9925	0.9947	0.9953
4	1.1051	1.0845	1.0853	1.0854	0.8950	0.7927	0.8035	0.8056

8.4.5 Comparison of the validity of 1D, 2D, and 3D dipolar GP equations

The validity of reduced-dimensional GP equations warrants careful examination, particularly for the one-dimensional (equation (8.20)) and two-dimensional (equation (8.27)) cases describing cigar-shaped and disk-shaped BECs, respectively. For cigar-shaped systems, the axial density profiles obtained from the 1D reduction can be quantitatively compared with the integrated axial densities derived from full 3D solutions through radial coordinate integration. Similar verification procedures apply to disk-shaped geometries, where the 2D reduction's planar density distributions should correspond to the transversely integrated densities from complete 3D calculations.

Table 8.4. Comparison of energy and chemical potential from numerical solutions of the 3D GP equation (8.10) with results from Bao *et al* [30] for different spatial and temporal discretizations. Calculations are shown for parameters $\lambda^2 = 0.25$, $4\pi a = 0.20716$, and $4\pi a_{dd} = 0.033\,146$ with varying atom numbers N. These nonlinearity parameters correspond to a ^{52}Cr dipolar BEC with $a \approx 100a_0$, $a_{dd} \approx 16a_0$, and oscillator length $l \approx 0.321$ μm. Results include variational approximations, TF solutions, and numerical data from [30]. Numerical simulations were performed using Crank–Nicolson discretization with: (A) $dx = dy = dz = 0.15$, $dt = 0.002$ $(NX = NY = NZ = 256)$; and (B) $dx = dy = dz = 0.3$, $dt = 0.003$ $(NX = NY = NZ = 128)$. Reproduced with permission from [4]. CC-BY-NC 4.0.

N	E var	E (B)	E (A)	E [30]	μ var	μ TF	μ (B)	μ (A)	μ [30]
100	1.579	1.567	1.567	1.567	1.840	1.323	1.813	1.813	1.813
500	2.287	2.224	2.224	2.225	2.951	2.518	2.835	2.835	2.837
1000	2.836	2.728	2.728	2.728	3.767	3.322	3.582	3.582	3.583
5000	5.036	4.744	4.744	4.745	6.935	6.324	6.485	6.486	6.488
10 000	6.563	6.146	6.146	6.147	9.100	8.345	8.475	8.475	8.479
50 000	12.34	11.46	11.46	11.47	17.23	15.87	15.96	15.96	15.98

Table 8.5. rms sizes $\langle x \rangle$ and $\langle z \rangle$ from numerical solutions of equation (8.10) with parameters $\lambda^2 = 0.25$, $4\pi a = 0.207\,16$, and $4\pi a_{dd} = 0.033\,146$ for varying atom numbers. Results are compared with variational approximations, TF solutions, and numerical data from Bao *et al* [30]. Numerical simulations employed: (A) $dx = dy = dz = 0.15$, $dt = 0.002$ $(NX = NY = NZ = 256)$; and (B) $dx = dy = dz = 0.3$, $dt = 0.003$ $(NX = NY = NZ = 128)$. Reproduced with permission from [4]. CC-BY-NC 4.0.

N	$\langle z \rangle$ TF	$\langle z \rangle$ var	$\langle z \rangle$ (B)	$\langle z \rangle$ (A)	$\langle z \rangle$ [30]	$\langle x \rangle$ TF	$\langle x \rangle$ var	$\langle x \rangle$ (B)	$\langle x \rangle$ (A)	$\langle x \rangle$ [30]
100	1.285	1.316	1.305	1.304	1.299	0.600	0.799	0.794	0.795	0.796
500	1.773	1.797	1.752	1.752	1.745	0.828	0.952	0.938	0.939	0.940
1000	2.037	2.079	2.015	2.014	2.009	0.951	1.054	1.034	1.035	1.035
5000	2.810	2.904	2.795	2.795	2.790	1.313	1.392	1.353	1.353	1.354
10 000	3.228	3.345	3.217	3.217	3.212	1.508	1.586	1.537	1.537	1.538
50 000	4.454	4.629	4.451	4.451	4.441	2.080	2.171	2.093	2.093	2.095

$$n(z) \equiv |\varphi(z)|^2 = \int |\psi(x, y, z)|^2 \, dx \, dy.$$

Figure 8.6(a) demonstrates this comparison between densities obtained from the 1D and 3D GP equations. For these comparisons, one uses the 3D dipolar model from table 8.4 with $N = 1000$ atoms. The 1D case employs trap parameters $\lambda = 1/4$. Similarly, the 2D GP equation yields radial density profiles that can be compared with the reduced radial density obtained by integrating the 3D solution over the axial coordinate

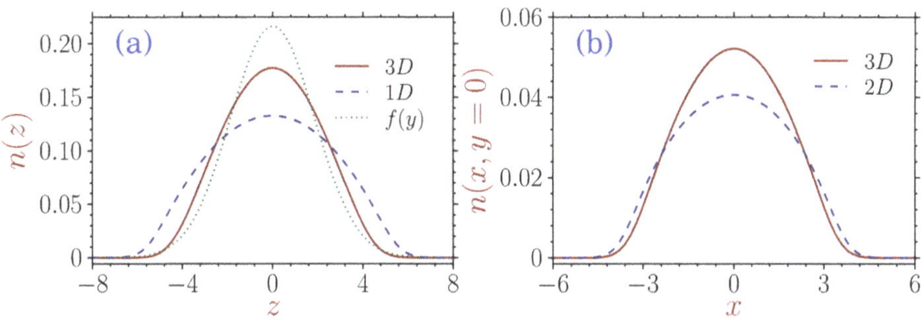

Figure 8.6. (a) Axial density profile ($\lambda = 0.25$) for a cigar-shaped BEC with $N = 1000$ atoms, comparing results from the 1D equation ((8.20)) with the integrated 3D density from equation (8.10). (b) Radial density profile ($\lambda = 4$) in the x–y plane ($n(x, y = 0)$) for a disk-shaped BEC with $N = 1000$ atoms, comparing results from the 2D equation (8.27) with the integrated 3D density from equation (8.10). Parameters: $4\pi a = 0.207\,16$, $4\pi a_{\text{dd}} = 0.033\,146$ (consistent with table 8.4). Reprinted with permission from [4]. CC-BY-NC 4.0.

$$n(x, y) \equiv |\varphi(x, y)|^2 = \int \mathrm{d}z\,|\psi(x, y, z)|^2.$$

Figure 8.6(b) shows this comparison between 2D and 3D results, specifically plotting the radial density section $n(x, 0)$. This comparison uses the same 3D dipolar model with $N = 1000$ atoms, but with trap parameters $\lambda = 4$. Figures 8.6(a) and (b) demonstrate satisfactory agreement between densities obtained from the full 3D GP equation and those from the reduced-dimensional equations.

8.5 Summary and future challenges

The dynamics of BECs with long-range anisotropic tunable DDIs were investigated in this chapter, contrasted with short-range contact interactions. Mean-field GP equations were derived for one-dimensional (cigar-shaped), two-dimensional (disk-shaped), and three-dimensional dipolar BECs. The SSCN method was employed to solve these equations numerically, while Gaussian variational and TF approximations facilitated the analysis of ground state properties. The dynamics across various trap geometries, including cigar- and disk-shaped configurations, were compared, revealing geometry-dependent behaviour. Energy, chemical potential, and rms sizes were calculated using numerical and analytical methods, and the validity of 1D, 2D, and 3D GP equations was assessed, highlighting dimensional effects on dipolar BEC stability and dynamics.

The stability and dynamics of nonlinear excitations, such as solitons and vortices, in dipolar BECs, warranted further exploration, particularly under time-varying DDIs. Extending the SSCN method to incorporate time-varying trap potentials or higher-order interactions requires investigation to capture complex dynamics. The interplay of DDIs with spin–orbit coupling in multi-dimensional dipolar BECs awaits exploration, offering opportunities to study novel quantum phases. These investigations will certainly deepen the understanding of long-range interactions in ultracold quantum systems.

8.6 Problems

Exercise 8.1

1. Compute the ground state of a 1D BEC of 10^4 ^{52}Cr atoms ($m = 8.635 \times 10^{-26}$ kg, $a = 15a_0$) in a harmonic trap $V(x) = \frac{1}{2}m\omega^2x^2$, $\omega = 2\pi \times 150$ Hz, including long-range dipolar interactions. Use a split-step Fourier method to solve the GP equation with a nonlocal interaction term $U_{\rm dd}(x - x') = g_{\rm dd}/|x - x'|^3$, where $g_{\rm dd} = \mu_0\mu^2/4\pi$, $\mu = 6\mu_B$ (Bohr magneton) [15]. Compare the density profile with and without dipolar interactions.
2. Solve the stationary 1D dipolar GP equation for a ^{52}Cr BEC (dipolar strength $a_{\rm dd} = 16a_0$, where a_0 is the Bohr radius) in a harmonic trap with frequency $\omega_z = 2\pi \times 200$ Hz and contact interaction scattering length $a_s = 100a_0$. Use imaginary-time propagation with the SSCN method to compute the ground-state wavefunction, chemical potential, and density profile. Compare results with the TF approximation [29, 31]. Refine the numerical solutions with the Newton Conjugate gradient method.
3. Using the ground state from the previous problem, simulate the real-time dynamics after a sudden quench of the trap frequency to $\omega_z = 2\pi \times 150$ Hz. Calculate the time evolution of the density profile and centre-of-mass oscillation over 50 ms and analyze the impact of dipolar interactions on oscillation frequency.
4. Solve the 2D dipolar GP equation for a ^{164}Dy BEC ($a_{\rm dd} = 131a_0, a_s = 90a_0$) in a harmonic trap with frequencies $\omega_x = \omega_y = 2\pi \times 100$ Hz. Use imaginary-time propagation to find the ground state with a single vortex, computing the energy, chemical potential, and 2D density profile.
5. Solve the 3D dipolar GP equation for a ^{168}Er BEC ($a_{\rm dd} = 66a_0, a_s = 70a_0$) in an anisotropic trap with frequencies $\omega_x = 2\pi \times 150$ Hz, $\omega_y = 2\pi \times 120$ Hz, $\omega_z = 2\pi \times 200$ Hz. Use imaginary-time propagation with the SSCN method to compute the ground-state wavefunction, energy, chemical potential, and 3D density profile.

References

[1] Trefzger C, Menotti C, Capogrosso-Sansone B and Lewenstein M 2011 Ultracold dipolar gases in optical lattices *J. Phys. B: At. Mol. Opt. Phys.* **44** 193001

[2] Griesmaier A, Werner J, Hensler S, Stuhler J and Pfau T 2005 Bose-Einstein condensation of chromium *Phys. Rev. Lett.* **94** 160401

[3] Koch T, Lahaye T, Metz J, Fröhlich B, Griesmaier A and Pfau T 2008 Stabilization of a purely dipolar quantum gas against collapse *Nat. Phys.* **4** 218–22

[4] Kishor Kumar R 2013 Properties of dipolar Bose-Einstein condensates *PhD Thesis* Bharathidasan University, Tiruchirappalli, India http://hdl.handle.net/10603/207647

[5] Marinescu M and You L 1998 Controlling atom-atom interaction at ultralow temperatures by dc electric fields *Phys. Rev. Lett.* **81** 4596–9

[6] Deb B and You L 2001 Low-energy atomic collision with dipole interactions *Phys. Rev.* A **64** 022717

[7] Yi S and You L 2000 Trapped atomic condensates with anisotropic interactions *Phys. Rev.* A **61** 041604
[8] Yi S and You L 2001 Trapped condensates of atoms with dipole interactions *Phys. Rev.* A **63** 053607
[9] Landau L D and Lifschitz E M 1959 *Course of Theoretical Physics: Statistical Physics* 3rd edn (Pergamon)
[10] Huang K 1987 *Statistical Mechanics* 2nd edn (Wiley)
[11] Huang K and Yang C N 1957 Quantum-mechanical many-body problem with hard-sphere interaction *Phys. Rev.* **105** 767–75
[12] Sriraman T 2017 Study on the ground state properties and vortex structures of dipolar Bose-Einstein condensates within the mean-field description *PhD Thesis* Bharathidasan University, Tiruchirappalli, India http://hdl.handle.net/10603/220987.
[13] Santos L, Shlyapnikov G V, Zoller P and Lewenstein M 2000 Bose-Einstein condensation in trapped dipolar gases *Phys. Rev. Lett.* **85** 1791–4
[14] Ronen S, Bortolotti D C E and Bohn J L 2006 Bogoliubov modes of a dipolar condensate in a cylindrical trap *Phys. Rev.* A **74** 013623
[15] Lahaye T, Menotti C, Santos L, Lewenstein M and Pfau T 2009 The physics of dipolar bosonic quantum gases *Rep. Progr. Phys.* **72** 126401
[16] Muruganandam P and Adhikari S K 2012 Numerical and variational solutions of the dipolar Gross-Pitaevskii equation in reduced dimensions *Laser Phys.* **22** 813–20
[17] Sinha S and Santos L 2007 Cold dipolar gases in quasi-one-dimensional geometries *Phys. Rev. Lett.* **99** 140406
[18] Deuretzbacher F, Cremon J C and Reimann S M 2010 Ground-state properties of few dipolar bosons in a quasi-one-dimensional harmonic trap *Phys. Rev.* A **81** 063616
[19] Góral K and Santos L 2002 Ground state and elementary excitations of single and binary Bose-Einstein condensates of trapped dipolar gases *Phys. Rev.* A **66** 023613
[20] Fischer U R 2006 Stability of quasi-two-dimensional Bose-Einstein condensates with dominant dipole-dipole interactions *Phys. Rev.* A **73** 031602
[21] Pedri P and Santos L 2005 Two-dimensional bright solitons in dipolar Bose-Einstein condensates *Phys. Rev. Lett.* **95** 200404
[22] Dalfovo F, Giorgini S, Pitaevskii L P and Stringari S 1999 Theory of Bose-Einstein condensation in trapped gases *Rev. Mod. Phys.* **71** 463–512
[23] O'Dell D H J, Giovanazzi S and Eberlein C 2004 Exact hydrodynamics of a trapped dipolar Bose-Einstein condensate *Phys. Rev. Lett.* **92** 250401
[24] Eberlein C, Giovanazzi S and O'Dell D H J 2005 Exact solution of the Thomas-Fermi equation for a trapped Bose-Einstein condensate with dipole-dipole interactions *Phys. Rev.* A **71** 033618
[25] Parker N G and O'Dell D H J 2008 Thomas-Fermi versus one- and two-dimensional regimes of a trapped dipolar Bose-Einstein condensate *Phys. Rev.* A **78** 041601
[26] Ticknor C, Parker N G, Melatos A, Cornish S L, O'Dell D H J and Martin A M 2008 Collapse times of dipolar Bose-Einstein condensates *Phys. Rev.* A **78** 061607
[27] Wilson R M, Ronen S, Bohn J L and Pu H 2008 Manifestations of the roton mode in dipolar Bose-Einstein condensates *Phys. Rev. Lett.* **100** 245302
[28] Ronen S, Bortolotti D C E and Bohn J L 2007 Radial and angular rotons in trapped dipolar gases *Phys. Rev. Lett.* **98** 030406

[29] Kumar R K, Young S L E, Vudragović D, Balaž A, Muruganandam P and Adhikari S K 2015 Fortran and C programs for the time-dependent dipolar Gross-Pitaevskii equation in an anisotropic trap *Comput. Phys. Commun.* **195** 117–28

[30] Bao W, Cai Y and Wang H 2010 Efficient numerical methods for computing ground states and dynamics of dipolar Bose-Einstein condensates *J. Comput. Phys.* **229** 7874–92

[31] Young S L E, Muruganandam P, Balaž A and Adhikari S K 2023 OpenMP Fortran programs for solving the time-dependent dipolar Gross-Pitaevskii equation *Comput. Phys. Commun.* **286** 108669

IOP Publishing

An Introduction to Ultracold Atoms with Analytical and Numerical Methods

Paulsamy Muruganandam and Ramaswamy Radha

Chapter 9

Collisionally inhomogenous Bose–Einstein condensates

In the last few chapters, the dynamics of scalar and vectorial Bose–Einstein condensates (BECs) have been explored in detail. In particular, the properties of BECs and the associated nonlinear excitations were analyzed allowing the short-range interactions (both binary and three-body interaction) to vary with time. At this juncture, it should be pointed out that the interaction between the atoms in a condensate in general can vary with both space and time. It was realized in recent times that the spatial variation of the laser field intensity by proper choice of resonance detuning can lead to the spatial dependence of the atomic scattering length leading to the so-called 'collisionally inhomogenous BECs'. Under these circumstances, the dynamics of such collisionally inhomogenous BECs is governed by the variable coefficient nonlinear Schrödinger (NLS) equation of the following form [1–3]

$$i\hbar\frac{\partial\Psi(\mathbf{r},\tau)}{\partial\tau} = \left[-\frac{\hbar^2}{2m}\nabla^2 + V(\mathbf{r}) + g(\mathbf{r})|\Psi(\mathbf{r},\tau)|^2\right]\Psi(\mathbf{r},\tau). \tag{9.1}$$

In the above equation, $V(\mathbf{r})$ represents the external trapping potential and $g(\mathbf{r})$ the binary spatially dependent interatomic interaction. Such spatial dependence of the scattering lengths which can be implemented utilizing a spatially inhomogeneous external magnetic field in the vicinity of a Feshbach resonance [4] renders the collisional dynamics inhomogeneous across the BECs. The resulting so-called *collisionally inhomogeneous environment* provides a variety of interesting and previously unexplored dynamical phenomena and potential applications like adiabatic compression of matter waves [5–7], atomic soliton emission and atom lasers [6], enhancement of the transmittivity of matter waves through barriers [8], dynamical trapping of matter-wave solutions [9] etc. The investigation of

doi:10.1088/978-0-7503-5447-9ch9

collisionally inhomogeneous interactions with linear and nonlinear lattices [10] and the recent identification of gap solitons in periodic media supported by localized nonlinearites [11] underscore the impact of a spatially inhomogeneous environment on optical solitons and BECs. The impact of spatially inhomogeneous interaction on the condensates in a monochromatic optical lattice [12] has been explored recently. This chapter is mainly devoted to identify the impact of collisionally inhomogeneous short-range interactions on the condensates. As a prelude, one first analyzes the combined impact of collisionally inhomogenous binary and collisionally inhomogeneous three-body interactions on the condensates of a dilute gas in a trapping potential, say in a bichormatic optical lattice potential. Another endeavour is to extract other nonlinear excitations arising out of spatially inhomogenous interactions like Faraday waves both in scalar and vector BECs, vortices, etc, and study their stability. The impact of higher-order collisionally inhomogenous nonlinearities like quintic and septimal nonlinearities on the nonlinear excitations is also being planned to be analyzed.

9.1 Model and evolution equation

This section has been reproduced with permission from [20].

The dynamics of a BEC in three dimensions incorporating both binary (two-body) and three-body interactions, is governed by the mean-field Gross–Pitaevskii (GP) equation

$$i\hbar\frac{\partial\Psi(\mathbf{r},\tau)}{\partial\tau} = \left[-\frac{\hbar^2\nabla^2}{2m} + U(\mathbf{r}) + g(\mathbf{r})|\Psi(\mathbf{r},\tau)|^2 + k(\mathbf{r})|\Psi(\mathbf{r},\tau)|^4\right]\Psi(\mathbf{r},\tau), \quad (9.2)$$

where $\Psi(\mathbf{r},\tau)$ is the condensate wavefunction, normalized by $\int_{-\infty}^{\infty}|\Psi(\mathbf{r},\tau)|^2\,d\mathbf{r} = N$, with N the number of atoms and τ the time. The potential $U(\mathbf{r})$ represents a bi-chromatic optical lattice, while $g(\mathbf{r})$ and $k(\mathbf{r})$ denote the spatially dependent strengths of two-body and three-body interactions, respectively.

The bi-chromatic optical lattice potential generated by two standing-wave polarized laser beams with incommensurate wavelengths is expressed as [5]:

$$U(\mathbf{r}) = \sum_{i=1}^{2} s_i E_i \sin^2(k_i\cdot\mathbf{r}), \quad (9.3)$$

where s_i (for $i = 1, 2$) are the amplitudes in units of recoil energies $E_i = 2\pi^2\hbar^2/(m\lambda_i'^2)$, $k_i = 2\pi/\lambda_i'$ are the wavenumbers, and λ_i' are the wavelengths. In experiments by Roati *et al* [13], the wavelengths are $\lambda_1' = 1032$ nm and $\lambda_2' = 862$ nm.

For a cigar-shaped trap with strong transverse confinement, the three-dimensional equation (9.2) is reduced to one dimension by freezing the transverse dynamics (in y and z) to the ground state and integrating over the transverse coordinates. The resulting quasi-one-dimensional dimensionless GP equation is given by:

$$i\frac{\partial\phi(x,t)}{\partial t} = \left[-\frac{1}{2}\frac{\partial^2}{\partial x^2} + V(x) + g(x)|\phi(x,t)|^2 + k(x)|\phi(x,t)|^4\right]\phi(x,t), \quad (9.4)$$

where $\phi(x, t)$ is the wavefunction, normalized by $\int_{-\infty}^{\infty} |\phi(x,t)|^2\, \mathrm{d}x = N$. Length is in units of $\sqrt{\hbar/m\omega_\perp}$, time in units of $\omega_\perp^{-1}$, and energy in units of $\hbar\omega_\perp$, with $\omega_\perp$ the transverse trap frequency. The potential $V(x)$ derived from equation (9.3) is written as:

$$V(x) = \sum_{i=1}^{2}\frac{4\pi^2 s_i}{\lambda_i^2}\sin^2\left(\frac{2\pi}{\lambda_i}x\right), \quad (9.5)$$

where $\lambda_i = \lambda_i'/\sqrt{\hbar/m\omega_\perp}$. The interaction strengths are $g(x) = 2a_x/a_\perp$, with a_x being the spatially dependent s-wave scattering length tunable via Feshbach resonance and $k(x)$ the three-body interaction strength. The scattering length a_x varies spatially as [14–16]:

$$a_x = a_0 + \frac{\alpha I(x)}{\delta + \beta I(x)}, \quad (9.6)$$

where $I(x)$ is the laser intensity, a_0 is the scattering length without light, and α and β are constants depending on the laser detuning δ. For a Gaussian laser beam with large detuning, the interaction strengths are [14–16]:

$$g(x) = \gamma_0 + \gamma_1\exp(-x^2/2), \quad (9.7a)$$

$$k(x) = \eta_0 + \eta_1\exp(-x^2/2), \quad (9.7b)$$

where γ_0, γ_1, η_0, and η_1 are constants. The three-body interaction $k(x)$ is typically weaker than $g(x)$, often set to $k(x) = 0.1g(x)$ [17], allowing control of three-body interactions via Feshbach resonance by tuning a_x.

Assuming a stationary solution $\phi(x, t) = \phi(x)\exp(-\mathrm{i}\mu t)$, the time-independent form of equation (9.4) is given by:

$$\mu\phi(x) = \left[-\frac{1}{2}\frac{\partial^2}{\partial x^2} + V(x) + g(x)|\phi(x)|^2 + k(x)|\phi(x)|^4\right]\phi(x), \quad (9.8)$$

where μ is the chemical potential. The Lagrangian for equation (9.8) is written as [17]:

$$L = \int\left[\mu|\phi|^2 - \frac{1}{2}\left|\frac{\partial\phi}{\partial x}\right|^2 - V(x)|\phi|^2 - \frac{g(x)}{2}|\phi|^4 - \frac{k(x)}{3}|\phi|^6\right]\mathrm{d}x - \mu. \quad (9.9)$$

For variational approximation, a Gaussian trial function is assumed to be of the following form:

$$\phi(x) = \pi^{-\frac{1}{4}}\sqrt{\frac{N}{w}}\exp\left(-\frac{x^2}{2w^2}\right), \quad (9.10)$$

where $\mathcal{N}$ is the norm and w is the width. Substituting equations (9.10) and (9.5) into equation (9.9) and integrating over space yields the effective Lagrangian of the following form

$$L_{\text{eff}} = \mu(\mathcal{N}-1) + \frac{\mathcal{N}}{4w^2} + \mathcal{N}\sum_{i=1}^{2}\frac{A_i}{2}\left[1-\exp(-\alpha_i^2 w^2)\right] + \frac{\gamma_0}{2\sqrt{2\pi}}\frac{\mathcal{N}^2}{w} + \frac{\gamma_1}{\sqrt{2\pi}}\frac{\mathcal{N}^2}{w\sqrt{4+w^2}} + \frac{\eta_0}{3\pi\sqrt{3}}\frac{\mathcal{N}^3}{w^2} + \frac{\eta_1\sqrt{2}}{3\pi}\frac{\mathcal{N}^3}{w^2\sqrt{6+w^2}}, \tag{9.11}$$

where $A_i = 4\pi^2 s_i/\lambda_i^2$, $\alpha_i = 2\pi/\lambda_i$. The variational equation $\partial L_{\text{eff}}/\partial\mu = 0$ gives $\mathcal{N} = 1$. The equation $\partial L_{\text{eff}}/\partial w = 0$ yields

$$2w^4\sum_{i=1}^{2}\left[A_i\alpha_i^2\exp(-\alpha_i^2 w^2)\right] - \frac{\gamma_0 w}{\sqrt{2\pi}} - \sqrt{\frac{2}{\pi}}\frac{\gamma_1 w^3}{(w^2+4)^{3/2}} - \sqrt{\frac{2}{\pi}}\frac{\gamma_1 w}{\sqrt{w^2+4}} - \frac{4\eta_0}{3\pi\sqrt{3}} - \frac{\eta_1 2\sqrt{2}w^2}{3\pi(w^2+6)^{3/2}} - \frac{4\sqrt{2}\eta_1}{3\pi\sqrt{w^2+6}} = 1, \tag{9.12}$$

determining w. The equation $\partial L_{\text{eff}}/\partial\mathcal{N} = 0$ gives the chemical potential:

$$\mu = \frac{1}{4w^2} + \sum_{i=1}^{2}\frac{A_i}{2}\left[1-\exp(-\alpha_i^2 w^2)\right] + \frac{\gamma_0}{w\sqrt{2\pi}} + \sqrt{\frac{2}{\pi}}\frac{\gamma_1}{w\sqrt{w^2+4}} + \frac{\eta_0}{\pi\sqrt{3}w^2} + \frac{\eta_1}{3\pi w^2}\frac{2\sqrt{2}}{\sqrt{w^2+6}}. \tag{9.13}$$

9.1.1 Numerical study

Numerical simulations employ the split-step Crank–Nicolson (SSCN) method [18] for real- and imaginary-time propagation, using space and time steps of 0.0025 and 0.00005, respectively, to ensure convergence. Consistency is verified by comparing real- and imaginary-time propagation results, which agree well. Numerical accuracy is confirmed by varying step sizes. The bi-chromatic optical lattice potential maintains a strength ratio $s_2/s_1 = 1$, with wavelengths $\lambda_1 = 5$ and $\lambda_2 = 0.864\lambda_1$ [13].

9.1.1.1 Impact of spatially inhomogeneous binary interactions ($k(x) = 0$)

The localization of BECs with constant repulsive binary interactions has been studied earlier [19]. Introducing repulsive spatially inhomogeneous two-body interactions (γ_1), density profiles for various γ_1 values are shown in figures 9.1(a)–(d). Increasing γ_1 induces instability in the condensates.

The central localized density amplitude can be controlled by adding a nearly equal constant attractive binary interaction (γ_0). Figure 9.2(a) shows the resulting density profile, representing a solitary mode, while figure 9.2(b) displays a phase plot indicating stable regions for repulsive γ_1 balanced by attractive γ_0.

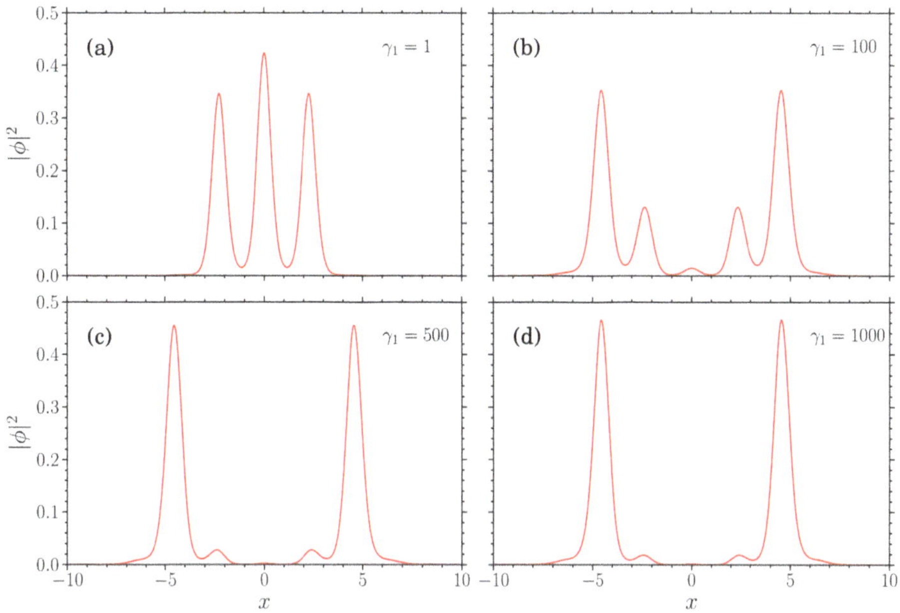

Figure 9.1. Density profiles for repulsive spatially inhomogeneous binary interaction strengths with potential (9.5): (a) $\gamma_1 = 1$, (b) $\gamma_1 = 100$, (c) $\gamma_1 = 500$, (d) $\gamma_1 = 1000$. Reprinted from [20]. Copyright 2013 IOP Publishing Ltd.

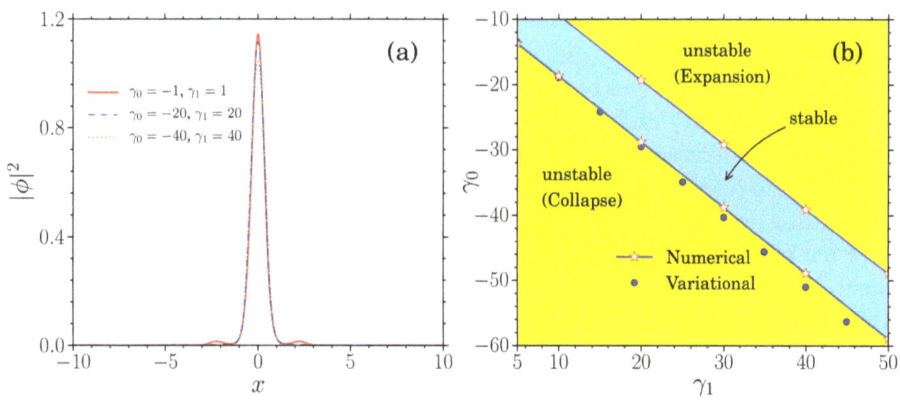

Figure 9.2. (a) Density profiles for potential (9.5). (b) Phase plot showing stability regions for repulsive spatially inhomogeneous binary interaction versus constant attractive binary interaction. Reprinted from [20]. Copyright 2013 IOP Publishing Ltd.

Figure 9.3 presents density profiles for $\gamma_0 = -15$, $\gamma_1 = 20$ over time, corresponding to the region above the stable domain in figure 9.2(b). Despite appearing dynamically stable, boundary fluctuations increase the root-mean-square size, leading to BEC expansion. Below the stable domain, the central density increases with γ_0, collapsing at a critical value.

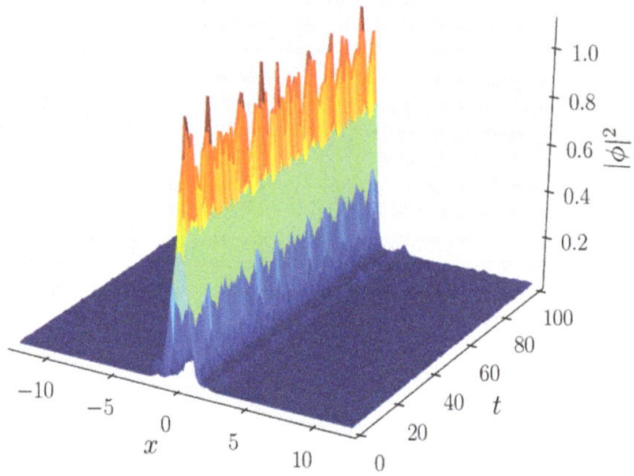

Figure 9.3. Density profiles for $\gamma_0 = -15$, $\gamma_1 = 20$ with potential (9.5) as a function of time. Rerpinted from [20]. Copyright 2013 IOP Publishing Ltd.

Alternatively, central localized density can be sustained by balancing constant repulsive binary interactions with attractive spatially inhomogeneous binary interactions, as shown in figure 9.4(a). Increasing constant repulsive interactions causes the density to occupy multiple sites, as in figure 9.4(b). Figure 9.4(c) provides the numerical stability domain for these interactions.

9.1.1.2 Impact of spatially inhomogeneous binary and three-body interactions

Introducing weak three-body interactions ($k(x) = 0.1g(x)$) alongside binary interactions, figures 9.5(a) and 9.5(b) show density and phase plots for repulsive spatially inhomogeneous binary and three-body interactions versus constant attractive binary and three-body interactions. The density in figure 9.5(a), corresponding to the stable region in figure 9.5(b), represents a solitary mode, nearly identical to figure 9.2 (a), indicating negligible impact from weak three-body interactions in the stable region. Above the stable region, condensates expand faster than with $k(x) = 0$, similar to figure 9.2(b).

Reversing the signs of γ_0 and γ_1, a central localized solitary mode is achieved with balanced γ_0 and γ_1, as in figure 9.6(a). Imbalance leads to multi-site occupation (figure 9.6(b)), above the stable region in figure 9.7 or condensate collapse (below the stable region). Comparing phase plots in figures 9.2(b), 9.4(c), 9.5(b), and 9.7, the combined effect of spatially inhomogeneous binary and three-body interactions accelerates instability [21].

9.2 Faraday and resonant waves in scalar Bose–Einstein condensates

9.2.1 Introduction

Over the past two decades, ultracold quantum gases have emerged as an ideal platform for nonlinear science due to their versatility and precise experimental control. BECs are particularly compelling [22, 23] for several reasons: highly tunable

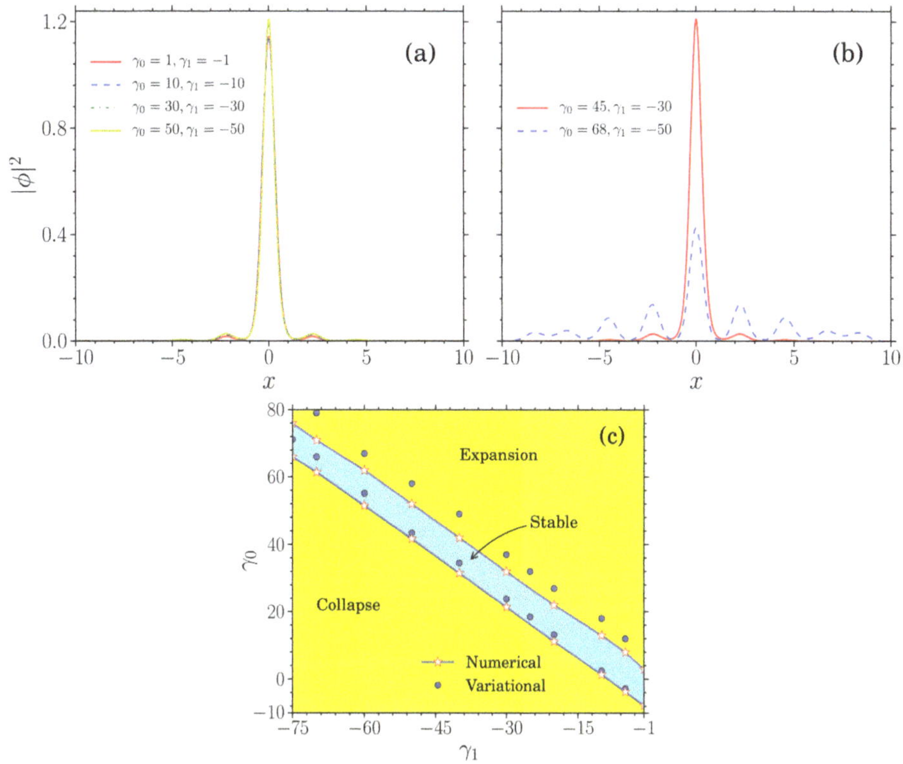

Figure 9.4. Density profiles for potential (9.5): (a) stable region, (b) expansion region. (c) Phase plot for attractive spatially inhomogeneous binary interaction versus constant repulsive binary interaction. Reprinted from [20]. Copyright 2013 IOP Publishing Ltd.

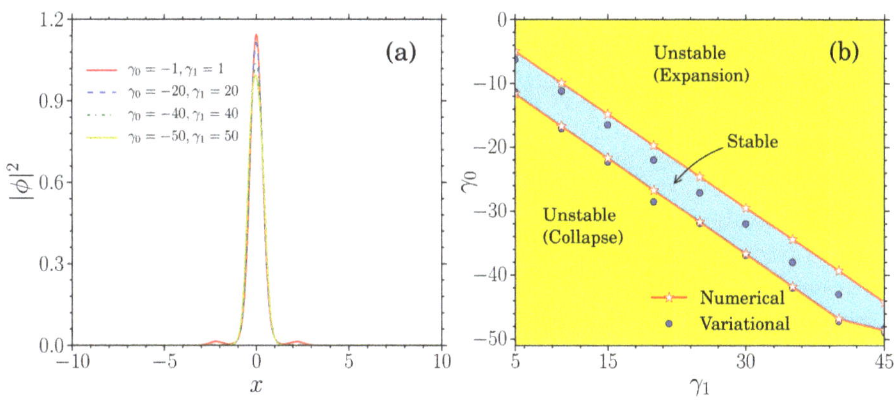

Figure 9.5. (a) Density profiles for potential (9.5) with varying γ_0 and γ_1. (b) Phase plot for repulsive spatially inhomogeneous binary and three-body interactions versus constant attractive binary and three-body interactions. Reprinted from [20]. Copyright 2013 IOP Publishing Ltd.

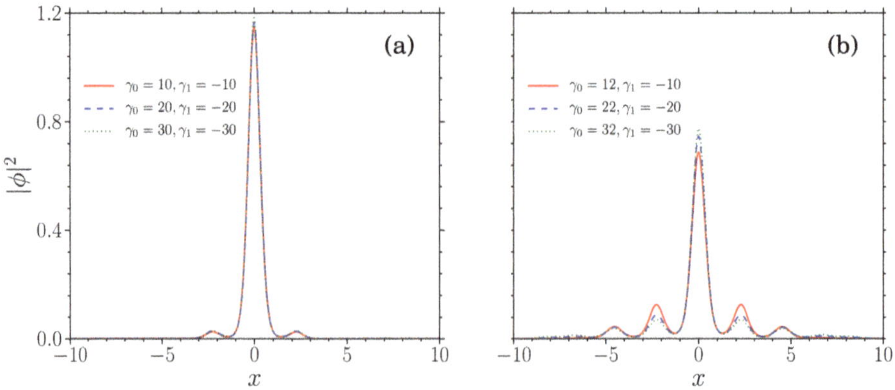

Figure 9.6. Density profiles for potential (9.5): (a) stable region, (b) expansion region. Reprinted from [20]. Copyright 2013 IOP Publishing Ltd.

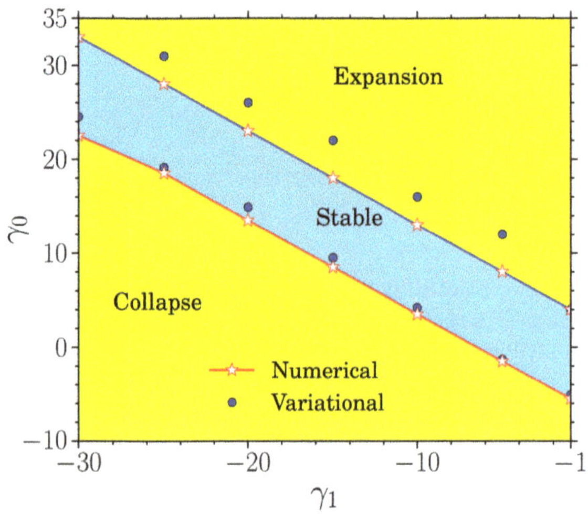

Figure 9.7. Phase plot for attractive spatially inhomogeneous binary and three-body interactions versus constant repulsive binary and three-body interactions. Reprinted from [20]. Copyright 2013 IOP Publishing Ltd.

short-range two-body interactions via magnetic or optical Feshbach resonances, the presence of long-range dipole–dipole interactions in certain atomic species, the realization of multi-component condensates (using one atomic species in multiple hyperfine states or distinct species) with adjustable inter-component interactions, and the ability to flexibly shape condensate geometries [23]. These experimental capabilities are complemented by accurate mean-field theoretical models, primarily the GP equation [1], enabling detailed studies of condensate dynamics at or near zero temperature.

Pattern formation in quantum fluids is a vibrant research area, with experimental observations of Faraday patterns in BECs [3] and liquid helium-4 cells [24] under parametric driving, alongside theoretical investigations of Faraday waves in condensates with short-range interactions [24, 25], dipolar condensates [26], binary condensates [27], Fermi–Bose mixtures [28], and superfluid Fermi gases [29]. Faraday waves can be suppressed using resonant parametric modulations [29] or spatiotemporally modulated potentials [30, 31], a topic extensively explored [32–39]. Additionally, density patterns have been studied in expanding ultracold Bose gases, both fully [40] and partially condensed [41, 42], and spontaneous density waves have been reported in antiferromagnetic BECs [43].

This section focuses on cigar-shaped condensates with radially inhomogeneous, Gaussian-modulated scattering lengths under periodic modulation of radial confinement, operating in the collisionally inhomogeneous regime [44]. This regime can be achieved magnetically or optically. Magnetic Feshbach resonances, used to study ultracold molecules [45], BEC–Bardeen–Cooper–Schrieffer (BCS) crossover [46], and Efimov trimer states [47], typically apply over scales larger than BEC samples, limiting their use for inhomogeneity. Optical Feshbach resonances, however, enable fine spatial control of the scattering length, with modulations on submicron scales [48]. The collisionally inhomogeneous regime supports diverse nonlinear phenomena, including adiabatic matter-wave compression [5, 49], Bloch oscillations of solitons [50], atomic soliton emission and atom lasers [51], dynamical soliton trapping [16, 52–55], enhanced matter-wave transmission through barriers [56–58], stable condensates with mixed interactions [6, 59, 60], delocalization in optical lattices [61], symmetry breaking in double-well potentials [62], competition between linear and nonlinear lattices [53, 63], soliton and vortex ring generation [64–66], and more.

Through numerical simulations and variational calculations, this study demonstrates that the spatial period of Faraday waves in collisionally inhomogeneous condensates depends strongly on the scattering length's spatial modulation. The period increases as inhomogeneity weakens, saturating when the Gaussian inhomogeneity's width approaches the condensate's radial width. This behaviour reflects an effective nonlinearity, where weaker inhomogeneity enhances nonlinearity, producing Faraday waves with longer periods and shorter instability onset times. Strongly inhomogeneous regimes yield hollow cylinder-like radial profiles, while weakly inhomogeneous ones produce cigar-shaped, Thomas–Fermi profiles. The latter is described by standard variational methods [67], while the former requires a tailored trial wavefunction capturing both bulk properties and density wave emergence.

9.2.2 Variational treatment of the Gross–Pitaevskii equation

The ground state and dynamics of a three-dimensional BEC at zero temperature are described by the time-independent GP equation:

$$\left(-\frac{\hbar^2}{2m}\Delta + V(\mathbf{r}) + g(\mathbf{r})N|\psi|^2\right)\psi = \mu\psi, \tag{9.14}$$

and its time-dependent counterpart:

$$i\hbar\frac{\partial\psi}{\partial t} = \left(-\frac{\hbar^2}{2m}\Delta + V(\mathbf{r}) + g(\mathbf{r}, t)N|\psi|^2\right)\psi, \tag{9.15}$$

where μ is the chemical potential, N is the atom number, and the trapping potential is given by:

$$V(\mathbf{r}) = \frac{m}{2}\left(\Omega_\rho^2(t)\rho^2 + \Omega_z^2(t)z^2\right). \tag{9.16}$$

The nonlinear interaction strength g relates to the s-wave scattering length a_s as

$$g = \frac{4\pi\hbar^2}{m}a_s, \tag{9.17}$$

and can be spatially or temporally modulated via optical Feshbach resonances or magnetic field variations near a Feshbach resonance. Equation (9.15) can be solved numerically using split-step Crank–Nicolson or Fourier transform methods (see chapter 3). For analytical insights, variational or hydrodynamic approaches are employed [23]. Variational methods reduce condensate dynamics to ordinary differential equations, enabling calculations of collective excitation frequencies, sound speed, and resonance positions. The GP Lagrangian density is given by:

$$\mathcal{L}(\mathbf{r}, t) = \frac{\hbar^2}{2m}|\nabla\psi|^2 + V(\mathbf{r}, t)|\psi|^2 + \frac{gN}{2}|\psi|^4, \tag{9.18}$$

which is minimized for a trial wavefunction tailored to the physics of the system.

Consider a longitudinally homogeneous, cigar-shaped condensate ($\Omega_z(t) = 0$) with harmonically modulated radial frequency of the form:

$$\Omega_\rho(t) = \Omega_{\rho 0}(1 + \epsilon\sin\omega t), \tag{9.19}$$

where ϵ is the modulation amplitude and ω is the frequency. The scattering length is radially modulated, yielding:

$$g(\rho) = \frac{4\pi\hbar^2 a(0)}{m}e^{-\frac{\rho^2}{2b^2}} = g_0 e^{-\frac{\rho^2}{2b^2}}, \tag{9.20}$$

where $a(0) = a_s|_{\rho=0}$ is the scattering length along the z-axis, and b is the modulation length scale.

The trial wavefunction for the collisionally inhomogeneous BEC is assumed to be of the following form:

$$\begin{aligned}\psi(\mathbf{r}, t) &= \phi(\mathbf{r}, t)\{1 + [u(t) + iv(t)]\cos kz\} \\ &= A(t)(1 + \gamma\rho^2)\exp\left(-\frac{\rho^2}{2w^2(t)} + i\rho^2\alpha(t)\right)\{1 + [u(t) + iv(t)]\cos kz\},\end{aligned} \tag{9.21}$$

where $A(t)$ ensures density normalization over one period $[-\pi/k, \pi/k]$. The radial envelope $\phi(\mathbf{r}, t)$ describes collective dynamics, while the periodic term captures longitudinal density waves. One-dimensional systems exhibit uniform wave patterns, unaffected by the inhomogeneity of the scattering length and hence the focus is on spatial periods and instability onset times. Higher-dimensional systems show pattern geometry changes (e.g., triangular to square) with scattering length variations.

The Lagrangian density is integrated over one spatial period, and the resulting time-dependent Lagrangian is minimized via Euler–Lagrange equations [68], yielding four differential equations for variational parameters $w(t)$, $\alpha(t)$, $u(t)$, and $v(t)$, and one algebraic equation for γ. Here, $w(t)$ is the condensate width, $\alpha(t)$ is the phase, $u(t) + iv(t)$ is the density wave's complex amplitude, and γ reflects collisional inhomogeneity. The wave's spatial period $2\pi/k$ is determined by conditions for density wave emergence which means that k is not treated as a variational parameter. The accuracy of the trial wavefunction affects the results, and alternative forms are explored elsewhere [69–71]. Simplified variational equations for weakly and strongly inhomogeneous regimes are analyzed, validated by numerical results in section 9.2.5. Natural units ($\hbar = m = 1$) are used henceforth.

9.2.3 Weakly inhomogeneous collisions

In the weakly inhomogeneous regime, the modulation length b is large, making the exponential in equation (9.20) nearly unity. The stationary density, obtained numerically from equation (9.14), shows strong atom localization near the symmetry axis, allowing a small γ. The variational equations for the bulk condensate are the following.

$$\gamma = \frac{4b^2 + \tilde{w}^2}{4\tilde{w}^2(8b^6E_1 + 16b^4E_2\tilde{w}^2 + 4b^2E_2\tilde{w}^4 + \pi\tilde{w}^6)} \times \left\{8b^4E_3 + 4b^2E_3\tilde{w}^2 + \pi\left(1 - 16b^4\Omega_{\rho 0}^2\right)\tilde{w}^4 + 8\pi b^2\Omega_{\rho 0}^2\tilde{w}^6 + \pi\Omega_{\rho 0}^2\tilde{w}^8\right\}, \tag{9.22}$$

$$\begin{aligned}\ddot{w}(t) = {} & \frac{2\pi + ng_0}{2\pi w(t)^3} - \frac{2\gamma(2\pi + ng_0)}{\pi w(t)} \\ & - \frac{w(t)}{2\pi(4b^2 + w(t)^2)^3}\{4b^2(ng_0 + 32\pi b^4\Omega_\rho(t)^2) + 2\pi\Omega_\rho(t)^2 w(t)^6 \\ & + [ng_0 + 96\pi b^4\Omega_\rho(t)^2]w(t)^2 - 4[\gamma ng_0 - 6\pi b^2\Omega_\rho(t)^2]w(t)^4\},\end{aligned} \tag{9.23}$$

For the density wave, the variational equations are given by:

$$\dot{u}(t) = \frac{k^2}{2}v(t), \tag{9.24}$$

$$\dot{v}(t) = -\left(\frac{k^2}{2} + \frac{4b^2ng_0}{\pi w(t)^2(4b^2 + w(t)^2)}\right)u(t), \tag{9.25}$$

where $\tilde{w}$ is the equilibrium width from equation (9.23) with $\epsilon = 0$, $E_1 = 8\pi + 3ng_0$, $E_2 = 3\pi + ng_0$, $E_3 = 2\pi + ng_0$, and n is the longitudinal density. These equations neglect terms of order $\mathcal{O}(\gamma^2)$.

Equations (9.24)–(9.25) form a Mathieu-like equation:

$$\ddot{u}(\tau) + u(\tau)[A_{\rm W}(k,\,\omega) + \epsilon B_{\rm W}(k,\,\omega)\sin 2\tau] = 0, \tag{9.26}$$

with $\omega t = 2\tau$, and:

$$A_{\rm W}(k,\,\omega) = \frac{2k^2}{\omega^2}\left(\frac{k^2}{2} + \frac{4b^2 n g_0}{\pi\tilde{w}^2(4b^2 + \tilde{w}^2)}\right). \tag{9.27}$$

The periodic modulation of the trapping potential drives oscillations in $w(t)$, acting as a parametric drive in equation (9.26). For small ϵ, this reduces to a standard Mathieu equation [72, 73]. The most unstable solutions, corresponding to observed Faraday waves [73–75], satisfy $A_{\rm W}(k,\,\omega) = 1$, yielding:

$$k_{\rm F,W} = \left\{\sqrt{\omega^2 + \frac{16b^4n^2g_0^2}{\pi^2\tilde{w}^4(4b^2 + \tilde{w}^2)^2}} - \frac{4b^2ng_0}{\pi\tilde{w}^2(4b^2 + \tilde{w}^2)}\right\}^{1/2}, \tag{9.28}$$

with spatial period $p = 2\pi/k_{\rm F,W}$.

9.2.4 Strongly inhomogeneous collisions

In the strongly inhomogeneous regime, small b values produce a hollow cylinder-like stationary density profile, as interaction energy peaks along the longitudinal axis while potential energy increases radially. The ground state balances these, with maximal density offset from the axis. For large γ, the variational equations for the bulk condensate are given by:

$$\begin{aligned}\gamma = {}& \frac{2}{\tilde{w}^2}\left[\frac{1024b^8}{\tilde{w}^2} + 1280b^6 + 160b^2\tilde{w}^4 + 8b^4\left(80 + \frac{3ng_0}{\pi}\right)\tilde{w}^2 + 20\tilde{w}^6 + \frac{\tilde{w}^8}{b^2}\right]\\ &\times\left[\frac{1024b^8}{\tilde{w}^2} + 32b^6\left(40 - \frac{3ng_0}{\pi}\right) + \frac{\tilde{w}^8}{b^2} + 128b^4\left(5 + 8b^4\Omega_{\rho0}^2\right)\tilde{w}^2 + \frac{\Omega_{\rho0}^2\tilde{w}^{12}}{b^2}\right.\\ &\left. + 160b^2\left(1 + \Omega_{\rho0}^2\tilde{w}^4 + 8b^4\Omega_{\rho0}^2\right)\tilde{w}^4 + 20\left(1 + \Omega_{\rho0}^2\tilde{w}^4 + 32b^4\Omega_{\rho0}^2\right)\tilde{w}^6\right]^{-1},\end{aligned} \tag{9.29}$$

$$\begin{aligned}\ddot{w}(t) = {}& \frac{1}{3w(t)^3} - \frac{2}{3\gamma w(t)^5} - \Omega_\rho(t)^2 w(t) + \frac{256b^{12}ng_0}{\pi E_4 w(t)^3} + \frac{128b^{10}ng_0}{\pi\gamma E_4 w(t)^3}\\ &+ \frac{384b^{10}ng_0}{\pi E_4 w(t)} + \frac{192b^8ng_0}{\pi\gamma E_4 w(t)},\end{aligned} \tag{9.30}$$

and for the density wave, one obtains:

$$\dot{u}(t) = \frac{k^2}{2}v(t), \tag{9.31}$$

$$\dot{v}(t) = -\left(\frac{k^2}{2} + \frac{384b^{10}ng_0}{\pi w(t)^2(4b^2 + w(t)^2)^5}\right)u(t), \tag{9.32}$$

where $E_4 = (4b^2 + w(t)^2)^6$. Equation (9.29) is accurate to $\mathcal{O}(\gamma^{-4})$, others to $\mathcal{O}(\gamma^{-2})$, and n is the longitudinal density.

Equations (9.31)–(9.32) yield a Mathieu-like equation:

$$\ddot{u}(\tau) + u(\tau)[A_S(k, \omega) + \epsilon B_S(k, \omega)\sin 2\tau] = 0, \tag{9.33}$$

with:

$$A_S(k, \omega) = \frac{2k^2}{\omega^2}\left(\frac{k^2}{2} + \frac{384b^{10}ng_0}{\pi\tilde{w}^2(4b^2 + \tilde{w}^2)^5}\right). \tag{9.34}$$

The most unstable solution satisfies $A_S(k, \omega) = 1$, giving:

$$k_{F,S} = \frac{1}{\sqrt{\pi\tilde{w}^2(4b^2 + \tilde{w}^2)^5}}\left[\sqrt{Cb^{20}n^2g_0^2 + \pi^2\tilde{w}^4(4b^2 + \tilde{w}^2)^{10}\omega^2} - 384b^{10}ng_0\right]^{1/2}, \tag{9.35}$$

where $C = 147, 456$, and $\tilde{w}$ is the equilibrium width from equation (9.30) with $\epsilon = 0$. The spatial period is $p = 2\pi/k_{F,S}$.

The dispersion relations (9.28) and (9.35) are the key results, accurately describing density wave properties despite variational simplifications, as shown in section 9.2.5.

9.2.5 Results [30]

This section compares variational results with numerical simulations for a condensate of $N = 2.5 \times 10^5$ ^{87}Rb atoms in a magnetic trap with $\Omega_{\rho 0} = 160 \times 2\pi$ Hz and $\Omega_z = 7 \times 2\pi$ Hz. Using imaginary-time propagation via the SSCN method [18], the ground state is computed for a constant scattering length $a_s = 100.4\ a_0$, yielding a radial width $b_0 = 1.86\ \mu\text{m}$, a reference for the inhomogeneity parameter b in equation (9.20). Ground states are then calculated for b from $b_0/4$ to $4b_0$, and the homogeneous limit ($b \to \infty$). Real-time dynamics are simulated [18] to monitor Faraday patterns in radially integrated density profiles under the following parametric driving:

$$\Omega_\rho(t) = \Omega_{\rho 0}(1 + \epsilon \sin \omega t). \tag{9.36}$$

Figures 9.8(a) and (b) show the radial density profile at $z = 0$ and the full ρ–z profile for $b = 4b_0$ (weak inhomogeneity), exhibiting a Thomas–Fermi profile. Figure 9.8(c) depicts the time evolution of the radially integrated longitudinal density with $\epsilon = 0.1$, $\omega = 250 \times 2\pi$ Hz, where Faraday waves emerge after 200 ms. Figure 9.8(d) shows the Fourier spectrum at 250 ms, with a peak at $k_{F,W} = 0.60\ \mu\text{m}^{-1}$, corresponding to a period $p = 2\pi/k_{F,W} = 10.5\ \mu\text{m}$. Variational results from equations (9.22) and (9.23) overestimate peak density and underestimate radial extent but capture density wave emergence.

For strong inhomogeneity ($b = b_0/4$), figures 9.9(a) and (b) show a hollow cylinder-like radial profile. Figure 9.9(c) shows the density evolution, with Faraday waves visible after 200 ms, reducing longitudinal extent by half compared to figure 9.8 (c). The Fourier spectrum in figure 9.9(d) at 250 ms peaks at $k_{F,S} = 1.16\ \mu\text{m}^{-1}$, giving

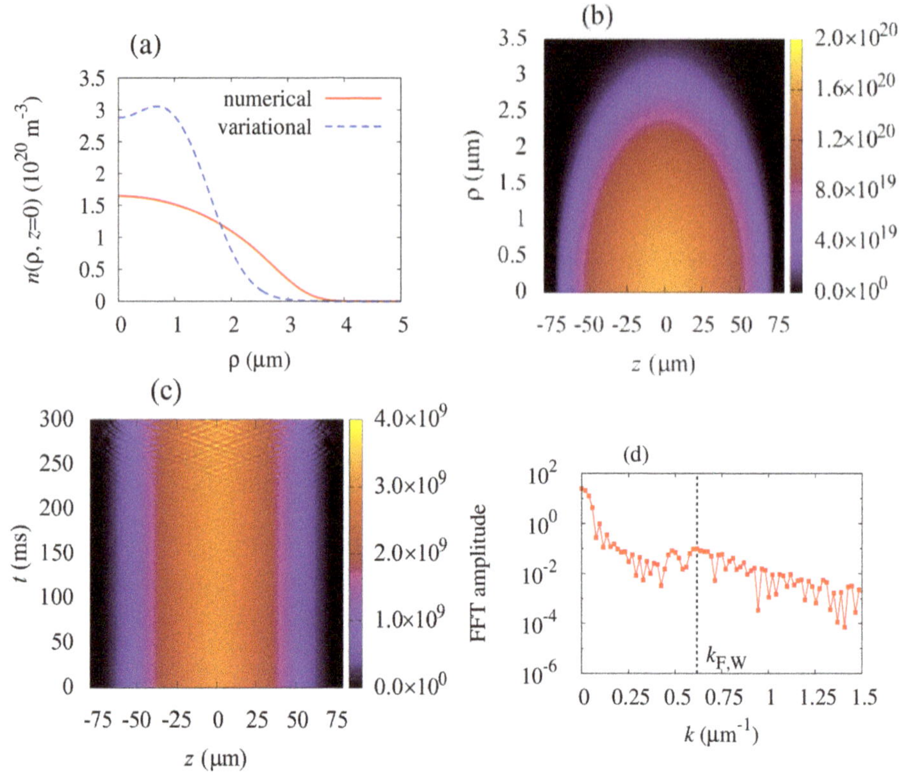

Figure 9.8. Weakly inhomogeneous collisions ($b = 4b_0$). (a) Radial density profile at $z = 0$ for the ground state; red line: numerical GP results, blue dashed line: variational results. (b) Full ρ–z density profile. (c) Time evolution of radially integrated longitudinal density with $\epsilon = 0.1$, $\omega = 250 \times 2\pi$ Hz; Faraday waves appear after 200 ms. (d) Fourier spectrum at 250 ms, with peak at $k_{\rm F,W} = 0.60\ \mu{\rm m}^{-1}$, yielding $p = 2\pi/k_{\rm F,W} = 10.5\ \mu{\rm m}$. Reprinted with permission from [30], Copyright (2014) by the American Physical Society.

$p = 2\pi/k_{\rm F,S} = 5.4\ \mu{\rm m}$. Fourier analysis confirms that density waves have a frequency of $\omega/2$, identifying them as Faraday waves across all b.

Figure 9.10 compares the spatial period of Faraday waves at $\omega = 250 \times 2\pi$ Hz versus b, with numerical results (black squares) from Fourier analysis and variational predictions from equations (9.28) (blue circles, weak inhomogeneity) and (9.35) (red triangles, strong inhomogeneity). A longitudinal Thomas–Fermi density of the form:

$$n(z) = \frac{3(L^2 - z^2)}{4L^3}, \quad z \in [-L, L], \tag{9.37}$$

is used, with $n(z) = 0$ elsewhere. The average wave vector is given by:

$$\bar{k} = \frac{1}{2L}\int_{-L}^{L} k(z)\, {\rm d}z, \tag{9.38}$$

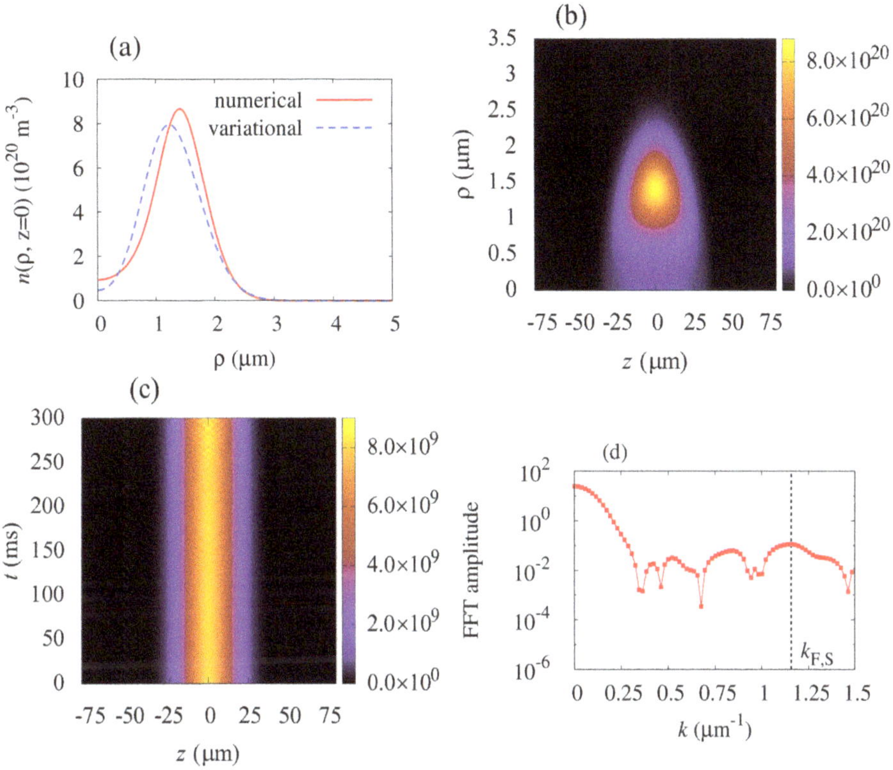

Figure 9.9. Strongly inhomogeneous collisions ($b = b_0/4$). (a) Radial density profile at $z = 0$; red line: numerical Gross–Pitaevskii results, blue dashed line: variational results. (b) Full ρ–z density profile. (c) Time evolution of radially integrated longitudinal density with $\epsilon = 0.1$, $\omega = 250 \times 2\pi$ Hz; Faraday waves appear after 200 ms. (d) Fourier spectrum at 250 ms, with peak at $k_{\mathrm{F,S}} = 1.16\ \mu\mathrm{m}^{-1}$, yielding $p = 2\pi/k_{\mathrm{F,S}} = 5.4\ \mu\mathrm{m}$. Reprinted from [30], Copyright (2014) by the American Physical Society.

where $2L$ is derived from a one-dimensional GP equation:

$$i\hbar\frac{\partial \tilde{f}}{\partial t} = \left(-\frac{\hbar^2}{2m}\frac{\partial^2}{\partial z^2} + \frac{1}{2}\Omega_z z^2 + g_{1\mathrm{D}} N |\tilde{f}|^2\right)\tilde{f}, \tag{9.39}$$

with:

$$g_{1\mathrm{D}} = g_0 \int_0^\infty 2\pi\rho\ \phi(\rho)^4 \mathrm{e}^{-\frac{\rho^2}{2b^2}}\ \mathrm{d}\rho. \tag{9.40}$$

Applying the Thomas–Fermi approximation, the longitudinal extent is given by:

$$2L = \left(\frac{12Ng_{1\mathrm{D}}}{\Omega_z^2}\right)^{1/3}, \tag{9.41}$$

using radial wavefunctions from equations (9.22), (9.23) (weak) or (9.29), (9.30) (strong).

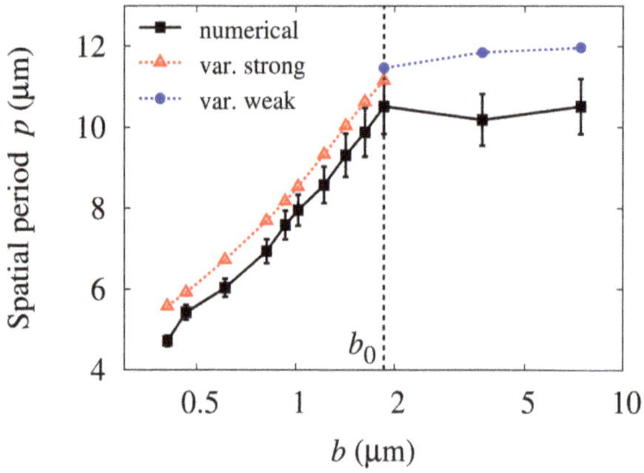

Figure 9.10. Spatial period of Faraday waves versus inhomogeneity scale b for $\epsilon = 0.1$, $\omega = 250 \times 2\pi$ Hz. Black squares: numerical results from Fourier analysis of the time-dependent GP equation. Red triangles: variational prediction from equation (9.35) (strong inhomogeneity). Blue circles: variational prediction from equation (9.28) (weak inhomogeneity). Error bars reflect peak widths in Fourier spectra. Reprinted from [30], Copyright (2014) by the American Physical Society.

Faraday waves appear for non-resonant driving frequencies ($\omega/\Omega_{\rho 0}$ non-integer). Resonant or near-resonant dynamics differ. Density waves couple with collective modes (stronger in weak inhomogeneity), and their frequency equals the drive's. Figure 9.11 shows resonant dynamics at $\omega = \Omega_{\rho 0} = 160 \times 2\pi$ Hz for $b = 4b_0$, b_0, and $b_0/4$, with longitudinal oscillations similar to experiments [76]. Radial extent remains nearly constant, with small oscillations from the drive, yielding one-dimensional longitudinal collective modes.

For $\omega = 2\Omega_{\rho 0} = 320 \times 2\pi$ Hz, figure 9.12 shows dynamics for $b = b_0$ and $b_0/4$. At $b = b_0$, the condensate destabilizes rapidly after the onset of the Faraday wave (figure 9.12(a)), while at $b = b_0/4$, the destabilization is slower, with clear wave evolution and collective mode excitation (figure 9.12(b)). This violent destabilization persists in weak inhomogeneity, including $b \to \infty$. The effective nonlinearity, given by:

$$g^* = \int_0^\infty 2\pi\rho\ g(\rho)\ d\rho = 2\pi g_0 b^2, \tag{9.42}$$

decreases with smaller b, reducing nonlinear effects like collective modes and density waves, increasing instability onset times, and decreasing longitudinal oscillation amplitudes. Larger b enhances nonlinearity, increasing Faraday wave periods and visibility, with shorter onset times. Exploring collisionally inhomogeneous interactions in vector BECs could reveal further insights.

9.3 Faraday and resonant waves in binary condensates

This section has been reproduced with permission from [94].

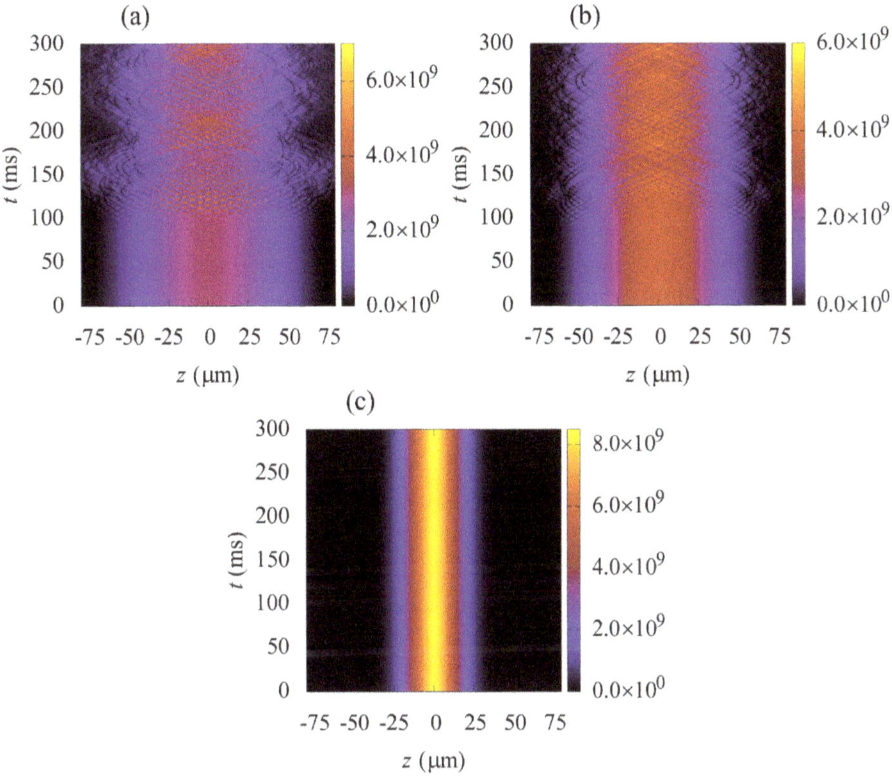

Figure 9.11. Time evolution of radially integrated longitudinal density with $\epsilon = 0.1$, $\omega = 160 \times 2\pi$ Hz: (a) $b = 4b_0$; (b) $b = b_0$; (c) $b = b_0/4$. Collective modes soften for smaller b. Reprinted from [30], Copyright (2014) by the American Physical Society.

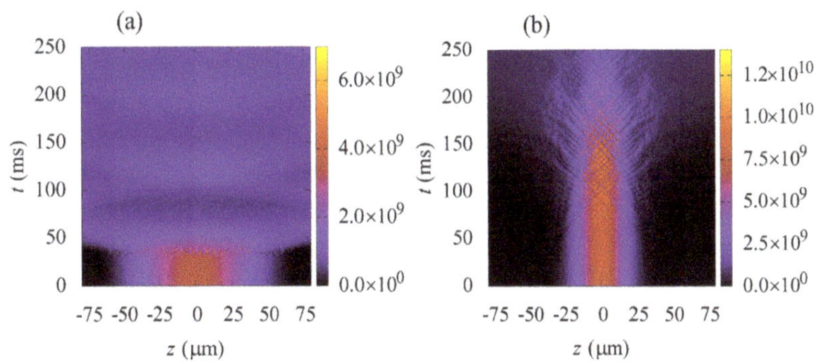

Figure 9.12. Time evolution of radially integrated longitudinal density with $\epsilon = 0.1$, $\omega = 320 \times 2\pi$ Hz: (a) $b = b_0$, rapid destabilization after Faraday wave onset; (b) $b = b_0/4$, slower destabilization with clear Faraday wave evolution and collective mode. Reprinted from [30], Copyright (2014) by the American Physical Society.

9.3.1 Introduction

This section explores pattern-forming modulational instabilities in binary cigar-shaped BECs with spatially varying collisional interactions. Extensive numerical simulations reveal that as binary collisions are located at the centre of the magnetic trap, the system approaches a nearly linear regime. Typically, binary cigar-shaped condensates exhibit two immiscible configurations: segregated, where components are spatially distinct, and symbiotic, where one component envelops the other. In this regime, these configurations transition to a miscible state, with components overlapping and their wavefunctions approximating Gaussians. Periodic modulation of the radial trap strength induces density waves, Faraday, or resonant types, but their emergence is notably delayed, as shown by extended instability onset times. The complexity of the wavefunctions limits the utility of variational methods which means that this study relies on numerical solutions of the GP equation. The following sections describe the numerical computation of stationary states, the shift to miscibility, and the condensate's dynamic behaviour.

9.3.2 Mean-field theory and numerical approach

Investigation of the properties of BECs often hinges on accurate numerical solutions of the mean-field GP equation. For the ground state of a two-component BEC, the equation is given by

$$\mu_j \psi_j = -\frac{1}{2}\Delta\psi_j + V(\mathbf{r}, t)\psi_j + N_j G_j(\mathbf{r})|\psi_j|^2\psi_j + N_{3-j}G_{12}(\mathbf{r})|\psi_{3-j}|^2\psi_j, \tag{9.43}$$

while the dynamics is described by

$$i\frac{\partial\psi_j}{\partial t} = -\frac{1}{2}\Delta\psi_j + V(\mathbf{r}, t)\psi_j + N_j G_j(\mathbf{r})|\psi_j|^2\psi_j + N_{3-j}G_{12}(\mathbf{r})|\psi_{3-j}|^2\psi_j, \tag{9.44}$$

where $j \in \{1, 2\}$, and N_j is the fixed number of atoms in component j. Lagrange multipliers μ_j, akin to eigenenergies, maintain the fixed atom number. Natural units $\hbar = m = 1$ are adopted for simplicity, and wavefunctions are normalized as

$$\int \mathrm{d}\mathbf{r}|\psi_j(\mathbf{r}, t)|^2 = 1. \tag{9.45}$$

The nonlinear interaction strengths $G_j(\mathbf{r})$ and $G_{12}(\mathbf{r})$ scale with the intra- and inter-component s-wave scattering lengths, which can be spatially modulated via optical Feshbach resonances, as demonstrated in [77].

The analysis involves two hyperfine states of ^{87}Rb (denoted A and B) in a harmonic trapping potential

$$V(\mathbf{r}, t) = \frac{1}{2}\Omega_\rho^2(t)\rho^2 + \frac{1}{2}\Omega_z^2 z^2, \tag{9.46}$$

where $\rho^2 = x^2 + y^2$, and the system is cylindrically symmetric, i.e., $\psi_j(\mathbf{r}, t) = \psi_j(\rho, z, t)$. Strong radial confinement ($\Omega_\rho(t) \gg \Omega_z$) is assumed, with scattering lengths modulated radially, yielding nonlinear interactions

$$G_j(\mathbf{r}) = G_j(\rho) = 4\pi a_j(0)\mathrm{e}^{-\rho^2/2b^2} = g_j\mathrm{e}^{-\rho^2/2b^2}, \tag{9.47}$$

$$G_{12}(\mathbf{r}) = G_{12}(\rho) = 4\pi a_{12}(0)\mathrm{e}^{-\rho^2/2b^2} = g_{12}\mathrm{e}^{-\rho^2/2b^2}, \tag{9.48}$$

where $a_j(0)$ and $a_{12}(0)$ are the constant s-wave scattering lengths along the z-axis for intra- and inter-component collisions, and b is the radial modulation length scale. Experimental scattering lengths from [78–80] are used:

$$a_1(0) = 100.4a_0, \quad a_2(0) = 98.98a_0, \quad a_{12}(0) = a_1(0), \tag{9.49}$$

where a_0 is the Bohr radius.

Numerical solution of the GP equation is a robust research field, with efficient algorithms for ground and excited states. Techniques include imaginary-time propagation [81], finite-difference schemes [72], time-splitting spectral methods [82], harmonic oscillator basis expansions [83], and symplectic shooting methods [18]. A versatile code package in Fortran and C [84–90], including MPI and CUDA parallelized versions, computes stationary states and dynamics for one- to three-dimensional BECs. OpenMP-parallelized C codes reduce execution times by an order of magnitude on multi-core CPUs compared to serial versions.

Simulations employ codes from [84] for a system with $N_1 = 2.5 \times 10^5$ atoms in state A and $N_2 = 1.25 \times 10^5$ atoms in state B, in a quasi-one-dimensional trap with frequencies $\Omega_{\rho 0} = 2\pi \times 160$ Hz and $\Omega_z = 2\pi \times 7$ Hz. The ground state is calculated using imaginary-time propagation with scattering lengths from equation (9.49). By setting $t = -i\tau$, the equation becomes a nonlinear diffusion equation, with wavefunctions normalized to unity after each time step to converge to a local energy minimum, as imaginary-time propagation does not preserve unitarity.

First, the ground state of a single-component BEC with a constant scattering length $a = 100.0a_0$ is computed, yielding a radial width $b_0 = 1.86\ \mu\mathrm{m}$, which serves as a reference for the inhomogeneity parameter b in equations (9.47) and (9.48). The two-component system is then studied with interaction parameters from equation (9.49), computing its ground state and dynamics for various b. The radial trap frequency is modulated as $\Omega_\rho(t) = \Omega_{\rho 0}(1 + \epsilon \sin(\omega t))$, with ϵ and ω as the modulation amplitude and frequency.

9.3.3 Stationary configurations

Two-component BECs with short-range interactions are either miscible or immiscible, based on intra- and inter-component scattering lengths. If $g_1 g_2 < g_{12}^2$, the system is immiscible, with components spatially separated [91–93]. Otherwise, miscibility occurs, with significant component overlap when inter-component interactions are weaker. Here, the condition indicates immiscibility.

In a quasi-one-dimensional trap, immiscible binary BECs exhibit two configurations [32]: segregated (components adjacent but distinct) or symbiotic (one component surrounds the other, resembling a soliton pair). Given the interaction strengths, the symbiotic configuration is the ground state, and the segregated is an excited state. Numerical methods compute both configurations using varied initial conditions.

Numerical results show that spatially modulated scattering lengths can induce miscibility. As the scattering length localizes at the trap centre (smaller b), components overlap more, becoming fully miscible when the scattering length profile is nearly delta-like, with wavefunctions approaching Gaussians. This occurs because effective nonlinearity decreases with b, as shown by the effective interactions:

$$\tilde{G}_j = \int_0^{\infty} \mathrm{d}\rho\ 2\pi\rho G_j(\rho) = 2\pi g_j b^2, \tag{9.50}$$

$$\tilde{G}_{12} = \int_0^{\infty} \mathrm{d}\rho\ 2\pi\rho G_{12}(\rho) = 2\pi g_{12} b^2. \tag{9.51}$$

The quadratic dependence on b implies that smaller b reduces nonlinearity, pushing the system towards a linear regime.

9.3.4 Symbiotic pair state–ground state

The symbiotic configuration has one component centred in the trap, surrounded by the other. Figure 9.13 shows radially integrated density profiles

$$n_j(z) = \int_0^{\infty} \mathrm{d}\rho\ 2\pi\rho|\psi_j(\rho, z)|^2, \tag{9.52}$$

obtained via imaginary-time propagation from identical Gaussian initial conditions

$$\psi_1(\rho, z, t = 0) = \psi_2(\rho, z, t = 0) = \frac{1}{\pi^{3/2}}\mathrm{e}^{-\frac{1}{2}(\rho^2 + z^2)}. \tag{9.53}$$

As b decreases, the immiscible configuration vanishes, and components become miscible, aligning with reduced effective interactions.

9.3.5 Segregated state

Figure 9.14 presents radially integrated density profiles for the segregated configuration, computed via imaginary-time propagation from separated Gaussian initial conditions

$$\psi_1(\rho, z, t = 0) = \frac{1}{\pi^{3/2}}\mathrm{e}^{-\frac{1}{2}[(\rho-\rho_0)^2 + (z-z_0)^2]}, \tag{9.54}$$

$$\psi_2(\rho, z, t = 0) = \frac{1}{\pi^{3/2}}\mathrm{e}^{-\frac{1}{2}[(\rho+\rho_0)^2 + (z+z_0)^2]}. \tag{9.55}$$

Figures 9.14(a)–(e) demonstrate the shift to miscibility as b decreases, with interactions localizing along the z-axis.

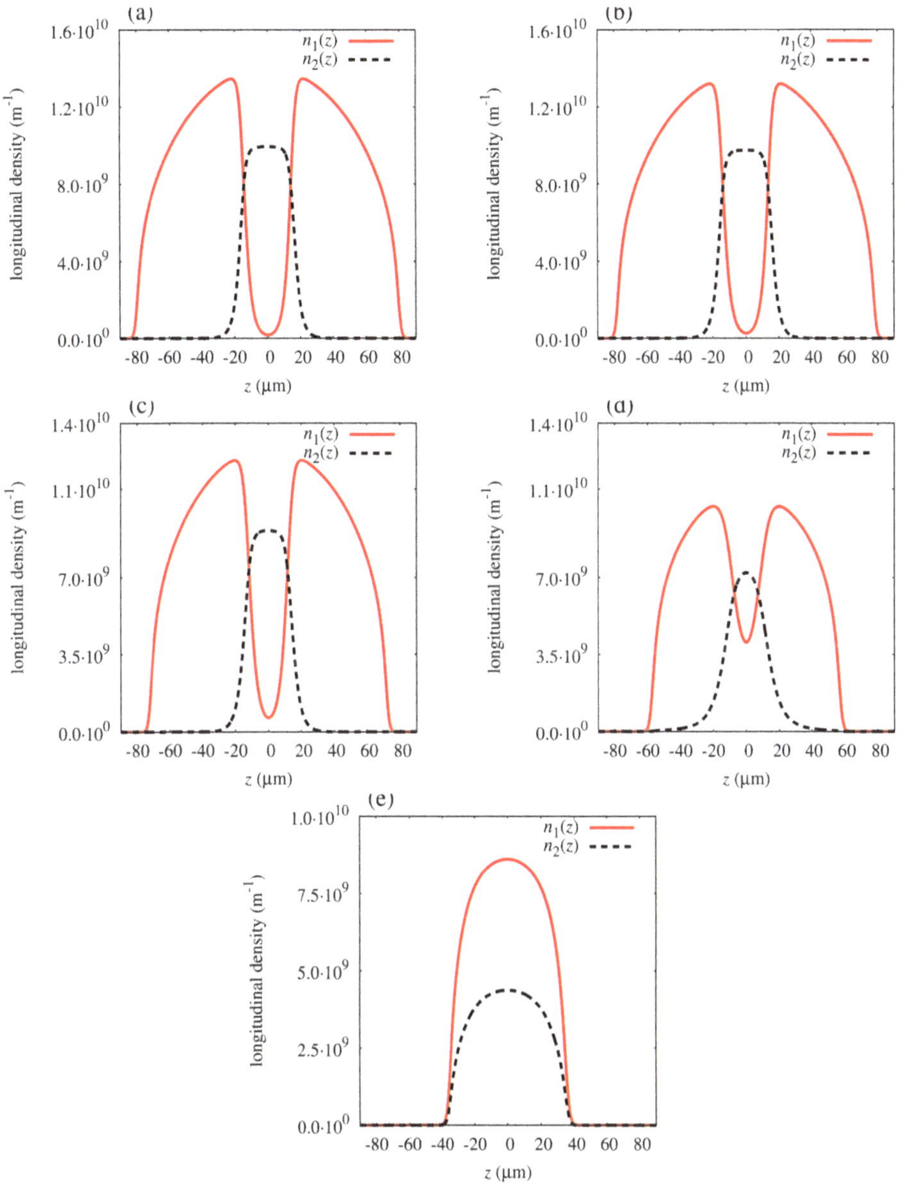

Figure 9.13. Stationary symbiotic pair states of a binary condensate with collisionally inhomogeneous interactions. Longitudinal density profiles $n_1(z)$ and $n_2(z)$ are shown for various inhomogeneity parameters: (a) $b = 4b_0$, (b) $b = 2b_0$, (c) $b = b_0$, (d) $b = b_0/2$, (e) $b = b_0/4$. Decreasing b increases component overlap, leading to a miscible state. Reprinted from [94]. Copyright 2016 IOP Publishing Ltd.

Comparing figures 9.13 and 9.14, one understands that the symbiotic configuration exhibits higher peak densities at weak inhomogeneity (large b) due to component separation, limiting available space. As inhomogeneity strengthens (smaller b), miscibility increases space, lowering peak density. In the segregated

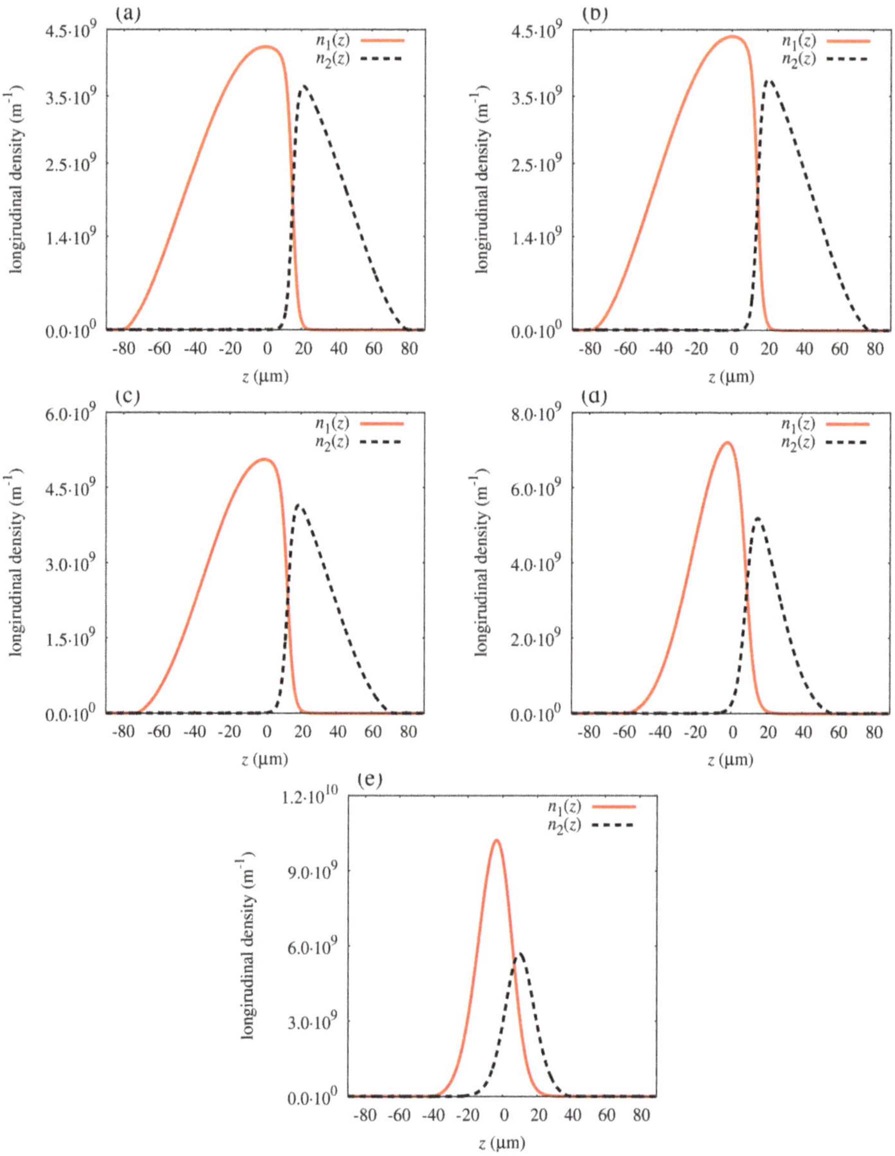

Figure 9.14. Stationary segregated states of a binary condensate with collisionally inhomogeneous interactions. Longitudinal density profiles $n_1(z)$ and $n_2(z)$ are shown for various inhomogeneity parameters: (a) $b = 4b_0$, (b) $b = 2b_0$, (c) $b = b_0$, (d) $b = b_0/2$, (e) $b = b_0/4$. Decreasing b increases component overlap, transitioning to a miscible state. Reprinted from [94]. Copyright 2016 IOP Publishing Ltd.

configuration, stronger inhomogeneity compresses components, reducing space and raising peak density, reflecting their distinct spatial structures.

Although miscibility typically requires $g_1 g_2 > g_{12}^2$, which is not satisfied here [91–93], spatially inhomogeneous interactions enable miscibility at strong inhomogeneity, suggesting a method to control miscibility in binary BECs.

9.3.6 Dynamical results [94]

This section analyzes excitations from harmonic modulation of the radial trapping potential, generating Faraday or resonant density waves [27, 31, 95]. Faraday waves have half the driving frequency, while resonant waves match it, growing exponentially via resonant energy transfer. The study examines how inhomogeneous interactions influence these waves in a two-component BEC.

Simulations use $N_1 = 2.5 \times 10^5$ atoms in state A and $N_2 = 1.25 \times 10^5$ atoms in state B, with the trap potential from equation (9.46), parameters $\Omega_{\rho 0} = 2\pi \times 160$ Hz, $\Omega_z = 2\pi \times 7$ Hz [25], and modulation $\Omega_\rho(t) = \Omega_{\rho 0}(1 + \epsilon \sin(\omega t))$. Two driving frequencies are used: $\omega = \omega_{\text{res}} = \Omega_{\rho 0}$, producing resonant waves, and $\omega = \omega_{\text{F}} = 2\pi \times 250$ Hz, producing Faraday waves, as it is off-resonance with $\Omega_{\rho 0}$ and $2\Omega_{\rho 0}$ [27, 96, 97]. The modulation amplitude is $\epsilon = 0.1$.

9.3.7 Symbiotic pair state

For the symbiotic ground state, resonant and Faraday waves are generated across various b.

Figure 9.15 illustrates the evolution of longitudinal density profiles for resonant waves ($\omega = \omega_{\text{res}}$) under weak (large b) and strong (small b) inhomogeneity.

Visibility, defined as the normalized standard deviation of the density profile from the ground state ([95]), quantifies resonant wave onset (figure 9.16). For weak inhomogeneity, resonant waves emerge after 80 ms, while for strong inhomogeneity, they appear after 350–400 ms, indicating softened excitations in the linear regime (figure 9.15). Figure 9.17 shows the spatial period, converging for strong inhomogeneity but diverging for weak inhomogeneity due to differing atom numbers. As in single-component BECs [32], the period increases with weaker inhomogeneity, saturating at the homogeneous limit. Figure 9.18 depicts Faraday wave dynamics ($\omega = \omega_{\text{F}}$). Visibility (figure 9.19) shows Faraday waves emerge after 150–200 ms for weak inhomogeneity and 350–400 ms for strong inhomogeneity, reflecting delayed onset due to cloud shape changes. Figure 9.20 shows the spatial period, diverging as inhomogeneity weakens, converging to the homogeneous case.

9.3.8 Segregated state

For the segregated configuration (section 9.3.5), figure 9.21 shows resonant wave dynamics ($\omega = \omega_{\text{res}}$). Visibility (figure 9.22) indicates resonant waves appear after 80–100 ms for weak inhomogeneity and 200 ms for strong inhomogeneity, a less pronounced delay than in the symbiotic case due to subtler cloud shape changes (figure 9.14 versus figure 9.13). Figure 9.23 shows the spatial period, mirroring symbiotic trends.

Figure 9.24 shows Faraday wave dynamics ($\omega = \omega_{\text{F}}$). Visibility (figure 9.25) indicates Faraday waves emerge after 150–200 ms for weak inhomogeneity and 200–250 ms for strong inhomogeneity, with a slight delay. Figure 9.26 shows the spatial period, consistent with symbiotic trends.

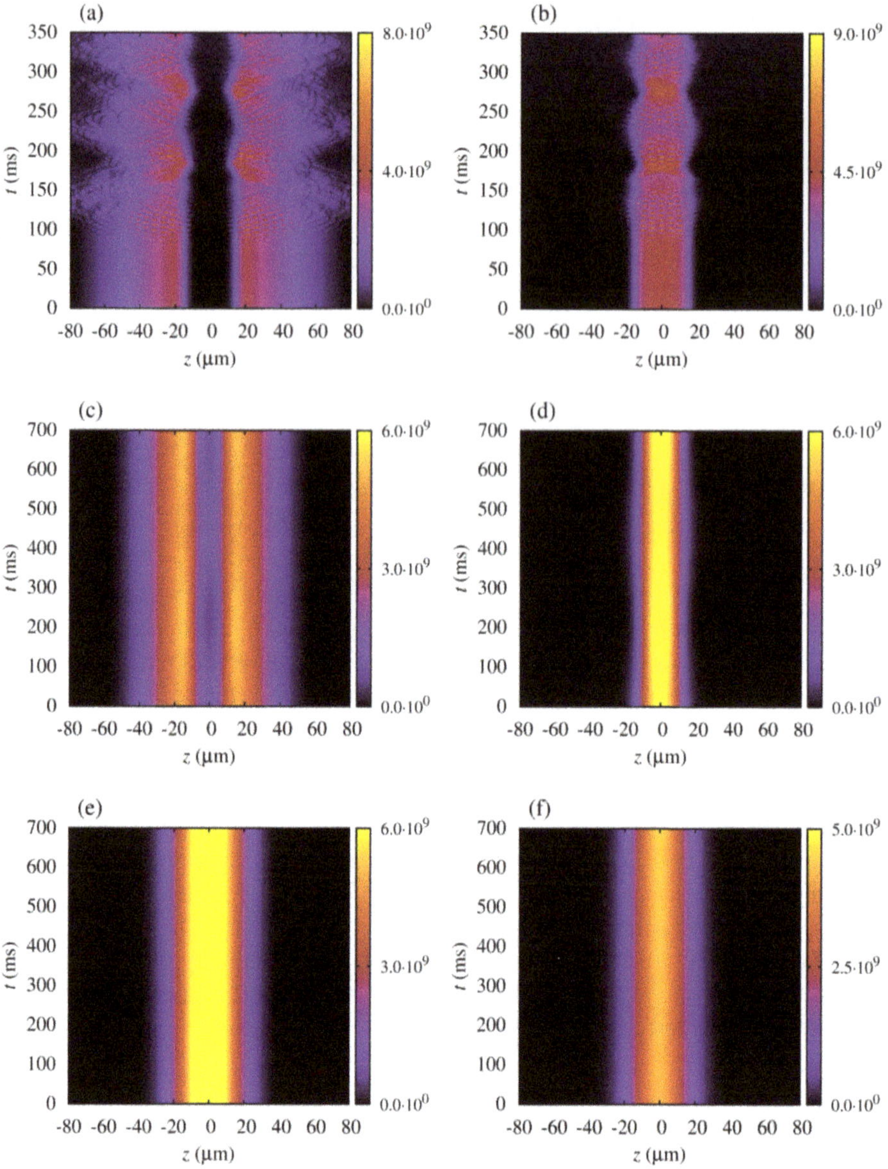

Figure 9.15. Time evolution of radially integrated density profiles for weak and strong inhomogeneous interactions in the symbiotic configuration with $\omega = \omega_{\rm res}$. Left (right) panels show component A (B) for: (a, b) $b = 4b_0$, (c, d) $b = b_0/2$, (e, f) $b = b_0/4$. Nonlinear excitations soften as b decreases, approaching a linear regime. Reprinted from [94]. Copyright 2016 IOP Publishing Ltd.

Comparing figures 9.17, 9.20, 9.23, and 9.26, spatial periods are similar for weak inhomogeneity in both configurations, indicating consistent wave behaviour.

Numerical simulations reveal that collisionally inhomogeneous binary condensates enter a linear regime as interactions localize at the trap centre, diminishing

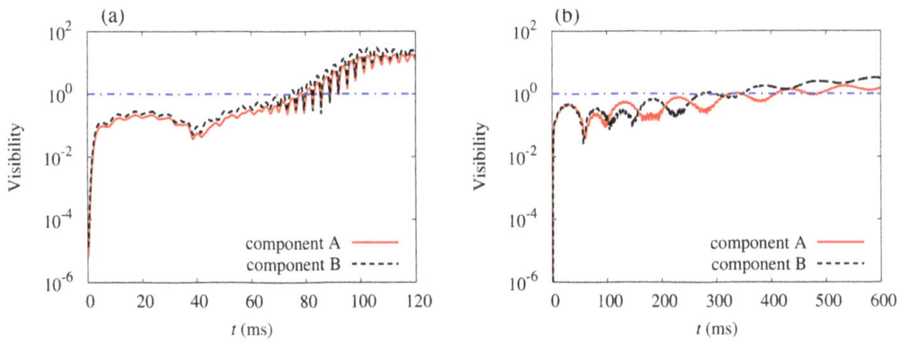

Figure 9.16. Time dependence of visibility for weak and strong inhomogeneous interactions in the symbiotic configuration with $\omega = \omega_{\rm res}$, for: (a) $b = 4b_0$, (b) $b = b_0/4$. The dashed-dotted line at visibility = 1 marks the onset of resonant waves. Reprinted from [94]. Copyright 2016 IOP Publishing Ltd.

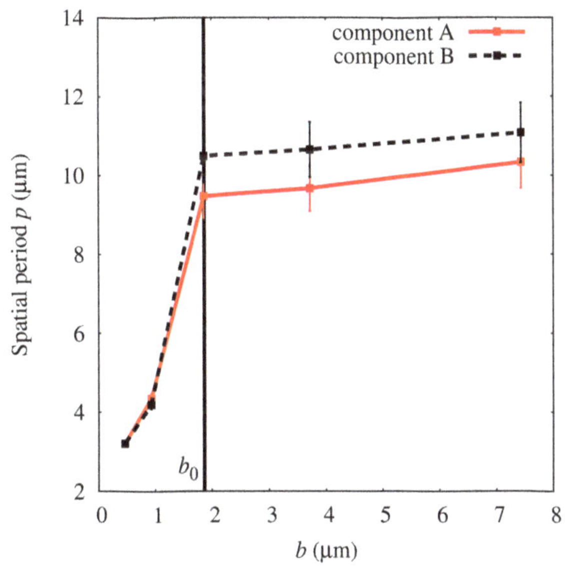

Figure 9.17. Spatial period of resonant waves versus inhomogeneity b in the symbiotic configuration for $\omega = \omega_{\rm res}$, obtained via FFT analysis. Reprinted from [94]. Copyright 2016 IOP Publishing Ltd.

nonlinear effects. This is evident in prolonged instability onset times for Faraday and resonant waves and the transition to miscibility, regardless of configuration. Stronger inhomogeneity reduces spatial periods of both waves. Resonant waves are excited by a driving frequency matching the radial trap frequency, while Faraday waves use an off-resonance frequency between the radial frequency and its first harmonic.

Symbiotic and segregated configurations show quantitative differences in the linear regime. Inhomogeneous interactions impact symbiotic states more, with longer onset times and stronger visibility fluctuations (figures 9.16 and 9.19) than

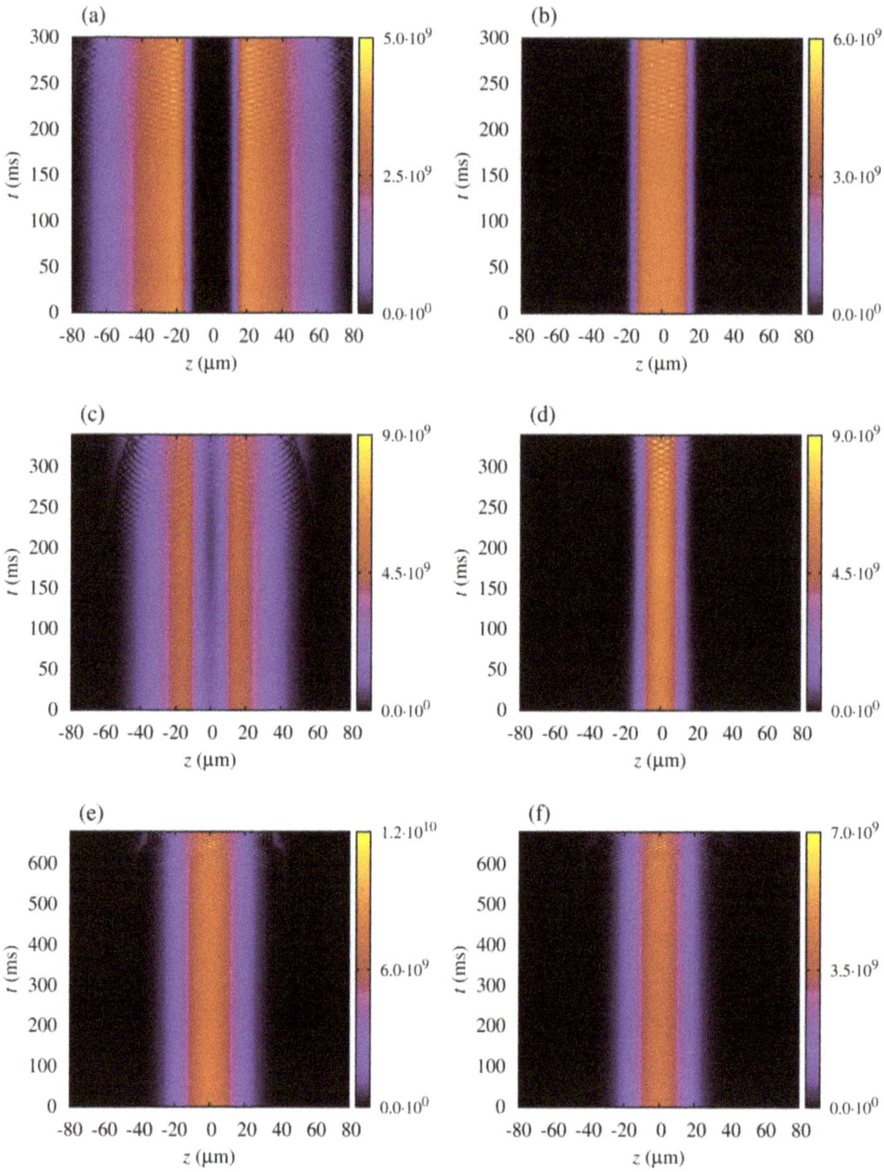

Figure 9.18. Time evolution of radially integrated density profiles for weak and strong inhomogeneous interactions in the symbiotic configuration with $\omega = \omega_F$. Left (right) panels show component A (B) for: (a, b) $b = 4b_0$, (c, d) $b = b_0/2$, (e, f) $b = b_0/4$. Nonlinear excitations soften as b decreases, approaching a linear regime. Reprinted from [94]. Copyright 2016 IOP Publishing Ltd.

segregated states (figures 9.22 and 9.25). Homogeneous interactions yield similar onset times for both.

Immiscible configurations transition to fully miscible states as the scattering length profile nears a delta function, with wavefunctions becoming nearly Gaussian.

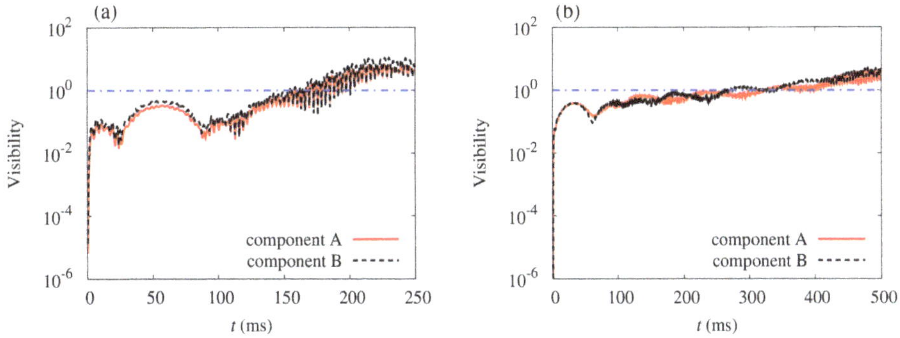

Figure 9.19. Time dependence of visibility for weak and strong inhomogeneous interactions in the symbiotic configuration with $\omega = \omega_F$, for: (a) $b = 4b_0$, (b) $b = b_0/4$. The dashed-dotted line at visibility = 1 marks the onset of Faraday waves. Reprinted from [94]. Copyright 2016 IOP Publishing Ltd.

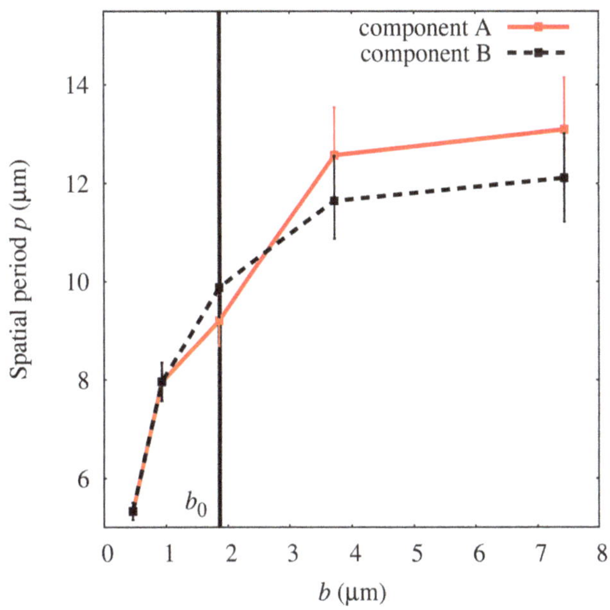

Figure 9.20. Spatial period of Faraday waves versus inhomogeneity b in the symbiotic configuration for $\omega = \omega_F$, obtained via FFT analysis. Reprinted from [94]. Copyright 2016 IOP Publishing Ltd.

This suggests that inhomogeneous interactions can experimentally control miscibility in binary BECs. Observing Faraday waves in BECs with inhomogeneous interactions encourages the exploration of other excitations, such as vortices, in similar systems.

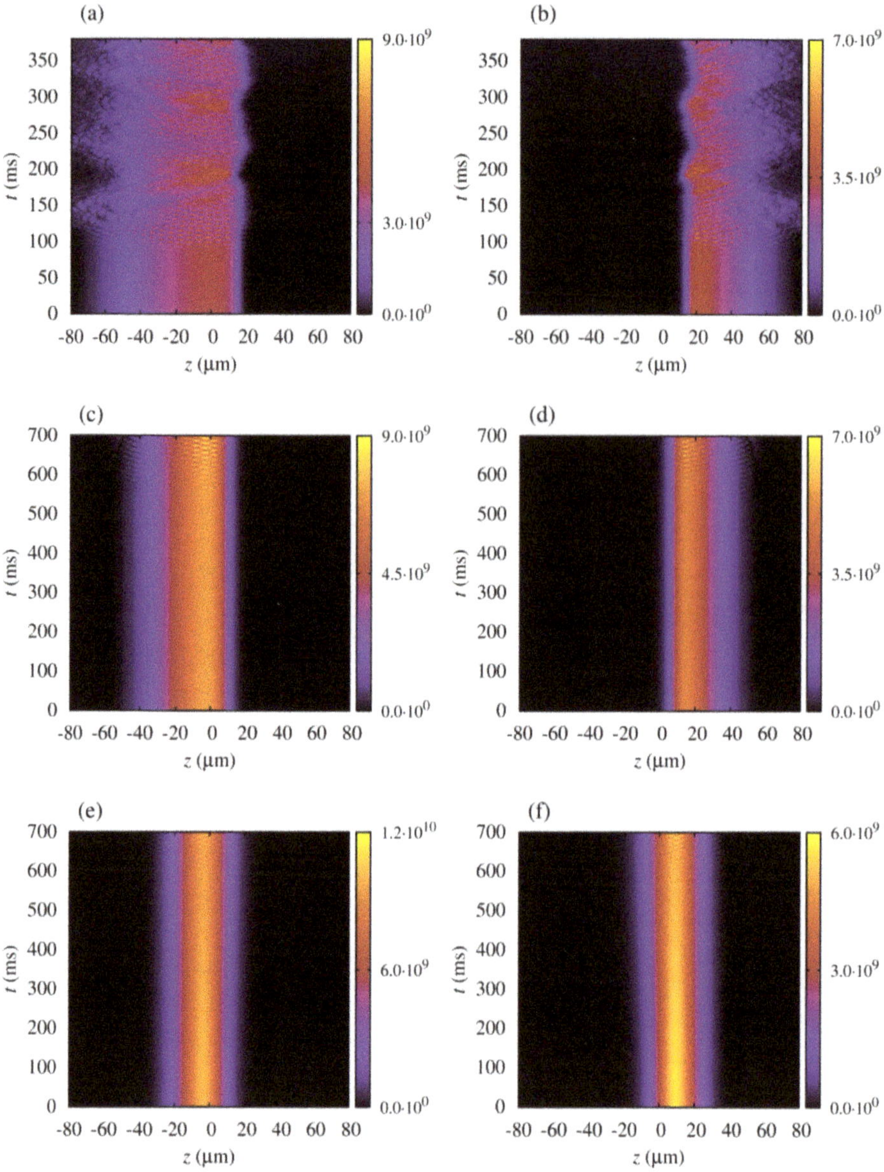

Figure 9.21. Time evolution of radially integrated density profiles for weak and strong inhomogeneous interactions in the segregated configuration with $\omega = \omega_{\text{res}}$. Left (right) panels show component A (B) for: (a, b) $b = 4b_0$, (c, d) $b = b_0/2$, (e, f) $b = b_0/4$. Nonlinear excitations soften as b decreases, approaching a linear regime. Reprinted from [94]. Copyright 2016 IOP Publishing Ltd.

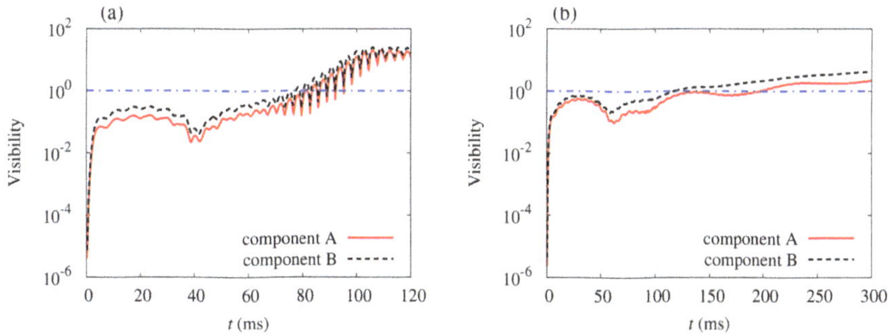

Figure 9.22. Time dependence of visibility for weak and strong inhomogeneous interactions in the segregated configuration with $\omega = \omega_{\text{res}}$, for: (a) $b = 4b_0$, (b) $b = b_0/4$. The dashed-dotted line at visibility = 1 marks the onset of resonant waves. Reprinted from [94]. Copyright 2016 IOP Publishing Ltd.

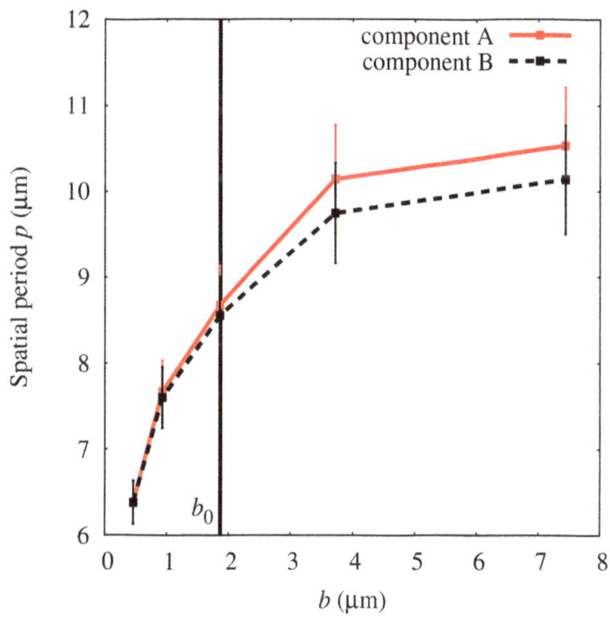

Figure 9.23. Spatial period of resonant waves versus inhomogeneity b in the segregated configuration for $\omega = \omega_{\text{res}}$, obtained via FFT analysis. Reprinted from [94]. Copyright 2016 IOP Publishing Ltd.

9.4 Stable multiple vortices in collisionally inhomogeneous attractive Bose–Einstein condensates

9.4.1 Introduction

The observation of robust nonlinear excitations, such as *dromions* [141, 142] (two-dimensional localized patterns formed by intersecting one-dimensional ghost solitons) and *lumps* (weakly localized two-dimensional solitons) [98], in various two-dimensional systems has inspired the search for analogous structures in BECs.

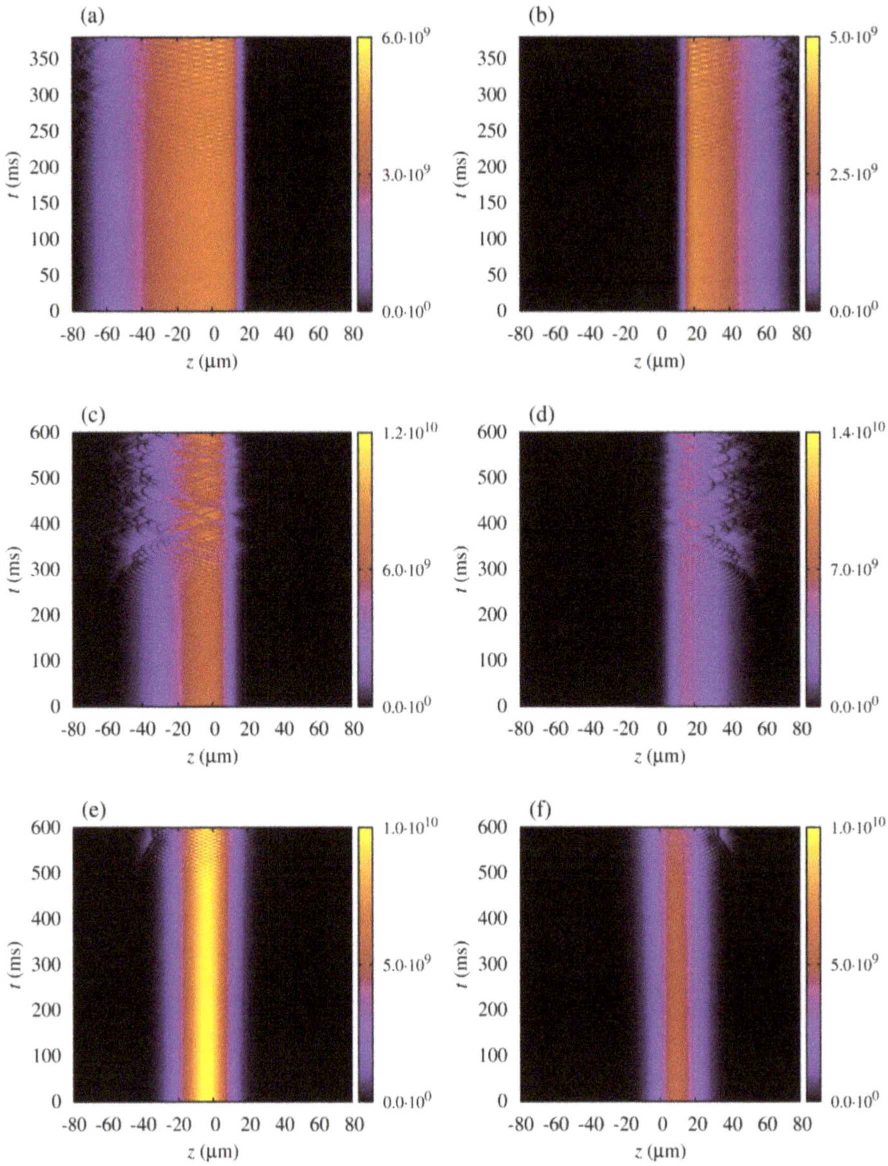

Figure 9.24. Time evolution of radially integrated density profiles for weak and strong inhomogeneous interactions in the segregated configuration with $\omega = \omega_F$. Left (right) panels show component A (B) for: (a, b) $b = 4b_0$, (c, d) $b = b_0/2$, (e, f) $b = b_0/4$. Nonlinear excitations soften as b decreases, approaching a linear regime. Reprinted from [94]. Copyright 2016 IOP Publishing Ltd.

Prior studies have focused on simpler nonlinear modes, including vortices [99] and Faraday waves [25, 30, 32, 35], in effectively two-dimensional, pancake-shaped BECs. Faraday waves arise from periodic modulation of the trapping potential, while vortices are generated through laser stirring, coherent angular momentum transfer

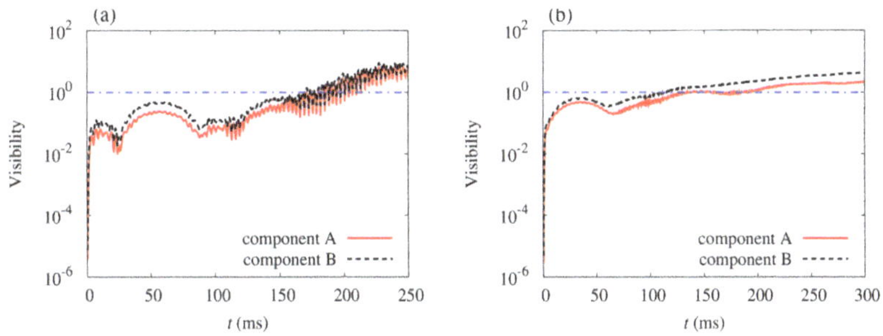

Figure 9.25. Time dependence of visibility for weak and strong inhomogeneous interactions in the segregated configuration with $\omega = \omega_F$, for: (a) $b = 4b_0$, (b) $b = b_0/4$. The dashed-dotted line at visibility = 1 marks the onset of Faraday waves. Reprinted from [94]. Copyright 2016 IOP Publishing Ltd.

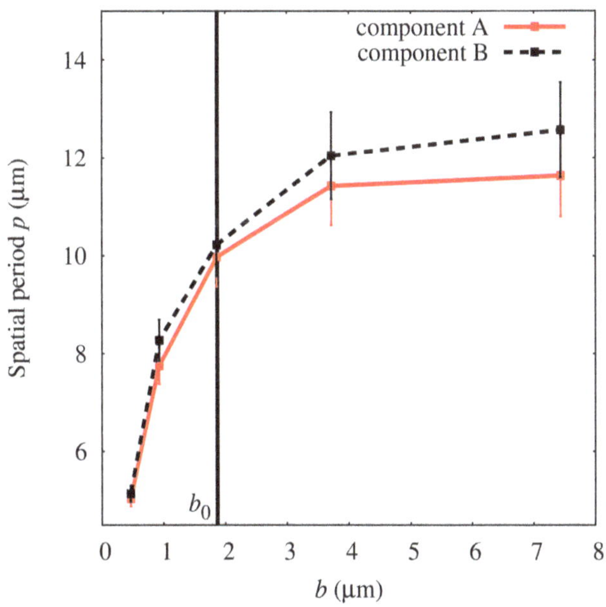

Figure 9.26. Spatial period of Faraday waves versus inhomogeneity b in the segregated configuration for $\omega = \omega_F$, obtained via FFT analysis. Reprinted from [94]. Copyright 2016 IOP Publishing Ltd.

via two-photon stimulated Raman processes [99, 100], or phase imprinting [101], as reviewed in [102]. In self-repulsive BECs, single vortices with topological charge $S = 1$ are typically stable, but in attractive BECs, self-attraction induces collapse and azimuthal instabilities, destabilizing solitary vortices [103, 104]. Even in non-collapsing media with quadratic nonlinearity, vortex solitons are often unstable [148]. Complex vortical structures, such as vortex dipoles [143, 144] and quadrupoles [145–147], have been identified primarily in self-repulsive BECs.

Pancake-shaped BECs, strongly confined along the z-axis and weakly confined in the transverse plane, provide an ideal platform for hosting vortices. While stable single vortices ($S = 1$) have been predicted in two-dimensional attractive

BECs with in-plane harmonic trapping [107, 108], higher-order vortices ($S \geqslant 2$) are unstable in such models, splitting into unitary vortices in both attractive and repulsive systems [100, 104, 106]. Stabilizing multiple vortices is a significant challenge [109]. This section predicts stable higher-order vortices ($S \geqslant 2$) in BECs with spatially localized attractive nonlinearity, which supports stable fundamental solitons ($S = 0$), but not vortices without trapping [8, 110]. This configuration appears to be the first to yield stable higher-order vortex states in trapped attractive BECs, with potential applications in nonlinear optics.

9.4.2 The model and key findings

In the mean-field framework, a self-attractive BEC, strongly confined along the z-axis and weakly trapped radially, is described by the three-dimensional GP equation for the single-atom wavefunction ([3, 23]):

$$i\hbar\frac{\partial \Psi}{\partial T} = \left(-\frac{\hbar^2}{2m}\nabla_{3\mathrm{D}}^2 + \frac{m}{2}\left[\Omega_r^2(X^2+Y^2) + \Omega_z^2 Z^2\right] + \frac{4\pi\hbar^2 a_s \mathcal{N}}{m}|\Psi|^2\right)\Psi, \quad (9.56)$$

where $\nabla_{3\mathrm{D}}^2$ operates on coordinates (X, Y, Z), m is the atomic mass, Ω_r and Ω_z are the radial and axial trapping frequencies, $a_s < 0$ is the attractive s-wave scattering length, and $\mathcal{N}$ is the number of atoms, with the wavefunction normalized to unity. By factorizing the wavefunction as $\Psi(X, Y, Z, T) = \pi^{-1/4} a_z^{-3/2} \exp(-i\Omega_z T/2 - Z^2/2a_z^2)\psi(x, y, t)$, where $a_z = \sqrt{\hbar/m\Omega_z}$ is the axial confinement length, and integrating over Z, the two-dimensional GP equation is derived and is of the following form [110–112]:

$$i\frac{\partial \psi}{\partial t} = \left[-\frac{1}{2}\nabla_{2\mathrm{D}}^2 + \frac{1}{2}\Omega_r^2 r^2 + gN|\psi|^2\right]\psi, \quad (9.57)$$

where $(x, y) = (X, Y)/a_z$, $t = \Omega_z T$, $r^2 = x^2 + y^2$, $g = 2a_s\sqrt{2\pi m\Omega_z/\hbar}$, and the two-dimensional wavefunction satisfies $\int|\psi(x, y, t)|^2\, dx\, dy = 1$. For low atomic densities, deviations from cubic nonlinearity are negligible [111].

Spatially inhomogeneous nonlinearity, achievable through optically controlled Feshbach resonances [8, 14, 53], modifies the two-dimensional equation to the following form:

$$\mathrm{i}\frac{\partial \psi}{\partial t} = \left(-\frac{1}{2}\nabla_{2\mathrm{D}}^2 + \frac{1}{2}r^2 + g(r)N|\psi|^2\right)\psi, \quad (9.58)$$

with $\Omega_r = 1$ via rescaling. Stationary vortex solutions are sought as:

$$\psi(r, \theta, t) = R(r)\exp(\mathrm{i}S\theta - \mathrm{i}\mu t), \quad (9.59)$$

where S is the integer vorticity and μ is the chemical potential. Substituting (9.59) into equation (9.58) yields the radial equation:

$$2\mu R + R'' + r^{-1}R' - (S^2 r^{-2} + r^2)R - 2gNR^3 = 0, \quad (9.60)$$

which is solved numerically. The nonlinearity is modelled as a Gaussian profile:

$$g(r) = -\exp(-b^2r^2/2), \tag{9.61}$$

where b^{-1} sets the nonlinearity radius, and N in equation (9.58) is proportional to the atom number, adjusted by absorbing a Gaussian coefficient. Thus, b and N are key parameters. For $b^2 \ll 1$, only $S = 0, 1$ states are stable [105, 107]. For $b^2 \gg 1$, the system is nearly linear, with vortices resembling stable linear harmonic oscillator modes. The critical question is the minimum $b_{\min}$ for stabilizing vortices with $S \geqslant 2$. Numerical results indicate $b_{\min} \approx 1.0$ for $S = 4$, consistent with the estimate $b_{\min} \approx \sqrt{2\ln 10/(1+S)} \approx 0.96$, derived from the linear harmonic oscillator's radial wavefunction $R_S(r) = (\pi S!)^{-1/2} r^S \exp(-r^2/2)$, where $\langle r^2 \rangle = 1 + S$, and nonlinearity is significantly attenuated at $r^2 = \langle r^2 \rangle$.

This model also applies to optics, where equation (9.58) describes spatial propagation in a waveguide with localized Kerr nonlinearity [8], and the harmonic potential represents refractive index modulation. Localized nonlinearity can be induced by doping with two-photon resonance-enhancing materials [113].

For small N, the model reduces to a harmonic oscillator with eigenvalues $\mu_{\max} = 1 + S$. The negative contribution of self-focusing lowers μ, making $\mu_{\max}$ an upper bound. Key results are summarized in figure 9.27 and tables 9.1 and 9.2, showing $\mu(N)$ curves for vortices at $b = 1.5$. Stable vortices exist for $\mu_{\min} < \mu < \mu_{\max}$, with instability (splitting and collapse) below $\mu_{\min}$. Higher S allows larger stable N.

For $S = 5, 6$, stability intervals may start at finite $N_{\min}$, corresponding to $\mu_{\max} < 1 + S$, a behaviour requiring numerical analysis. For ^{7}Li BECs with $\Omega_z/2\pi \approx 1$ kHz, $a_z \approx 3$ μm, and radial frequency ~10 Hz, the physical atom number is ~$500N$, and vortex sizes (figures 9.29–9.30) are scaled by ~30 μm. Stable atom numbers range from ~5000 ($S = 1$) to ~5×10^7 ($S = 6$).

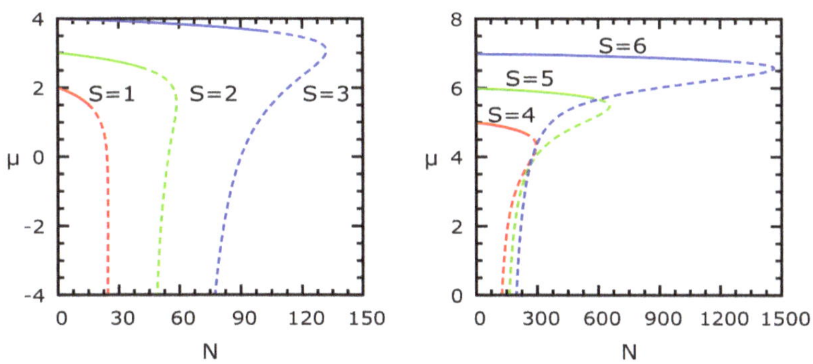

Figure 9.27. Chemical potential μ versus atom number N for vortices with vorticities S, obtained from equation (9.60) with $b = 1.5$. Solid lines indicate stable vortices; dashed lines indicate unstable ones.Reprinted with permission from [114], Copyright (2015) by the American Physical Society.

Table 9.1. Stability intervals (μ_{min}, μ_{max}) for vortices at various S and b. For $S = 1, 3, 4$, $\mu_{\text{max}} = 1 + S$. Adapted from [114], Copyright (2015) by the American Physical Society.

	$S = 1$	$S = 2$	$S = 3$	$S = 4$	$S = 5$	$S = 6$
$b = 0.75$	1.21	(2.16, 2.27)	unstable	unstable	unstable	unstable
$b = 1.0$	1.29	(2.30, 3)	unstable	4.82	(5.51, 5.84)	(6.51, 6.60)
$b = 1.5$	1.51	(2.58, 3)	3.64	4.69	(5.73, 6)	(6.76, 7)
$b = 2.0$	1.66	(2.73, 3)	3.79	4.83	(5.85, 6)	(6.84, 6.96)

Table 9.2. Stability intervals (N_{min}, N_{max}) for vortices at various S and b. For $S = 1, 3, 4$, $N_{\text{min}} = 0$. Adapted from [114], Copyright (2015) by the American Physical Society.

	$S = 1$	$S = 2$	$S = 3$	$S = 4$	$S = 5$	$S = 6$
$b = 0.75$	11.0	(18, 20.3)	unstable	unstable	unstable	unstable
$b = 1.0$	12.2	(0, 25)	unstable	27.4	(42.5, 103)	(150, 172)
$b = 1.5$	15.1	(0, 41)	101	237	(0, 550)	(0, 1274)
$b = 2.0$	20.7	(0, 83)	299	1066	(0, 3829)	(2160, 13 418)

9.4.2.1 Linear stability analysis

Stability is assessed using perturbed solutions ([105]):

$$\psi(r, t) = \left[R(r) + \varepsilon \sum_L u_L(r) e^{iL\theta - i\omega_L t} + \varepsilon \sum_L v_L^*(r) e^{-iL\theta + i\omega_L^* t}\right] e^{iS\theta - i\mu t}, \tag{9.62}$$

where (u_L, v_L) and ω_L are perturbation eigenmodes and eigenfrequencies for azimuthal index L. Linearization yields the Bogoliubov–de Gennes equations [23]:

$$\begin{pmatrix} \hat{D}_+ & gR^2 \\ -gR^2 & -\hat{D}_- \end{pmatrix} \begin{pmatrix} u_L \\ v_L \end{pmatrix} = \omega_L \begin{pmatrix} u_L \\ v_L \end{pmatrix}, \tag{9.63}$$

where $\hat{D}_\pm = -\frac{1}{2}\left[\frac{\partial^2}{\partial r^2} + \frac{1}{r}\frac{\partial}{\partial r} - \frac{(S \pm L)^2}{r^2}\right] + \frac{1}{2}\Omega_r^2 r^2 + 2gNR^2 - \mu$, with boundary conditions $u_L, v_L \sim r^{|S \pm L|}$ at $r \to 0$ and decaying exponentially at $r \to \infty$. Instability occurs if any ω_L is complex. Figure 9.28 shows instability growth rates (largest $\text{Im}(\omega_L)$) versus μ, confirming stability regions for $S = 1, 2, 3, 6$.

9.4.2.2 Perturbed vortex evolution

Simulations of equation (9.58) using split-step and D'yakonov methods [115] with a 400 $\times$ 400 grid ($\Delta x = \Delta y = 0.025, \Delta t = 0.000\,05$) and 5% random perturbations confirm stability predictions. Figure 9.29 shows a stable vortex ($S = 3$, $\mu = 3.8$, $N = 69$) self-cleaning perturbations by $t = 0.8$. Unstable vortices, as in figure 9.30

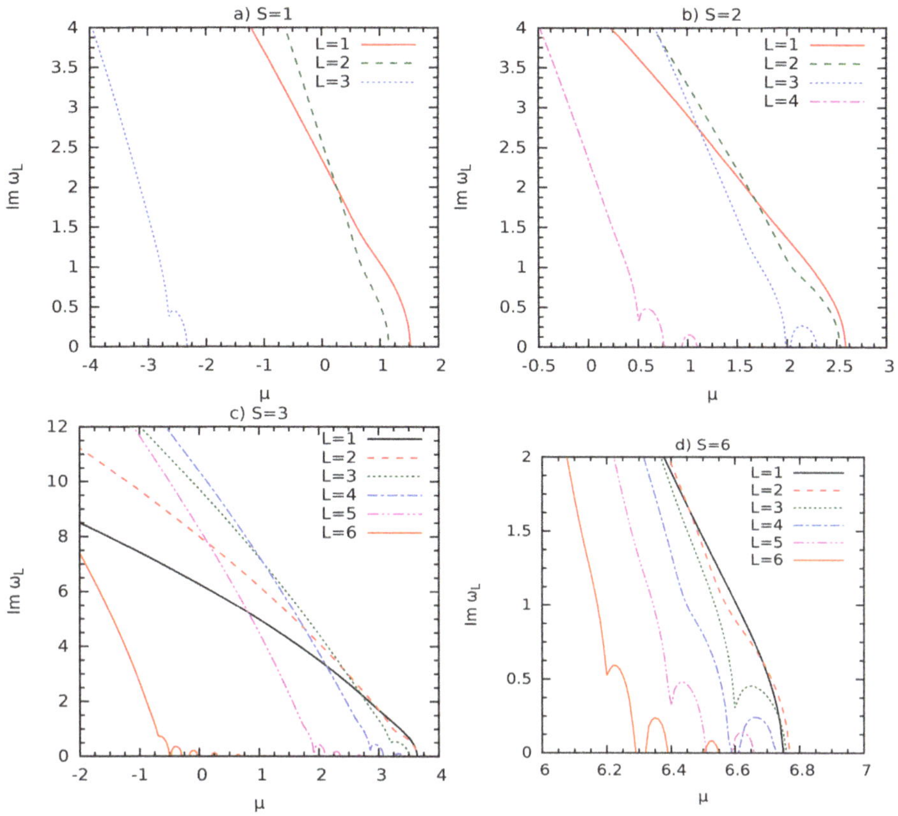

Figure 9.28. Imaginary part of eigenfrequencies for perturbation modes with azimuthal index L, for vortices with $S = 1, 2, 3, 6$. Reprinted from [114], Copyright (2015) by the American Physical Society.

($S = 3$, $\mu = 0$, $N = 90$), fragment into four parts (dominant mode $L = 4$) and collapse, consistent with linear stability analysis [114].

Unlike uniform nonlinearity models [107], no semi-stable region exists where vortices split and recombine. Future studies could explore higher-order nonlinearities (e.g., quintic, septimal) in collisionally inhomogeneous BECs.

9.5 Solitons under spatially localized cubic–quintic–septimal nonlinearities

This section has been reproduced with permission from [139].

9.5.1 Introduction

Spatial solitons, sustained by balancing diffraction and self-focusing, are pivotal in photonics [7, 8, 30, 94, 116–119]. The cubic NLS equation models Kerr self-focussing, supporting stable solitons in fibres and waveguides [117]. Higher-order nonlinearities are essential for phenomena like harmonic generation [120], filamentation [121], Kerr saturation [122], modulational instability in metamaterials [123],

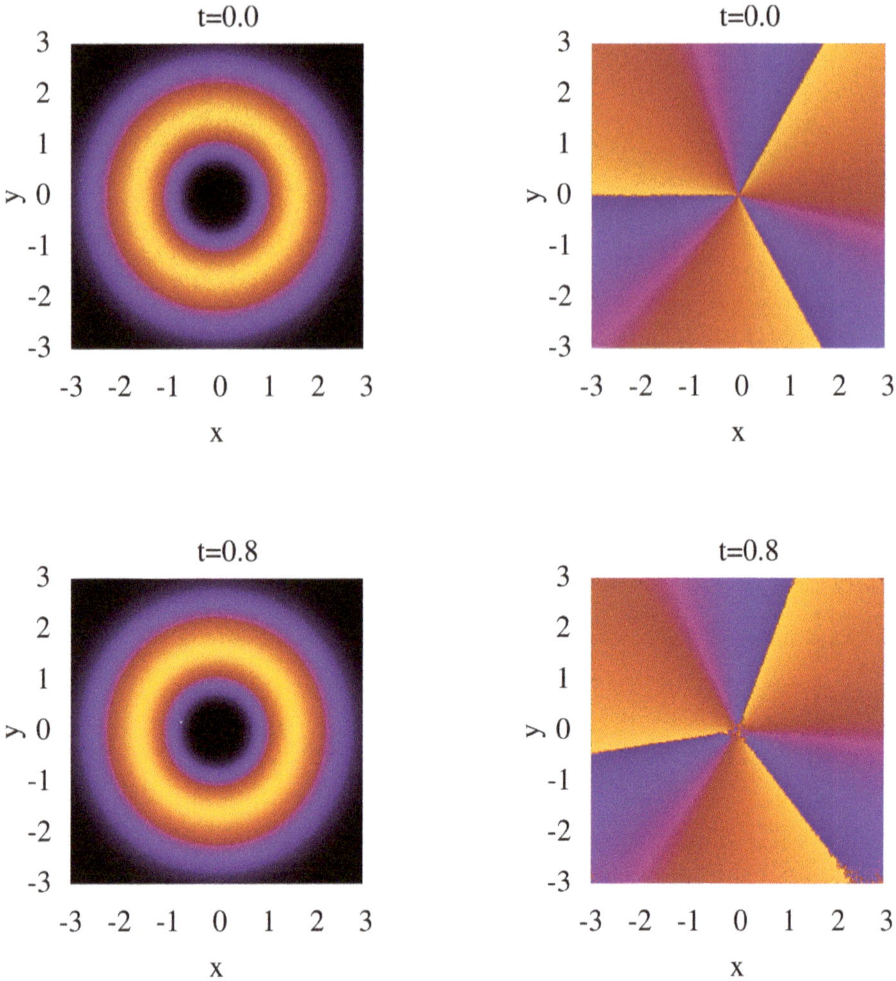

Figure 9.29. Density and phase evolution for a stable vortex with $S = 3$, $\mu = 3.8$, $N = 69$. Initial perturbations vanish by $t = 0.8$. Brighter regions indicate higher density; phase varies from $-\pi$ (dark) to π (bright). Reprinted from [114], Copyright (2015) by the American Physical Society.

and quasi-stable vortical beams [124]. Cubic–quintic (CQ) [125] and quintic–septimal [126] nonlinearities are the classic examples showing the propagation of stable two- and three-dimensional solitons [127, 149, 150]. Colloidal metallic nanoparticles enable tunable CQ and septimal nonlinearities [128], prompting analysis of one-dimensional solitons under cubic–quintic–septimal (CQS) nonlinearities [124]. Despite critical (quintic) and supercritical (septimal) collapse risks [129, 130], CQS combinations with focussing terms can yield stable solitons [131].

Spatially modulated nonlinearities enhance soliton stabilization [8]. Stepwise cubic nonlinearity stabilizes two-dimensional Townes solitons [132], and modulation accelerates Faraday and resonant wave onset [30, 94]. Localized cubic nonlinearity

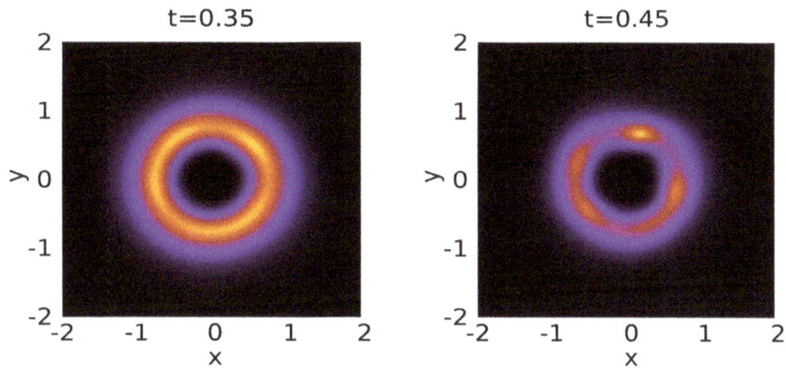

Figure 9.30. Density evolution of an unstable vortex with $S = 3$, $\mu = 0$, $N = 90$, fragmenting into four parts. Reprinted from [114], Copyright (2015) by the American Physical Society.

with harmonic trapping supports stable vortices up to $S = 6$ [114], unlike uniform nonlinearity [8, 118]. This section investigates stable one-dimensional solitons under spatially modulated CQS nonlinearities in planar waveguides, where modulation arises from varying waveguide thickness or dopant density [133]. The focus is on a 'sandwich' configuration with focusing cubic and septimal terms and a variable-sign quintic term, addressing stability amidst competing nonlinearities and potential collapse.

9.5.2 The model

The one-dimensional NLS equation for the electromagnetic wave amplitude $\Psi(x, z)$ with spatially localized CQS nonlinearities is given by:

$$i\frac{\partial \Psi}{\partial z} + \frac{1}{2}\frac{\partial^2 \Psi}{\partial x^2} + [G_3(x)P|\Psi|^2 + G_5(x)P^2|\Psi|^4 + G_7(x)P^3|\Psi|^6]\Psi = 0, \quad (9.64)$$

with normalization

$$\int_{-\infty}^{\infty} |\Psi(x, z)|^2 \, \mathrm{d}x = 1, \quad (9.65)$$

and total power P. Here, z is the propagation distance, x is the transverse coordinate, and nonlinearity coefficients are assumed to be of the following form

$$G_n(x) = g_n \exp\left(-\frac{b^2x^2}{2}\right), \quad n = 3, 5, 7, \quad (9.66)$$

where b^{-1} defines the nonlinearity region's width. Focussing on cubic and septimal terms allow rescaling

$$z = \sqrt{\frac{g_7}{g_3^3}}z', \quad x = \left(\frac{g_7}{g_3^3}\right)^{1/4} x', \quad \Psi = \left(\frac{g_3^3}{g_7}\right)^{1/8} \Psi', \quad P = (g_3 g_7)^{-1/4}P', \quad b = \left(\frac{g_3^3}{g_7}\right)^{1/4} b', \quad (9.67)$$

fixing $g_3 = g_7 = 1$, with b, g_5, and P as control parameters ($g_5 > 0$ for focussing, $g_5 < 0$ for defocussing quintic terms). This model also approximates the GP equation for dense BECs ([111]), with spatial modulation via Feshbach resonances ([53, 114]).

Stationary solutions $\Psi(x, z) = \psi(x)e^{ikz}$ satisfy

$$k\psi = \frac{1}{2}\frac{d^2\psi}{dx^2} + \exp\left(-\frac{b^2x^2}{2}\right)P\psi^3(1 + g_5P\psi^2 + P^2\psi^4). \tag{9.68}$$

Stability is analyzed using perturbed solutions $\Psi = [\psi + \delta\psi]e^{ikz}$, with $\delta\psi = u(x)e^{i\lambda z} + v^*(x)e^{-i\lambda * z}$, yielding:

$$\begin{aligned}
&\frac{1}{2}\frac{d^2u}{dx^2} + \exp\left(-\frac{b^2x^2}{2}\right)P\psi^2[(2u + v) + g_5P\psi^2(3u + 2v) + P^2\psi^4(4u + 3v)] - ku = \lambda u,\\
&\frac{1}{2}\frac{d^2v}{dx^2} + \exp\left(-\frac{b^2x^2}{2}\right)P\psi^2[(2v + u) + g_5P\psi^2(3v + 2u) + P^2\psi^4(4v + 3u)] - kv = -\lambda v.
\end{aligned} \tag{9.69}$$

Stability requires all λ to be real. Equation (9.68) is solved using a relaxation algorithm [134] ($\Delta x = 0.01$) and verified with Runge–Kutta shooting [135]. Time evolution uses Crank–Nicolson [18] ($\Delta t = 0.001$), and equation (9.69) is diagonalized on a 1000-point grid using LAPACK [151].

9.5.3 Numerical results

Figure 9.31 compares numerically computed solitons with analytical approximations, showing peakon-like shapes akin to Camassa–Holm or Salerno models [136–138].

Figure 9.32 shows $k(P)$ curves for $g_5 = -1$ and $g_5 = +1$, with stable (solid) and unstable (dashed) segments. For $g_5 < 0$, stable solitons exist in $P_{\min} < P < P_{\max}$. For $g_5 > 0$, stability occurs in specific P ranges. Narrower nonlinearity regions (larger b) expand stability.

For $g_5 = 1.5$, figure 9.33 shows stable segments between unstable ones near the stability transition.

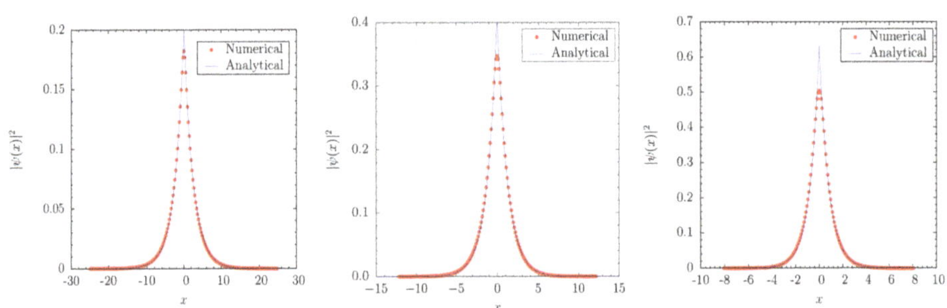

Figure 9.31. Numerical (dots) and analytical (lines) soliton profiles for $b = 3$, $g_5 = -1$: (a) stable, $k = 0.02$, $P = 1.89$; (b) stability boundary, $k = 0.085$, $P = 2.22$; (c) unstable, $k = 0.2$, $P = 2.13$. Reprinted from [139]. Copyright 2017 IOP Publishing Ltd.

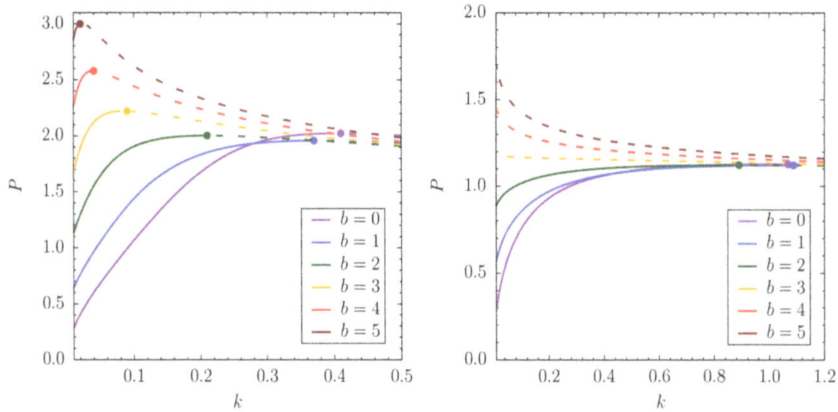

Figure 9.32. Propagation constant k versus power P for solitons with (a) $g_5 = -1$, (b) $g_5 = +1$, at various b. Solid lines are stable; dashed lines are unstable. Bold dots mark P_{min} and P_{max}. Reprinted from [139]. Copyright 2017 IOP Publishing Ltd.

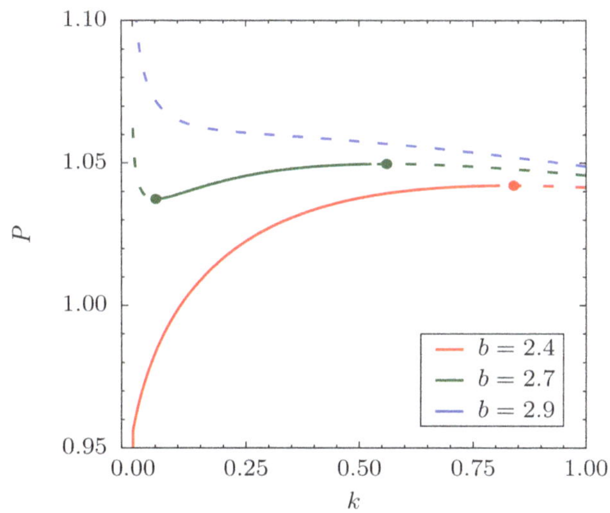

Figure 9.33. Same as figure 9.32, for $g_5 = 1.5$, showing stability transitions at $b \approx 2.7$. Reprinted from [139]. Copyright 2017 IOP Publishing Ltd.

Figure 9.34 plots P_{max} versus b for $g_5 = -1$, matching analytical predictions. Figure 9.35 maps P_{max} across (b, g_5), highlighting stability enhancement with narrower nonlinearity.

The Vakhitov–Kolokolov criterion ($dk/dP > 0$) [129, 140] accurately predicts stability. Simulations (figures 9.36, 9.37) confirm: at $g_5 = 1.5$, $b = 2.7$, unstable solitons ($k = 0.04$) form breathers, stable ones ($k = 0.25$) persist, and others ($k = 0.7$) collapse.

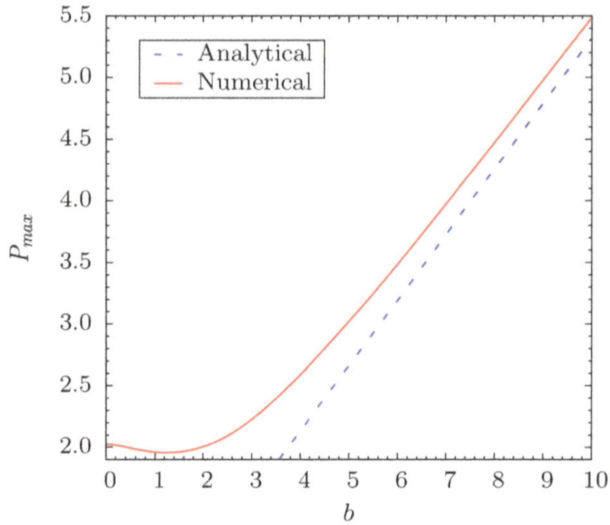

Figure 9.34. Maximum stable power $P_{\max}$ versus b for $g_5 = -1$. Red: numerical; blue: analytical (delta-function limit). Reprinted from [139]. Copyright 2017 IOP Publishing Ltd.

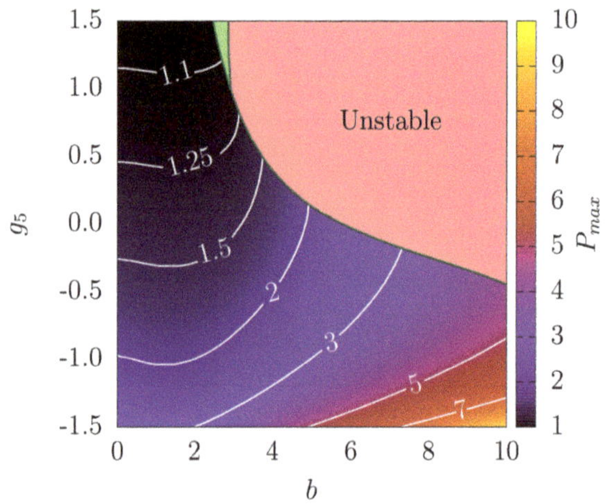

Figure 9.35. Map of $P_{\max}$ in the (b, g_5) plane. White curves indicate constant $P_{\max}$. Green wedge shows stability structure as in figure 9.33. Reprinted from [139]. Copyright 2017 IOP Publishing Ltd.

9.5.4 Analytical solution for delta-function modulation

For large b, the Gaussian in equation (9.66) approximates to a delta-function:

$$\exp\left(-\frac{b^2x^2}{2}\right) \approx \frac{\sqrt{2\pi}}{b}\delta(x). \tag{9.70}$$

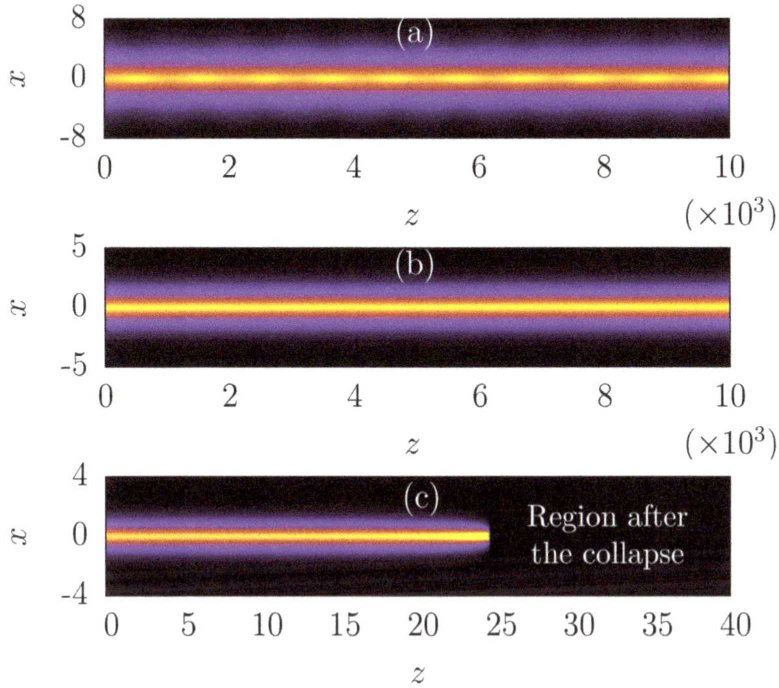

Figure 9.36. Evolution for $g_5 = 1.5$, $b = 2.7$: (a) unstable, $k = 0.04$, $P = 1.038$; (b) stable, $k = 0.25$, $P = 1.045$; (c) unstable, $k = 0.7$, $P = 1.049$. Reprinted from [139]. Copyright 2017 IOP Publishing Ltd.

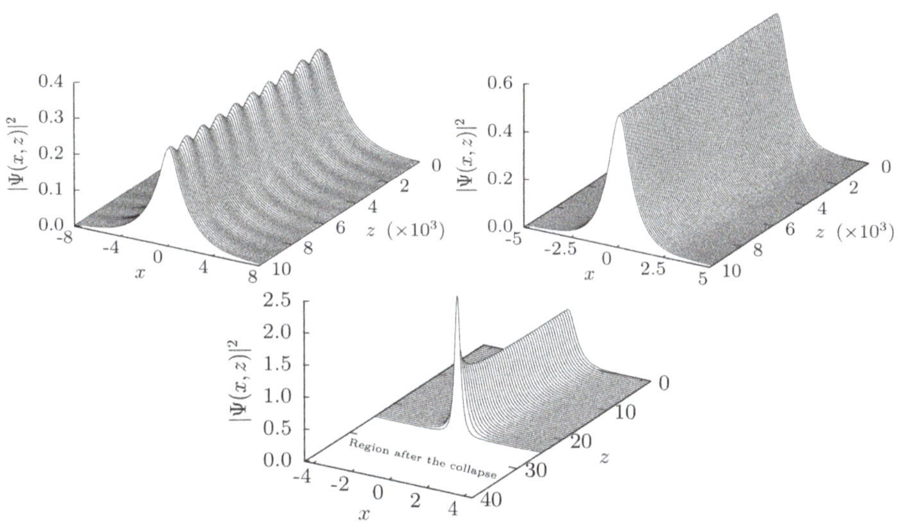

Figure 9.37. Evolution for: (a) unstable breather, $k = 0.04$, $P = 1.038$; (b) stable, $k = 0.2$, $P = 1.045$; (c) collapsing, $k = 0.7$, $P = 1.049$. Reprinted from [139]. Copyright 2017 IOP Publishing Ltd.

Equation (9.68) becomes:

$$k\psi = \frac{1}{2}\frac{\mathrm{d}^2\psi}{\mathrm{d}x^2} + \frac{\sqrt{2\pi}}{b}\delta(x)P\psi^3(1 + g_5P\psi^2 + P^2\psi^4). \tag{9.71}$$

For $x \neq 0$, $\psi'' = -2k\psi$, with solution given by:

$$\psi(x) = (2k)^{1/4}\exp(-\sqrt{2k}\,|x|). \tag{9.72}$$

Integrating at $x = 0$:

$$\frac{1}{2}\frac{\mathrm{d}\psi}{\mathrm{d}x}\bigg|_{x=0} = -\frac{\sqrt{2\pi}}{b}P\psi^3(1 + g_5P\psi^2 + P^2\psi^4)\bigg|_{x=0}, \tag{9.73}$$

yielding a quadratic equation for $\sqrt{k}$:

$$2P^3k + g_5P^2\sqrt{2k} + \left(P - \frac{b}{\sqrt{2\pi}}\right) = 0. \tag{9.74}$$

For $g_5 > 0$:

$$P_{\max}(g_5 > 0; b) = \frac{b}{\sqrt{2\pi}}, \tag{9.75}$$

with unstable solitons for $P < P_{\max}$. For $g_5 < 0$, stable solitons exist in the interval:

$$P_{\min} = \frac{b}{\sqrt{2\pi}} < P < P_{\max} = \frac{b}{\sqrt{2\pi}}\left(1 - \frac{g_5^2}{4}\right)^{-1}, \tag{9.76}$$

for $|g_5| < 2$, matching numerical results in figure 9.34.

9.5.5 Summary and future challenges

This chapter explored the dynamics of scalar and binary BECs with collisionally inhomogeneous short-range interactions. A mean-field model and evolution equation were developed, supported by numerical studies, to analyze nonlinear excitations. Faraday and resonant waves were investigated in scalar BECs using variational treatments, examining weak and strong inhomogeneous collisions. In binary condensates, these waves were studied across symbiotic and segregated states, revealing distinct dynamical behaviours. Stable multiple vortices were analyzed, highlighting their robustness under inhomogeneous interactions. Solitons arising from collisionally inhomogeneous cubic–quintic–septimal nonlinearities were also examined, with numerical and analytical solutions in the delta-functional limit, confirming their stability. These findings elucidated the role of spatially inhomogeneous interactions in shaping nonlinear excitations in BECs. The stability of Faraday and resonant waves in higher-dimensional BECs largely remains unaddressed, posing challenges for numerical and analytical studies.

It would be interesting to study the nature of such nonlinear excitations in nonequilibrium condensates generated by collisionally inhomogenous interactions in excitons–polaritons and their stability.

References

[1] Gross E P 1961 Structure of a quantized vortex in boson systems *Nuovo Cimento* **20** 454–77

[2] Gross E P 1963 Hydrodynamics of a superfluid condensate *J. Math. Phys.* **4** 195–207

[3] Pitaevskii L P 1961 Vortex lines in an imperfect Bose gas *Sov. Phys. JETP* **13** 451–4

[4] Xiong H, Liu S, Zhang W and Zhan M 2005 Ultracold two-component fermionic gases with a magnetic field gradient near a Feshbach resonance *Phys. Rev. Lett.* **95** 120401

[5] Theocharis G, Schmelcher P, Kevrekidis P and Frantzeskakis D 2005 Matter-wave solitons of collisionally inhomogeneous condensates *Phys. Rev.* A **72** 033614

[6] Garnier J and Abdullaev F K 2006 Transmission of matter-wave solitons through nonlinear traps and barriers *Phys. Rev.* A **74** 013604

[7] Kartashov Y V, Vysloukh V A and Torner L 2009 Soliton shape and mobility control in optical lattices *Prog. Opt.* **52** 63–148

[8] Kartashov Y V, Malomed B A and Torner L 2011 Solitons in nonlinear lattices *Rev. Mod. Phys.* **83** 247–305

[9] Sivan Y, Fibich G and Weinstein M I 2006 Waves in nonlinear lattices: ultrashort optical pulses and Bose-Einstein condensates *Phys. Rev. Lett.* **97** 193902

[10] Kartashov Y V, Vysloukh V A and Torner L 2016 Asymmetric soliton mobility in competing linear-nonlinear parity-time-symmetric lattices *Opt. Lett.* **41** 4348

[11] Dror N and Malomed B A 2011 Solitons supported by localized nonlinearities in periodic media *Phys. Rev.* A **83** 033828

[12] Sekh G A 2012 Effects of spatially inhomogeneous atomic interactions on Bose-Einstein condensates in optical lattices *Phys. Lett.* A **376** 1740–7

[13] Roati G, D'Errico C, Fallani L, Fattori M, Fort C, Zaccanti M, Modugno G, Modugno M and Inguscio M 2008 Anderson localization of a non-interacting Bose-Einstein condensate *Nature* **453** 895–8

[14] Fedichev P O, Kagan Y, Shlyapnikov G V and Walraven J T M 1996 Influence of nearly resonant light on the scattering length in low-temperature atomic gases *Phys. Rev. Lett.* **77** 2913–6

[15] Theis M, Thalhammer G, Winkler K, Hellwig M, Ruff G, Grimm R and Denschlag J H 2004 Tuning the scattering length with an optically induced Feshbach resonance *Phys. Rev. Lett.* **93** 123001

[16] Sakaguchi H and Malomed B A 2005 Matter-wave solitons in nonlinear optical lattices *Phys. Rev.* E **72** 046610

[17] Gammal A, Frederico T, Tomio L and Chomaz P 2000 Atomic Bose-Einstein condensation with three-body interactions and collective excitations *J. Phys. B: At. Mol. Opt. Phys.* **33** 4053–67

[18] Muruganandam P and Adhikari S 2009 Fortran programs for the time-dependent Gross-Pitaevskii equation in a fully anisotropic trap *Comput. Phys. Commun.* **180** 1888–912

[19] Adhikari S and Salasnich L 2009 Localization of a Bose-Einstein condensate in a bichromatic optical lattice *Phys. Rev.* A **80** 023606

[20] Sudharsan J B, Radha R and Muruganandam P 2013 Collisionally inhomogeneous Bose-Einstein condensates with binary and three-body interactions in a bichromatic optical lattice *J. Phys. B: At. Mol. Opt. Phys.* **46** 155302

[21] Sudharsan J B, Radha R and Muruganandam P 2013 Collisionally inhomogeneous Bose-Einstein condensates with binary and three-body interactions in a bichromatic optical lattice *J. Phys. B: At. Mol. Opt. Phys.* **46** 155302

[22] Kevrekidis P G, Carretero-González R and Frantzeskakis D J 2008 *Emergent Nonlinear Phenomena in Bose–Einstein Condensates* (Springer) (includes bibliographical references and index)

[23] Pethick C and Smith H 2008 *Bose–Einstein Condensation in Dilute Gases* (Cambridge University Press) (ncludes bibliographical references and index)

[24] Abe H, Ueda T, Morikawa M, Saitoh Y, Nomura R and Okuda Y 2007 Faraday instability of superfluid surface *Phys. Rev.* E **76** 046305

[25] Engels P, Atherton C and Hoefer M A 2007 Observation of Faraday waves in a Bose-Einstein condensate *Phys. Rev. Lett.* **98** 095301

[26] Staliunas K, Longhi S and de Valcárcel G J 2002 Faraday patterns in Bose-Einstein condensates *Phys. Rev. Lett.* **89** 210406

[27] Nicolin A I, Carretero-González R and Kevrekidis P G 2007 Faraday waves in Bose-Einstein condensates *Phys. Rev.* A **76** 063609

[28] Nicolin A I and Raportaru M C 2010 Faraday waves in high-density cigar-shaped Bose–Einstein condensates *Physica* A **389** 4663–7

[29] Nicolin A I and Raportaru M C 2011 Faraday waves in one-dimensional Bose-Einstein condensates *Proc. Rom. Acad.* A **12** 209

[30] Balaž A, Paun R, Nicolin A I, Balasubramanian S and Ramaswamy R 2014 Faraday waves in collisionally inhomogeneous Bose-Einstein condensates *Phys. Rev.* A **89** 023609

[31] Nath R and Santos L 2010 Faraday patterns in two-dimensional dipolar Bose-Einstein condensates *Phys. Rev.* A **81** 033626

[32] Balaž A and Nicolin A I 2012 Faraday waves in binary nonmiscible Bose-Einstein condensates *Phys. Rev.* A **85** 023613

[33] Abdullaev F K, Ögren M and Sørensen M P 2013 Faraday waves in quasi-one-dimensional superfluid Fermi-Bose mixtures *Phys. Rev.* A **87** 023616

[34] Capuzzi P and Vignolo P 2008 Faraday waves in elongated superfluid fermionic clouds *Phys. Rev.* A **78** 043613

[35] Nicolin A I 2011 Resonant wave formation in Bose-Einstein condensates *Phys. Rev.* E **84** 056202

[36] Gaul C, Díaz E, Lima R P A, Domínguez-Adame F and Müller C A 2011 Stability and decay of Bloch oscillations in the presence of time-dependent nonlinearity *Phys. Rev.* A **84** 053627

[37] Díaz E, García Mena A, Asakura K and Gaul C 2013 Super-Bloch oscillations with modulated interaction *Phys. Rev.* A **87** 015601

[38] Brouzos I and Schmelcher P 2012 Controlled excitation and resonant acceleration of ultracold few-Boson systems by driven interactions in a harmonic trap *Phys. Rev.* A **85** 033635

[39] Diakonos F K, Kalozoumis P A, Karanikas A I, Manifavas N and Schmelcher P 2012 Geometric-phase-propagator approach to time-dependent quantum systems *Phys. Rev.* A **85** 062110

[40] Vidanović I, Balaž A, Al-Jibbouri H and Pelster A 2011 Nonlinear Bose-Einstein-condensate dynamics induced by a harmonic modulation of the s-wave scattering length *Phys. Rev.* A **84** 013618
[41] Kobyakov D, Bezett A, Lundh E, Marklund M and Bychkov V 2012 Quantum swapping of immiscible Bose-Einstein condensates as an alternative to the Rayleigh-Taylor instability *Phys. Rev.* A **85** 013630
[42] Hu W H, Jin L and Song Z 2013 Dynamics of one-dimensional tight-binding models with arbitrary time-dependent external homogeneous fields *Quantum Inf. Process.* **12** 3569–85
[43] Sakhel R R, Sakhel A R and Ghassib H B 2013 Nonequilibrium dynamics of a Bose-Einstein condensate excited by a red laser inside a power-law trap with hard walls *J. Low Temp. Phys.* **173** 177–206
[44] Balaž A, Vidanović I, Bogojević A, Belić A and Pelster A 2011 Fast converging path integrals for time-dependent potentials: I. Recursive calculation of short-time expansion of the propagator *J. Stat. Mech: Theory Exp.* **2011** P03004
[45] Vidanović I, Bogojević A, Balaž A and Belić A 2009 Properties of quantum systems via diagonalization of transition amplitudes. II. Systematic improvements of short-time propagation *Phys. Rev.* E **80** 066706
[46] Salasnich L, Manini N, Bonelli F, Korbman M and Parola M 2007 Self-induced density modulations in the free expansion of Bose-Einstein condensates *Phys. Rev.* A **75** 043616
[47] Imambekov A, Mazets I, Petrov D, Gritsev V, Manz S, Hofferberth S, Schumm T, Demler E and Schmiedmayer J 2009 Density ripples in expanding low-dimensional gases as a probe of correlations *Phys. Rev.* A **80** 033604
[48] Manz S *et al* 2010 Two-point density correlations of quasicondensates in free expansion *Phys. Rev.* A **81** 031610
[49] Kronjäger J, Becker C, Soltan-Panahi P, Bongs K and Sengstock K 2010 Spontaneous pattern formation in an antiferromagnetic quantum gas *Phys. Rev. Lett.* **105** 090402
[50] Köhler T, Góral K and Julienne P S 2006 Production of cold molecules via magnetically tunable Feshbach resonances *Rev. Mod. Phys.* **78** 1311–61
[51] Altmeyer A, Riedl S, Kohstall C, Wright M, Denschlag J H and Grimm R 2006 Note on 'Collective excitations of a degenerate gas at the BEC-BCS crossover *Phys. Rev. Lett.* **92** 203201
[52] Kraemer T *et al* 2006 Evidence for Efimov quantum states in an ultracold gas of caesium atoms *Nature* **440** 315–8
[53] Yamazaki R, Taie S, Sugawa S and Takahashi Y 2010 Submicron spatial modulation of an interatomic interaction in a Bose-Einstein condensate *Phys. Rev. Lett.* **105** 050405
[54] Abdullaev F K and Salerno M 2003 Adiabatic compression of soliton matter waves *J. Phys. B: At. Mol. Opt. Phys.* **36** 2851
[55] Rodas-Verde M I, Michinel H and Pérez-García V M 2005 Controllable soliton emission from a Bose-Einstein condensate *Phys. Rev. Lett.* **95** 153903
[56] Sakaguchi H and Malomed B A 2010 Solitons in combined linear and nonlinear lattice potentials *Phys. Rev.* A **81** 013624
[57] Theocharis G, Schmelcher P, Kevrekidis P G and Frantzeskakis D J 2006 Dynamical trapping and transmission of matter-wave solitons in a collisionally inhomogeneous environment *Phys. Rev.* A **74** 053614
[58] Mateo A M and Delgado V 2013 Effective equations for matter-wave gap solitons in higher-order transversal states *Phys. Rev.* E **88** 042916

[59] Lee C, Huang J, Deng H, Dai H and Xu J 2012 Nonlinear quantum interferometry with Bose condensed atoms *Front. Phys.* **7** 109–30
[60] Dong G, Hu B and Lu W 2006 Ground-state properties of a Bose-Einstein condensate tuned by a far-off-resonant optical field *Phys. Rev.* A **74** 063601
[61] Young S L E and binary S K 2012 Mixing, demixing, and structure formation in a binary dipolar Bose–Einstein condensate *Phys. Rev.* A **86** 063611
[62] Garcia-March M A and Busch T 2013 Quantum gas mixtures in different correlation regimes *Phys. Rev.* A **87** 063633
[63] Bludov Y V, Brazhnyi V A and Konotop V V 2007 Delocalizing transition in one-dimensional condensates in optical lattices due to inhomogeneous interactions *Phys. Rev.* A **76** 023603
[64] Mayteevarunyoo T, Malomed B A and Dong G 2008 Spontaneous symmetry breaking in a nonlinear double-well structure *Phys. Rev.* A **78** 053601
[65] Abdullaev F, Abdumalikov A and Galimzyanov R 2007 Gap solitons in Bose-Einstein condensates in linear and nonlinear optical lattices *Phys. Lett.* A **367** 149–55
[66] Wang C, Kevrekidis P, Horikis T and Frantzeskakis D 2010 Collisional-inhomogeneity-induced generation of matter-wave dark solitons *Phys. Lett.* A **374** 3863–8
[67] Mithun T, Porsezian K and Dey B 2013 Vortex dynamics in cubic-quintic Bose-Einstein condensates *Phys. Rev.* E **88** 012904
[68] Pinsker F, Berloff N G and Pérez-García V M 2013 Nonlinear quantum piston for the controlled generation of vortex rings and soliton trains *Phys. Rev.* A **87** 053624
[69] Madarassy E J and Toth V T 2013 Numerical simulation code for self-gravitating Bose-Einstein condensates *Comput. Phys. Commun.* **184** 1339–43
[70] Marojević V, Göklü E and Lämmerzahl C 2013 Energy eigenfunctions of the 1D Gross-Pitaevskii equation *Comput. Phys. Commun.* **184** 1920–30
[71] Trallero-Giner C, Cipolatti R and Liew T C H 2013 One-dimensional cubic-quintic Gross-Pitaevskii equation for Bose-Einstein condensates in a trap potential *Eur. Phys. J.* D **67** 1–7
[72] Bao W, Jaksch D and Markowich P A 2003 Numerical solution of the Gross-Pitaevskii equation for Bose-Einstein condensation *J. Comput. Phys.* **187** 318–42
[73] McLachlan N W 1947 *Theory and Application of Mathieu Functions* (Clarendon)
[74] Caplan R M 2013 NLSEmagic: nonlinear Schrödinger equation multi-dimensional Matlab-based GPU-accelerated integrators using compact high-order schemes *Comput. Phys. Commun.* **184** 1250–71
[75] Grišins P and Mazets I E 2014 Metropolis-Hastings thermal state sampling for numerical simulations of Bose-Einstein condensates *Comput. Phys. Commun.* **185** 1926–31
[76] Gubeskys A, Malomed B A and Merhasin I M 2005 Alternate solitons: nonlinearly managed one- and two-dimensional solitons in optical lattices *Stud. Appl. Math.* **115** 255–77
[77] Verhaar B J, van Kempen E G M and Kokkelmans S J J M F 2009 Predicting scattering properties of ultracold atoms: adiabatic accumulated phase method and mass scaling *Phys. Rev.* A **79** 032711
[78] Middelkamp S, Chang J, Hamner C, Carretero-González R, Kevrekidis P, Achilleos V, Frantzeskakis D, Schmelcher P and Engels P 2011 Dynamics of dark-bright solitons in cigar-shaped Bose-Einstein condensates *Phys. Lett.* A **375** 642–6
[79] Hamner C, Chang J J, Engels P and Hoefer M A 2011 Generation of dark-bright soliton trains in superfluid-superfluid counterflow *Phys. Rev. Lett.* **106** 065302

[80] Chiofalo M L, Succi S and Tosi M P 2000 Ground state of trapped interacting Bose-Einstein condensates by an explicit imaginary-time algorithm *Phys. Rev.* E **62** 7438–44
[81] Cerimele M M, Chiofalo M L, Pistella F, Succi S and Tosi M P 2000 Numerical solution of the Gross-Pitaevskii equation using an explicit finite-difference scheme: an application to trapped Bose-Einstein condensates *Phys. Rev.* E **62** 1382–9
[82] Tiwari R P and Shukla A 2006 A basis-set based Fortran program to solve the Gross-Pitaevskii equation for dilute bose gases in harmonic and anharmonic traps *Comput. Phys. Commun.* **174** 966–82
[83] Hua W, Liu X and Ding P 2006 Numerical solution for the Gross-Pitaevskii equation *J. Math. Chem.* **40** 243–55
[84] Vudragović D, Vidanović I, Balaž A, Muruganandam P and Adhikari S K 2012 C programs for solving the time-dependent Gross-Pitaevskii equation in a fully anisotropic trap *Comput. Phys. Commun.* **183** 2021–5
[85] Kumar R K, Young S L E, Vudragović D, Balaž A, Muruganandam P and Adhikari S K 2015 Fortran and C programs for the time-dependent dipolar Gross-Pitaevskii equation in an anisotropic trap *Comput. Phys. Commun.* **195** 117–28
[86] Lončar V, Balaž A, Bogojević A, Škrbić S, Muruganandam P and Adhikari S K 2016 CUDA programs for solving the time-dependent dipolar Gross-Pitaevskii equation in an anisotropic trap *Comput. Phys. Commun.* **200** 406–10
[87] Lončar V, Young S L E, Škrbić S, Muruganandam P, Adhikari S K and Balaž A 2016 OpenMP, OpenMP/MPI, and CUDA/MPI C programs for solving the time-dependent dipolar Gross-Pitaevskii equation *Comput. Phys. Commun.* **209** 190–6
[88] Satarić B, Slavnić V, Belić A, Balaž A, Muruganandam P and Adhikari S K 2016 Hybrid OpenMP/MPI programs for solving the time-dependent Gross-Pitaevskii equation in a fully anisotropic trap *Comput. Phys. Commun.* **200** 411–7
[89] Young S L E, Vudragović D, Muruganandam P, Adhikari S K and Balaž A 2016 OpenMP Fortran and C programs for solving the time-dependent Gross-Pitaevskii equation in an anisotropic trap *Comput. Phys. Commun.* **204** 209–13
[90] Ho T L and Shenoy V 1996 Binary mixtures of Bose condensates of alkali atoms *Phys. Rev. Lett.* **77** 3276
[91] Ao P and Chui S T 1998 Binary Bose-Einstein condensate mixtures in weakly and strongly segregated phases *Phys. Rev.* A **58** 4836–40
[92] Egorov M, Opanchuk B, Drummond P, Hall B V, Hannaford P and Sidorov A I 2013 Measurement ofs-wave scattering lengths in a two-component Bose-Einstein condensate *Phys. Rev.* A **87** 053614
[93] Vidanović I, van Druten N J and Haque M 2013 Spin modulation instabilities and phase separation dynamics in trapped two-component Bose condensates *New J. Phys.* **15** 035008
[94] Sudharsan J B, Radha R, Raportaru M C, Nicolin A I and Balaž A 2016 Faraday and resonant waves in binary collisionally-inhomogeneous Bose-Einstein condensates *J. Phys. B: At. Mol. Opt. Phys.* **49** 165303
[95] Pollack S E, Dries D, Hulet R G, Magalhães K M F, Henn E A L, Ramos E R F, Caracanhas M A and Bagnato V S 2010 Collective excitation of a Bose-Einstein condensate by modulation of the atomic scattering length *Phys. Rev.* A **81** 053627
[96] Staliunas K, Longhi S and de Valcárcel G J 2004 Faraday patterns in low-dimensional Bose-Einstein condensates *Phys. Rev.* A **70** 011601

[97] Radha R and Vinayagam P S 2015 An analytical window into the world of ultracold atoms *Rom. Rep. Phys.* **67** 89

[98] Satsuma J and Ablowitz M J 1979 Two-dimensional lumps in nonlinear dispersive systems *J. Math. Phys.* **20** 1496–503

[99] Matthews M R, Anderson B P, Haljan P C, Hall D S, Wieman C E and Cornell E A 1999 Vortices in a Bose–Einstein condensate *Phys. Rev. Lett.* **83** 2498

[100] Fetter A L 2009 Rotating trapped Bose-Einstein condensates *Rev. Mod. Phys.* **81** 647–91

[101] Leanhardt A, Görlitz A, Chikkatur A, Kielpinski D, Shin Y I, Pritchard D and Ketterle W 2002 Imprinting vortices in a Bose-Einstein condensate using topological phases *Phys. Rev. Lett.* **89** 190403

[102] White A C, Anderson B P and Bagnato V S 2014 Vortices and turbulence in trapped atomic condensates *Proc. Natl. Acad. Sci.* **111** 4719–26

[103] Dalfovo F and Stringari S 1996 Bosons in anisotropic traps: ground state and vortices *Phys. Rev.* A **53** 2477

[104] Neu J C 1990 Vortices in complex scalar fields *Physica* D **43** 385–406

[105] Mihalache D, Mazilu D, Malomed B A and Lederer F 2006 Vortex stability in nearly-two-dimensional Bose-Einstein condensates with attraction *Phys. Rev.* A **73** 043615

[106] Pu H, Law C K, Eberly J H and Bigelow N P 1999 Coherent disintegration and stability of vortices in trapped Bose condensates *Phys. Rev.* A **59** 1533–7

[107] Dodd R 1996 Approximate solutions of the nonlinear Schrödinger equation for ground and excited states of Bose-Einstein condensates *J. Res. Natl. Inst. Stand. Technol.* **101** 545

[108] Brtka M, Gammal A and Malomed B A 2010 Hidden vorticity in binary Bose-Einstein condensates *Phys. Rev.* A **82** 053610

[109] Kawaguchi Y and Ohmi T 2004 Splitting instability of a multiply charged vortex in a Bose-Einstein condensate *Phys. Rev.* A **70** 043610

[110] Pérez-García V M, Michinel H and Herrero H 1998 Bose-Einstein solitons in highly asymmetric traps *Phys. Rev.* A **57** 3837–42

[111] Salasnich L, Parola A and Reatto L 2002 Effective wave equations for the dynamics of cigar-shaped and disk-shaped Bose condensates *Phys. Rev.* A **65** 043614

[112] Mateo A M and Delgado V 2008 Effective mean-field equations for cigar-shaped and disk-shaped Bose-Einstein condensates *Phys. Rev.* A **77** 013617

[113] Hukriede J, Runde D and Kip D 2003 Fabrication and application of holographic Bragg gratings in lithium niobate channel waveguides *J. Phys. Appl. Phys.* **36** R1

[114] Sudharsan J B, Radha R, Fabrelli H, Gammal A and Malomed B A 2015 Stable multiple vortices in collisionally inhomogeneous attractive Bose-Einstein condensates *Phys. Rev.* A **92** 053601

[115] D'yakonov E G 1962 Difference schemes with a 'disintegrating' operator for multidimensional problems *Zh. Vychisl. Mat. Mat. Fiz.* **2** 549–68

[116] Buryak A V, Di Trapani P, Skryabin D V and Trillo S 2002 Optical solitons due to quadratic nonlinearities: from basic physics to futuristic applications *Phys. Rep.* **370** 63–235

[117] Kivshar Y S and Agrawal G P 2003 *Optical Solitons: From Fibers to Photonic Crystals* (Academic Press)

[118] Malomed B A, Mihalache D, Wise F and Torner L 2005 Spatiotemporal optical solitons *J. Opt. B: Quantum Semiclass. Opt.* **7** R53–72

[119] Chen Z, Segev M and Christodoulides D N 2012 Optical spatial solitons: historical overview and recent advances *Rep. Progr. Phys.* **75** 086401

[120] Moll K, Homoelle D, Gaeta A L and Boyd R W 2002 Conical harmonic generation in isotropic materials *Phys. Rev. Lett.* **88** 153901
[121] Béjot P, Kasparian J, Henin S, Loriot V, Vieillard T, Hertz E, Faucher O, Lavorel B and Wolf J-P 2010 Higher-order Kerr terms allow ionization-free filamentation in gases *Phys. Rev. Lett.* **104** 103903
[122] Brée C, Demircan A and Steinmeyer G 2011 Saturation of the all-optical Kerr effect *Phys. Rev. Lett.* **106** 183902
[123] Saha M and Sarma A K 2013 Modulation instability in nonlinear metamaterials induced by cubic-quintic nonlinearities and higher order dispersive effects *Opt. Commun.* **291** 321–5
[124] Reyna A S, Malomed B A and de Araújo C B 2015 Stability conditions for one-dimensional optical solitons in cubic-quintic-septimal media *Phys. Rev.* A **92** 033810
[125] Falcão-Filho E L, de Araújo C B, Boudebs G, Leblond H and Skarka V 2013 Robust two-dimensional spatial solitons in liquid carbon disulfide *Phys. Rev. Lett.* **110** 013901
[126] Reyna A S, Jorge K C and de Araújo C B 2014 Two-dimensional solitons in a quintic-septimal medium *Phys. Rev.* A **90** 063835
[127] Mihalache D, Mazilu D, Crasovan L C, Towers I, Buryak A, Malomed B A, Torner L, Torres J and Lederer F 2002 Stable spinning optical solitons in three dimensions *Phys. Rev. Lett.* **88** 073902
[128] Reyna A S and de Araújo C B 2014 Spatial phase modulation due to quintic and septic nonlinearities in metal colloids *Opt. Express* **22** 22456
[129] Bergé L 1998 Wave collapse in physics: principles and applications to light and plasma waves *Phys. Rep.* **303** 259–370
[130] Rasmussen J J and Rypdal K 1986 Blow-up in nonlinear Schroedinger equations–I. A general review *Phys. Scr.* **33** 481
[131] Pelinovsky D E, Kivshar Y S and Afanasjev V V 1998 Internal modes of envelope solitons *Physica* D **116** 121–42
[132] Sakaguchi H and Malomed B A 2012 Stable two-dimensional solitons supported by radially inhomogeneous self-focusing nonlinearity *Opt. Lett.* **37** 1035–7
[133] Zubairy M S, Matsko A B and Scully M O 2002 Resonant enhancement of high-order optical nonlinearities based on atomic coherence *Phys. Rev.* A **65** 043804
[134] Brtka M, Gammal A and Tomio L 2006 Relaxation algorithm to hyperbolic states in Gross-Pitaevskii equation *Phys. Lett.* A **359** 339–44
[135] Gammal A, Frederico T and Tomio L 1999 Improved numerical approach for the time-independent Gross-Pitaevskii nonlinear Schrödinger equation *Phys. Rev.* E **60** 2421–4
[136] Camassa R and Holm D D 1993 An integrable shallow water equation with peaked solitons *Phys. Rev. Lett.* **71** 1661
[137] Salerno M 1992 Quantum deformations of the discrete nonlinear Schrödinger equation *Phys. Rev.* A **46** 6856–9
[138] Gomez-Gardeñes J, Malomed B A, Floría L M and Bishop A R 2006 Solitons in the Salerno model with competing nonlinearities *Phys. Rev.* E **73** 036608
[139] Fabrelli H, Sudharsan H B, Radha R, Gammal A and Malomed B A 2017 Solitons under spatially localized cubic-quintic-septimal nonlinearities *J. Opt.* **19** 075501
[140] Vakhitov N G and Kolokolov A A 1973 Stationary solutions of the wave equation in the medium with nonlinearity saturation Radiophys *Quantum Electron.* **16** 783–9
[141] Radha R and Lakshmanan M 1994 Singularity analysis and localized coherent structures in (2+1)-dimensional generalized Korteweg–de Vries equations *J. Math. Phys.* **35** 4746

[142] Radha R and Lakshmanan M 1995 Dromion-like structures in the (2+1)-dimensional breaking soliton equation *Phys. Lett.* A **197** 7

[143] Crasovan L C, Vekslerchik V, Pérez-García V M, Torres J P, Mihalache D and Torner L 2003 Stable vortex dipoles in nonrotating Bose–Einstein condensates *Phys. Rev.* A **68** 063609

[144] Neely T W, Samson E C, Bradley A S, Davis M J and Anderson B P 2010 Observation of vortex dipoles in an oblate Bose–Einstein condensate *Phys. Rev. Lett.* **104** 160401

[145] Crasovan L C, Molina-Terriza G, Torres J P, Torner L, Pérez-García V M and Mihalache D 2002 Globally linked vortex clusters in trapped wave fields *Phys. Rev.* E **66** 036612

[146] Lashkin V M 2007 Two-dimensional nonlocal vortices, multipole solitons, and rotating multisolitons in dipolar Bose–Einstein condensates *Phys. Rev.* A **75** 043607

[147] Middelkamp S, Kevrekidis P G, Frantzeskakis D J, Carretero-González R and Schmelcher P 2010 Bifurcations, stability, and dynamics of multiple matter-wave vortex states *Phys. Rev.* A **82** 013646

[148] Firth W J and Skryabin D V 1997 Optical solitons carrying orbital angular momentum *Phys. Rev. Lett.* **79** 2450–3

[149] Quiroga-Teixeiro M and Michinel H 1997 Stable azimuthal stationary state in quintic nonlinear optical media *J. Opt. Soc. Am.* B **14** 2004–9

[150] Pego R L and Warchall J 2002 Spectrally stable encapsulated vortices for nonlinear Schrödinger equations . *Nonlinear Sci.* **12** 347–94

[151] Anderson E, Bai Z, Bischof C, Blackford S, Demmel J, Dongarra J, Du Croz J, Greenbaum A, Hammarling S, McKenney A *et al* 1999 *LAPACK Users' Guide* 3rd edn (Philadelphia: Society for Industrial and Applied Mathematics)

IOP Publishing

An Introduction to Ultracold Atoms with Analytical and Numerical Methods

Paulsamy Muruganandam and Ramaswamy Radha

Chapter 10

Quantum vortices in Bose–Einstein condensates

Quantum vortices in Bose–Einstein condensates (BECs) exhibit unique properties, including irrotational and resistance-free flow, analogous to superfluid helium. These properties lead to vortex formation, characterized by quantized circulation with phase changes in integer multiples of 2π. Theoretical models, employing the Gross–Pitaevskii (GP) equation, describe vortex nucleation and dynamics in single vortices and lattices. Vortex lattices, observed in weakly interacting Bose gases, represent highly excited collective states with finite angular momentum, similar to magnetic flux lines in superconductors. These structures appear across domains like astrophysics and condensed matter physics, underscoring their universal significance.

A range of experimental and numerical techniques have been developed to generate and study vortices in BECs. Rotating traps are commonly used, where vortices form once the rotation exceeds a critical frequency, often resulting in ordered vortex lattices. Vortices can also arise from turbulence or dynamical instabilities within the condensate. In merging BECs, interference between matter waves leads to the formation of solitons that decay into vortex–antivortex pairs. Numerical methods, such as phase engineering and the use of time-dependent potentials, have been employed to simulate vortex formation and lattice crystallization. To accurately capture energy dissipation into thermal (noncondensed) modes, these simulations often include dissipative terms in the GP equation. Together, these approaches offer valuable insight into the complex dynamics of vortex generation in quantum fluids.

In two-dimensional BECs trapped in harmonic potentials, vortices with charge $s = 1$ are found to be more stable compared to those with higher charges, as confirmed by solutions to the Bogoliubov equations. Vortices in the Thomas–Fermi (TF) regime and the lowest Landau level (LLL) exhibit distinct behaviours, with phase transitions to highly correlated states revealing key quantum effects. In dipolar BECs, the anisotropic dipole–dipole interactions (DDIs) play a crucial

doi:10.1088/978-0-7503-5447-9ch10

role in shaping vortex structure and stability, leading to second-order-like transitions in both straight and helical vortex lines. These results highlight the complex interplay of interparticle interactions, trapping geometries, and quantum mechanics in vortex dynamics.

10.1 Single-vortex dynamics

The GP equation is primarily developed to describe a single straight vortex in a uniform BEC with particle density n. The condensate wavefunction in hydrodynamic form is given by

$$\Psi(\mathbf{r}, t) = |\Psi(\mathbf{r}, t)|\exp(\mathrm{i}S(\mathbf{r}, t)) \tag{10.1}$$

where $|\Psi(\mathbf{r}, t)|$ is the magnitude and S is the condensate phase. The condensate density is $n(\mathbf{r}, t) = |\Psi(\mathbf{r}, t)|^2$, and the condensate-current density is

$$\mathbf{j} = \frac{\hbar}{2Mi}(\Psi^* \nabla \Psi - \Psi \nabla \Psi^*) = |\Psi|^2 \frac{\hbar \nabla S}{M} \tag{10.2}$$

Using the hydrodynamic relation, the velocity field for a BEC is written as

$$v(\mathbf{r}, t) = \frac{\hbar}{M} \nabla S(\mathbf{r}, t) = \nabla \phi(\mathbf{r}, t)$$

where $\phi = \hbar S/M$ is the velocity potential or phase.

10.1.1 Unbounded condensates

The vortex structure in an unbounded, uniform dilute Bose gas is analyzed by assuming a single straight vortex. The condensate wavefunction has been defined as

$$\Psi(\mathbf{r}) = \sqrt{n}\chi(\mathbf{r}), \quad \chi(\mathbf{r}) = \mathrm{e}^{\mathrm{i}\phi} f(r/\xi), \tag{10.3}$$

where $\mathbf{r} = (r, \phi)$ represents 2D polar coordinates, n is the uniform density, ξ is the healing length, and $f(r/\xi) \to 1$ as $r \to \infty$. The phase $\mathrm{e}^{\mathrm{i}\phi}$ induces a velocity field $\mathbf{v} = (\hbar/Mr)\hat{\phi}$, which diverges as $r \to 0$. The circulation around the vortex, given by

$$\kappa = \oint \mathbf{v} \cdot d\mathbf{l} = \frac{2\pi\hbar}{M}, \tag{10.4}$$

is derived using the phase gradient. By applying Stokes' theorem, the vorticity is localized at the core:

$$\nabla \times \mathbf{v} = \frac{2\pi\hbar}{M}\delta^{(2)}(\mathbf{r})\hat{z}. \tag{10.5}$$

The kinetic energy per unit length is computed using the wavefunction in equation (10.3) as:

$$E_k = \frac{\hbar^2}{2M}\int \mathrm{d}^2\mathbf{r} |\nabla \Psi|^2 = \frac{\hbar^2 n}{2M}\int \mathrm{d}^2\mathbf{r}\left[\left(\frac{\mathrm{d}f}{\mathrm{d}r}\right)^2 + \frac{f^2}{r^2}\right]. \tag{10.6}$$

The first term reflects density variations near the vortex core, and the second term represents the incompressible kinetic energy of the circulating flow. The Euler–Lagrange equation, derived by minimizing the GP energy functional with this kinetic term, yields a nonlinear differential equation for the radial profile $f(r)$. The profile satisfies $f(r) \to 0$ as $r \to 0$, vanishes within the core ($r \lesssim \xi$), and approaches $f(r) \to 1$ for $r \gg \xi$, defining the vortex core size.

The compressible nature of the condensate introduces a speed of sound, $s = \sqrt{gn/M}$, where g is the interaction strength. By relating the healing length to interactions via $\xi = \hbar/\sqrt{2Mgn}$, the speed of sound is expressed as $s = \hbar/(\sqrt{2}\, M\xi)$. The circulating flow velocity $v = \hbar/(Mr)$ exceeds s for $r < \xi$, indicating supersonic flow within the vortex core, consistent with the divergent velocity field near the singularity.

10.1.2 Bounded condensates

In classical viscous fluids, rotation within a container at angular velocity Ω causes the fluid to match the container's velocity due to microscopically rough walls. In contrast, superfluids and BECs in bounded, rotating traps exhibit quantized rotation, supporting vortices only at discrete frequencies despite continuous changes in the container's rotation. The rotating walls impose a time-dependent potential, necessitating a transformation to the rotating frame, where the potential becomes stationary. The Hamiltonian in this frame is expressed as $H_{\rm rot} = H - \Omega \cdot \mathbf{L}$, where H is the laboratory-frame Hamiltonian and $\mathbf{L} = \mathbf{r} \times \mathbf{p}$ is the angular momentum. Using the GP equation, the energy is minimized to yield states with non-zero, positive angular momentum corresponding to vortex configurations.

In the TF, dominant repulsive interactions result in a large condensate radius R_0 compared to the oscillator length d_0, with the vortex core size ξ satisfying $\xi/d_0 = d_0/R_0$. Central vortices induce a toroidal density profile due to zero density at the core, while off-centre vortices at position (x_0, y_0) alter the phase to the azimuthal angle around the shifted origin. The TF density remains largely unperturbed by a few vortices. This regime, common in experimental vortex studies, facilitates the analysis of vortex dynamics without significant density changes.

Energy analysis reveals vortex stability thresholds. The formation energy of an off-centre vortex, $\Delta E(r_0, \Omega)$, relative to a vortex-free state has been derived by Svidzinsky and Fetter [1] as a function of radial position r_0, rotation frequency Ω, and axial width $R_\perp$. In non-rotating condensates, the energy decreases as r_0 approaches $R_\perp$, driving vortices to the edge via spiral trajectories under weak dissipation. At a metastable frequency $\Omega_{\rm m} = \frac{3}{5}\Omega_{\rm c}$, where $\Omega_{\rm c}$ is the critical frequency for vortex creation, the energy curvature at the centre becomes zero, marking the onset of metastability. For $\Omega > \Omega_{\rm m}$, vortices near the centre are locally stable, while for $\Omega > \Omega_{\rm c}$, central vortices are both locally and globally stable, defining the thermodynamic threshold for quantized vortex formation in the TF limit.

10.1.3 Motion of trapped vortex

The dynamics of a single vortex in a trapped BEC is analyzed using the time-dependent GP equation, which serves as the Euler–Lagrange equation for the Lagrangian functional $\mathcal{L}[\Psi] = \int dV \frac{i\hbar}{2}(\Psi^* \frac{\partial \Psi}{\partial t} - \frac{\partial \Psi^*}{\partial t}\Psi) - E[\Psi]$, where $E[\Psi]$ is the GP energy functional. For a straight vortex in a non-rotating, disk-shaped condensate in the TF limit, the Lagrangian formalism predicts angular precession. The off-centre vortex position r_0 parameterizes the motion, with the precession rate proportional to the radial derivative of the energy functional. This has yielded the phase evolution $\dot{\phi} = \frac{\Omega_m}{1 - r_0^2/R_\perp^2}$, where $\Omega_m = \frac{3\hbar\omega_\perp^2}{4\mu}\ln\left(\frac{R_\perp}{\xi}\right)$ is the critical rotation frequency for metastability, μ is the chemical potential, $R_\perp = \sqrt{2\mu/(M\omega_\perp^2)}$ is the TF radius, and ξ is the healing length. The factor $1 - r_0^2/R_\perp^2$ reflects the parabolic TF density, indicating faster precession near the condensate boundary.

An alternative approach employs matched asymptotic expansions to derive the vortex's translational velocity in the TF limit, where the vortex core size is small compared to the trap. The vortex at position r_0 moves with velocity $v(r_0)$ in the x–y plane, and the condensate is transformed into a comoving frame where the vortex is stationary. Near the core ($\xi \ll |r - r_0|$), the trap potential exerts a force proportional to $\nabla_\perp V_{tr}(r_0)$, capturing the core's structure. Far from the core, the vortex is approximated as a line singularity. Matching these solutions in the overlap region has yielded $v(r_0) = \frac{3\hbar}{4M\mu}\ln\left(\frac{R_\perp}{\xi}\right)\hat{\mathbf{z}} \times \nabla_\perp V_{tr}(r_0)$. For the axisymmetric trap $V_{tr} = \frac{1}{2} M(\omega_\perp^2 r^2 + \omega_z^2 z^2)$, this velocity becomes $v(r_0) = \Omega_m r\hat{\Phi}$, matching the Lagrangian result near the trap centre, indicating motion along equipotential lines and conserving energy.

A density-based approach further confirms the vortex's precessional velocity. Mass conservation, $\nabla \cdot (nv) = 0$, is applied in the comoving frame, with velocity $v \propto \nabla S$ and TF density $n(r)$. Perturbing the phase S with an azimuthal term proportional to $\ln|r - r_0|$ accounts for the vortex's circulation. The resulting velocity is $v(r_0) = \frac{\hbar}{M}\frac{\hat{\mathbf{z}} \times (-\nabla n)}{2n}\ln\left(\frac{R_\perp}{\xi}\right)$, evaluated at r_0. Since $\nabla n \propto -\nabla V_{tr}$ in the TF approximation, this matches the asymptotic velocity, reinforcing that the vortex's precession arises from the trap's potential gradient or equivalent density gradient, driven by a Magnus-like force perpendicular to the applied force.

10.2 Vortex lattices in BECs

Vortex lattices in BECs form when rotation frequencies exceed a critical threshold, stabilizing multiple vortices in a harmonically confined condensate. At moderate rotation frequencies, vortices arrange into triangular arrays, analogous to flux-line lattices in type-II superconductors and rotating superfluid ^{4}He. The TF approximation describes these lattices, capturing their uniform vorticity distribution. As the rotation frequency Ω approaches the radial trap frequency $\omega_\perp$, the condensate expands, entering a two-dimensional superfluid regime described by the LLL, where vortex lattices exhibit enhanced quantization effects. The lattices achieve an average vorticity matching a rigidly rotating body, with circulation quantized in units of $\kappa = h/M$ and vortex density given by $n_v = 2\Omega/\kappa$, yielding an angular momentum per particle of $N_v\hbar/2$, where N_v is the number of vortices.

Experiments produce vortex lattices by rotating condensates along their long axis using blue-detuned laser beams at wavelengths around 500 nm. Resonant absorption imaging reveals highly ordered triangular lattices, termed Abrikosov lattices, containing over 100 vortices. These lattices exhibit Tkachenko oscillations, characteristic transverse modes arising from quantized vorticity, distinct from classical fluids with uniform vorticity. Surface instabilities, observed when stirring frequencies exceed a critical value dependent on trap aspect ratio, disrupt lattice stability by exciting high-angular-momentum modes, as analyzed through Bogoliubov stability methods.

At finite initial temperatures, vortex lattice formation is modelled without explicit damping terms. A classical field $\Psi = \sqrt{N}\psi + \Psi_\perp$, combining the zero-temperature condensate wavefunction ψ and a thermal noise component $\Psi_\perp$, describes the system. The noise term, expressed via Bogoliubov modes with Gaussian-distributed amplitudes scaled by thermal energy, captures finite-temperature effects. This approach, validated at both zero and finite temperatures, demonstrates that weakly interacting BECs form stable vortex lattices, highlighting the robustness of quantized vorticity under thermal fluctuations.

10.2.1 Zero temperature initial state

At zero temperature, a BEC in a symmetric harmonic trap is subjected to a rotating frequency that increases from zero to a final value Ω_f. An eccentricity parameter, introduced by adjusting one radial trapping frequency, abruptly induces trap anisotropy. As the condensate evolves under this rotating anisotropy, its angular momentum approaches a steady state. For $\Omega_f < 0.7$, the condensate adiabatically follows a steady state, with minor surface mode excitations causing small angular momentum oscillations. At higher frequencies ($\Omega_f \approx 0.75$), the condensate deforms elliptically and then becomes S-shaped before undergoing turbulent motion. This leads to a ring of vortices surrounding the condensate, which does not enter the high-density region until the onset of turbulence, when multiple vortices penetrate and form a lattice.

As the lattice stabilizes in the rotating frame, vortices exhibit small random motions around equilibrium positions, accompanied by density fluctuations. The damping of vortex motion transfers initial energy irreversibly to other degrees of freedom. To estimate the system's effective temperature, the final state is evolved in imaginary time with the trap rotating at Ω_f, minimizing its energy to a local minimum associated with the vortex lattice. The energy difference ΔE between the real-time final state and the minimum state is calculated. Using Bogoliubov theory, ΔE is equated to the thermal energy of weakly coupled harmonic oscillator modes, given by $\Delta E = \sum_n \langle b_n^* b_n \rangle \varepsilon_n = (N-1)k_B T$, where b_n and ε_n are mode amplitudes and energies, and N is the number of modes. The resulting temperature is negligible compared to the chemical potential.

10.2.2 Finite initial temperature

At finite initial temperatures, a BEC in a rotating trap contains both condensate and thermally occupied noncondensed modes, unlike the zero-temperature case, where only the condensate is occupied. Below a critical rotation frequency, termed the Landau frequency, the condensate remains undeformed, and vortices do not enter. The noncondensed modes facilitate particle, energy, and angular momentum exchange with the condensate. Above the Landau frequency, the condensate minimizes its energy by nucleating vortices, which enter the high-density region one at a time as angular momentum increases. A single vortex from the surrounding cloud spirals slowly towards the centre, forming a stable configuration.

Once formed, the vortex lattice becomes stationary in the rotating frame, without additional vortices entering. As the rotation frequency increases further, additional vortices enter sequentially, contrasting with the zero-temperature case, where multiple vortices enter simultaneously in a ring configuration, with only some forming a lattice. The sequential entry at finite temperatures reflects thermal fluctuations, which reduce the energy barrier for vortex nucleation, enabling a gradual lattice buildup compared to the abrupt, turbulent lattice formation at zero temperature.

10.3 Vortices in Thomas–Fermi regime

In the TF limit, rapidly rotating BECs exhibit vortices with distinct dynamics, modelled using the GP equation. A single vortex at position r_0 displays a 2π-phase singularity, as described by the condensate wavefunction. For a rotating condensate with angular momentum quantum number $m \geqslant 0$, the wavefunction is expressed as a linear combination, $\Psi = \sum_{m\geqslant 0} c_m \psi_m$. The expectation value of the angular momentum operator $\hat{L} = -i\hbar\partial/\partial\phi$ yields $\langle L \rangle = \sum_{m\geqslant 0} m|c_m|^2$, and the one-body Hamiltonian's energy is $\langle H_0 \rangle = \hbar\omega_\perp(\langle L \rangle + 1)$. This energy is degenerate for states with the same $\langle L \rangle$, but interactions break the degeneracy, allowing the determination of coefficients c_m and vortex positions as functions of angular momentum. Studies, such as Butts and Rokhsar [2], identify vortex configurations with a single vortex on the symmetry axis for $\langle L \rangle = 1$, while multiple vortices form for higher angular momenta, yielding continuous angular momentum values.

Vortex lattices in rapidly rotating condensates approximate uniform classical vorticity. The total circulation around a contour enclosing $\mathcal{N}_\text{v}$ vortices in area $\mathcal{A}$ is $\Gamma = \kappa\mathcal{N}_v$, where $\kappa = 2\pi\hbar/M$ is the single-vortex circulation. Equating this to the classical circulation $\Gamma^{\text{cl}} = 2\Omega\mathcal{A}$ via Stokes' theorem, the areal vortex density is derived as

$$n_\text{v} = \frac{\mathcal{N}_\text{v}}{\mathcal{A}} = \frac{M\Omega}{\pi\hbar}. \tag{10.7}$$

The inverse density defines the area per vortex, yielding a cell radius $l = \sqrt{\hbar/M\Omega}$, with intervortex separation approximately $2l$. As angular velocity Ω increases, l decreases, and vortex density grows linearly per the Feynman relation. Centrifugal forces expand the condensate radially, increasing the number of vortices faster than linearly, while particle conservation causes axial compression.

The TF energy functional in the rotating frame, assuming negligible density gradient energy for length scales much larger than the healing length ξ, is expressed as

$$E'[\Psi] = \int \mathrm{d}V\left[\frac{1}{2}Mv^2 + V_{\mathrm{tr}} - M(\mathbf{\Omega}\cdot(\mathbf{r}\times\mathbf{v}))|\Psi|^2 + \frac{1}{2}g|\Psi|^4\right], \tag{10.8}$$

where $V_{\mathrm{tr}} = \frac{1}{2}M(\omega_\perp^2 r_\perp^2 + \omega_z^2 z^2)$. For a dense vortex lattice, the velocity approximates solid-body rotation, $\mathbf{v} = \mathbf{\Omega}\times\mathbf{r}$, reducing the functional to a non-rotating TF form with effective radial frequency $\omega_\perp^2 - \Omega^2$. The condensate density is

$$n(r, z) = |\Psi(r, z)|^2 = n_0\left(1 - \frac{r^2}{R_\perp^2} - \frac{z^2}{R_z^2}\right), \tag{10.9}$$

with $R_\perp^2 = 2\mu/[M(\omega_\perp^2 - \Omega^2)]$ and $R_z^2 = 2\mu/(M\omega_z^2)$. The condition $\Omega < \omega_\perp$ ensures radial confinement, while increasing Ω reduces central density n_0 and chemical potential $\mu \approx gn_0$. The TF approximation holds for sparse lattices but breaks down when vortex cores ($\xi \approx l$) overlap, increasing kinetic energy contributions.

10.4 Vortices in lowest Landau level

In the TF regime, the kinetic energy from density variations is neglected, and vortex cores remain small at moderate angular velocities. As the dimensionless angular velocity $\bar{\Omega} = \Omega/\omega_\perp$ approaches 1, increased vortex nucleation reduces the intervortex spacing, causing vortex cores to overlap. This overlap invalidates the TF approximation, as kinetic energy from density variations near vortex cores becomes significant. The centrifugal force expands the condensate radially into a disk-shaped, quasi-two-dimensional form, with axial contraction conserving particle number. The one-body oscillator Hamiltonian in the rotating frame yields eigenvalues

$$\varepsilon_{nm} = \hbar[\omega_\perp + n(\omega_\perp + \Omega) + m(\omega_\perp - \Omega)], \tag{10.10}$$

where, in the limit $\Omega \to \omega_\perp$, the eigenvalues become independent of the angular momentum quantum number m, resulting in large degeneracy. The quantum number n defines the Landau-level index, with the LLL ($n = 0$) separated by a gap of approximately $2\hbar\omega_\perp$. The low central density due to radial expansion reduces the interaction energy $gn(0)$ below this gap.

The LLL ground state wavefunction is a Gaussian, $\Psi_{00} \propto \exp(-r^2/2d_\perp^2)$, with $d_\perp = \sqrt{\hbar/M\omega_\perp}$. General LLL eigenfunctions take the form $\Psi_{0m} \propto \zeta^m \exp(-r^2/2d_\perp^2)$, where $\zeta = x + \mathrm{i}y$. The condensate wavefunction is expressed as

$$\Psi_{\mathrm{LLL}}(r) = \sum_{m\geqslant 0} c_m\Psi_{0m}(r) = f(\zeta)\exp(-r^2/2d_\perp^2), \tag{10.11}$$

where $f(\zeta) = \sum_{m\geqslant 0} c_m \zeta^m$. The zeros of $f(\zeta)$ at positions $\{\zeta_j\}$ mark singly quantized vortices, each with a 2π phase winding, and their core sizes approximate the intervortex spacing. The quasi-2D wavefunction is transformed as $\Psi(x, y, z) = \Psi(x, y)/\sqrt{2\pi d_z}$, where $d_z = \sqrt{\hbar/M\omega_z}$ is the axial confinement length. The energy functional is

$$E[\psi] = \int \mathrm{d}^2 r\ \psi^* \left(\frac{p^2}{2M} + \frac{1}{2} M\omega_\perp^2 r^2 - \Omega L_z + \frac{1}{2} g_{2D} |\psi|^2 \right) \psi, \tag{10.12}$$

with $g_{2D} = g/\sqrt{2\pi d_z}$ as the scaled 2D interaction strength and $p = Mv$ as the momentum.

Experimental realization of LLL BECs requires rotation at $\Omega > 0.99\omega_\perp$, near the centrifugal limit, with reduced atom numbers to minimize interaction energy. Three regimes are identified based on the filling factor $\nu = \mathcal{N}/\mathcal{N}_v$, where $\mathcal{N}$ is the atom number and $\mathcal{N}_v$ the vortex number. At high ν, the condensate forms an ordered vortex lattice in the mean-field quantum-Hall regime. As ν decreases, the vortex lattice's elastic shear strength, probed by Tkachenko oscillations, diminishes. For $\nu \approx 10$, quantum fluctuations melt the lattice, signalling a transition to correlated states. The condensate's density profile shifts from a parabolic TF form to a Gaussian, assuming uniform vortex density and weak interactions ($\mathcal{N}a/R_z \ll 1$).

10.5 Phase transitions to highly correlated states

As the rotation frequency Ω approaches $\omega_\perp$, the LLL regime transitions to highly correlated, non-superfluid states, marking a quantum phase transition. The filling factor $\nu = \mathcal{N}/\mathcal{N}_v$ governs this transition, with lower ν favouring correlated many-body states analogous to the fractional quantum-Hall effect.

A key example is the bosonic Laughlin state,

$$\Psi_{\mathrm{Laughlin}}(r_1, r_2, \ldots, r_N) \propto \prod_{n<n'} (z_n - z_{n'})^2 \exp\left(-\sum_{n=1}^{N} \frac{|z_n|^2}{2d_\perp^2} \right), \tag{10.13}$$

where $z_n = x_n + iy_n$ denotes the nth particle's position. The pair product $(z_n - z_{n'})^2$ introduces correlations by reducing energy for short-range repulsive interactions, distinguishing it from the coherent GP wavefunction. Other correlated states, such as the Pfaffian state, are predicted for specific ν, exhibiting non-Abelian statistics.

Achieving these states requires reducing ν, which is proposed through elongated condensates with large vortex arrays or synthetic gauge fields generated by lasers to mimic rotational effects. These methods lower the vortex density relative to the atom number, driving the system towards strongly correlated regimes. The transition alters the condensate's collective behaviour, eliminating superfluidity and introducing quantum-Hall-like properties.

10.6 Vortex stability and Bogoliubov equations

Theoretical studies analyze the collective-mode spectrum of a non-rotating, vortex-free BEC using the GP equation for the condensate wavefunction and the Bogoliubov equations for small-amplitude normal modes. These models accurately predict experimental observations, validating the approach. The analysis extends to a single quantized vortex in a non-rotating condensate, positioned on the symmetry axis, where the wavefunction carries one unit of angular momentum along the z-axis, conserving this quantity.

The quasiparticle excitation spectrum of the vortex reveals an anomalous mode with a positive normalization integral, negative angular momentum ($m_a = -1$), and negative frequency ($\omega_a = -|\omega_a|$). This negative frequency indicates energetic instability, as the vortex occupies an excited state rather than the ground state. In the weak-coupling limit, the vortex wavefunction is proportional to $re^{i\phi}\exp\left(-\frac{1}{2}r^2\right)$, with excitation energy $\hbar\omega_\perp$ (where $\omega_\perp$ is the trap frequency), compared to the ground state's zero energy and angular momentum. The anomalous mode's density perturbation, $\delta n_a \propto \exp[i(-\phi + |\omega_a|t)]$, precesses positively around the symmetry axis at rate $|\omega_a|$, a phenomenon also observed in the TF limit where ω_a equals the mechanical rotation frequency.

In rotating condensates, the mode frequency shifts in the rotating frame to $\omega'(\Omega) = \omega_j - m_j\Omega$, where Ω is the rotation frequency and m_j the angular momentum quantum number. For the anomalous mode, this shift yields $\omega_a(\Omega) = -|\omega_a| + \Omega$, achieving energetic stability when $\Omega > |\omega_a|$. In the TF limit, this condition confirms a critical rotation frequency for vortex metastability, ensuring stability against spiraling out in the presence of dissipation. These findings elucidate the role of anomalous modes in vortex stability across different interaction regimes.

10.7 Vortices in dipolar BECs

In rotating BECs with dominant dipolar interactions, vortex dynamics exhibit distinct characteristics compared to conventional BECs with only contact interactions. Mean-field models, incorporating the GP equation with DDIs, reveal that dipolar BECs nucleate more vortices due to the anisotropic nature of DDIs. When rotated along the trap's symmetry axis (z-axis), the condensate elongates, and its surface becomes unstable, developing ripples that facilitate vortex entry above a critical rotation frequency (Ω_c). A dissipative mechanism, implemented by modifying the time derivative in the GP equation to $(i - \gamma)\partial_t$, where γ represents a decay term, is essential for stabilizing vortex lattices. These lattices, often off-centred due to DDI-induced axial symmetry breaking, form as vortices and relax into stable configurations.

The interplay of dipolar and contact interactions further shapes vortex dynamics. Stronger DDI lowers Ω_c, enabling vortex nucleation at smaller rotation frequencies, and increases vortex numbers compared to contact-dominated BECs. Repulsive contact interactions enhance condensate stability and further increase vortex populations, though DDI plays a dominant role in reducing Ω_c. At high rotation frequencies, the condensate's density profile deviates from the parabolic TF shape, approaching the LLL regime, driven by DDI's anisotropic effects. These findings highlight the critical role of dipolar interactions in enhancing vortex formation and altering lattice geometries in rotating BECs.

10.8 Vortex formation via merging BECs

The merger of spatially separated, identical BECs produces vortices through matter-wave interference. When two condensates overlap, their interference generates patterns that evolve into nonlinear excitations. In two-dimensional condensates, these patterns form dark soliton stripes, while in three-dimensional condensates they create ring solitons. These solitons emerge with low initial velocities, propagate towards the condensate boundaries, and increase their velocity and width during propagation.

The soliton instability leads to vortex formation via the snake instability mechanism. Dark soliton stripes and ring solitons decay into vortex–antivortex pairs, which, in the absence of rotation, annihilate each other and dissipate as sound waves. In rotating condensates, vortices with circulation matching the direction of rotation persist, while those with opposite circulation spiral out, consistent with critical rotation frequencies observed in vortex lattice studies. This process highlights the role of rotational effects in stabilizing vortex configurations.

The vortex population depends on several factors. Slower mergers produce wider solitons that decay more rapidly, reducing vortex pair lifetimes compared to faster mergers. Larger condensates, with higher atom numbers or stronger repulsive interactions, sustain more solitons and, thus, more vortices. The initial relative phases of the merging condensates also influence vortex formation, with effects more pronounced in slower mergers. Furthermore, angular momentum transfer between condensates, facilitated by dark soliton stripes, enhances vortex dynamics, offering insights into collective excitations in merged BECs.

10.9 Vortex dynamics in dipolar BECs

Numerical studies model vortex dynamics in rotating dipolar BECs using the GP equation, incorporating DDIs. For a BEC with N atoms of mass m at absolute zero in a rotating frame, the three-dimensional GP equation is given by [3]

$$\begin{aligned} \mathrm{i}\frac{\partial\psi(\mathbf{r}, t)}{\partial t} = \Bigg[& -\frac{1}{2}\nabla^2 + V(\mathbf{r}) + 4\pi a N|\psi(\mathbf{r}, t)|^2 - \Omega L_z \\ & + N\int U_{\mathrm{dd}}(\mathbf{r}-\mathbf{r}')|\psi(\mathbf{r}', t)|^2 d^3r' \Bigg]\psi(\mathbf{r}, t), \end{aligned} \tag{10.14}$$

where $L_z = -\mathrm{i}(x\partial_y - y\partial_x)$ represents the z-component of angular momentum due to rotation about the z-axis with angular velocity Ω, expressed in units of the radial trap frequency ω_ρ. The quasi-two-dimensional GP equation for a disk-shaped BEC is given by:

$$\begin{aligned} \mathrm{i}\frac{\partial\psi_{2D}(\vec{\rho}, t)}{\partial t} = \Bigg[& -\frac{\nabla_\rho^2}{2} + V_{2\mathrm{D}} - \Omega L_z + \frac{4\pi a N}{\sqrt{2\pi}\,d_z}|\psi_{2D}(\vec{\rho}, t)|^2 \\ & + \frac{4\pi a_{\mathrm{dd}} N}{\sqrt{2\pi}\,d_z}\int \frac{\mathrm{d}^2k_\rho}{(2\pi)^2}\mathrm{e}^{\mathrm{i}\mathbf{k}_\rho\cdot\vec{\rho}}\tilde{n}(\mathbf{k}_\rho)h_{2\mathrm{D}}\left(\frac{k_\rho d_z}{\sqrt{2}}\right)\Bigg]\psi_{2\mathrm{D}}(\vec{\rho}, t), \end{aligned} \tag{10.15}$$

with the trap potential

$$V_{2D} = \frac{1}{2}[(1+\epsilon)x^2 + (1-\epsilon)y^2], \tag{10.16}$$

where ϵ introduces anisotropy to nucleate vortices [4, 5], and h_{2D} describes the Fourier-transformed DDI. Simulations use spatial steps $dx = dy = 0.2$ and time step $dt = 0.005$. The ground state is prepared via imaginary time propagation of equation (10.15) without rotation ($\Omega = 0$) or anisotropy ($\epsilon = 0$). Vortices form by evolving the ground state in real time with $\Omega \neq 0$, anisotropy $\epsilon = 0.06$, and phenomenological dissipation ($\gamma \sim 10^{-5}$), implemented by replacing i with $\mathrm{i} - \gamma$ in equation (10.15), mimicking thermal losses [4–6].

10.9.1 Vortex formation in pure dipolar BECs

In pure dipolar BECs ($a = 0$), vortices form above a critical angular velocity (Ω_c) in a disk-shaped geometry ($N = 10\,000$, trap aspect ratio $\lambda = 30$). For a dipolar strength $a_{dd} = 15a_0$, rotation elongates the condensate, destabilizing its boundary. Surface ripples develop, and vortices enter at $\Omega_c \approx 0.38$. For $\Omega = 0.6$, seven vortices form a stable pattern after $t = 20,000\omega^{-1}$, as shown in figure 10.1(a)–(d). The phase distribution, varying from 0 to 2π, reveals bifurcations at vortex cores (figure 10.1

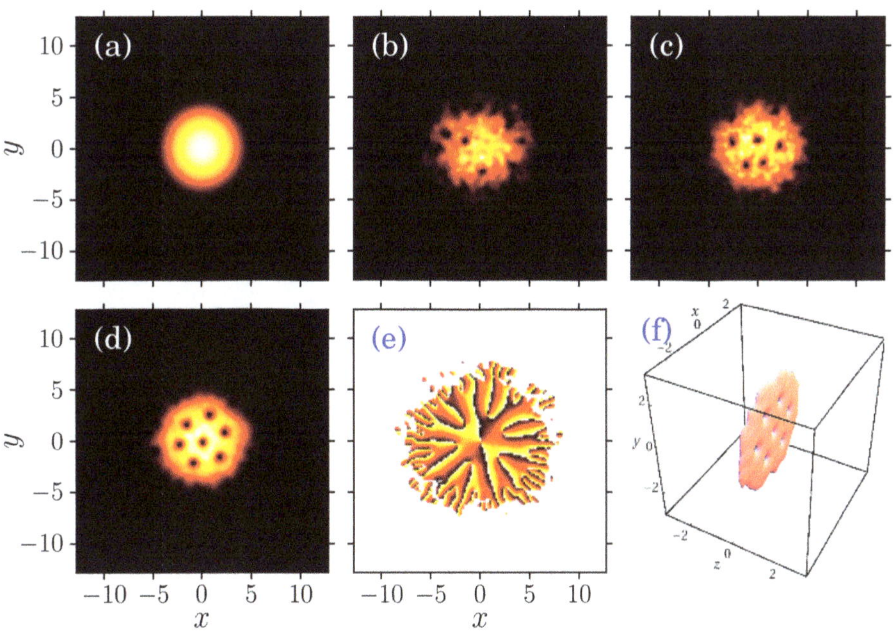

Figure 10.1. Contour plots of the density $|\psi_{2D}(\vec{\rho}, t)|^2$ showing the development of vortices in a pure dipolar BEC ($a = 0$, $a_{dd} = 15a_0$, $N = 10\,000$, $\lambda = 30$, $\Omega = 0.6$): (a) $t = 0$, (b) $t = 1,000\omega^{-1}$, (c) $t = 2,000\omega^{-1}$, (d) $t = 20,000\omega^{-1}$. (e) Phase distribution of ψ_{2D} for the steady vortex state in (d). (f) Three-dimensional contour plot of vortices from equation (10.14). Reproduced from [7]. Copyright 2012 IOP Publishing Ltd. All rights reserved.

(e)). Three-dimensional simulations using equation (10.14) confirm these results (figure 10.1(f)). The angular momentum expectation value given by:

$$\langle L_z \rangle = \mathrm{i} \int \psi^*(\vec{\rho}, t)(y\partial_x - x\partial_y)\psi(\vec{\rho}, t)\, \mathrm{d}\vec{\rho}, \tag{10.17}$$

increases with periodic oscillations, stabilizing at a steady value (figure 10.2). Vortex patterns at $\Omega = 0.38, 0.44, 0.47$ show increasing vortex numbers (figures 10.3(a)–(c)), with corresponding phase distributions (figures 10.3(d)–(f)).

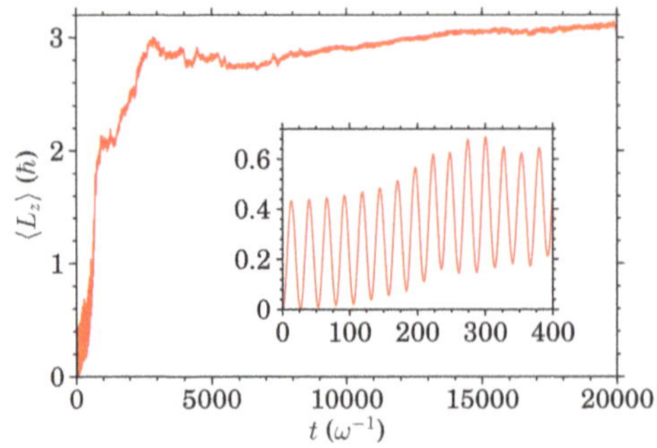

Figure 10.2. Time evolution of $\langle L_z \rangle$ during vortex development in figure 10.1. Reproduced from [7]. Copyright 2012 IOP Publishing Ltd. All rights reserved.

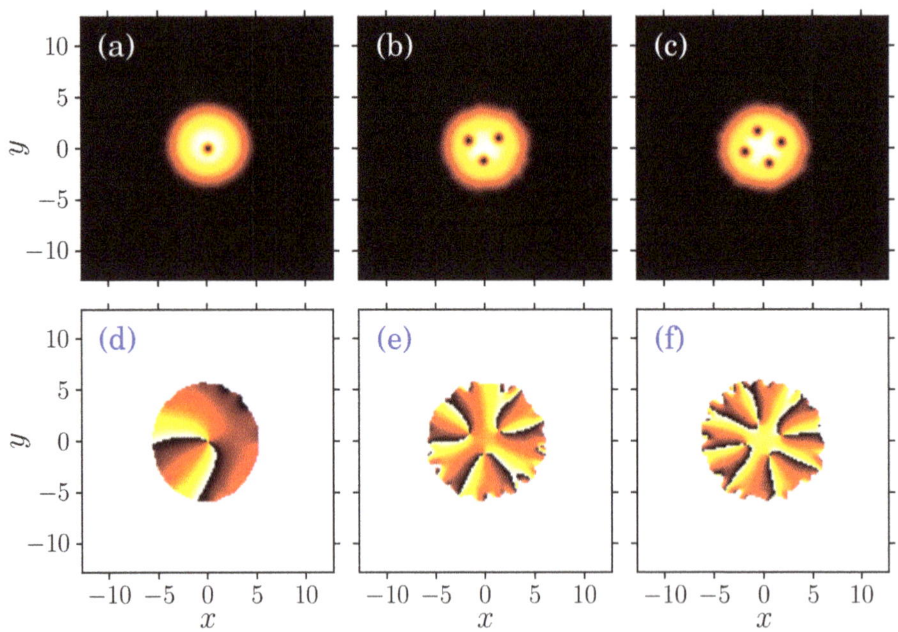

Figure 10.3. Contour plots of $|\psi_{2D}|^2$ showing steady-state vortices in a pure dipolar BEC ($a = 0$, $a_{dd} = 15a_0$, $N = 10, 000$, $\lambda = 30$): (a) $\Omega = 0.38$, (b) $\Omega = 0.44$, (c) $\Omega = 0.47$. (d–f) Corresponding phase distributions. Reproduced from [7]. Copyright 2012 IOP Publishing Ltd. All rights reserved.

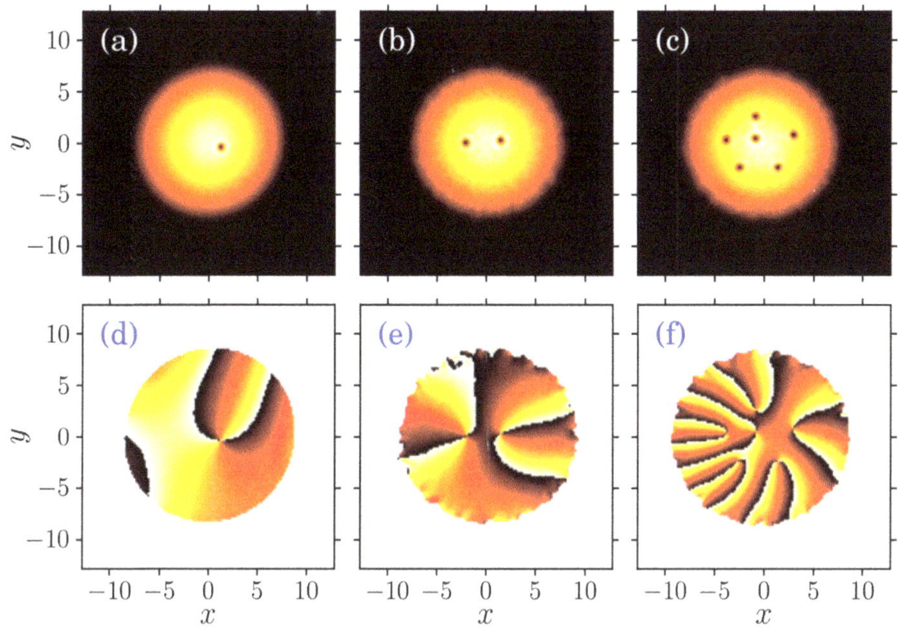

Figure 10.4. Contour plots of $|\psi_{2D}|^2$ showing vortices in a pure dipolar BEC ($a = 0$, $a_{\rm dd} = 130a_0$, $N = 10, 000$, $\lambda = 30$): (a) $\Omega = 0.271$, (b) $\Omega = 0.29$, (c) $\Omega = 0.33$. (d–f) Corresponding phase distributions. Reproduced from [7].

For a stronger dipolar strength ($a_{\rm dd} = 130a_0$), the critical velocity decreases to $\Omega_c \approx 0.271$. A single vortex forms at Ω_c (figure 10.4(a)), with more vortices at $\Omega = 0.29, 0.33$ (figure 10.4(b)–(c)), and phase distributions showing 2π windings (figures 10.4(d)–(f)). The condensate destabilizes at $\Omega > 0.62$. At $\Omega \approx 0.35$, cyclic dynamics emerge: rotation deforms the condensate, vortices enter, stabilize briefly, then destabilize as more vortices enter, repeating over long times ($t = 100\ 000\omega^{-1}$), as shown in figures 10.5(a)–(f). This behaviour is absent for lower $a_{\rm dd}$. Vortex formation occurs in a few tens of milliseconds, an order of magnitude faster than the few hundred milliseconds in non-dipolar BECs, due to DDI anisotropy breaking axial symmetry.

The vortex number ($N_{\rm v}$) follows Feynman's rule:

$$N_{\rm v} = \frac{m\Omega}{\hbar} R_\rho^2(\Omega), \tag{10.18}$$

where the TF radius under rotation is:

$$R_\rho(\Omega) = R_\rho(0)\left(1 - \frac{\Omega^2}{\Omega_\rho^2}\right)^{-1/4}, \tag{10.19}$$

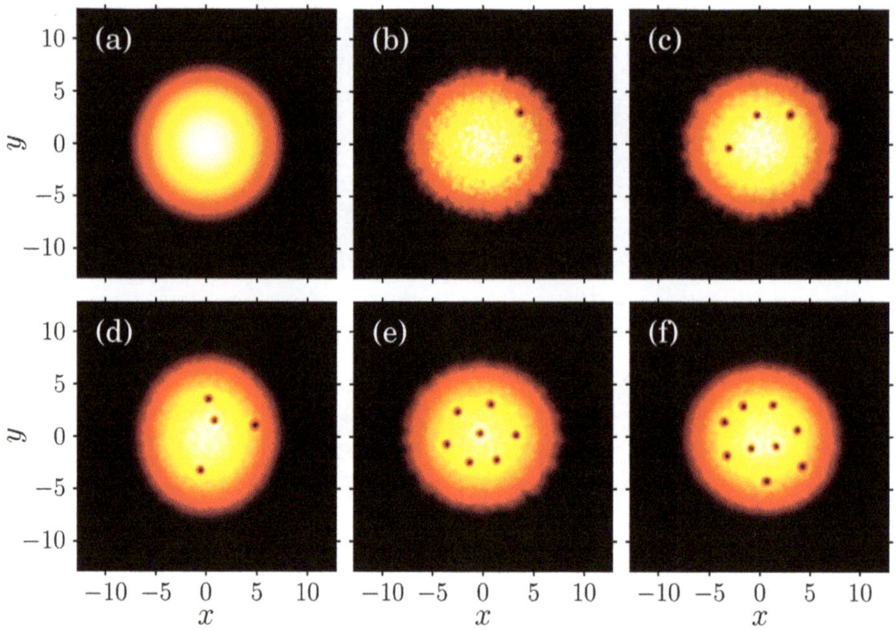

Figure 10.5. Contour plots of $|\psi_{2D}(\vec{\rho}, t)|^2$ showing cyclic dynamics in a pure dipolar BEC ($a = 0$, $a_{dd} = 130a_0$, $N = 10\ 000$, $\lambda = 30$, $\Omega = 0.35$): (a) $t = 0$, (b) $t = 30\ 000\omega^{-1}$, (c) $t = 40\ 000\omega^{-1}$, (d) $t = 50\ 000\omega^{-1}$, (e) $t = 80\ 000\omega^{-1}$, (f) $t = 100\ 000\omega^{-1}$. Reproduced from [7].

and the non-rotating radius is:

$$R_\rho^{\mathrm{TF}}(0) = \left[15N\frac{\lambda}{\sqrt{2\pi}}(a + 2a_{\mathrm{dd}})l^4\right]^{1/5}, \tag{10.20}$$

with $R_\rho(0) \approx 0.6R_\rho^{\mathrm{TF}}(0)$. Combining these, N_v is given by:

$$N_v = \frac{m\Omega}{\hbar}R_\rho^2(0)\left(1 - \frac{\Omega^2}{\Omega_\rho^2}\right)^{-1/2}. \tag{10.21}$$

Numerical simulations of equation (10.15) over $150\,000$–$200\,000\omega^{-1}$ show that N_v agrees with equation (10.21) for $a_{dd} = 15a_0$, but deviates for $a_{dd} = 130a_0$ due to anisotropic interactions (figure 10.6(a)). For a higher trap aspect ratio ($\lambda = 100$), similar trends appear, but the condensate with $a_{dd} = 130a_0$ remains stable at higher Ω (figure 10.6(b)).

10.9.2 Effect of contact interactions

Contact interactions enhance vortex stability and number [8]. For $a = 10a_0$, $a_{dd} = 15a_0$, vortices form at $\Omega = 0.5, 0.6, 0.65$, with eight vortices at $\Omega = 0.6$ (figures 10.7(a–c)), compared to seven without contact interactions, showing better-organized lattices

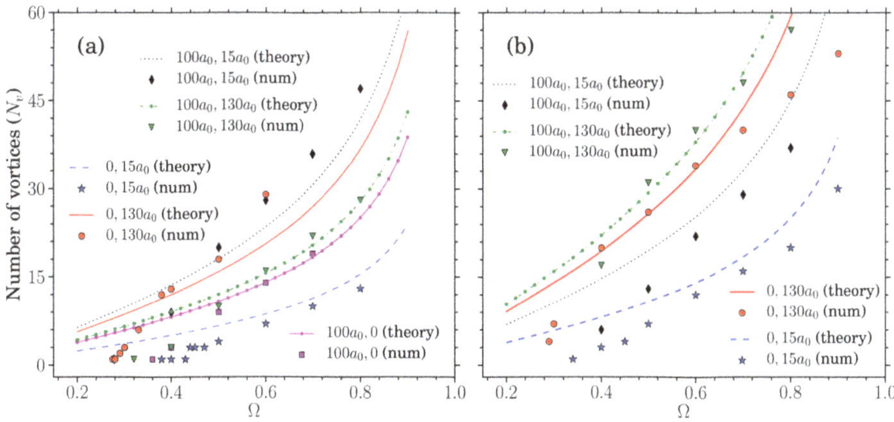

Figure 10.6. Equilibrium vortex number (N_v) versus Ω for pure dipolar BECs ($a = 0$, $a_{dd} = 15a_0$, $130a_0$): 'theory' used equation (10.21), 'num' was numerical. (a) $\lambda = 30$, (b) $\lambda = 100$. Repoduced from [7].

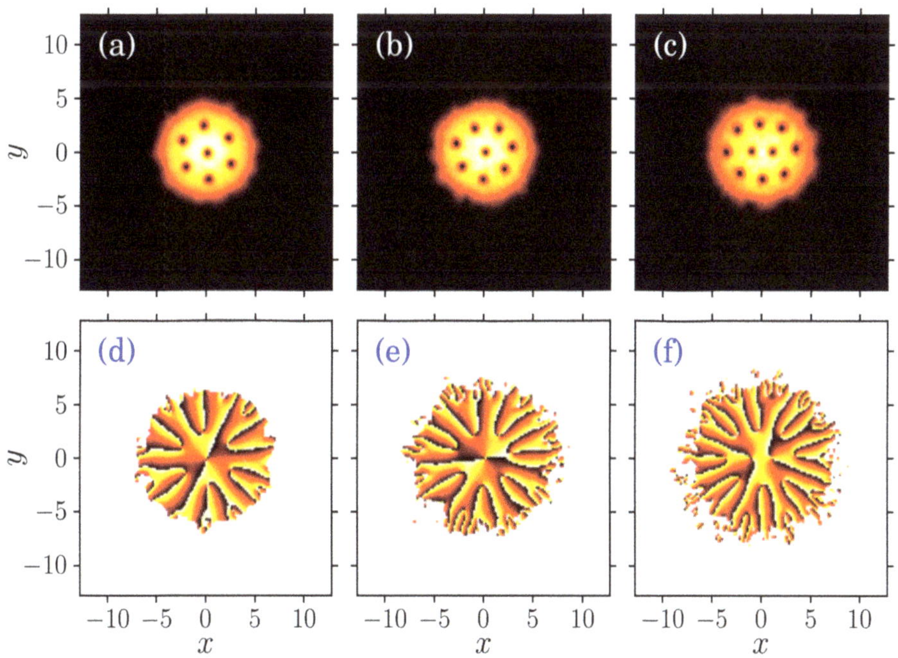

Figure 10.7. Contour plots of $|\psi_{2D}|^2$ showing vortices in a dipolar BEC ($a = 10a_0$, $a_{dd} = 15a_0$, $N = 10\,000$, $\lambda = 30$): (a) $\Omega = 0.5$, (b) $\Omega = 0.6$, (c) $\Omega = 0.65$. (d)–(f) Corresponding phase distributions. Reproduced from [7].

(figure 10.7(d–f)). For $a_{dd} = 130a_0$, 18, 30, and 35 vortices form at $\Omega = 0.5$, 0.6, 0.65, respectively (figures 10.8(a)–(c)), with phase distributions in figures 10.8(d)–(f). The condensate remains stable beyond $\Omega = 0.62$, unlike the pure dipolar case. For fixed $a = 20a_0$, $\Omega = 0.6$, the vortex number increases from 5 ($a_{dd} = 0$) to 10 ($a_{dd} = 15a_0$) and

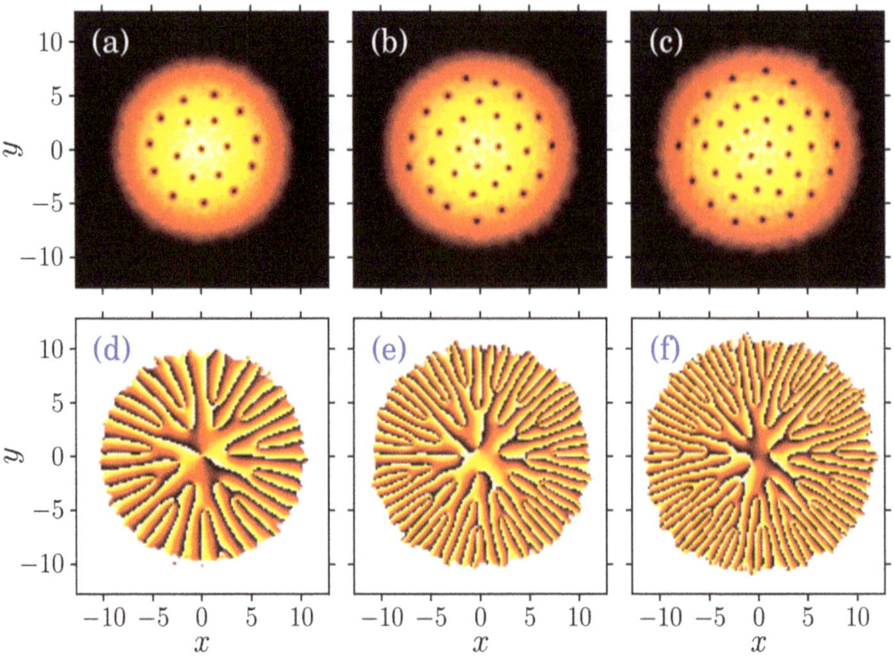

Figure 10.8. Contour plots of $|\psi_{2D}|^2$ showing vortices in a dipolar BEC ($a = 10a_0$, $a_{\rm dd} = 130a_0$, $N = 10\ 000$, $\lambda = 30$): (a) $\Omega = 0.5$, (b) $\Omega = 0.6$, (c) $\Omega = 0.65$. (d–f) Corresponding phase distributions. Reproduced from [7].

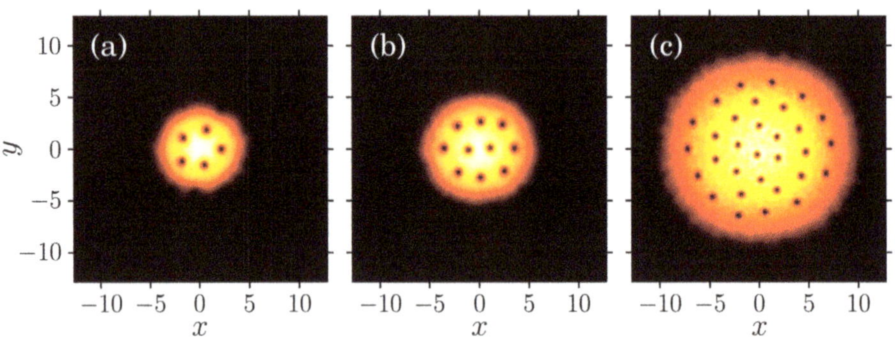

Figure 10.9. Contour plots of $|\psi_{2D}|^2$ showing vortices in a rotating BEC ($a = 20a_0$, $\Omega = 0.6$, $N = 10,\ 000$, $\lambda = 30$): (a) $a_{\rm dd} = 0$, (b) $a_{\rm dd} = 15a_0$, (c) $a_{\rm dd} = 130a_0$. Reproduced from [7].

then to 31 ($a_{\rm dd} = 130a_0$) (figures 10.9(a)–(c)). Critical angular velocities (Ω_c) vary with interaction strengths (table 10.1). Stronger DDIs lower Ω_c (e.g., 0.380 to 0.271 for $a = 0$, $a_{\rm dd} = 15a_0$ to $130a_0$), while contact interactions increase Ω_c (e.g., 0.500 for $a = 20a_0$, $a_{\rm dd} = 0$). These trends highlight the complementary roles of contact and dipolar interactions in stabilizing and nucleating vortices.

Table 10.1. Critical angular velocities (Ω_c) for vortex formation in a dipolar BEC with varying contact (a) and dipolar ($a_{\rm dd}$) interaction strengths. Reproduced from [7].

a/a_0	$a_{\rm dd}/a_0$	Ω_c
0	15	0.380
0	130	0.271
20	0	0.500
20	15	0.390
20	130	0.275
100	0	0.360
100	15	0.320

10.10 Summary and future challenges

This chapter explored quantum vortices in BECs, focussing on their quantized circulation and collective dynamics across various regimes. Single-vortex dynamics was explored in unbounded and trapped systems, revealing precession and stability governed by trap geometry and interactions. Vortex lattices, analyzed at zero and finite temperatures, formed as highly excited states with finite angular momentum and their stability is influenced by rotation and interaction strengths. In the TF regime, vortices exhibited robust structures due to strong interactions, while in the LLL, quantum fluctuations dominated vortex configurations. Phase transitions to highly correlated states highlighted quantum effects, such as vortex melting into strongly correlated phases. Stability analysis using the Bogoliubov equations identified anomalous modes with negative frequency, indicating energetic instability for singly quantized ($s = 1$) vortices in non-rotating condensates, through rotation stabilized above a critical angular velocity. In dipolar BECs, numerical simulations of the GP equation showed that DDI anisotropy enabled vortex formation faster than non-dipolar BECs. Stronger dipolar interactions lowered the critical angular velocity and increased vortex numbers, with deviations from Feynman's rule for large dipolar strengths. Contact interactions enhanced vortex stability and lattice organization. Vortex formation via merging BECs demonstrated alternative nucleation mechanisms driven by interference and dissipation.

The dynamics and stability of vortices in SOC BECs under realistic experimental conditions, including thermal fluctuations and trap imperfections, require advanced theoretical and numerical investigation. In dipolar BECs, modelling strong DDIs in three-dimensional systems poses challenges, particularly for predicting vortex lattice stability and cyclic dynamics observed at high rotation frequencies. The formation of vortex lattices in complex trapping geometries, such as optical lattices or anharmonic traps, remains unexplored, necessitating new analytical and computational approaches to capture lattice defects and pinning effects. The impact of higher-order interactions on vortex stability in three-dimensional BECs warrants further investigation to understand collective excitations beyond the mean-field approximation. Additionally, the interplay of dissipation and quantum fluctuations in vortex nucleation via merging BECs warrants investigation to optimize experimental protocols. These challenges highlight the need for integrated theoretical, numerical, and experimental efforts to advance the understanding of quantum vortices in diverse BEC systems.

References

[1] Svidzinsky A A and Fetter A L 2000 Stability of a vortex in a trapped Bose-Einstein condensate *Phys. Rev. Lett.* **84** 5919–23
[2] Butts D A and Rokhsar D S 1999 Predicted signatures of rotating Bose-Einstein condensates *Nature* **397** 327–9
[3] Yi S and Pu H 2006 Vortex structures in dipolar condensates *Phys. Rev.* A **73** 061602
[4] Tsubota M, Kasamatsu K and Ueda M 2002 Vortex lattice formation in a rotating Bose-Einstein condensate *Phys. Rev.* A **65** 023603
[5] Kasamatsu K, Machida M, Sasa N and Tsubota M 2005 Three-dimensional dynamics of vortex-lattice formation in Bose-Einstein condensates *Phys. Rev.* A **71** 063616
[6] García-Ripoll J J and Pérez-García V M 2001 Vortex bending and tightly packed vortex lattices in Bose-Einstein condensates *Phys. Rev.* A **64** 053611
[7] Kumar R K and Muruganandam P 2012 Vortex dynamics of rotating dipolar Bose-Einstein condensates *J. Phys. B: At. Mol. Opt. Phys.* **45** 215301
[8] Koch T, Lahaye T, Metz J, Fröhlich B, Griesmaier A and Pfau T 2008 Stabilization of a purely dipolar quantum gas against collapse *Nat. Phys.* **4** 218–22

IOP Publishing

An Introduction to Ultracold Atoms with Analytical and Numerical Methods

Paulsamy Muruganandam and Ramaswamy Radha

Chapter 11

Nascent outlook of exciton–polariton condensates

In addition to atomic Bose–Einstein condensates (BECs) formed from alkali atoms, BECs can also be realized with bosonic quasi-particles, which are collective excitations in many-body systems such as solids, liquids, or gases. Quasi-particles, behaving as bosons, possess finite lifetimes and typically have effective masses significantly lower than those of alkali atoms used in atomic BECs. Examples of bosonic quasi-particles include phonons (quantized lattice vibrations), magnons (quantized spin waves), excitons (bound electron–hole pairs), and various polaritons. Polaritons are hybrid quasi-particles resulting from the strong coupling of photons with other excitations. For instance, phonon–polaritons arise from the coupling of infrared photons with optical phonons, magnon–polaritons from the coupling of magnons with light, exciton–polaritons from the coupling of visible photons with excitons, and plasmon–polaritons from the coupling of plasmons with excitons.

Exciton–polaritons, in particular, are bosonic quasi-particles formed as a coherent superposition of excitons and photons, with their composition tunable by experimental conditions. Their effective mass, typically of the order of $10^{-4}m_e$ (where m_e is the electron mass), is dominated by the photonic component, making them significantly lighter than atomic particles. Exciton–polaritons are short-lived, with lifetimes ranging from 1 to 10 ps, reflecting their transient nature. Their low effective mass enables the creation of exciton–polariton BECs at critical temperatures much higher than those required for atomic BECs, often approaching room temperature since the critical temperature for BEC formation is inversely proportional to the effective mass, as described by the relation $T_c \propto (m_{\text{eff}})^{-1}$, where m_{eff} is the effective mass, allowing lighter particles to condense at elevated temperatures.

doi:10.1088/978-0-7503-5447-9ch11

Recent advancements in nanophotonics and optical computing have been driven by the robust interaction between photons and excitons forming composite quasi-particles known as polaritons [1]. Excitons arise as bound states of electron–hole pairs held together by Coulomb interactions. Polaritons, in turn, emerge as hybrid light–matter quasi-particles due to the strong coupling between excitons and photons [1–5]. This strong coupling induces nonlinear interactions among polaritons, establishing a unique platform for exploring nonlinear collective phenomena, such as Bose–Einstein condensation [1, 5, 6], superfluidity [7], and quantized vortices [8, 9].

The exceptionally small effective mass of polaritons, approximately 10^{-5} times that of a free electron, originates primarily from their photonic component. Combined with their large coherence length on the millimetre scale and the pronounced nonlinearity stemming from their excitonic component, polaritons exhibit a strong nonlinear optical response. This response holds significant potential for applications in optical switching [10], optical computing [11], and photonic neural networks [12]. Quantum microcavities hosting polaritons provide a distinctive environment for investigating quantum collective phenomena such as polariton BECs (pBECs) in non-equilibrium systems across a temperature range from a few kelvins to room temperature [4]. The ability to operate at room temperature enhances the practical utility of these systems, enabling the development of novel coherent light sources and optical switching mechanisms leveraging Kerr nonlinearity [10].

As already pointed out, polariton condensates constitute non-equilibrium systems, necessitating continuous replenishment from a reservoir to maintain a steady state. This steady state emerges from a dynamic equilibrium between pumping and decay processes. The primary source of decay stems from the recombination of electrons and holes, which leads to photon emission and the dissipation of excitons. To address this challenge, a polariton reservoir is employed from which polaritons are cooled and injected into the condensate to offset the decay rate. Simultaneously, the reservoir maintains a low polariton density to minimize interactions among them, thereby preserving the stability of the condensate.

Polariton BECs are described by the complex Gross–Pitaevskii (GP) equation, as noted in Keeling and Berloff [9], which captures their non-equilibrium dynamics through terms accounting for pumping and decay. This contrasts with the conventional GP equation, widely applied to equilibrium systems of ultracold atoms.

The nonlinear nature of pBECs, arising from interparticle interactions and the interplay between intrinsic nonlinearity and loss/gain mechanisms, give rise to a diverse array of phenomena. These include bright solitons [13, 14], dark solitons [15, 16], gap solitons [17, 18], oblique dark solitons [19], dissipative solitons [20], spin–orbit-coupled pBECs [21] and dipolar pBECs [22]. These observations underscore the rich physics displayed by the nonlinear and dissipative nature of the pBECs.

11.1 Theoretical model

The polariton system is modelled using an open-dissipative GP equation coupled to a rate equation for the density $n_R(x, t)$ of the uncondensed reservoir of high-energy, near-excitonic polaritons, as detailed in [9, 23–25]. The coupled system is

$$i\hbar\frac{\partial\Psi}{\partial t} = \left[-\frac{\hbar^2}{2m}\frac{\partial^2}{\partial x^2} + V_{\text{imp}} + U(x, t) + \frac{i\hbar}{2}(Rn_{\text{R}} - \gamma_C)\right]\Psi + P_{\text{ad}}(x, t)\Psi, \tag{11.1a}$$

$$\frac{\partial n_{\text{R}}}{\partial t} = P_{\text{incoh}}(x, t) - (\gamma_{\text{R}} + R|\Psi|^2)n_{\text{R}}, \tag{11.1b}$$

where $\Psi(x, t)$ is the polariton condensate wavefunction, and $n_{\text{R}}(x, t)$ is the reservoir density. The effective potential is $U(x, t) = g_{\text{C}}|\Psi|^2 + g_{\text{R}}n_{\text{R}}$, with g_{C} representing the strength of polariton–polariton interactions and g_{R} the condensate–reservoir coupling. The parameter R denotes the stimulated scattering rate from reservoir to condensate, while γ_{C} and γ_{R} are the loss rates for condensate and reservoir polaritons, respectively. The impurity potential is $V_{\text{imp}} = -V_0\delta(x)$ [9], with V_0 as its strength. The effective mass of lower polaritons is m, and $P_{\text{ad}}(x, t)$ and $P_{\text{incoh}}(x, t)$ represent adiabatic and incoherent continuous-wave pumping rates, respectively.

Equations (11.1) describe the dynamics of a one-dimensional polariton condensate interacting with a reservoir, influenced by an impurity and dissipative processes. The GP equation includes terms for kinetic energy, impurity potential, nonlinear interactions, gain/loss mechanisms, and external pumping, while the rate equation balances incoherent pumping with losses from decay and scattering into the condensate.

11.1.1 Adiabatic approximation

To establish the density background for nonlinear excitations, the steady-state of equations (11.1) is determined. When the incoherent pumping rate P_{incoh} exceeds the threshold $P_{\text{th}} = \gamma_{\text{R}}\gamma_{\text{C}}/R$, a stable condensate forms with density $|\Psi|^2 = (P_{\text{incoh}}/\gamma_{\text{C}}) - (\gamma_{\text{R}}/R)$. The homogeneous steady-state densities for the condensate and reservoir are given by

$$n_{\text{C}}^0 = \frac{P_{\text{incoh}} - P_{\text{th}}}{\gamma_{\text{C}}}, \quad n_{\text{R}}^0 = \frac{\gamma_{\text{C}}}{R}, \tag{11.2}$$

with $P_{\text{incoh}} = P_{\text{stat}}$. The equations are nondimensionalized by scaling the wavefunction as $\Psi \to \Psi/\sqrt{n_{\text{C}}^0}$, reservoir density as $\bar{n}_{\text{R}} = n_{\text{R}}/n_{\text{C}}^0$, length by the healing length $l = \hbar^2/(mg_{\text{C}}n_{\text{C}}^0)$, and time by $\tau_0 = \hbar/(g_{\text{C}}n_{\text{C}}^0)$. Substituting these into equation (11.1a), the dimensionless system becomes

$$i\frac{\partial\bar{\Psi}}{\partial\bar{t}} = \left[-\frac{1}{2}\frac{\partial^2}{\partial\bar{x}^2} - \gamma\delta(\bar{x}) + |\bar{\Psi}|^2 + \bar{g}_{\text{R}}\bar{n}_{\text{R}} + \frac{i}{2}(\bar{R}\bar{n}_{\text{R}} - \bar{\gamma}_{\text{C}})\right]\bar{\Psi} + \bar{P}_{\text{ad}}(\bar{x}, \bar{t})\bar{\Psi}, \tag{11.3a}$$

$$\frac{\partial\bar{n}_{\text{R}}}{\partial\bar{t}} = \bar{P}_{\text{incoh}}(\bar{x}, \bar{t}) - (\bar{\gamma}_{\text{R}} + \bar{R}|\bar{\Psi}|^2)\bar{n}_{\text{R}}, \tag{11.3b}$$

where the dimensionless parameters are $\bar{g}_R = g_R/g_C$, $\bar{\gamma}_C = \gamma_C\bar{\gamma}_R/\gamma_R$, $\bar{P}_{ad}(\bar{x}, \bar{t}) = P_{ad}/(g_C n_C^0)$, $\bar{P}_{incoh} = (P_{incoh} - P_{stat})/(g_C n_C^0)$, $\bar{R} = \hbar R/(g_C n_C^0)$, and $\gamma = V_0/(g_C n_C^0)$. Time and space are normalized as $\bar{t} = t/\tau_0, \bar{x} = x/l$.

The steady-state reservoir density deviation is

$$m_R = n_R - n_R^0. \tag{11.4}$$

Rewriting equation (11.3b) in terms of $\bar{m}_R = m_R/n_C^0$, one obtains

$$\frac{\partial \bar{m}_R}{\partial \bar{t}} = \bar{P}_{incoh}(\bar{x}) + \bar{\gamma}_C(1 - |\bar{\Psi}|^2) - \bar{\gamma}_R\bar{m}_R - \bar{R}|\bar{\Psi}|^2\bar{m}_R. \tag{11.5}$$

The simplified dimensionless system is given by

$$\mathrm{i}\frac{\partial\psi}{\partial\bar{t}} + \frac{1}{2}\frac{\partial^2\psi}{\partial\bar{x}^2} + |\psi|^2\psi + \gamma\delta(\bar{x})\psi = 2|\psi|^2\psi + \bar{P}_{ad}(\bar{x})\psi + \left(\bar{g}_R\bar{m}_R + \frac{i}{2}\bar{R}\bar{m}_R\right)\psi, \tag{11.6a}$$

$$\frac{\partial \bar{m}_R}{\partial \bar{t}} = \bar{P}_{incoh}(\bar{x}) + \bar{\gamma}_C(1 - |\psi|^2) - \bar{\gamma}_R\bar{m}_R - \bar{R}|\psi|^2\bar{m}_R, \tag{11.6b}$$

where $\psi = \bar{\Psi}$, and variables are in dimensionless units. In the fast reservoir limit, the reservoir density in equation (11.6b) is

$$\bar{m}_R = \frac{\bar{P}_{incoh}(\bar{x})}{\bar{\gamma}_R} + \frac{\bar{\gamma}_C}{\bar{\gamma}_R}(1 - |\psi|^2), \tag{11.7}$$

with $\bar{P}_{incoh}(\bar{x}) = \bar{P}^c_{incoh} + \bar{P}^v_{incoh}(\bar{x})$, comprising a constant $\bar{P}^c_{incoh}$ and a spatially varying term $\bar{P}^v_{incoh}(\bar{x})$. Substituting equation (11.7) into equation (11.6a) and following [24] yields

$$\mathrm{i}\frac{\partial\psi}{\partial\bar{t}} + \frac{1}{2}\frac{\partial^2\psi}{\partial\bar{x}^2} + |\psi|^2\psi + \gamma\delta(\bar{x})\psi = \frac{\mathrm{i}}{2}[P(\bar{x}) - \sigma - \chi|\psi|^2]\psi, \tag{11.8}$$

where $P(\bar{x}) = \bar{R}\bar{P}^v_{incoh}(\bar{x})/\bar{\gamma}_R$ assumes a Gaussian profile $P(\bar{x}) = P_0 e^{-\bar{x}^2/\omega^2}$ with amplitude P_0 and width ω, and parameters are $\sigma = -(\bar{P}^c_{incoh} + \bar{\gamma}_C)\bar{R}/\bar{\gamma}_R$, $\chi = \bar{R}\bar{\gamma}_C/\bar{\gamma}_R$, representing loss rate and gain saturation, respectively. The adiabatic pumping is constrained as $\bar{P}_{ad}(\bar{x}) = -2|\psi|^2 - \bar{g}_R\bar{m}_R$. This framework enables analysis of bright soliton interactions with the impurity governed by γ.

11.2 Variational approach

The dissipative effects in equation (11.8) are modelled by $D(\psi) = \frac{\mathrm{i}}{2}[P(\bar{x}) - \sigma - \chi|\psi|^2]\psi$. Without the impurity ($\gamma = 0$), equation (11.8) supports a bright soliton solution. A Lagrangian variational approach within perturbation theory analyzes these effects. For the unperturbed case ($D(\psi) = 0$), equation (11.8) reduces to a nonlinear Schrödinger equation, or a linear Schrödinger equation without nonlinearity. A trial wavefunction is taken as

$$\psi(\bar{x}, \bar{t}) = [\eta(\bar{t})\mathrm{sech}[\eta(\bar{t})(\bar{x} - z(\bar{t}))]e^{\mathrm{i}\kappa(\bar{t})\bar{x}} + a(\bar{t})\sqrt{\lambda(\bar{t})}e^{-\lambda(\bar{t})|\bar{x}|+\mathrm{i}\varphi(\bar{t})}]e^{\mathrm{i}\phi(\bar{t})}, \tag{11.9}$$

with time-dependent parameters $\eta(\bar{t})$ (soliton amplitude), $z(\bar{t})$ (centre position), $\phi(\bar{t})$ (global phase), $\kappa(\bar{t})$ (wavenumber), $a(\bar{t})$ and $\lambda(\bar{t})$ (amplitude and decay rate of the impurity-induced component), and $\varphi(\bar{t})$ (relative phase).

The Lagrangian approach assumes that the soliton and impurity-induced components retain their functional forms, with slowly varying parameters. The time evolution of parameters $q_i = \{\eta, z, \phi, \kappa, a, \lambda, \varphi\}$ is governed by the Euler–Lagrange equations for dissipative systems [26–29]

$$\frac{\partial L}{\partial q_i} - \frac{\mathrm{d}}{\mathrm{d}\bar{t}}\left(\frac{\partial L}{\partial \dot{q}_i}\right) = 2\mathrm{Re}\left(\int_{-\infty}^{\infty} D^*(\psi)\frac{\partial \psi}{\partial q_i}\,\mathrm{d}\bar{x}\right), \tag{11.10}$$

where $\dot{q}_i = \mathrm{d}q_i/\mathrm{d}\bar{t}$ and $L = \int_{-\infty}^{\infty} \mathcal{L}\,\mathrm{d}\bar{x}$ is the average Lagrangian for the unperturbed equation (11.8) ($D(\psi) = 0$) with Lagrangian density being given by

$$\mathcal{L} = \frac{\mathrm{i}}{2}\left(\psi^*\frac{\partial \psi}{\partial \bar{t}} - \psi\frac{\partial \psi^*}{\partial \bar{t}}\right) - \frac{1}{2}\left|\frac{\partial \psi}{\partial \bar{x}}\right|^2 + \frac{1}{2}|\psi|^4 + \gamma|\psi|^2\delta(\bar{x}). \tag{11.11}$$

Substituting equation (11.9) into equation (11.10), the averaged Lagrangian becomes

$$\begin{aligned} L = &- 2\eta\dot{\phi} - 2\dot{\kappa}z - a^2(\dot{\phi} + \dot{\varphi}) + \frac{\eta^3}{3} - \kappa^2\eta - \frac{a^2\lambda^2}{2} + \gamma a^2\lambda + \gamma\eta^2\mathrm{sech}^2(z) \\ &+ 2\gamma\eta a\sqrt{\lambda}\,\mathrm{sech}(z)\cos(\varphi) + \mathcal{O}(a^4), \end{aligned} \tag{11.12}$$

neglecting higher-order terms $\mathcal{O}(a^4)$ and following [30] which ignores direct soliton-impurity mode interactions except through energy exchange at the defect. This approximation is validated by comparing analytical results from equation (11.12) with numerical simulations of equation (11.8).

Applying equation (11.12) to equation (11.10), the equations of motion for the parameters are

$$\dot{\eta} = \frac{1}{12}[(6a^2 + 12\eta)(P_0 - \sigma) - 12a\dot{a} - 8\chi\eta^3 - 3\chi a^4\lambda], \tag{11.13a}$$

$$\dot{z} = \frac{1}{3}[3z(P_0 - \sigma) - 2\chi z\eta^2 + 3\eta\kappa], \tag{11.13b}$$

$$\dot{a} = \frac{1}{4}[2a(P_0 - \sigma) + 4\gamma\mathrm{sech}(z)\sin(\varphi)\eta\sqrt{\lambda} - \chi a^3\lambda], \tag{11.13c}$$

$$\dot{\kappa} = -\gamma\mathrm{sech}^2(z)\tanh(z)\eta^2 - \gamma a\cos(\varphi)\mathrm{sech}(z)\tanh(z)\eta\sqrt{\lambda}, \tag{11.13d}$$

$$\begin{aligned} \dot{\varphi} = &- \gamma\mathrm{sech}^2(z)\eta - \frac{\eta^2}{2} + \frac{\kappa^2}{2} - \gamma a\cos(\varphi)\mathrm{sech}(z)\sqrt{\lambda} \\ &+ \gamma a^{-1}\cos(\varphi)\mathrm{sech}(z)\eta\sqrt{\lambda} + \gamma\lambda - \frac{\lambda^2}{2}, \end{aligned} \tag{11.13e}$$

$$\dot{\phi} = \gamma \text{sech}^2(z)\eta + \frac{\eta^2 - \kappa^2}{2} + \gamma a \cos(\varphi)\text{sech}(z)\sqrt{\lambda}, \tag{11.13f}$$

$$0 = a(\gamma - \lambda) + \gamma\eta\lambda^{-1/2}\cos(\varphi)\text{sech}(z). \tag{11.13g}$$

11.3 Collisions of bright solitons with an unexcited impurity

The equation for $\dot{\phi}$, derived from varying η, governs the phase evolution but is less critical to soliton dynamics. The primary equations describe the evolution of the soliton's position z, amplitude η, and the localized mode amplitude $a\sqrt{\lambda}$, with $\dot{a}$ characterizing soliton excitations.

Figures 11.1, 11.2, and 11.3 illustrate bright soliton interactions with impurities of strengths $\gamma = 0.02$, 0.05, and 0.14, corresponding to effective impurity masses $m_{\text{eff}} = 1.04$, 1.10, and 1.28, respectively [30]. The system of equations (11.13) is solved using a fifth-order Runge–Kutta method, with the algebraic equation (11.13g) handled iteratively via root-finding.

There exist three interaction scenarios which depend on γ:

(i) **Transmission**: For a weak impurity ($\gamma = 0.02$, $m_{\text{eff}} = 1.04$), the soliton passes through with minimal disturbance in the absence of dissipation ($P_0 = \sigma = \chi = 0$), as shown in figure 11.1(a). The impurity mode (dashed

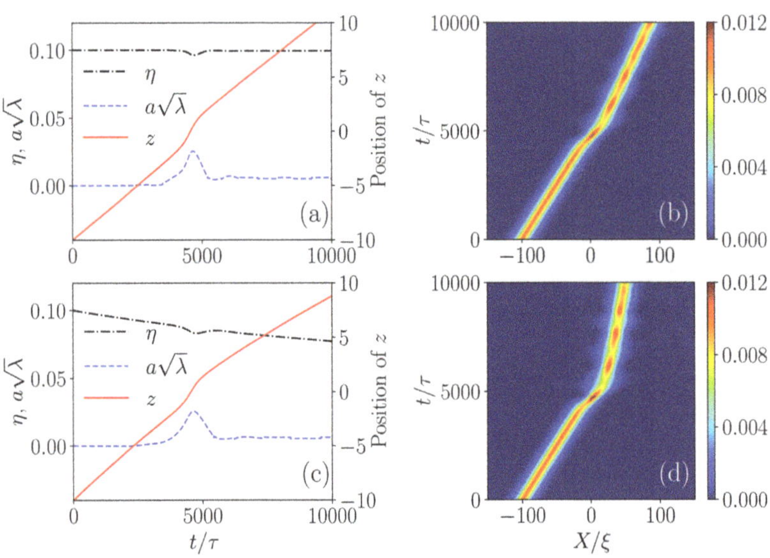

Figure 11.1. Plots depicting the transmission of a bright soliton with an initial amplitude $\eta = 0.1$ through an impurity of strength $\gamma = 0.02$ [25]. The left column shows results from solving the variational equations (11.13a) using the Runge–Kutta method with adaptive step size control, while the right column presents numerical simulations of dissipative GP equation (11.14) via the split-step Crank–Nicolson method [31]. Panels (a) and (b) illustrate amplitude and position without dissipation; panels (c) and (d) include dissipative effects, demonstrating reduced amplitude after collision. Reproduced from [25]. CC BY 4.0.

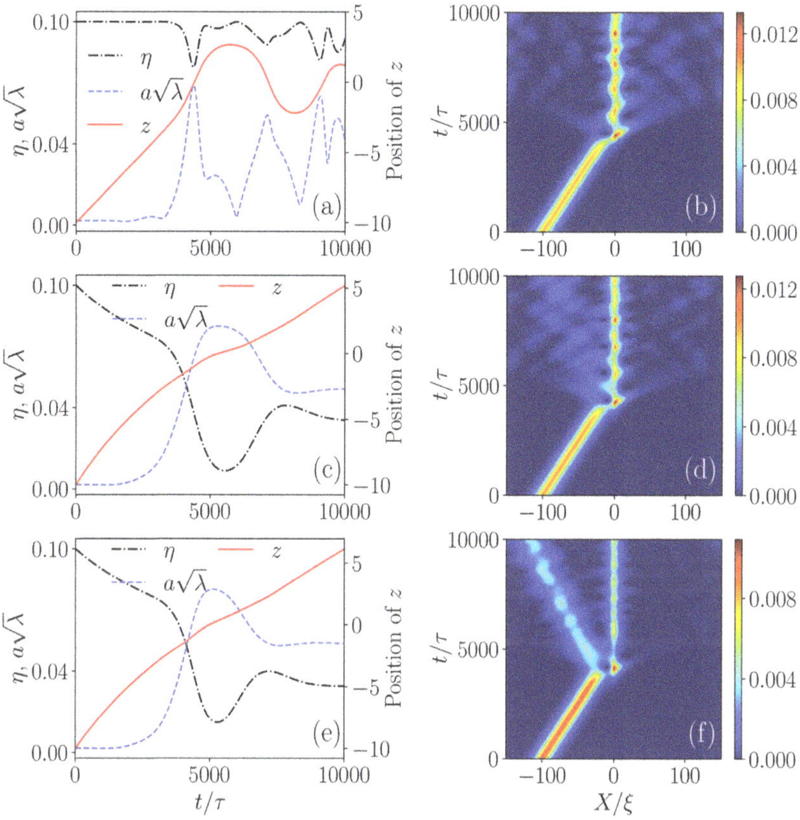

Figure 11.2. Plots depicting the trapping of a bright soliton with an initial amplitude $\eta = 0.1$ through an impurity of strength $\gamma = 0.05$ [25]. The left column shows results from solving the variational equations (11.13a) using the Runge–Kutta method with adaptive step size control, while the right column presents numerical simulations of dissipative GP equation (11.14) via the split-step Crank–Nicolson method [31]. Panels (a) and (b) show trapping without dissipation; (c), (d), (e), and (f) show reduced oscillations with dissipation. Reproduced from [25]. CC BY 4.0.

blue lines) is excited during collision but diminishes afterwards. Numerical simulations using the Crank–Nicolson method for equation (11.8) (figure 11.1(b)) confirm this observation. With dissipation ($D(\psi) \neq 0$), the soliton amplitude decreases post-collision (figure 11.1(c)), consistent with simulations (figure 11.1(d)) while higher dissipation accelerates amplitude decay.

(ii) **Trapping**: For a moderate impurity ($\gamma = 0.05$, $m_{\text{eff}} = 1.10$), the soliton is trapped without dissipation, as shown in figures 11.2(a) and (b). The impurity mode oscillates strongly, and the soliton amplitude drops significantly [figure 11.2(c)], verified by simulations (figure 11.2(d)). With dissipation (figures 11.2(c)–(f)), trapping persists, but oscillations diminish, indicating that dissipation suppresses low-energy excitations from soliton-impurity collisions.

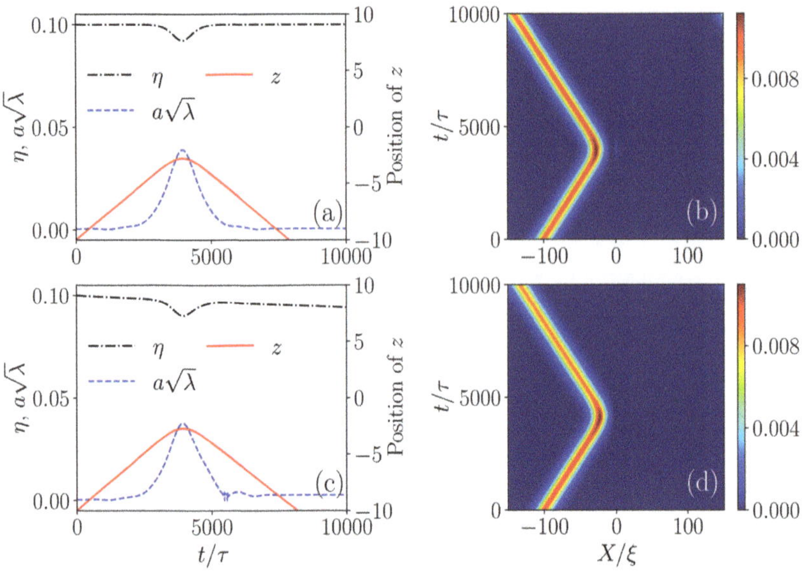

Figure 11.3. Plots depicting the reflection of a bright soliton with an initial amplitude $\eta = 0.1$ through an impurity of strength $\gamma = 0.14$ [25]. The left column shows results from solving the variational equations (11.13a) using the Runge–Kutta method with adaptive step size control, while the right column presents numerical simulations of dissipative GP equation (11.14) via the split-step Crank–Nicolson method [31]. Panels (a) and (b) show reflection without dissipation; (c) and (d) show minimal amplitude reduction with dissipation. Reproduced from [25].

(iii) **Reflection**: For a strong impurity ($\gamma = 0.14$, $m_{\text{eff}} = 1.28$), the soliton is reflected without dissipation (figures 11.3(a) and (b)). Dissipation has minimal impact, causing slight amplitude reduction (figures 11.3(c) and (d)), unlike transmission and trapping, due to the heavier impurity reducing interaction intensity.

11.4 Stability window of trapless polariton Bose–Einstein condensates

To formulate the complex GP equation for polariton Bose–Einstein condensates, non-resonant pumping is considered, introducing stimulated scattering into the condensate, described by $\partial_t \psi \mid_{\text{gain}} = \gamma\psi$. Particle decay, or loss, is modelled by $\partial_t \psi \mid_{\text{loss}} = -\kappa\psi$. If gain exceeds loss, the condensate grows indefinitely; if loss exceeds gain, it vanishes. In most non-resonantly pumped systems, the gain is saturable, aiming to equilibrate the condensate with an external particle density. The density-dependent gain rate, $\partial_t \rho \mid_{\text{gain}} = (\gamma - \Gamma\rho)\rho$, leads to an equilibrium density $\rho = \gamma/\Gamma$.

11.4.1 Theoretical model

Under these conditions, non-resonantly pumped pBECs are modelled by a generalized complex GP equation, capturing the dynamics of polariton condensates with ground-state populations replenished from a reservoir of incoherent excitons. The classical field ψ, the macroscopic wavefunction or order parameter, replaces the field operator, neglecting quantum and thermal fluctuations of the non-condensate fraction. Assuming a weakly interacting polariton gas at low temperatures, the condensate dynamics are described by a GP type equation, excluding microscopic polariton physics [9]

$$i\hbar\frac{\partial\psi(r,t)}{\partial t} = \left[-\frac{\hbar^2}{2m_p}\nabla^2 + V_{3\,D} + i\Gamma_{\text{eff}}\right]\psi(r,t) + [g|\psi(r,t)|^2 - ig_1|\psi(r,t)|^2]\psi(r,t), \tag{11.14}$$

where the harmonic trapping potential is $V_{3D} = \frac{1}{2}m_p\omega^2 r^2$, with ω as the angular frequency, m_p the polariton mass, and r the radial coordinate. The parameter g represents the interaction energy, and $\Gamma_{\text{eff}} = \gamma - \kappa$ is the net linear gain, balancing stimulated scattering into the condensate and polariton decay out of the cavity. Equation (11.14) satisfies the normalization condition

$$N = \int |\psi|^2\, \mathrm{d}r, \tag{11.15}$$

where N is the total number of polaritons, and $\mathrm{d}r$ is the volume element. Due to gain and loss terms in equation (11.14), N is not conserved.

For convenience, equation (11.14) is expressed in dimensionless form using the transformations $\bar{r} = r/l$, $\bar{t} = t\omega$, and $\bar{\psi} = \psi/\sqrt{\hbar\omega}$, where $l = \sqrt{\hbar/(m_p\omega)}$ is the harmonic oscillator length. Dropping overbars, the dimensionless equation becomes

$$i\frac{\partial\psi(r,t)}{\partial t} = \left[-\frac{1}{2}\nabla^2 + \frac{\Omega^2 r^2}{2} + i\gamma_{\text{eff}}\right]\psi(r,t) + [g|\psi(r,t)|^2 - ig_1|\psi(r,t)|^2]\psi(r,t), \tag{11.16}$$

where $\gamma_{\text{eff}} = \Gamma_{\text{eff}}/\hbar\omega$. Experimental parameters from references [1, 6] yield $\gamma_{\text{eff}} = 0 \sim 20\ \hbar\omega$, $g = 0 \sim 50\ \hbar\omega$, and $g_1 = 0 \sim 50\ \hbar\omega$.

For strong radial confinement, polariton motion is free axially but restricted radially, allowing the dynamics to be described by a one-dimensional GP equation

$$\left[i\frac{\partial}{\partial t} + \frac{1}{2}\frac{\partial^2}{\partial x^2} - V_{1\text{D}} - g|\psi(x,t)|^2\right]\psi(x,t) = i[\gamma_{\text{eff}} - g_1|\psi(x,t)|^2]\psi(x,t) \equiv W \tag{11.17}$$

where $V_{1\text{D}} = \frac{1}{2}d(t)\Omega^2 x^2$, with Ω^2 as the trap frequency and $d(t)$ the trap strength, varying from 1 to 0 as the trap is removed.

11.4.2 Variational results

To derive the governing equations for system parameters, a variational approach is employed using a Gaussian trial wavefunction

$$\psi(x,t) = A(t)\exp\left[-\frac{x^2}{2R(t)^2} + i\beta(t)x^2 + i\alpha(t)\right], \tag{11.18}$$

where $A(t)$, $R(t)$, $\beta(t)$, and $\alpha(t)$ are the time-dependent amplitude, width, chirp, and phase, respectively. As dissipative and amplifying terms do not affect the centre-of-mass coordinate in the Gaussian ansatz, they are neglected here. The variational method is applied to the averaged Lagrangian of the conservative system

$$L = \int \mathcal{L} \, \mathrm{d}x, \tag{11.19}$$

with the Lagrangian density taking the form

$$\mathcal{L}(x, t) = \frac{\mathrm{i}}{2}(\dot{\psi}\psi^* - \dot{\psi}^*\psi) - \frac{1}{2}|\nabla \psi|^2 - V_{1\mathrm{D}}|\psi|^2 - \frac{1}{2}g|\psi|^4. \tag{11.20}$$

Substituting equation (11.18) into equations (11.20) and (11.19) yields the averaged Lagrangian

$$L = -\frac{\sqrt{\pi}A(t)^2R(t)}{2}\left[\frac{1}{R(t)^2} + \frac{g}{\sqrt{2}}A(t)^2 + 2\dot{\alpha} + \frac{d(t)\Omega^2}{2}R(t)^2 + R(t)^2\left(\beta(t)^2 + \frac{\dot{\beta}}{2}\right)\right]. \tag{11.21}$$

A dissipative term $\mathcal{L}_W$, satisfying $\delta\mathcal{L}_W/\delta\psi^* = -W$, where W is given by equation (11.17), is added to equation (11.20). Applying the Euler–Lagrange equations to $\mathcal{L}' = \mathcal{L} + \mathcal{L}_W$ with respect to ψ^*, one obtains

$$\left[\frac{\partial\mathcal{L}'}{\partial\psi^*} - \frac{\mathrm{d}}{\mathrm{d}t}\frac{\partial\mathcal{L}'}{\partial\dot{\psi}^*}\right] = \left[\frac{\partial\mathcal{L}}{\partial\psi^*} - \frac{\mathrm{d}}{\mathrm{d}t}\frac{\partial\mathcal{L}}{\partial\dot{\psi}^*}\right] - W(\psi, \psi^*) = 0, \tag{11.22}$$

which recovers equation (11.17) (the conjugate equation follows similarly).

The variational principle is

$$\delta\int_0^t L' \, \mathrm{d}t = \delta\int_0^t (L + L_W) \, \mathrm{d}t = 0, \tag{11.23}$$

where $L_W = \int \mathcal{L}_W \, \mathrm{d}x$. For a small variation $\delta\eta$ in a variational parameter η, defined by

$$f(\eta + \delta\eta) = f(\eta) + \delta\eta\frac{\partial f}{\partial\eta}, \tag{11.24}$$

where $f = L$ or L_W, a system of equations for the variational parameters $\eta_i \in \{A, R, \beta, \alpha\}$ is obtained as

$$\frac{\partial L}{\partial\eta_i} - \frac{\mathrm{d}}{\mathrm{d}t}\frac{\partial L}{\partial\dot{\eta}_i} = \int \mathrm{d}x\left[W\frac{\partial\psi^*}{\partial\eta_i} + W^*\frac{\partial\psi}{\partial\eta_i}\right]. \tag{11.25}$$

Substituting equations (11.18) and (11.21) into equation (11.25) yields the coupled nonlinear ordinary differential equations [32]

$$\dot{A} = (\gamma_{\mathrm{eff}} - \beta)A - \frac{5g_1}{4\sqrt{2}}A^3, \tag{11.26a}$$

$$\dot{R} = 2\beta R + \frac{g_1}{2\sqrt{2}}A^2R, \tag{11.26b}$$

$$\dot{\beta} = -\frac{d(t)\Omega^2}{2} + \frac{1}{2R^4} + \frac{gA^2}{2\sqrt{2}R^2} - 2\beta^2, \tag{11.26c}$$

$$\dot{\alpha} = -\frac{1}{2R^2} - \frac{5g}{4\sqrt{2}}A^2. \tag{11.26d}$$

The normalization condition (11.15) gives:

$$\dot{N} = 2\sqrt{\pi}\gamma_{\text{eff}}A^2R - \sqrt{2\pi}\,g_1A^4R. \tag{11.26e}$$

This equation reflects the non-equilibrium nature of the system, where the particle number varies due to reservoir interactions.

Multiplying equation (11.26a) by $2A$, equation (11.26b) by $2R$, and defining $X = A^2$, $Y = R^2$, the system becomes

$$\dot{X} = 2(\gamma_{\text{eff}} - \beta)X - \frac{5g_1}{2\sqrt{2}}X^2, \tag{11.27a}$$

$$\dot{Y} = 4\beta Y + \frac{g_1}{\sqrt{2}}XY, \tag{11.27b}$$

$$\dot{\beta} = -\frac{d(t)\Omega^2}{2} + \frac{1}{2Y^2} + \frac{g}{2\sqrt{2}}\frac{X}{Y} - 2\beta^2. \tag{11.27c}$$

A stationary solution is

$$X^{*} = \frac{\sqrt{2}\gamma_{\text{eff}}}{g_1}, \tag{11.28a}$$

$$Y^{*} = \frac{2g\gamma_{\text{eff}} \pm 2\sqrt{g_1^2\gamma_{\text{eff}}^2 + g^2\gamma_{\text{eff}}^2 + 4g_1^2\Omega}}{g_1\gamma_{\text{eff}}^2 + 4g_1\Omega}, \tag{11.28b}$$

$$\beta^{*} = -\frac{\gamma_{\text{eff}}}{4}. \tag{11.28c}$$

Since $X = A^2$ and $Y = R^2$, both X^* and Y^* must be positive. Linear stability analysis of the equilibrium point (X^*, Y^*, β^*) identifies stable regions in the $\gamma_{\text{eff}} - g_1$ plane.

Figure 11.4 shows stability regions for (X^*, Y^*, β^*) (with $X^*>0$, $Y^*>0$) in the $\gamma_{\text{eff}} - g_1$ plane for a repulsive pBEC with $g = 5$ under varying trap strengths. Figure 11.5 illustrates the time evolution of amplitude A and width R in stable and unstable regimes for $\Omega = 0.01$, $g = 5$, $g_1 = 10^{-2}$, with $\gamma_{\text{eff}} = 10^{-3}$ (stable, dashed-blue) and $\gamma_{\text{eff}} = 10^{-2}$ (unstable, solid-red).

Figure 11.6 depicts the amplitude A and width R for trapless ($d(t) = 0$) attractive pBECs: (a) $g = -10.0$, $g_1 = 0.0$, $\gamma_{\text{eff}} = 0.0$; (b) $g = -5.0$, $g_1 = 0.0$, $\gamma_{\text{eff}} = 0.0$; (c) $g = -10.0$, $g_1 = 0.0142$, $\gamma_{\text{eff}} = 0.01$.

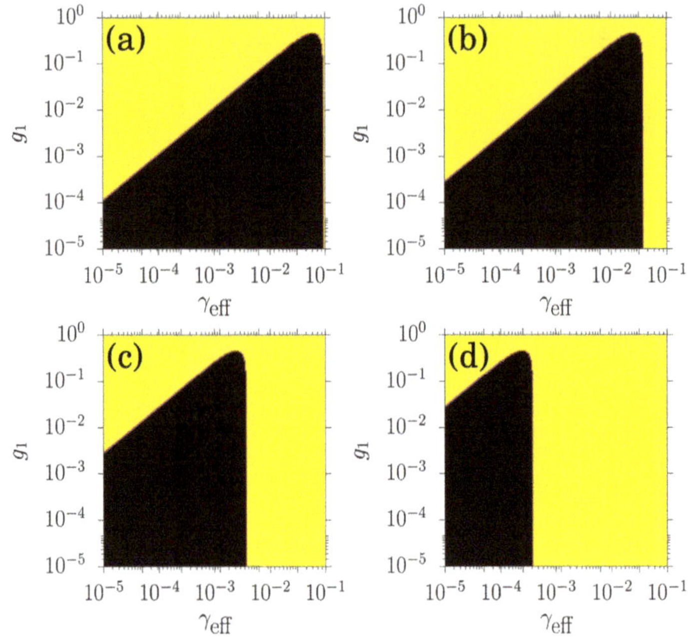

Figure 11.4. Stability regions (black) where the equilibrium point (X^*, Y^*, β^*) is stable, with $g = 5$ and $d(t) = 1$: (a) $\Omega = 0.25$, (b) $\Omega = 0.1$, (c) $\Omega = 0.01$, (d) $\Omega = 0.001$. Reprinted figure with permission from [33], Copyright (2022) by the American Physical Society.

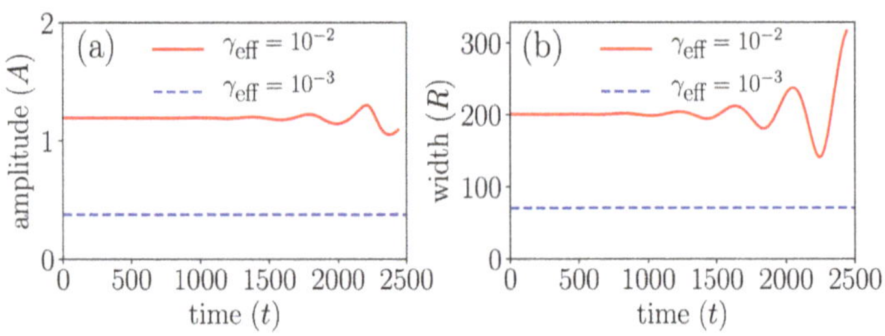

Figure 11.5. Time evolution of: (a) amplitude A, (c) width R, in stable and unstable regimes with $\Omega = 0.01$, $g = 5$, $g_1 = 10^{-2}$: (i) $\gamma_{\text{eff}} = 10^{-3}$ (stable, dashed-blue), (ii) $\gamma_{\text{eff}} = 10^{-2}$ (unstable, solid-red).Reprinted with permission from [33], Copyright (2022) by the American Physical Society.

The stability domain shrinks as trap strength decreases, as evident in figure 11.4. An unstable equilibrium, such as $(g_1, \gamma_{\text{eff}}) = (10^{-2}, 10^{-4})$ at $\Omega = 0.25$ in figure 11.4 (a), becomes stable at $\Omega = 0.001$ in figure 11.4(d). The maximum stable nonlinear loss/gain g_1 remains constant, independent of the trap strength. Stability is verified by numerically solving equations (11.27) using the fourth-order Runge–Kutta method with initial conditions near the equilibrium point. Figures 11.5(a) and (c) show sustained amplitude and width in the stable regime ($\gamma_{\text{eff}} = 10^{-3}$) versus oscillations in the unstable regime ($\gamma_{\text{eff}} = 10^{-2}$), confirming stability in the repulsive regime.

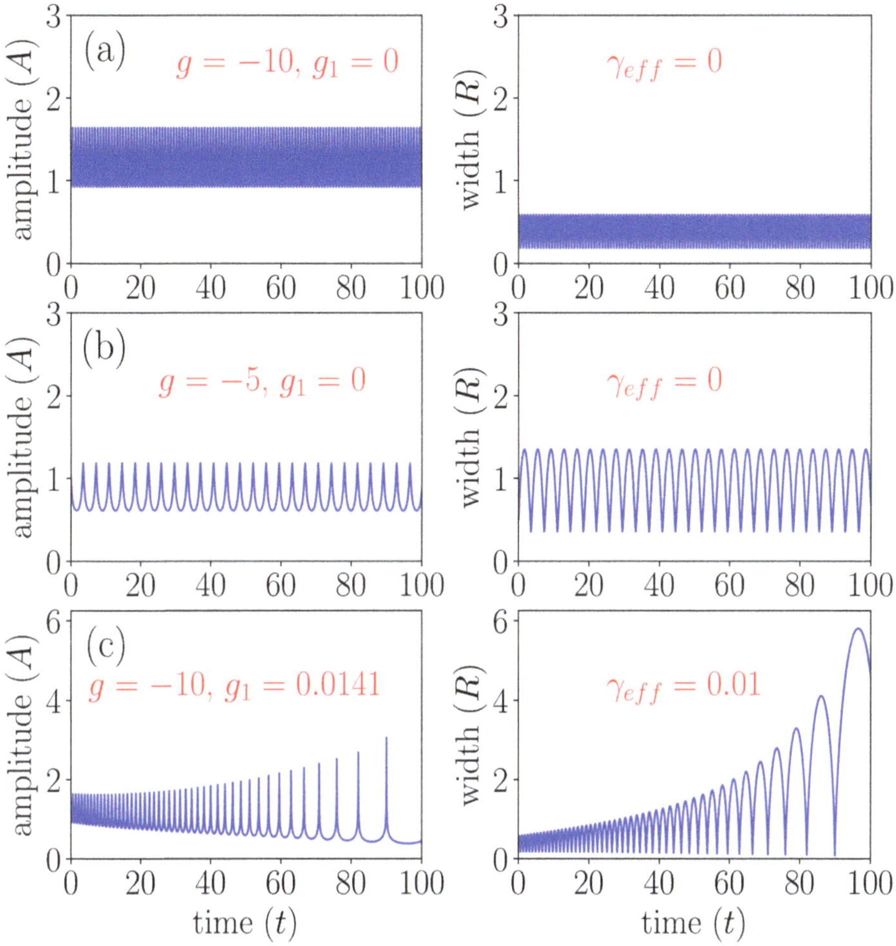

Figure 11.6. Amplitude A (left) and width R (right) for trapless ($d(t) = 0$) attractive pBECs: (a) $g = -10.0$, $g_1 = 0.0$, $\gamma_{\text{eff}} = 0.0$; (b) $g = -5.0$, $g_1 = 0.0$, $\gamma_{\text{eff}} = 0.0$; (c) $g = -10.0$, $g_1 = 0.0142$, $\gamma_{\text{eff}} = 0.01$. Reprinted with permission from [33], Copyright (2022) by the American Physical Society.

11.4.2.1 Attractive polariton condensates

The stability of trapless attractive pBECs is analyzed by solving equations (11.26a)–(11.26c) using the fourth-order Runge–Kutta method. Figure 11.6 shows the dynamics of amplitude A (left) and width R (right). For $g = -10.0$, $g_1 = 0$, $\gamma_{\text{eff}} = 0$, $d(t) = 0$ (panel a), the amplitude stabilizes when the interaction strength is reduced to $g = -5.0$ (panel b), with constant width. Reintroducing $g = -10.0$ with $g_1 = 0.0142$, $\gamma_{\text{eff}} = 0.01$ (panel c) induces small fluctuations in both amplitude and width, indicating instability due to cubic nonlinearity reinforced by nonlinear loss/gain terms.

11.4.2.2 Repulsive polariton condensates

The stability of repulsive pBECs with a weak trap ($\Omega = 0.01$, $d(t) = 1$) is examined. Figure 11.7 shows amplitude A (left) and width R (right): (a) $g = 5.0$, $g_1 = 0.0$,

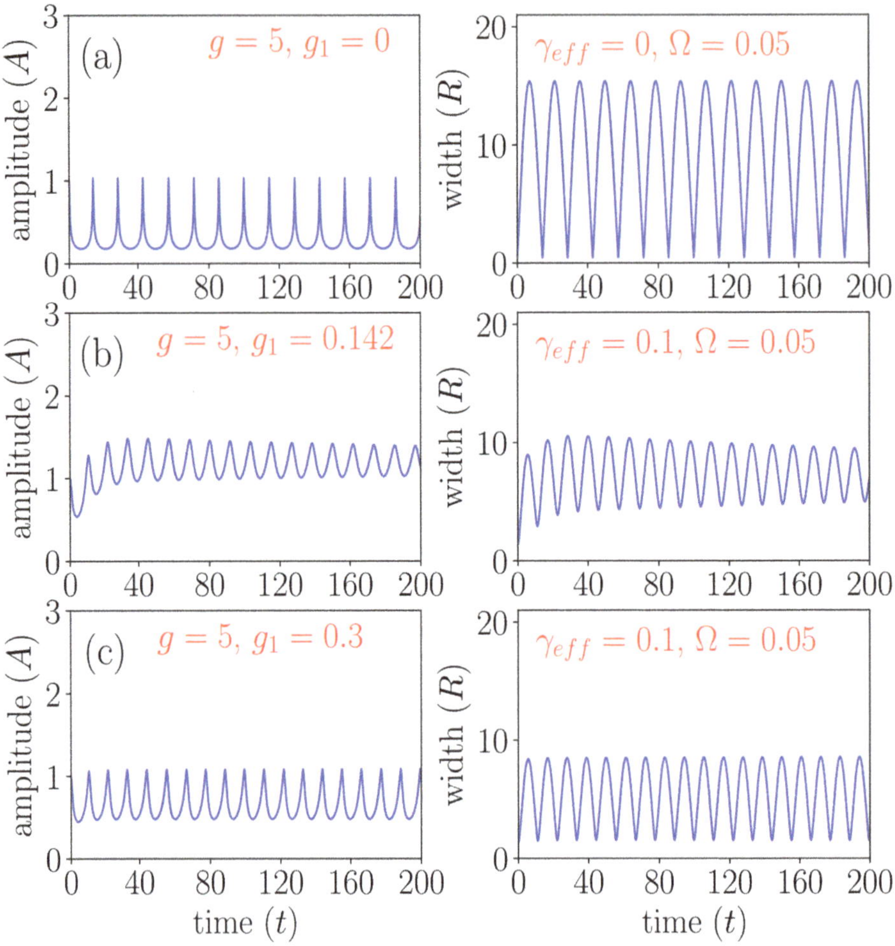

Figure 11.7. Amplitude A (left) and width R (right) for repulsive pBECs with weak trap ($\Omega = 0.01, d(t) = 1$): (a) $g = 5.0$, $g_1 = 0.0$, $\gamma_{\text{eff}} = 0.0$; (b) $g = 5.0$, $g_1 = 0.142$, $\gamma_{\text{eff}} = 0.1$; (c) $g = 5.0$, $g_1 = 0.3$, $\gamma_{\text{eff}} = 0.1$. Reprinted with permission from [33], Copyright (2022) by the American Physical Society.

$\gamma_{\text{eff}} = 0.0$, with stable dynamics up to $t = 200$; (b) $g = 5.0$, $g_1 = 0.142$, $\gamma_{\text{eff}} = 0.1$, showing slight variations; (c) $g = 5.0$, $g_1 = 0.3$, $\gamma_{\text{eff}} = 0.1$, where increased nonlinear loss/gain stabilizes both amplitude and width. Stabilization is achievable in the repulsive domain, unlike the inevitable instability in attractive pBECs (figure 11.6(c)), attributed to nonlinear loss/gain reinforcing cubic nonlinearity in non-equilibrium systems.

11.4.3 Numerical results

The time-dependent GP equation (11.17) is solved numerically using the split-step Crank–Nicolson method [31, 34]. Stabilizing a trapless system with nonlinearity above a critical value requires specific initial conditions. Choosing a system size close to the desired value stabilizes the system after a finite time, a technique applicable to

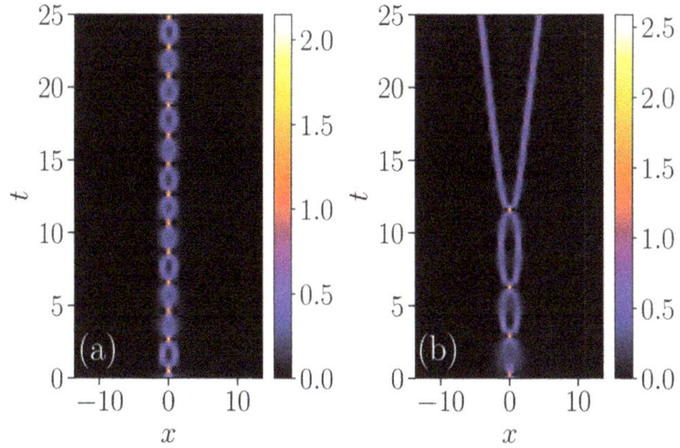

Figure 11.8. Density profiles for attractive pBECs: (a) $g = -10.0$, $g_1 = 0.0$, $\gamma_{\text{eff}} = 0.0$; (b) $g = -10.0$, $g_1 = 0.0142$, $\gamma_{\text{eff}} = 0.01$. Reprinted with permission from [33], Copyright (2022) by the American Physical Society.

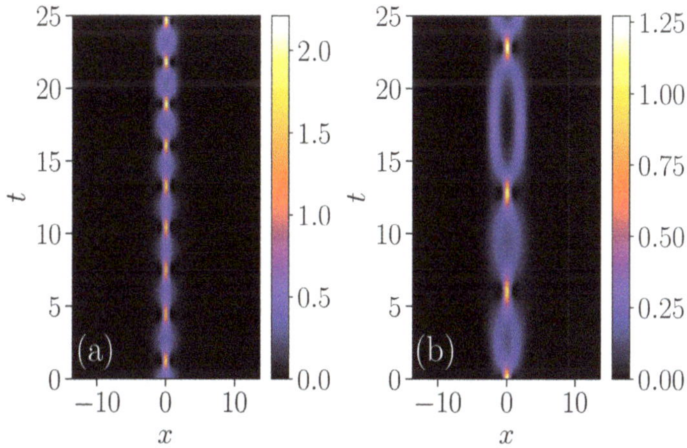

Figure 11.9. Density profiles for attractive pBECs: (a) $g = -5.0$, $g_1 = 0.01$, $\gamma_{\text{eff}} = 0.01$; (b) $g = -5.0$, $g_1 = 0.03$, $\gamma_{\text{eff}} = 0.01$. Reprinted with permission from [33], Copyright (2022) by the American Physical Society.

experimental setups. The external trap is removed while increasing nonlinearity to achieve stability. During time iteration, nonlinear coefficients increase as $g(t) = f(t)g$, where $f(t) = t/\tau$ for $0 \leqslant t \leqslant \tau$, and $f(t) = 1$ for $t \geqslant \tau$. The trap is removed by varying $d(t) = 1 - f(t)$ from 1 to 0, achieving full nonlinearity g at time τ.

Density profiles for attractive pBECs are shown in figures 11.8–11.10: (a) $g = -10.0$, $g_1 = 0.0$, $\gamma_{\text{eff}} = 0.0$; (b) $g = -10.0$, $g_1 = 0.0142$, $\gamma_{\text{eff}} = 0.01$; (a) $g = -5.0$, $g_1 = 0.01$, $\gamma_{\text{eff}} = 0.01$; (b) $g = -5.0$, $g_1 = 0.03$, $\gamma_{\text{eff}} = 0.01$; (a) $g = -5.0$, $g_1 = 0.03$, $\gamma_{\text{eff}} = -0.01$; (b) $g = -5.0$, $g_1 = 0.1$, $\gamma_{\text{eff}} = -0.01$. Figure 11.8(a) exhibits a 'beating effect', with periodic superposition of condensates, controllable by linear and nonlinear loss/gain terms (figures 11.8(b), 11.9, 11.10).

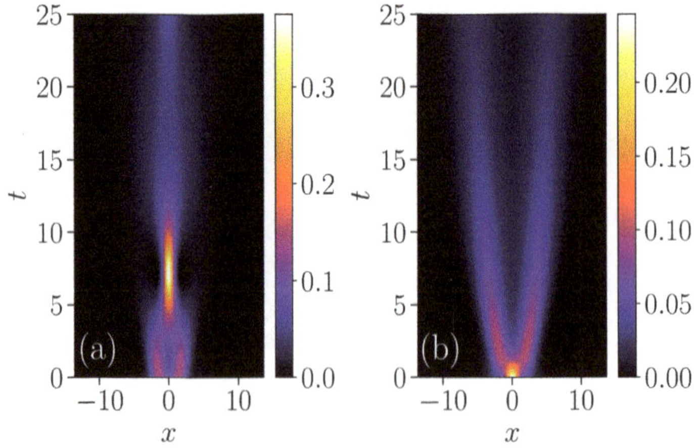

Figure 11.10. Density profiles for attractive pBECs: (a) $g = -5.0$, $g_1 = 0.03$, $\gamma_{\mathrm{eff}} = -0.01$; (b) $g = -5.0$, $g_1 = 0.1$, $\gamma_{\mathrm{eff}} = -0.01$. Reprinted with permission from [33], Copyright (2022) by the American Physical Society.

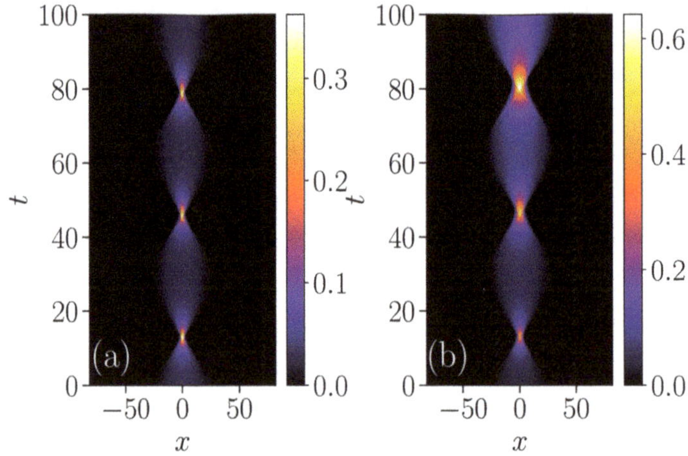

Figure 11.11. Density profiles for repulsive pBECs: (a) $g = 5.0$, $g_1 = 0.0$, $\gamma_{\mathrm{eff}} = 0.0$; (b) $g = 5.0$, $g_1 = 0.05$, $\gamma_{\mathrm{eff}} = 0.05$. Reprinted with permission from [33], Copyright (2022) by the American Physical Society.

Density profiles for repulsive pBECs with a weak trap are shown in figures 11.11–11.13: (a) $g = 5.0$, $g_1 = 0.0$, $\gamma_{\mathrm{eff}} = 0.0$; (b) $g = 5.0$, $g_1 = 0.05$, $\gamma_{\mathrm{eff}} = 0.05$; (a) $g = 5.0$, $g_1 = 0.1$, $\gamma_{\mathrm{eff}} = 0.1$; (b) $g = 5.0$, $g_1 = 0.2$, $\gamma_{\mathrm{eff}} = 0.1$; (a) $g = 5.0$, $g_1 = 0.3$, $\gamma_{\mathrm{eff}} = 0.1$; (b) $g = 5.0$, $g_1 = 0.5$, $\gamma_{\mathrm{eff}} = 0.1$. These exhibit a beating effect, with density stretching as nonlinear loss/gain increases, resembling a classical stretched string with nodes and antinodes, maintaining constant loop periodicity.

Attractive pBECs exhibit periodic superposition in density profiles, with periodicity adjustable via linear and nonlinear loss/gain terms. Repulsive pBECs show density stretching with constant loop periodicity, resembling a classical stretched string.

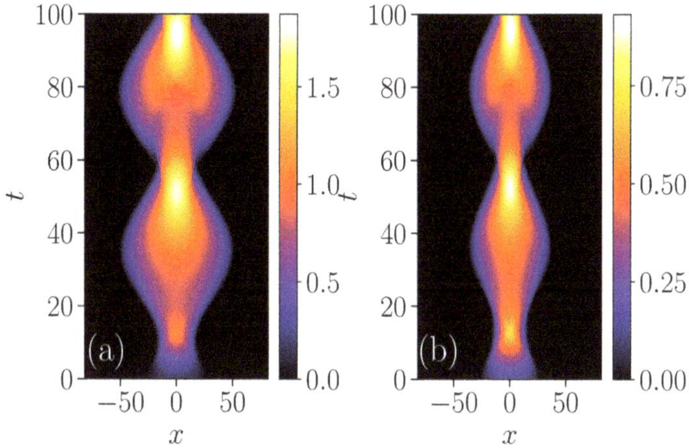

Figure 11.12. Density profiles for repulsive pBECs: (a) $g = 5.0$, $g_1 = 0.1$, $\gamma_{\text{eff}} = 0.1$; (b) $g = 5.0$, $g_1 = 0.2$, $\gamma_{\text{eff}} = 0.1$. Reprinted with permission from [33], Copyright (2022) by the American Physical Society.

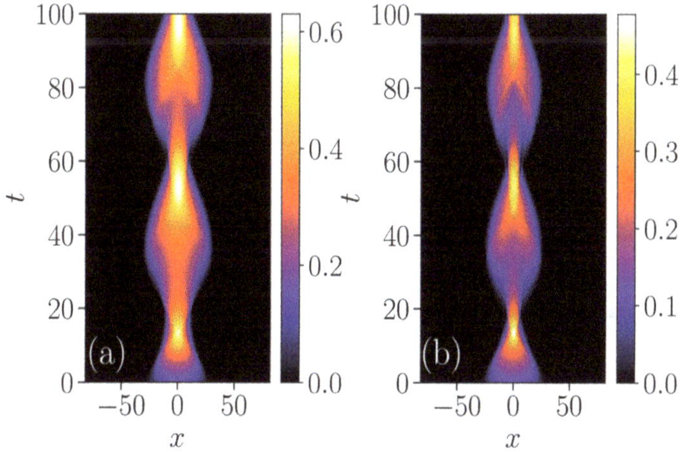

Figure 11.13. Density profiles for repulsive pBECs: (a) $g = 5.0$, $g_1 = 0.3$, $\gamma_{\text{eff}} = 0.1$; (b) $g = 5.0$, $g_1 = 0.5$, $\gamma_{\text{eff}} = 0.1$. Reprinted with permission from [33], Copyright (2022) by the American Physical Society.

11.5 Summary and future challenges

This chapter presented an overview of exciton–polariton condensation within semiconductor microcavities and an associated theoretical model based on an open-dissipative GP framework is derived. The dynamics arising out of the interaction of impurities with bright solitons in polariton condensates, revealing transmission, reflection, and trapping scenarios is explored. The stability window for trapless polariton condensation under non-resonant pumping is also characterized. These results underscore the significance of polaritons as tunable quasi-particles in microcavity physics, offering insights into nonlinear dynamics under driven-dissipative conditions.

At this juncture, it should be noted that several challenges remain unresolved, offering opportunities for further research. The interaction between an impurity and nonlinear excitations such as dark solitons and breathers in exciton–polariton condensates and their impact on microcavity dynamics warrants further investigation. The influence of three-body interactions on condensate stability and how the associated nonlinear excitations interact with the impurity remains open. Despite the experimental realization of polariton condensates near room temperature, their short lifetimes (typically of the order of picoseconds) highlight the need to study two-component exciton–polariton condensates for enhanced stability.

References

[1] Balili R, Hartwell V, Snoke D, Pfeiffer L and West K 2007 Bose-Einstein condensation of microcavity polaritons in a trap *Science* **316** 1007–10

[2] Byrnes T, Kim N Y and Yamamoto Y 2014 Exciton-polariton condensates *Nat. Phys.* **10** 803–13

[3] Plumhof J D, Stöferle T, Mai L, Scherf U and Mahrt R F 2014 Room-temperature Bose-Einstein condensation of cavity exciton-polaritons in a polymer *Nat. Mater.* **13** 247–52

[4] Christopoulos S *et al* 2007 Room-temperature polariton lasing in semiconductor microcavities *Phys. Rev. Lett.* **98** 126405

[5] Carusotto I and Ciuti C 2013 Quantum fluids of light *Rev. Mod. Phys.* **85** 299–366

[6] Kasprzak J *et al* 2006 Bose-Einstein condensation of exciton polaritons *Nature* **443** 409–14

[7] Amo A, Lefrère J, Pigeon S, Adrados C, Ciuti C, Carusotto I, Houdré R, Giacobino E and Bramati A 2009 Superfluidity of polaritons in semiconductor microcavities *Nat. Phys.* **5** 805–10

[8] Lagoudakis K G, Wouters M, Richard M, Baas A, Carusotto I, André R, Dang L S and Deveaud-Plédran B 2008 Quantized vortices in an exciton-polariton condensate *Nat. Phys.* **4** 706–10

[9] Keeling J and Berloff N G 2008 Spontaneous rotating vortex lattices in a pumped decaying condensate *Phys. Rev. Lett.* **100** 250401

[10] Jabri H and Eleuch H 2020 Optical Kerr nonlinearity in quantum-well microcavities: from polariton to dipolariton *Phys. Rev.* A **102** 063713

[11] Ghosh S and Liew T C H 2020 Quantum computing with exciton-polariton condensates *npj Quantum Inf.* **6** 16

[12] Xu H, Ghosh S, Matuszewski M and Liew T C 2020 Universal self-correcting computing with disordered exciton-polariton neural networks *Phys. Rev. Appl.* **13** 064074

[13] Sich M, Krizhanovskii D N, Skolnick M S, Gorbach A V, Hartley R, Skryabin D V, Cerda-Méndez E A, Biermann K, Hey R and Santos P V 2012 Observation of bright polariton solitons in a semiconductor microcavity *Nat. Photonics* **6** 50–5

[14] Egorov O A, Skryabin D V, Yulin A V and Lederer F 2009 Bright cavity polariton solitons *Phys. Rev. Lett.* **102** 153904

[15] Ma X, Egorov O A and Schumacher S 2017 Creation and manipulation of stable dark solitons and vortices in microcavity polariton condensates *Phys. Rev. Lett.* **118** 157401

[16] Grosso G, Nardin G, Morier-Genoud F, Léger Y and Deveaud-Plédran B 2011 Soliton instabilities and vortex street formation in a polariton quantum fluid *Phys. Rev. Lett.* **107** 245301

[17] Tanese D *et al* 2013 Polariton condensation in solitonic gap states in a one-dimensional periodic potential *Nat. Commun.* **4** 1749

[18] Ostrovskaya E A, Abdullaev J, Fraser M D, Desyatnikov A S and Kivshar Y S 2013 Self-localization of polariton condensates in periodic potentials *Phys. Rev. Lett.* **110** 170407

[19] El G A, Gammal A and Kamchatnov A M 2006 Oblique dark solitons in supersonic flow of a Bose-Einstein condensate *Phys. Rev. Lett.* **97** 180405

[20] Ostrovskaya E A, Abdullaev J, Desyatnikov A S, Fraser M D and Kivshar Y S 2012 Dissipative solitons and vortices in polariton Bose-Einstein condensates *Phys. Rev.* A **86** 013636

[21] Whittaker C E *et al* 2018 Exciton polaritons in a two-dimensional Lieb lattice with spin-orbit coupling *Phys. Rev. Lett.* **120** 097401

[22] Rosenberg I, Liran D, Mazuz-Harpaz Y, West K, Pfeiffer L and Rapaport R 2018 Strongly interacting dipolar-polaritons *Sci. Adv.* **4** eaat8880

[23] Wertz E *et al* 2010 Spontaneous formation and optical manipulation of extended polariton condensates *Nat. Phys.* **6** 860–4

[24] Smirnov L A, Smirnova D A, Ostrovskaya E A and Kivshar Y S 2014 Dynamics and stability of dark solitons in exciton-polariton condensates *Phys. Rev.* B **89** 235310

[25] Jia C and Liang Z 2022 Interaction between an impurity and nonlinear excitations in a polariton condensate *Entropy* **24** 1789

[26] Kivshar Y S and Królikowski W 1995 Lagrangian approach for dark solitons *Opt. Commun.* **114** 353–62

[27] Theocharis G, Schmelcher P, Kevrekidis P and Frantzeskakis D 2005 Matter-wave solitons of collisionally inhomogeneous condensates *Phys. Rev.* A **72** 033614

[28] Zhang Y, Jia C and Liang Z 2022 Dynamics of two dark solitons in a polariton condensate *Chin. Phys. Lett.* **39** 020501

[29] Xu X, Chen L, Zhang Z and Liang Z 2018 Dark-bright solitons in spinor polariton condensates under nonresonant pumping *J. Phys. B: At. Mol. Opt. Phys.* **52** 025303

[30] Forinash K, Peyrard M and Malomed B 1994 Interaction of discrete breathers with impurity modes *Phys. Rev.* E **49** 3400–11

[31] Muruganandam P and Adhikari S 2009 Fortran programs for the time-dependent Gross-Pitaevskii equation in a fully anisotropic trap *Comput. Phys. Commun.* **180** 1888–912

[32] Filho V S, Abdullaev F K, Gammal A and Tomio L 2001 Autosolitons in trapped Bose-Einstein condensates with two- and three-body inelastic processes *Phys. Rev.* A **63** 053603

[33] Sabari S, Kumar R K, Radha R and Muruganandam P 2022 Stability window of trapless polariton Bose-Einstein condensates *Phys. Rev.* B **105** 224315

[34] Kumar R K, Young S L E, Vudragović D, Balaž A, Muruganandam P and Adhikari S 2015 Fortran and C programs for the time-dependent dipolar Gross-Pitaevskii equation in an anisotropic trap *Comput. Phys. Commun.* **195** 117–28

IOP Publishing

An Introduction to Ultracold Atoms with Analytical and Numerical Methods

Paulsamy Muruganandam and Ramaswamy Radha

Chapter 12

Roadmap ahead

This chapter explores the captivating phenomena of supersolids, quantum droplets, and quantum turbulence, which have garnered significant interest in condensed matter physics. It elucidates how these quantum states, characterized by unique combinations of superfluidity, coherence, and turbulence, can be realized and studied in ultracold atomic systems. Through precise experimental control of ultracold atoms, researchers uncover the fundamental properties of these states, shedding light on their microscopic origins. The chapter also highlights the potential of these phenomena to advance quantum computing, quantum simulation, and materials science, where their exotic properties may enable novel computational paradigms and material designs.

12.1 Supersolids

Supersolidity represents a fascinating quantum phase that simultaneously exhibits the properties of superfluidity and crystalline order within a single, coherent quantum state. This exotic state of matter, first hypothesized in the context of solid helium by theorists such as Andreev and Lifshitz [1], has captivated researchers due to its unique combination of phase coherence and spatial periodicity. The advent of ultracold atomic gases, particularly Bose–Einstein condensates, has provided a highly controllable experimental platform to realize and study supersolids, reinvigorating interest in this phenomenon. Unlike conventional superfluids, which are phase-coherent but spatially homogeneous, supersolids spontaneously break both global gauge symmetry, leading to superfluidity, and translational symmetry, resulting in a periodic density modulation, typically forming a discrete lattice structure. This dual symmetry breaking makes supersolids an ideal system for investigating fundamental questions about quantum order, emergent phenomena, and the interplay of competing interactions.

doi:10.1088/978-0-7503-5447-9ch12

The theoretical foundation for supersolidity in Bose–Einstein condensates (BECs) hinges on mechanisms that induce long-range correlations while preserving global phase coherence. Two primary mechanisms have been extensively explored: dipolar interactions and spin–orbit coupling. In dipolar quantum gases, such as those composed of dysprosium or erbium atoms, the anisotropic, long-range dipole–dipole interaction can stabilize a periodic array of high-density regions, often termed quantum droplets, interconnected by a low-density superfluid background [2]. These quantum droplets arise from a balance between attractive dipolar forces, repulsive contact interactions, and quantum fluctuations, which prevent collapse and maintain phase coherence across the system [3]. Alternatively, spin–orbit coupling, engineered through Raman laser fields, can induce stripe phases in BECs, where the condensate density exhibits periodic modulations while retaining superfluid properties [4]. In such systems, the interplay between spin–orbit coupling and interatomic interactions creates a density-striped ground state, a distinct manifestation of supersolidity. These theoretical frameworks have guided experimental efforts to realize supersolids in controlled settings.

Experimental breakthroughs in ultracold atomic gases have provided compelling evidence for supersolidity. In 2016, Kadau *et al* [5] reported signatures of supersolidity in a dysprosium BEC, where the long-range dipolar interactions led to the formation of a droplet array with measurable phase coherence, confirming the coexistence of crystalline order and superfluidity. Subsequent experiments in 2019 by Tanzi *et al* [6] with erbium BECs further validated these findings, demonstrating robust supersolid behaviour through the observation of droplet lattices. Around the same time, Böttcher *et al* [7] achieved a supersolid phase in a spin–orbit-coupled sodium BEC, where Raman-induced spin–orbit coupling produced a stripe phase with periodic density modulations and superfluid characteristics. These milestones, achieved through precise control of interactions and external fields, have confirmed long-standing theoretical predictions and opened new avenues for studying the static and dynamic properties of supersolids.

Supersolids exhibit distinctive properties that set them apart from other quantum phases. Their excitation spectrum features two gapless modes: a phonon mode associated with the superfluid component, reflecting broken gauge symmetry, and a softened roton-like mode linked to the periodic density modulation, indicative of partial translational symmetry breaking. The critical velocity for superflow in supersolids is lower than in uniform superfluids due to the presence of the density lattice, which introduces additional energy scales and competes with kinetic energy. However, supersolids are inherently delicate, often existing as metastable states that are sensitive to parameters such as temperature, interaction strength, and trap geometry. For instance, imbalances between attractive and repulsive interactions or insufficient quantum stabilization can lead to collapse or melting of the supersolid phase. Controlling these instabilities is a critical challenge for achieving robust supersolids in larger systems or higher dimensions.

Significant open questions persist in the study of supersolids, driving ongoing research in both theory and experiment. The stability of supersolids in two- and three-dimensional systems, where competing interactions may lead to complex phase diagrams, remains an active area of exploration. The role of topological defects, such as vortices or domain walls, in supersolid dynamics is another frontier area of research, potentially revealing connections to other symmetry-broken phases. Non-equilibrium properties, including the response of supersolids to time-dependent perturbations, are also of great interest, as they probe the interplay between superfluid and crystalline degrees of freedom. Moreover, supersolids hold promise for applications in quantum simulation, where they can emulate exotic condensed matter systems, and in precision measurement, leveraging their sensitivity to external fields [8]. While connections to high-temperature superconductivity remain speculative, the study of supersolids may inspire analogies to other quantum phases with competing orders, such as fermionic supersolids or charge-density-wave systems.

By bridging superfluidity and crystalline order, supersolids offer profound insights into the principles of quantum coherence and symmetry breaking. As theoretical models refine and experimental techniques advance, the interplay between these disciplines will continue to unveil new aspects of this remarkable phase, paving the way for both fundamental discoveries and innovative quantum technologies.

12.2 Quantum droplets

Quantum droplets represent a fascinating state of matter within the domain of BECs. Unlike conventional BECs, which require external trapping potentials, such as magnetic or optical confinement, these self-bound quantum states form in ultracold bosonic mixtures through a precise balance between attractive and repulsive forces, without any confinement [3, 9]. They are characterized by significantly lower density than typical condensates. This delicate interplay of forces allows them to exist at ultralow density. Initially predicted through theoretical studies, quantum droplets have since been experimentally realized in two distinct systems: dipolar quantum gases (notably dysprosium and erbium), which exhibit strong magnetic dipole–dipole interactions [10], and binary BEC mixtures involving different hyperfine states of alkali atoms, such as potassium and rubidium [11, 12]. It is worth noting a key difference between solitons and quantum droplets: solitons are localized states that arise from a delicate balance between dispersion (the kinetic energy term) and nonlinearity in the mean-field approximation, whereas quantum droplets arise as a result of the equilibrium between the mean-field dynamics and the beyond-mean-field effect, which is eventually represented by the well-known Lee–Huang–Yang (LHY) correction which provides an additional repulsive term that stabilizes the system against collapse.

12.2.1 Mean-field description: Gross–Pitaevskii equation and Lee–Huang–Yang correction

In the mean-field framework, quantum droplets are described using an extended version of the GP equation, which governs the dynamics of the condensate wavefunction $\psi(\mathbf{r}, t)$. The standard GP equation accounts for a contact interaction term proportional to $g|\psi|^2$, where $g = 4\pi\hbar^2 a/m$ is the interaction strength, a is the s-wave scattering length, and m is the atomic mass. For quantum droplets, however, the mean-field description must incorporate beyond-mean-field corrections, such as the LHY term, which arises from quantum fluctuations.

The energy functional for a BEC in the mean-field approximation, including the LHY correction, can be written as

$$E = \int \left[\frac{\hbar^2}{2m} | \nabla \psi |^2 + V(\mathbf{r})|\psi|^2 + \frac{g}{2}|\psi|^4 + \frac{2}{3} g_{\mathrm{LHY}} |\psi|^3 \right] \mathrm{d}^3\mathbf{r}, \tag{12.1}$$

where $V(\mathbf{r})$ is the external potential (often zero for droplets), $g|\psi|^4/2$ is the mean-field contact interaction energy, $g_{\mathrm{LHY}}|\psi|^3$ is the LHY correction, with $g_{\mathrm{LHY}} \propto \sqrt{a^3}$ for a single-component BEC or adjusted for mixtures.

For a quantum droplet to form, the mean-field interaction (g) must be slightly attractive ($a < 0$), which would typically lead to collapse. However, the repulsive LHY term, scaling as $|\psi|^3$, counteracts this collapse, stabilizing the droplet at a finite density. This balance results in a self-bound state with liquid-like properties, such as a constant density in the bulk and a sharp surface.

12.3 Quantum turbulence in BECs

Turbulence, characterized by spatially and temporally disordered flow with numerous nonlinearly interacting degrees of freedom, poses a significant challenge in classical physics. In classical fluids, nonlinearity typically arises from the convective term $(\mathbf{v} \cdot \nabla)\mathbf{v}$ in the Navier–Stokes equation, where $\mathbf{v}$ represents the velocity field. Despite extensive study, classical turbulence remains poorly understood. In contrast, quantum turbulence (QT), a phenomenon pioneered by Vinen (1957) [13], emerges in quantum fluids such as superfluid helium and BECs. Unlike classical turbulence, which features a continuous distribution of eddies, QT is characterized by discrete, quantized vortex lines. Superfluid flow, being inviscid, prevents vortex decay through vorticity diffusion, as it occurs in classical fluids. Instead, energy dissipation in QT occurs via phonon emission or interactions with thermal clouds in BECs or the normal fluid component in superfluid helium. Superfluid helium exhibits a two-fluid nature, comprising an inviscid superfluid and a normal fluid component with a short mean free path at higher temperatures. This normal component introduces mutual friction between the superfluid and normal fluids, which, when stirred, can induce a complex, doubly turbulent state [14]. The interplay of quantum and classical behaviours highlights the unique complexity of quantum turbulence and its significance in advancing the understanding of turbulent phenomena.

QT is more closely related to the quantization of vortices than to the absence of viscosity in superfluids. In quantum fluids, at locations far from the vortex cores and on length scales larger than the intervortex spacing, the detailed structure of the vortex core becomes negligible, and quantized vortex lines behave similarly to vortex filaments in an ideal Euler fluid. Under certain conditions, QT exhibits striking similarities to classical turbulence. However, in other regimes, it displays fundamentally different behaviour. The identification of these connections provides a strong motivation for further investigation into the nature of QT and for exploring the emergence of classical behaviour in out-of-equilibrium quantum systems.

BECs, which constitute viscosity-free fluid, are a novel platform for QT studies due to their compressibility, weak interatomic interactions, tunability, and availability of new experimental methods for probing and studying superfluid flow. QT has been restricted to superfluid systems due to their accessibility and ability to generate hundreds of thousands of vortices with core diameters $\xi \sim 10^{-10}$ m and intervortex spacing $\ell_0 \sim 10^{-4}$ m indicating well-separated vortices. However, in superfluid systems, as the atomic interaction strength is fixed, fluid distribution is homogeneous, and controlling single-vortex dynamics in a turbulent flow is very difficult.

In contrast, atomic BECs are inherently inhomogeneous and offer a narrower range of accessible length scales for probing vortices. This inhomogeneity arises from the geometry of the trapping potential and the compressibility of BECs, which also allows for easier manipulation of the homogeneity and interatomic interactions of the underlying system. Turbulent BECs typically contain fewer vortices with smaller spatial separation. However, the positions and dynamics of individual vortices can be precisely controlled and observed using a wide range of advanced imaging techniques. Moreover, BECs provide a versatile platform for studying two-dimensional quantum turbulence and allow for the deterministic preparation of initial states, facilitating investigations into the chaotic dynamics of few-vortex systems. Atomic BECs also open pathways for exploring turbulence in spinor condensates and degenerate Fermi gases. Despite their intrinsic differences from classical fluids, quantum turbulence in atomic condensates was first experimentally observed by Henn *et al* [15], who reported hallmark signatures such as self-similar expansion and the emergence of energy cascades.

Experimental challenges in studying quantum turbulence in BECs include tracking vortices, probing vortex dynamics and interactions, measuring quantized circulation, and determining energy spectra. The tunability of BECs facilitates advanced techniques to address these issues. *In situ* imaging is being optimized to visualize vortices and atomic interferometry is being developed to measure circulation precisely. Novel forcing methods are under exploration to suppress phonon excitations, enabling accurate kinetic energy spectra measurements. These advancements enhance the experimental probing of quantum turbulence, deepening insights into superfluid dynamics.

12.3.1 Classical turbulence

Classical turbulence arises from a continuous distribution of eddies with varying vorticity and scales, sustained by constant kinetic energy injection and viscous dissipation at small scales. Governed by the Navier–Stokes equation, homogeneous isotropic turbulence (HIT) provides an idealized framework for studying these dynamics and serves as a reference for quantum turbulence. The energy cascade, conceptualized by Richardson, involves energy transfer from large scales D to small scales η, where viscosity dissipates it. Kolmogorov formalized this self-similar process, demonstrating that in the inertial range, the kinetic energy spectrum follows

$$E_{\text{kin}}(k) = C\epsilon^{2/3}k^{-5/3}, \tag{12.2}$$

where $k = 2\pi/r$ is the wavenumber, $C \approx 1$ is a constant, and ϵ is the energy dissipation rate. This scaling applies in the range $k_D = 2\pi/D$ to $k_\eta = 2\pi/\eta$. The enstrophy, defined as the squared vorticity $\Omega(k)$, exhibits a k^{-3} scaling in the inertial range, peaking at k_η. These universal properties of classical turbulence provide a foundation for understanding turbulent behaviour in quantum fluids.

Turbulent intensity is quantified by the Reynolds number, Re$=UD/\nu$, where U is the characteristic velocity at length scale D, and ν is the kinematic viscosity. This dimensionless ratio measures the dominance of inertial forces over viscous forces and indicates the range of length scales, from the largest eddies to the smallest dissipative scales. In classical turbulence, velocity components follow a Gaussian distribution, and the kinetic energy dissipation rate approaches a nonzero constant as $\nu \to 0$ (i.e., Re$\to\infty$), unlike laminar flows where dissipation vanishes. Small-scale motions with high vorticity but low energy sustain this dissipation. In contrast, quantum turbulence in superfluids, which exhibit zero kinematic viscosity due to their inviscid nature, relies on quantized vortices rather than viscous dissipation, highlighting fundamental differences from classical turbulence.

In classical turbulence, an inverse energy cascade can occur, particularly in two-dimensional flows, where energy transfers from smaller to larger length scales, yielding an energy spectrum scaling as $k^{-5/3}$. This process involves the coalescence of smaller vortices into larger structures, contrasting with the direct cascade, where energy flows to smaller scales. Unlike quantum turbulence, where vorticity is confined to discrete, quantized vortex cores, classical turbulence lacks such well-defined vorticity, complicating the study of vortex dynamics. Quantum turbulence in BECs offers a controlled platform to probe turbulent behaviour, revealing similarities and differences with classical systems. However, the confined geometry of trapped BECs introduces finite-size effects, challenging direct comparisons between classical and quantum turbulence.

12.3.2 Spectral analysis

Spectral analysis is a key tool for studying QT, decomposing kinetic energy into wavenumber k modes to reveal dynamics across multiple length scales. For compressible quantum fluids, classical spectral methods have been adapted by introducing a regularized velocity field to mitigate singularities at vortex cores. However, a comprehensive spectral formulation capturing all quantum phase effects, particularly the interplay between velocity and density in turbulent flows, remains an unresolved challenge. The Gross–Pitaevskii (GP) equation accurately models dilute BECs at low temperatures, serving as a platform for spectral analysis of turbulence induced experimentally via stirring or laser-driven perturbations. These studies elucidate the complex dynamics of QT in controlled settings.

The wavefunction of a superfluid, expressed as $\psi(\mathbf{r}, t) = \sqrt{n(\mathbf{r}, t)}\,\mathrm{e}^{\mathrm{i}\theta(\mathbf{r}, t)}$, encodes the particle density $n = |\psi|^2$ and phase θ. The superfluid velocity is defined as $\mathbf{v}(\mathbf{r}, t) = \frac{\hbar}{m}\nabla\theta$, reflecting phase gradients. Within the GP framework, the Hamiltonian for a BEC comprises kinetic, quantum pressure, external potential, and interaction energies as

$$H = \int \mathrm{d}^3\mathbf{r}\left[\frac{m}{2}n|\mathbf{v}|^2 + \frac{\hbar^2}{2m}|\nabla\sqrt{n}|^2 + Vn + \frac{gn^2}{2}\right], \tag{12.3}$$

where V is the external potential and g is the interaction strength. The quantum pressure term, $\frac{\hbar^2}{2m}|\nabla\sqrt{n}|^2$, arises from sharp density gradients, such as vortex cores or dark solitons. In two-dimensional BECs, this term scales with the number of vortices, while in three-dimensional systems, it correlates with vortex line length, providing a measure of topological defects in quantum turbulence.

To analyze the spectral properties of quantum turbulence in BECs, the kinetic energy is decomposed into incompressible, compressible, and quantum pressure components, revealing the distribution of energy from vortex lines and sound waves across wavenumber k scales. A density-weighted velocity field, $\mathbf{u} = \sqrt{n(\mathbf{r})}\,\mathbf{v}$, where $\mathbf{v} = \frac{\hbar}{m}\nabla\theta$ is the superfluid velocity, regularizes the $1/r$ singularity near vortex cores. Applying Helmholtz decomposition, this field is expressed as $\mathbf{u} = \mathbf{u}^{\mathrm{i}} + \mathbf{u}^{\mathrm{c}}$, with divergence-free ($\nabla\cdot\mathbf{u}^{\mathrm{i}} = 0$) incompressible ($\mathbf{u}^{\mathrm{i}}$) and irrotational ($\nabla\times\mathbf{u}^{\mathrm{c}} = 0$) compressible ($\mathbf{u}^{\mathrm{c}}$) components. Including the quantum pressure velocity, $\mathbf{u}^{\mathrm{q}} = \frac{\hbar}{m}\nabla\sqrt{n}$, the total kinetic energy becomes

$$E_{\mathrm{kin}} = \int \mathrm{d}^3\mathbf{r}\left[\frac{m}{2}n(|\mathbf{u}^{\mathrm{i}}|^2 + |\mathbf{u}^{\mathrm{c}}|^2) + \frac{\hbar^2}{2m}|\nabla\sqrt{n}|^2\right], \tag{12.4}$$

corresponding to $E^{\mathrm{i}}_{\mathrm{kin}}$, $E^{\mathrm{c}}_{\mathrm{kin}}$, and $E^{\mathrm{q}}_{\mathrm{kin}}$. This decomposition enables spectral analysis of turbulent energy distributions. The total kinetic energy of each component in d-dimensions can be expressed as

$$E^{\mathrm{i,c,q}}_{\mathrm{kin}} = \frac{m}{2}\int d^d\mathbf{r}\,|\mathbf{u}^{\mathrm{i,c,q}}(\mathbf{r})|^2 = \int \mathrm{d}k\,|\tilde{\mathbf{u}}^{\mathrm{i,c,q}}(k)|^2 = \int_0^\infty \mathrm{d}k\,\varepsilon^{\mathrm{i,c,q}}_{\mathrm{kin}}(k), \tag{12.5}$$

where $\tilde{\mathbf{u}}^{\mathrm{i,c,q}}(k)$ is the Fourier transform of the velocity field $\mathbf{u}^{\mathrm{i,c,q}}(\mathbf{r})$. The velocity power spectra $\varepsilon^{\mathrm{i,c,q}}_{\mathrm{kin}}(k)$, derived from the magnitude of $\tilde{\mathbf{u}}^{\mathrm{i,c,q}}(k)$, fully characterize the kinetic energy distribution across scales via Parseval's theorem. In d-dimensional space, the relationship between the two-point correlation function and its Fourier transform is given by

$$\int d^d\mathbf{r}\,\mathbf{u}^{\mathrm{i,c,q}}(\mathbf{r})^*\cdot\mathbf{u}^{\mathrm{i,c,q}}(\mathbf{r}+\mathbf{r}') \leftrightarrow (2\pi)^{d/2}|\tilde{\mathbf{u}}^{\mathrm{i,c,q}}(\mathbf{k})|^2. \tag{12.6}$$

Here, the left-hand side represents the spatial correlation of the velocity field, while the right-hand side encodes the same information in Fourier (scale) space. These quantities, often termed 'energy spectra', provide critical insights into the turbulent dynamics of BECs.

Conventionally, computing $\varepsilon_{\text{kin}}^{\text{i,c,q}}(k)$ for turbulent BECs involves discretizing the velocity field on Cartesian grids and binning Fourier modes into annular regions $k_1 < k < k_2$. However, this approach introduces numerical artifacts, particularly at small k, where sparse sampling leads to poor resolution. Modern methods address this limitation by analytically evaluating k-space integrals, decoupling spectral resolution from grid discretization, and enabling accurate angle-averaged correlations. One such technique leverages the angle-averaged Wiener–Khinchin theorem [16] for d-dimensional systems as:

$$\varepsilon_{\text{kin}}^{\text{i,c,q}}(k) = \frac{m}{2}\int d^d\mathbf{x}\;\Lambda_d(k,\,|\mathbf{x}|)C[\mathbf{u}^{\text{i,c,q}},\,\mathbf{u}^{\text{i,c,q}}](\mathbf{x}), \tag{12.7}$$

where $C[\mathbf{u}^{\text{i,c,q}},\,\mathbf{u}^{\text{i,c,q}}](\mathbf{x})$ is the two-point correlation function of the velocity field, and $\Lambda_d(k,\,r)$ is a dimension-dependent kernel given by

$$\Lambda_d(k,\,r) = \begin{cases} (2\pi)^{-1}kJ_0(kr) & \text{for}\, d = 2, \\ (2\pi^2)^{-1}k^2\text{sinc}(kr) & \text{for}\, d = 3. \end{cases} \tag{12.8}$$

These spectral decomposition methods, rooted in classical turbulence studies, are now standard tools for probing energy cascades in QT. They reveal energy transport across scales not only in incompressible quantum fluids but also in compressible systems, dipole gases, plasmas, and clustered quantum vortices. In particular, the kinetic energy flux across wavenumbers, central to both QT and CT, illuminates the role of quantized vortices in 2D and 3D dynamics.

A key similarity between classical and quantum turbulence is the emergence of the Kolmogorov energy spectrum in the inertial range of three-dimensional turbulence. For non-equilibrium steady-state systems with forced turbulence, the incompressible kinetic energy spectrum exhibits the characteristic scaling $E(k) \sim k^{-5/3}$ at length scales intermediate between the energy injection scale and the intervortex separation scale. This power-law behaviour originates from the Richardson cascade process, where larger vortices successively break down into smaller ones, ultimately dissipating energy at the smallest scales. This quintessential turbulence phenomenon, first established in classical fluids, has also been observed experimentally in turbulent superfluids, including ^{4}He and ^{3}He-B [17–19]. Moreover, theoretical and numerical studies have confirmed its presence in trapped Bose–Einstein condensates [20, 21], demonstrating the universal nature of this scaling across both classical and quantum turbulent systems.

12.3.3 Kolmogorov quantum turbulence

Kolmogorov quantum turbulence refers to a regime in which quantum turbulence exhibits behaviour analogous to that of classical turbulence. However, unlike classical Kolmogorov turbulence, the quantum case involves an additional characteristic length scale: the average spacing between quantized vortex lines, known as the intervortex spacing ℓ_0. This spacing can be estimated by the relation $\ell_0 \approx L^{-1/2}$, where L denotes the vortex line density, defined as the total vortex length per unit volume. Experimentally, vortex line density can be measured in superfluid helium using various techniques, such as second sound attenuation, ion trapping, and Andreev reflection [22]. Although vortex line density is a useful indicator of quantum turbulence intensity, a more informative dimensionless parameter is D/ℓ_0, where D represents the size of the condensate [23].

At higher temperatures, mutual friction becomes significant, facilitating energy exchange between the normal and superfluid components. This mutual friction depends on the relative velocity of the two components, the vortex line density, and temperature-dependent friction coefficients, typically denoted by α. In some regimes, the normal component is highly viscous and, therefore, dynamically passive, acting only to provide frictional resistance to vortex motion.

Kolmogorov-type turbulence in quantum fluids can be generated by injecting vortex rings into the condensate or by oscillating grids. The resulting turbulence is predominantly incompressible, as the average vortex velocities are much smaller than the speed of sound and, hence, do not significantly excite density (acoustic) modes. Notably, the Kolmogorov scaling law in quantum fluids emerges only over 'classical length scales', corresponding to wavenumbers satisfying $k_D \ll k \ll k_{\ell_0}$, which is the range over which vortex lines can effectively polarize. In contrast, for wavenumbers $k \gg k_{\ell_0}$, the dynamics deviate from classical behaviour due to the intrinsic quantization of circulation, a phenomenon with no classical counterpart. In this high-wavenumber regime, quantum turbulence may exhibit an energy spectrum scaling as k^{-1}, characteristic of isolated straight vortices.

12.3.3.1 3D cases

Feynman first proposed that QT could emerge from disordered vortex line configurations [24]. Three-dimensional QT provides valuable insights into how quantized vortices bend, interact, and form complex tangles. For studying 3D QT in BECs, one typically considers a spherically symmetric trap with comparable axial and radial trapping frequencies ($\omega_z \sim \omega_r$). A key feature of continuously forced 3D turbulence is the Kolmogorov cascade, where energy transfers from large to small length scales. This process suggests that vortex bundles can transfer energy to progressively smaller structures until reaching individual vortices. Experimental verification remains challenging due to the limited size of BECs (typically $<100\ \mu\text{m}$) and the restricted range of accessible length scales. Numerical simulations, however, enable vortex tracking and have confirmed both the Richardson cascade process and the $k^{-5/3}$ energy spectrum scaling. In both classical and quantum turbulence, this cascade originates from the nonlinear term $(\mathbf{v} \cdot \nabla)\mathbf{v}$ in the Euler equation. Simulations reveal that Kolmogorov-spectrum QT contains transient vortex bundles alongside randomly oriented vortices (see figure 12.1).

At scales smaller than the intervortex spacing ℓ_0, QT differs fundamentally from classical turbulence. The velocity statistics in trapped and homogeneous turbulent BECs follow power-law distributions rather than the Gaussian statistics of classical turbulence [26, 27]. This distinction stems from the singular nature of quantum vorticity and the $1/r$ velocity field of individual vortices. However, in the 'quasi-classical' limit where microscopic vortex structure becomes unresolved, the velocity statistics approach Gaussian distributions.

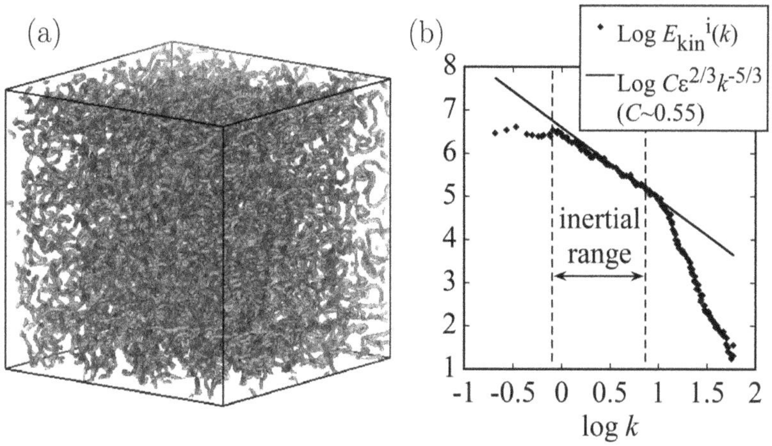

Figure 12.1. (a) 3D visualization of vortex tangles and reconnections demonstrating Kolmogorov turbulence in a GP simulation [18]. (b) The characteristic $k^{-5/3}$ scaling in the inertial range of the incompressible kinetic energy spectrum. Reproduced with permission from [25]. Copyright 2008 The Physical Society of Japan.

In 3D QT, the Kolmogorov cascade couples to energy dissipation through Kelvin wave excitations at small scales. These helical perturbations propagate along vortex lines with frequency $\omega \approx \beta k^2$, where $\beta = \kappa/(4\pi)\ln(k\xi)$ and $\xi \approx 10^{-10}$ m is the core radius. The waves decay exponentially as $\exp(-\alpha\beta k^2 t)$, with α being a temperature-dependent friction coefficient. Consequently, vortex tangle geometry evolves with temperature: high temperatures yield smooth vortex lines, while lower temperatures produce cusps and high-frequency Kelvin waves [28].

Kelvin waves can be generated through vortex-sound wave interactions or suppressed by dimensional confinement [29]. Theoretical studies suggest vortex loop fragmentation could create a self-sustaining cascade to microscopic scales [30]. This Kelvin wave cascade represents the primary dissipation mechanism in zero-temperature 3D BECs, though its dependence on system size, vortex density, and tangle polarity remains unresolved. In steady-state QT, vortex reconnections occur at a rate $\approx \kappa L^{5/2}$ per unit time and volume.

Stationary QT requires continuous energy injection to balance cascade dissipation. In decaying turbulence (without energy input), vortex reconnections drive the decay process, with line density L scaling as $t^{-3/2}$ [31, 32] and transitioning to t^{-2} at late times. The decay dynamics depend strongly on vortex density and trapping potential inhomogeneity.

Experimental studies have generated turbulent vortex tangles in cigar-shaped BECs through surface mode instabilities [15]. Controlled perturbation strength and duration can produce configurations ranging from few vortices to dense tangles [33]. While sparse vortex distributions allow individual vortex imaging, dense tangles make single-vortex identification challenging in absorption measurements.

12.3.3.2 2D cases

Understanding quantum turbulence in two dimensions requires examining how quantized vortices move within a plane. BECs provide the ideal platform for studying 2D QT due to their inherent inhomogeneity and the ability to create highly oblate trapping configurations [34, 35]. Such configurations are achieved by setting the axial trapping frequency much greater than the radial frequency ($\omega_z \gg \omega_r$) in the GP equation (section 1.3), effectively suppressing superfluid flow along the tightly confined direction and restricting vortex dynamics to the radial plane.

While 3D superfluids exhibit a direct Kolmogorov cascade, 2D quantum turbulence can display an inverse energy cascade (IEC), where energy flows towards larger length scales in the inertial range. This phenomenon, well-known in classical 2D turbulence, produces a kinetic energy spectrum $E^{\mathrm{i}}_{\mathrm{kin}}(k) \propto k^{-5/3}$ through vortex merging and growth. The 2D geometry enforces vorticity normal to the plane, leading to an additional conserved quantity, the enstrophy (net squared vorticity). This conservation enables simultaneous energy flux to larger scales and enstrophy flux to smaller scales, with the latter producing a k^{-3} spectrum in the energy spectrum.

Key research questions in 2D QT include:

(i) Conditions for observing IECs and enstrophy cascades in forced or decaying turbulence;
(ii) Superfluid dynamics accompanying cascade processes, particularly vortex clustering;
(iii) Fundamental vortex dynamics, including vortex–antivortex annihilation.

Unlike classical systems, enstrophy conservation in 2D QT is imperfect due to vortex–antivortex annihilation [36, 37]. Additional challenges arise in harmonically trapped BECs, where vortex energy can dissipate through radial precession towards the condensate boundary or through phonon emission from vortex motion and recombination events. As suggested by Onsager's arguments, achieving IECs may require minimum vortex densities or forcing rates.

The energy flux $\Pi(k) = -\mathrm{d}E^{<}(k)/\mathrm{d}t$, where $E^{<}(k) = \int_0^k E(k')\mathrm{d}k'$, characterizes cascade directions with $\Pi(k) > 0$ for direct and $\Pi(k) < 0$ for inverse cascades. While challenging to measure experimentally, the integral scale given by

$$L_{\mathrm{flow}} = 2\pi \frac{\int k^{-1} E^{\mathrm{i}}_{\mathrm{kin}}(k)\mathrm{d}k}{\int E^{\mathrm{i}}_{\mathrm{kin}}(k)\mathrm{d}k} \tag{12.9}$$

provides insight into energy distribution. When L_{flow} approaches the system size L_x, spectral condensation occurs, indicating large-scale energy concentration.

Compressibility effects complicate 2D QT studies, as vortex–antivortex annihilation and acoustic energy cascades compete with vortex dynamics. Introducing damping in the GP equation can approximate incompressible vortex dynamics [38], with studies showing vortex clustering and IEC emergence under appropriate damping (see figure 12.2).

Experimental realizations of 2D QT remain limited. The first observations by Neely *et al* [34] used laser stirring to generate vortex distributions in oblate annular BECs, though definitive IEC evidence remains elusive. Alternative methods,

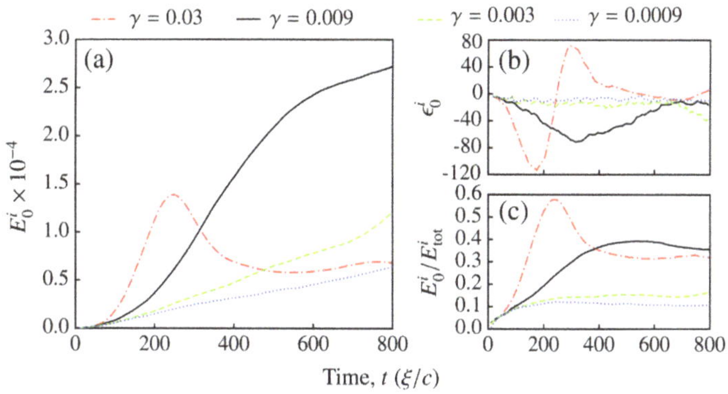

Figure 12.2. (a) Incompressible energy E_0^{i} at system scale. (b) Negative incompressible kinetic energy flux (ϵ_0^{i}) showing IEC onset for varying damping strengths γ. (c) System-scale energy fraction of total incompressible kinetic energy. Reprinted with permission from [38], copyright (2013) by the American Physical Society, demonstrating how increased damping suppresses sound waves to facilitate IEC.

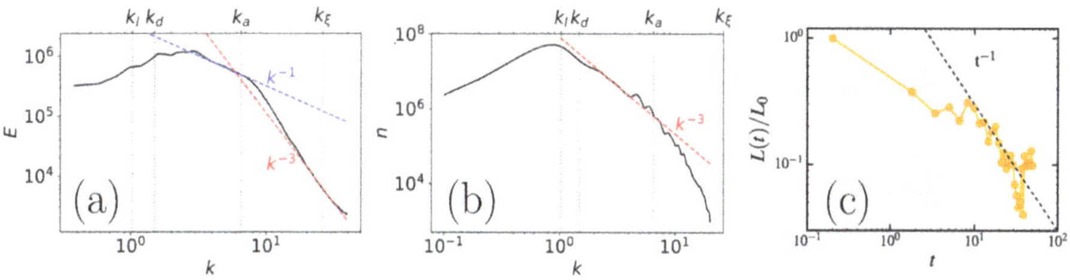

Figure 12.3. (a) Incompressible kinetic energy of a trapped BEC exhibiting k^{-1} scaling consistent with Vinen scaling and (b) shows corresponding momentum spectrum with a k^{-3} scaling in inertial range as shown by Marino and colleagues [40], reproduced with permission from Springer Nature. (c) Decay of vortex line density with respect to time obeying the $L \sim t^{-1}$, which is a characteristic of Vinen turbulence as shown by Cidrim *et al* [41] CC BY 4.0.

including trap modulation and localized repulsive potentials [39], can generate vortex turbulence but often excite unwanted density modes. Numerical studies of decaying turbulence show both $k^{-5/3}$ and k^{-3} spectra, while analytical approaches derive these scalings from vortex core dynamics of homogeneous systems.

12.3.4 Vinen or ultra-quantum turbulence

Ultra-quantum models of turbulence describe random distributions of vortices in superfluid He turbulence, with vortex decay facilitated by a Kelvin wave cascade process corresponding to decay of vortex line density L as $L \sim t^{-1}$ (see figure 12.3) which is the large-t decaying solution of the equation $\mathrm{d}L/\mathrm{d}t \sim -L^2$ proposed by Vinen on simple physical arguments to model random like flow. This was first envisaged by Volovik [42] and experimentally identified by Walmsley and Golov [43] and Bradley *et al* [32]. This line density decay was also observed numerically [31] in simulations of turbulence driven by a uniform normal flow (using vortex filament models) and also observed in thermal quenching of Bose gases where topological defects created by the Kibble–Zurek mechanism evolve into a turbulent vortex tangle. Decay of multicharged vortices in trapped BECs reducing to surface oscillations via vortex reconnections leading to disordered vortex lattice state was also shown to exhibit Vinen turbulence [41] (see figure 12.4). Numerical simulations of counterflow turbulence driven by a uniform normal fluid produced a superfluid energy spectrum peaking at mesoscales.

Besides the $L \sim t^{-1}$ decay, Vinen quantum turbulence is starkly different from that of Kolmogorov quantum turbulence in two respects:

(i) It lacks the concentration of energy at large length scales near k_D, which is typical of classical turbulence, peaks near $k \approx k_{\ell_0}$ and from there scales k^{-1}

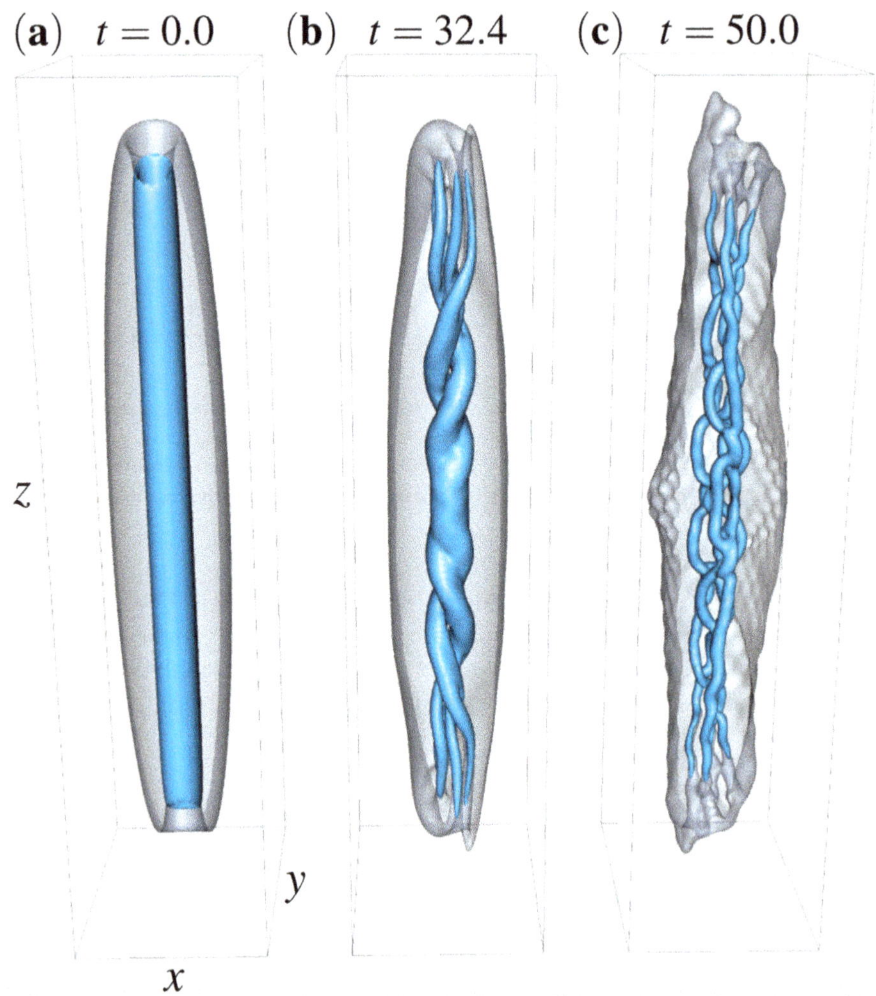

Figure 12.4. Isosurface plots of vortex decay exhibiting Vinen turbulence as reported by Cidrim *et al* [41] CC BY 4.0.

for $k \gg k_{\ell_0}$. Thus, the Vinen spectrum is reminiscent of the energy spectrum of a random gas of vortex rings. The absence of $k^{-5/3}$ scaling suggests that in Vinen quantum turbulence, the energy transfer from small wavenumbers $\sim k_D$ to wavenumbers k_{ℓ_0} is weak or essentially absent.

(ii) The lack of polarization of vortex lines is apparent when the coarse-grained vorticity is calculated as the vortex lines tend to be randomly oriented with respect to each other. This leads to vortices interacting weakly with each other, which is reflected in the rapid decay with distance of the velocity correlation function [44].

Numerical simulations of trapped BECs have provided important insights into momentum spectra during Vinen turbulence [40]. These studies consistently reveal a power-law behaviour $n(k) \propto k^{-3}$ in the momentum distribution, which aligns with theoretical predictions for a wave-turbulent regime characterized by particle cascade. Interestingly, homogeneous BEC systems under comparable conditions exhibit a different scaling behaviour, with momentum spectra following a steeper $n(k) \sim k^{-7/2}$ dependence. The emergence of these power-law spectra occurs in specific wavenumber ranges that notably lack self-similar dynamics, providing strong evidence for the simultaneous presence of wave-turbulence phenomena alongside quantum turbulence. This dual regime originates from nonlinear interactions between various wave excitations generated during vortex reconnection events. Such reconnections can produce both density fluctuations in the condensate and Kelvin wave excitations propagating along vortex lines, contributing to the complex turbulent behaviour.

Further analysis of energy fluxes offers an additional understanding of the underlying dynamics. The simulations show a clear direct particle cascade superimposed on the chaotic motion of the vortex tangle. Most significantly, the flux calculations reveal a distinct peak at length scales comparable to the characteristic intervortex spacing ℓ_0. This peak serves as a direct indicator of vortex reconnection processes dominating the turbulent behaviour at these intermediate scales, highlighting the importance of reconnections in the overall turbulent dynamics. The combination of these features provides a comprehensive picture of how energy transfers across different scales in quantum turbulent systems.

Models that generate both Kolmogorov and Vinen turbulence using energy fluxes at classical and quantum length scales were presented by Zmeev *et al* [45]. The key ingredient to create Vinen turbulence is either a lack of forcing at large length scales or an initial condition that lacks sufficient forcing at large length scales.

12.3.5 Quantum turbulence at small length scales

The zero-viscosity nature of superfluids renders the classical Reynolds number inapplicable. However, a superfluid Reynolds number $\mathrm{Re_s}$ can be defined through the ratio of inertial to frictional forces, analogous to classical fluids. This dimensionless parameter takes the form $Re_s = (1 - \alpha')/\alpha$, where α and α' represent the dissipative and non-dissipative mutual friction coefficients, respectively. Unlike its classical counterpart, Re_s exhibits temperature dependence but remains independent of flow velocity or system size D. As temperature approaches absolute zero ($T \to 0$), both friction coefficients vanish (α, $\alpha' \to 0$), causing $\mathrm{Re_s}$ to diverge ($\mathrm{Re_s} \to \infty$).

At ultralow temperatures, Kelvin waves propagate more freely, leading to increased vortex line curvature and cusp formation. Recent numerical studies [46] demonstrate that the kinetic energy dissipation rate through mutual friction initially decreases with increasing $\mathrm{Re_s}$, eventually reaching a plateau. This saturation occurs because intensive turbulent flows generate small-scale vortex structures that maintain finite dissipation. Remarkably, this behaviour mirrors classical turbulence and persists across both Kolmogorov and Vinen turbulence regimes at small scales ($k \gg k_\xi$).

As temperatures decrease further, an alternative dissipation mechanism emerges. Nonlinear interactions among finite-amplitude Kelvin waves generate progressively shorter wavelengths that rotate more rapidly. This Kelvin wave cascade efficiently transfers energy to scales much smaller than the intervortex spacing ℓ_0 until sound radiation becomes the dominant dissipation channel. Theoretical estimates for ^{4}He suggest this crossover occurs near $T \approx 0.5$ K for vortex line densities $L \approx 10^{10}\ \mathrm{m}^{-2}$.

Low-temperature quantum turbulence can simultaneously support two distinct energy cascades: (1) a three-dimensional Kolmogorov cascade of vortex bundles at intermediate scales ($k_D \ll k \ll k_{\ell_0}$), and (2) a one-dimensional Kelvin wave cascade along individual vortices at small scales ($k \gg k_{\ell_0}$). Large-scale GP simulations [47] reveal that these cascades are separated by a spectral bottleneck near $k \approx k_{\ell_0}$, arising from the disparity in energy transfer rates between the faster 3D Kolmogorov cascade and the slower 1D Kelvin wave cascade.

12.3.6 Rotating quantum turbulence

Research on 2D and 3D quantum turbulence has traditionally focused on the GP equation without rotational terms, instead introducing rotational effects through anharmonic and time-varying trapping potentials. A significant departure from this approach was presented in the work of Amette Estrada and colleagues [48], who examined 3D condensates in rotating frames, revealing behaviour fundamentally distinct from classical rotating turbulence. Their findings demonstrate that such systems develop an inverse energy cascade through quasi-two-dimensionalization of the flow, accompanied by self-organization of kinetic energy into quantized vortex structures. The study also identified an anisotropic dissipation mechanism and a non-Kolmogorovian energy scaling at small scales that align with Vinen turbulence characteristics. Notably, this Vinen turbulence regime emerges without visible thermal counterflow and exhibits minimal energy transfer to small scales.

Rotational quantum turbulence was further studied in 2D harmonically confined BECs by injecting a local perturbation into the condensate and gradually increasing its rotation frequency [50]. With appropriate initial conditions, the turbulent behaviour exhibited $k^{-5/3}$ Kolmogorov scaling or k^{-1} Vinen turbulence (see figure 12.5). Surprisingly, energy flux calculations revealed a direct energy cascade, contrary to previous studies on 2D and quasi-2D flows, which showed inverse energy

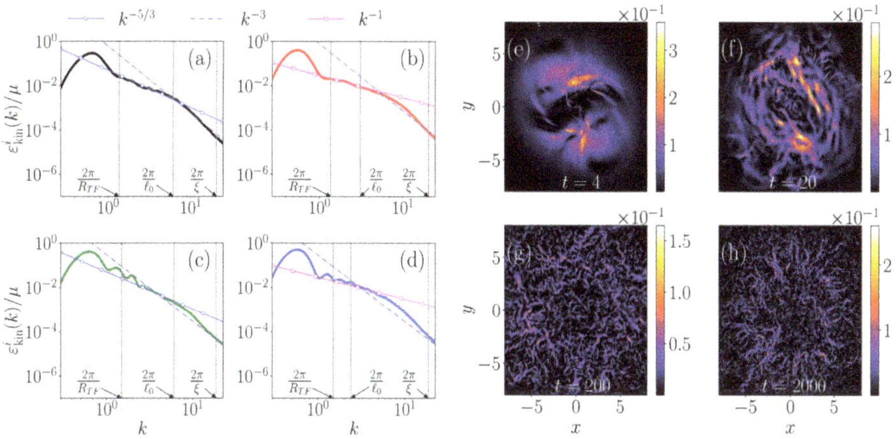

Figure 12.5. Time-averaged incompressible kinetic energy spectra for a condensate formed via rotational mergers: (a) and (b) correspond to a merging time of $t_m = 4$, and (c) and (d) correspond to a merging time of $t_m = 6$. In both cases, the initial spectra (a) and (c) exhibit Kolmogorov scaling ($k^{-5/3}$), which evolves into k^{-1} at later times (b) and (d), indicating Vinen turbulence. (e)–(h) Time evolution of compressible energy density for rotating condensates. At later times in the turbulent condensate (h), the compressible energy forms strand-like structures. Reprinted from reference [49] with the permission of AIP Publishing.

cascades. Furthermore, the onset of turbulence in collisions of rotating BECs was analyzed [49]. As the condensates merge, they generate interference patterns that evolve into dark solitons. These solitons propagate through the condensate, reflect off trap boundaries, and decay into vortex–antivortex pairs. Thus, kinetic energy injection via vortices is driven by both rotation and soliton decay. Faster collisions increase soliton numbers, thereby increasing the vortex population, displaying Kolmogorov scaling $k^{-5/3}$ over a wider range of length scales. As vortex pairs annihilate, turbulent dynamics become dominated by individual vortices at smaller length scales, exhibiting Vinen turbulence with k^{-1} scaling. Notably, as the rotation frequency increases, the depth of formed solitons decreases due to off-axis condensate merging, reducing vortices produced via soliton decay, as illustrated in figure 12.6.

Apart from the above, this study also elucidates the distribution of kinetic energy components in real space within a BEC. Notably, the compressible kinetic energy distribution reveals the evolution of sound and density waves in a turbulent BEC. The confining harmonic potential causes these density waves to reflect off the trap boundaries, collide, and fragment into smaller waves. This process leads to a uniform distribution of compressible kinetic energy, signalling the onset of thermalization in the BEC, which manifests as a k^{-1} scaling in the compressible kinetic energy spectrum. Furthermore, rotational effects cause these energy fragments to aggregate into strand-like structures. Rotational effects also elevate the critical collision speed required for condensates to merge without generating turbulence. This increase occurs because rotational forcing sustains density waves and vortices, preventing their rapid decay.

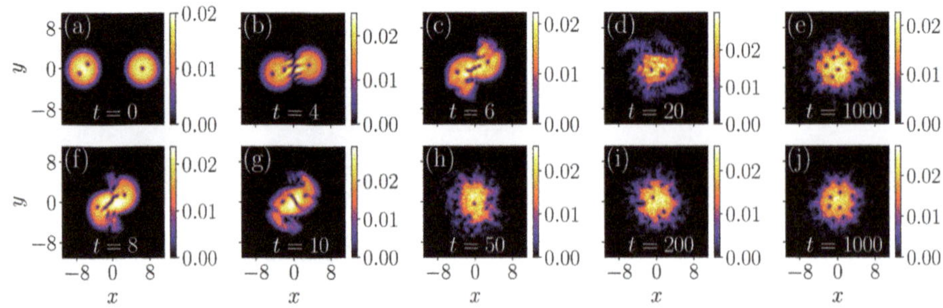

Figure 12.6. Snapshots of the density for the condensates rotating at frequency $\Omega = 0.45$. The top panel (a)–(e) and bottom panel (f)–(j) for merging times $t_m = 6$ and $t_m = 10$, respectively. (b),(f) and (g) show off-axis collisions forming weaker bent soliton structures. Reprinted from reference [49] with the permission of AIP Publishing.

12.3.7 Quantum turbulence in self-gravitating BECs

In recent years, BECs have been proposed as potential candidates for ultralight dark matter haloes [51–56], a concept supported by various cosmological simulations. The weakly interacting bosons are modelled using the GP equation. When considering interactions within the Newtonian gravitational framework, the GP equation is coupled with Poisson's equation for the gravitational field, forming the Gross–Pitaevskii–Poisson (GPP) model, expressed as

$$\mathrm{i}\frac{\partial\psi(\mathbf{r}, t)}{\partial t} = \left(-\frac{1}{2}\nabla^2 + \Phi(\mathbf{r}, t) + g|\psi(\mathbf{r}, t)|^2\right)\psi(\mathbf{r}, t) \tag{12.10}$$

$$\nabla^2\Phi(\mathbf{r}, t) = 4\pi|\psi(\mathbf{r}, t)|^2 \tag{12.11}$$

where $\psi(\mathbf{r}, t)$ is the condensate wavefunction, $\Phi(\mathbf{r}, t)$ is the gravitational potential replacing the harmonic trap, and g is the interatomic interaction strength (in dimensionless units). Compared to atomic BECs, self-gravitating BECs do not require external potentials for stabilization, relying instead on the balance between repulsive interatomic interactions and quantum kinetic pressure.

Collisions of self-gravitating BECs have attracted significant interest, as galactic collisions involving dark and luminous matter may reveal insights into the nature of dark matter. During such collisions, luminous matter is expelled from the gravitational potential while simultaneously generating gravitational waves. Self-gravitating BECs have been shown to survive and maintain their structural integrity post-collision or merger [58, 59]. Nikolaieva *et al* [60], using the GPP model, investigated collisions of solitonic BECs with vortex structures. Their findings highlight the dominance of the superfluid nature of bosonic dark matter, characterized by quantized vortex lines and vortex rings. The study also demonstrates that interference patterns influence the emission of gravitational waves, reinforcing prior conclusions that vortex structures remain intact even after head-on collisions.

Compared to advances in turbulence studies for conventional atomic BECs, quantum turbulence in dark matter BECs has been minimally explored. It has been studied in the fuzzy dark matter regime, where condensates lack self-interaction [61, 62]. For instance, Mocz *et al* [61] demonstrated that energy spectra follow $E(k) \sim k^{-1.1}$, resembling a thermally driven counterflow rather than a $k^{-5/3}$ scaling. However, few studies explore rotation-induced instabilities that could provide insights into generating turbulence in dark matter BECs. Recently, turbulence was investigated in the collision of two self-gravitating BECs, each with a central vortex [57]. Similar to atomic BECs, interference fringes appeared shortly after the collision, evolving into dark soliton rings that decay into vortex–antivortex pairs. Unlike conventional BECs, these vortices do not dissipate or annihilate but instead migrate to the periphery of the condensate. This instability is triggered when the total vortex circulation exceeds a critical value ξ_c, as shown by Nikolaieva *et al* [63].

Before the onset of this instability and just after the collision, the condensate displays turbulent behaviour with a Kolmogorov scaling $k^{-5/3}$ in the inertial range, similar to atomic BECs. During these stages, the incompressible kinetic energy is predominant. However, in later stages, the quantum pressure energy dominates over energies associated with condensate flow. This shift is attributed to the conversion of incompressible and compressible kinetic energies into the quantum pressure component (see figure 12.7). Since quantum pressure is a measure of condensate stability, this trend suggests that the condensate strives to stabilize itself by reducing condensate flow via trap deformations. This transfer of energy is not only due to the expulsion of vortices to the condensate edge, as mentioned earlier but also the suppression of fragmentation of density waves. In atomic condensates, where density waves can reflect off the trap walls and fragment, density waves in self-gravitating condensates instead deform the trap and transfer energy to the

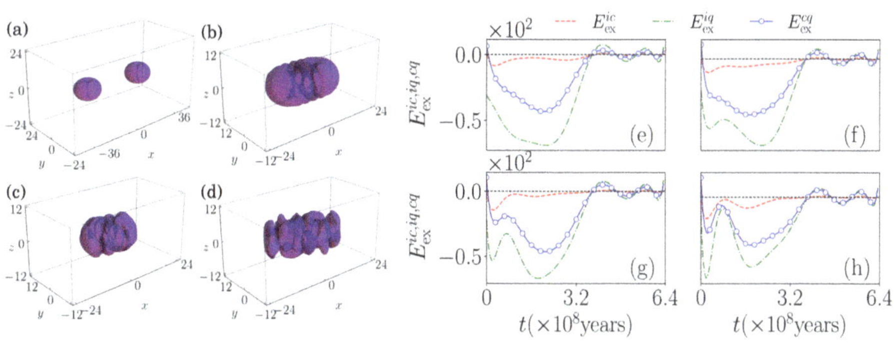

Figure 12.7. Three-dimensional density contours of colliding BECs at a velocity of $v = 9 \times 10^{-4}c$, shown at various times to illustrate their merger: (a) $t = 0$, (b) $t = 32 \times 10^6$ years, (c) $t = 38 \times 10^6$ years, and (d) $t = 180 \times 10^6$ years. Kinetic energy exchange profiles between incompressible and compressible components as a function of time for collisional velocities: (e) $v = 6 \times 10^{-4}c$, (f) $v = 7 \times 10^{-4}c$, (g) $v = 8 \times 10^{-4}c$, and (h) $v = 9 \times 10^{-4}c$. Negative exchange energy indicates transfer from the incompressible to the compressible component, while positive exchange energy indicates the reverse. Both incompressible and compressible kinetic energies contribute to the quantum pressure energy. Reprinted with permission from [57], Copyright (2025) by the American Physical Society.

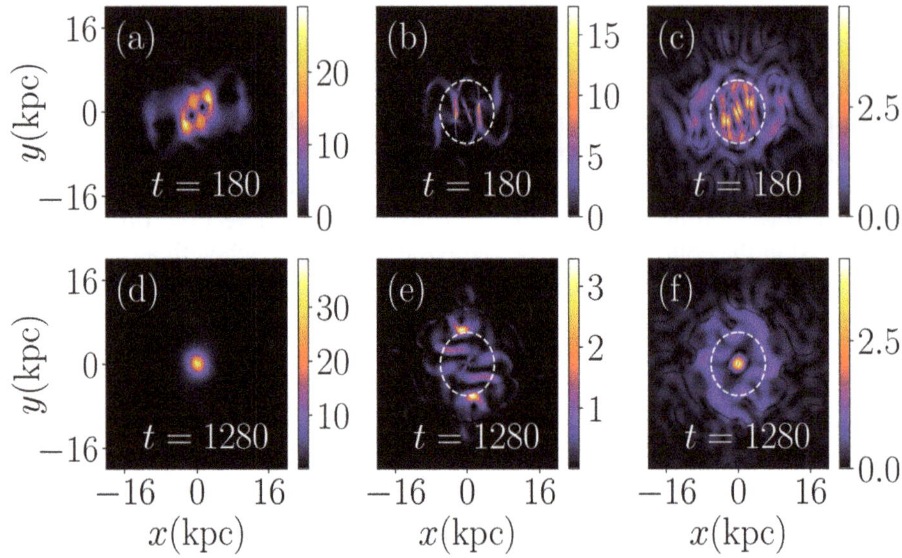

Figure 12.8. Energy density in the x–y plane for (a)–(b) compressible kinetic energy, (c)–(d) incompressible kinetic energy, and (e)–(f) quantum pressure contribution, before and after the turbulent phase. The dashed white circle indicates the Thomas–Fermi radius of the condensate. Post-turbulence, the incompressible kinetic energy extends beyond the Thomas–Fermi radius, while the compressible kinetic energy distribution lacks the fragmentation observed in atomic BECs, as shown in figure 12.5. Reprinted with permission from [57], Copyright (2025) by the American Physical Society.

condensate edges due to the trap being dependent on the condensate density. This process highlights a novel mechanism of turbulence decay without an explicit decay term.

Figure 12.8 illustrates the spatial distribution of the incompressible and compressible kinetic energy components during different stages of a collision between BECs. For the case with velocity $v = 9 \times 10^{-4}c$, figure 12.8(a), the incompressible and compressible energy density are concentrated in peripheral regions of the condensate. In post-collision stages, the compressible energy accumulates predominantly in the central region, as depicted in figures 12.8(b)–(c). The incompressible energy density, which indicates the presence of vortex cores, is primarily confined within the condensate, as delineated by the Thomas–Fermi radius (white-dashed line) in figures 12.8(c)–(d). Quantum pressure, arising from steep density gradients, exhibits localized nucleation around the edges of vortex cores, as shown in figures 12.8(e) and (f).

12.3.8 Summary and future challenges

This chapter explored exotic quantum phases and dynamics in BECs, focusing on supersolids, quantum droplets, and quantum turbulence. It introduced supersolids, where superfluidity and crystalline order coexist, focussing on their formation and stability in ultracold atomic systems. Quantum droplets, stabilized by quantum fluctuations, were analyzed using the GP equation with the LHY correction, which accounts for beyond-mean-field effects in dilute BECs. The chapter then examined quantum turbulence, contrasting it with classical turbulence, characterized by chaotic vortex motion and energy cascades. Spectral analysis techniques, including energy decomposition into incompressible and compressible components, were presented to study quantum turbulence's energy spectra across length scales. Specific regimes were discussed: Kolmogorov turbulence, exhibiting classical-like $k^{-5/3}$ scaling; Vinen (ultra-quantum) turbulence, dominated by random vortex tangles; small-scale turbulence, focussing on vortex core dynamics; and strong turbulence, involving intense vortex interactions. These analyses highlighted the unique role of quantized vortices in quantum turbulence compared to classical fluids, offering insights into nonlinear dynamics in quantum systems. A glimpse of quantum turbulence in self-gravitating BECs is also shown.

Several open questions remain to be explored further giving scope for further research. The stability and experimental realization of supersolids in diverse atomic systems, particularly with long-range interactions, remained unexplored. For quantum droplets, the impact of higher-order quantum corrections and external potentials on droplet stability and dynamics warranted further study. In quantum turbulence, developing advanced spectral analysis methods to capture transient vortex dynamics posed a significant challenge. The interplay between Kolmogorov and Vinen turbulence regimes, especially in finite systems like trapped BECs, required deeper investigation to clarify crossover conditions. Understanding small-scale quantum turbulence, including vortex reconnection and annihilation, demanded improved imaging and simulation techniques. Finally, exploring strong quantum turbulence's relevance to extreme quantum systems, such as neutron stars or quark-gluon plasmas, offered a pathway to bridge ultracold atomic physics with high-energy phenomena, necessitating interdisciplinary approaches.

12.4 Problems

Exercise 12.1

1. Derive and implement a numerical solver for the 1D spin–orbit-coupled GP equations for a binary BEC with the LHY correction term for studying quantum droplets in a 1D spin–orbit-coupled binary BEC. Solve the modified GP equations to find the ground state of a quantum droplet. Use the split-step Fourier transform method discussed in chapter 7 to simulate the ground state of a binary BEC with equal intra- and inter-species interactions.
2. Investigate the stability of quantum droplets by varying the spin–orbit coupling strength and the LHY correction strength. Use the numerical solver to compute the chemical potential and analyze droplet stability.
3. Generate quantum turbulence using an oscillating potential in the form specified in reference [64] in the rotating frame for various rotating frequencies. Observe how Kolmogorov and enstrophy scaling vary with and without rotation and at different rotation frequencies. Also note the critical rotation frequency which breaks enstrophy conservation and explain why it happens.

4. Solve the GPP system numerically using the Crank–Nicolson scheme for the GP equation. For the gravitational Poisson equation, use pseudospectral and relaxation schemes to obtain density profiles for vortex circulations $s = 1, 2, 3, 4$ and compare these densities with the results from [63].
5. For the turbulent condensate obtained in exercise 3, find the energy flux distribution across various Fourier modes k. Also, analyze how the energy flux at a particular mode varies with time for a turbulent condensate under harmonic confinement and box trap.

References

[1] Andreev A F and Lifshitz I M 1969 Quantum theory of defects in crystals *Sov. Phys. -JETP* **29** 1107–13

[2] Santos L, Shlyapnikov G V and Lewenstein M 2003 Roton-Maxon spectrum and stability of trapped dipolar Bose-Einstein condensates *Phys. Rev. Lett.* **90** 250403

[3] Petrov D S 2015 Quantum mechanical stabilization of a collapsing Bose-Bose mixture *Phys. Rev. Lett.* **115** 155302

[4] Li J -R, Lee J, Huang W, Burchesky S, Shteynas B, Top F C, Jamison A O and Ketterle W 2017 A stripe phase with supersolid properties in spin-orbit-coupled Bose-Einstein condensates *Nature* **543** 91–4

[5] Kadau H, Schmitt M, Wenzel M, Wink C, Maier T, Ferrier-Barbut I and Pfau T 2016 Observing the Rosensweig instability of a quantum ferrofluid *Nature* **530** 194–7

[6] Tanzi L, Roccuzzo S M, Lucioni E, Famà F, Fioretti A, Gabbanini C, Modugno G, Recati A and Stringari S 2019 Supersolid symmetry breaking from compressional oscillations in a dipolar quantum gas *Nature* **574** 382–5

[7] Böttcher F, Schmidt J N, Wenzel M, Hertkorn J, Guo M, Langen T and Pfau T 2019 Transient supersolid properties in an array of dipolar quantum droplets *Phys. Rev.* X **9** 011051

[8] Blakie P B, Baillie D, Chomaz L and Ferlaino F 2020 Supersolidity in an elongated dipolar condensate *Phys. Rev. Res.* **2** 043318

[9] Böttcher F, Schmidt J N, Hertkorn J, Ng K S H, Graham S D, Guo M, Langen T and Pfau T 2020 New states of matter with fine-tuned interactions: quantum droplets and dipolar supersolids *Rep. Progr. Phys.* **84** 012403

[10] Ferrier-Barbut I, Kadau H, Schmitt M, Wenzel M and Pfau T 2016 Observation of quantum droplets in a strongly dipolar Bose gas *Phys. Rev. Lett.* **116** 215301

[11] Cabrera C R, Tanzi L, Sanz J, Naylor B, Thomas P, Cheiney P and Tarruell L 2018 Quantum liquid droplets in a mixture of Bose-Einstein condensates *Science* **359** 301–4

[12] Semeghini G, Ferioli G, Masi L, Mazzinghi C, Wolswijk L, Minardi F, Modugno M, Modugno G, Inguscio M and Fattori M 2018 Self-bound quantum droplets of atomic mixtures in free space *Phys. Rev. Lett.* **120** 235301

[13] Vinen W 1957 Mutual friction in a heat current in liquid helium II I. Experiments on steady heat currents *Proc. R. Soc. Lond.* A **240** 114–27

[14] Guo W, Cahn S B, Nikkel J A, Vinen W F and McKinsey D N 2010 Visualization study of counterflow in superfluid using metastable helium molecules *Phys. Rev. Lett.* **105** 045301

[15] Henn E A L, Seman J A, Roati G, Magalhães K M F and Bagnato V S 2009 Emergence of turbulence in an oscillating Bose-Einstein condensate *Phys. Rev. Lett.* **103** 045301

[16] Bradley A S, Kumar R K, Pal S and Yu X 2022 Spectral analysis for compressible quantum fluids *Phys. Rev.* A **106** 043322
[17] Maurer J and Tabeling P 1998 Local investigation of superfluid turbulence *Europhys. Lett.* **43** 29–34
[18] Nore C, Abid M and Brachet M E 1997 Kolmogorov turbulence in low-temperature superflows *Phys. Rev. Lett.* **78** 3896–9
[19] Araki T, Tsubota M and Nemirovskii S K 2002 Energy spectrum of superfluid turbulence with no normal-fluid component *Phys. Rev. Lett.* **89** 145301
[20] Kobayashi M and Tsubota M 2007 Quantum turbulence in a trapped Bose-Einstein condensate *Phys. Rev.* A **76**
[21] Kobayashi M and Tsubota M 2008 Quantum turbulence in a trapped Bose-Einstein condensate under combined rotations around three axes *J. Low Temp. Phys.* **150** 587–92
[22] Skrbek L 2011 Quantum turbulence *J. Phys.: Conf. Ser.* **318** 012004
[23] Tough J 1982 Superfluid turbulence *Progress in Low Temperature Physics*vol 8 *(Elsevier) ch 3 pp 133–219*
[24] Feynman R P 1955 Chapter II Application of quantum mechanics to liquid helium *Prog. Low Temp. Phys.* **1** 17–53
[25] Tsubota M 2008 Quantum turbulence *J. Phys. Soc. Japan* **77** 111006
[26] Adachi H and Tsubota M 2011 Numerical study of velocity statistics in steady counterflow quantum turbulence *Phys. Rev.* B **83** 132503
[27] White A C, Barenghi C F, Proukakis N P, Youd A J and Wacks D H 2010 Nonclassical velocity statistics in a turbulent atomic Bose-Einstein condensate *Phys. Rev. Lett.* **104** 075301
[28] Tsubota M, Araki T and Nemirovskii S K 2000 Dynamics of vortex tangle without mutual friction in superfluid ^{4}He *Phys. Rev.* B **62** 11751–62
[29] Rooney S J, Blakie P B, Anderson B P and Bradley A S 2011 Suppression of Kelvon-induced decay of quantized vortices in oblate Bose-Einstein condensates *Phys. Rev.* A **84** 023637
[30] Simula T P 2011 Crow instability in trapped Bose-Einstein condensates *Phys. Rev.* A **84** 021603
[31] Baggaley A W, Barenghi C F and Sergeev Y A 2012 Quasiclassical and ultraquantum decay of superfluid turbulence *Phys. Rev.* B **85** 060501
[32] Bradley D I, Clubb D O, Fisher S N, Guénault A M, Haley R P, Matthews C J, Pickett G R, Tsepelin V and Zaki K 2006 Decay of pure quantum turbulence in superfluid *Phys. Rev. Lett.* **96** 035301
[33] Seman J A *et al* 2010 Three-vortex configurations in trapped Bose-Einstein condensates *Phys. Rev.* A **82** 033616
[34] Neely T W, Bradley A S, Samson E C, Rooney S J, Wright E M, Law K J H, Carretero-González R, Kevrekidis P G, Davis M J and Anderson B P 2013 Characteristics of two-dimensional quantum turbulence in a compressible superfluid *Phys. Rev. Lett.* **111**
[35] Abbott B P, Abbott R, Abbott T, Acernese F, Ackley K, Adams C, Adams T, Addesso P, Adhikari R X and Adya V B 2017 GW170817: observation of gravitational waves from a binary neutron star inspiral *Phys. Rev. Lett.* **119** 161101
[36] Numasato R and Tsubota M 2010 Possibility of inverse energy cascade in two-dimensional quantum turbulence *J. Low Temp. Phys.* **158** 415–21
[37] Numasato R, Tsubota M and L'vov V S 2010 Direct energy cascade in two-dimensional compressible quantum turbulence *Phys. Rev.* A **81** 063630

[38] Reeves M T, Billam T P, Anderson B P and Bradley A S 2013 Inverse energy cascade in forced two-dimensional quantum turbulence *Phys. Rev. Lett.* **110** 104501
[39] Wilson K E, Samson E C, Newman Z L, Neely T W and Anderson B P 2013 *Experimental Methods for Generating Two-Dimensional Quantum Turbulence in Bose-Einstein Condensates* (World Scientific) pp 261–98
[40] Marino A V M, Madeira L, Cidrim A, dos Santos F E A and Bagnato V S 2021 Momentum distribution of Vinen turbulence in trapped atomic Bose-Einstein condensates *Eur. Phys. J. Spec. Top.* **230** 809–12
[41] Cidrim A, White A C, Allen A J, Bagnato V S and Barenghi C F 2017 Vinen turbulence via the decay of multicharged vortices in trapped atomic Bose-Einstein condensates *Phys. Rev.* A **96** 023617
[42] Volovik G 2004 On developed superfluid turbulence *J. Low Temp. Phys.* **136** 309–27
[43] Walmsley P M and Golov A I 2008 Quantum and quasiclassical types of superfluid turbulence *Phys. Rev. Lett.* **100** 245301
[44] Stagg G W, Parker N G and Barenghi C F 2016 Ultraquantum turbulence in a quenched homogeneous Bose gas *Phys. Rev.* A **94** 053632S
[45] Zmeev D E, Walmsley P M, Golov A I, McClintock P V E, Fisher S N and Vinen W F 2015 Dissipation of quasiclassical turbulence in superfluid he4 *Phys. Rev. Lett.* **115** 155303
[46] Galantucci L, Rickinson E, Baggaley A W, Parker N G and Barenghi C F 2023 Dissipation anomaly in a turbulent quantum fluid *Phys. Rev. Fluids* **8** 034605
[47] di Leoni P C, Mininni P D and Brachet M E 2017 Dual cascade and dissipation mechanisms in helical quantum turbulence *Phys. Rev.* A **95** 053636
[48] Amette Estrada J, Brachet M E and Mininni P D 2022 Turbulence in rotating Bose-Einstein condensates *Phys. Rev.* A **105** 063321
[49] Sivakumar A, Mishra P K, Hujeirat A A and Muruganandam P 2024 Dynamic instabilities and turbulence of merged rotating Bose-Einstein condensates *Phys. Fluids* **36** 117121
[50] Sivakumar A, Mishra P K, Hujeirat A A and Muruganandam P 2024 Energy spectra and fluxes of turbulent rotating Bose-Einstein condensates in two dimensions *Phys. Fluids* **36** 027149
[51] Böhmer C G and Harko T 2007 Can dark matter be a Bose-Einstein condensate? *J. Cosmol. Astropart. Phys.* **2007** 025
[52] Chavanis P H 2011 Mass-radius relation of Newtonian self-gravitating Bose-Einstein condensates with short-range interactions. I. Analytical results *Phys. Rev.* D **84** 043531
[53] Chavanis P H and Harko T 2012 Bose-Einstein condensate general relativistic stars *Phys. Rev.* D **86** 064011
[54] Chavanis P H 2017 Dissipative self-gravitating Bose-Einstein condensates with arbitrary nonlinearity as a model of dark matter halos *Eur. Phys. J. Plus* **132**
[55] Hui L, Ostriker J P, Tremaine S and Witten E 2017 Ultralight scalars as cosmological dark matter *Phys. Rev.* D **95** 043541
[56] Hui L 2021 Wave dark matter *Annu. Rev. Astron. Astrophys.* **59** 247–89
[57] Sivakumar A, Mishra P K, Hujeirat A A and Muruganandam P 2025 Revealing turbulent dark matter via merging of self-gravitating condensates *Phys. Rev.* D **111** 083511
[58] Choi D I 2002 Collision of gravitationally bound Bose-Einstein condensates *Phys. Rev.* A **66** 063609
[59] Cotner E 2016 Collisional interactions between self-interacting nonrelativistic boson stars: effective potential analysis and numerical simulations *Phys. Rev.* D **94** 063503

[60] Nikolaieva Y, Bidasyuk Y, Korshynska K, Gorbar E, Jia J and Yakimenko A 2023 Stable vortex structures in colliding self-gravitating Bose-Einstein condensates *Phys. Rev.* D **108** 023503

[61] Mocz P, Vogelsberger M, Robles V H, Zavala J, Boylan-Kolchin M, Fialkov A and Hernquist L 2017 Galaxy formation with BECDM–I. Turbulence and relaxation of idealized haloes *Mon. Not. R. Astron. Soc.* **471** 4559–70

[62] Liu I K, Proukakis N P and Rigopoulos G 2023 Coherent and incoherent structures in fuzzy dark matter haloes *Mon. Not. R. Astron. Soc.* **521** 3625–47

[63] Nikolaieva Y O, Olashyn A O, Kuriatnikov Y I, Vilchynskii S I and Yakimenko A I 2021 Stable vortex in Bose-Einstein condensate dark matter *Low Temp. Phys.* **47** 684–92

[64] Middleton-Spencer H A J, Orozco A D G, Galantucci L, Moreno M, Parker N G, Machado L A, Bagnato V S and Barenghi C F 2023 Strong quantum turbulence in Bose-Einstein condensates *Phys. Rev. Res.* **5** 043081

Paulsamy Muruganandam and Ramaswamy Radha

Appendix A

Note on computational tools for the Gross–Pitaevskii equation

A.1 An overview of the computational tools

A comprehensive suite of computational codes in Fortran, C, and CUDA enable numerical solutions of the time-dependent Gross–Pitaevskii (GP) equation, modeling Bose–Einstein condensates (BECs) with contact, dipolar, spin–orbit-coupled, and spinor interactions in one-dimensional (1D), two-dimensional (2D), and three-dimensional (3D) trap geometries. These codes primarily use the split-step Crank–Nicolson algorithm or time-splitting Fourier spectral methods, compute stationary and non-stationary solutions, leveraging OpenMP, MPI, and CUDA for performance. Available via the CPC Program Library, they calculate energy, chemical potentials, density profiles, vortex dynamics, and spin correlations, serving as essential tools for ultracold quantum gas research. The following subsections detail the algorithms and codes, citing relevant studies.

A.1.1 Conventional BECs

For BECs with short-range contact interactions, reference [1] provides Fortran 77 and 90/95 codes for solving GP equations in 1D, 2D circularly-symmetric, 3D spherically-symmetric, and fully anisotropic traps, using split-step Crank–Nicolson for real- and imaginary-time propagation to compute dynamics and stationary states. C versions enhance portability, adding OpenMP-parallelized codes for multicore efficiency [2]. OpenMP Fortran and C programs for solving the time-dependent GP equation in an anisotropic trap [3] use experimental inputs (e.g., atom number, scattering length), while OpenMP programs for solving the time-dependent GP equation [4] offer compiler-specific optimizations. Hybrid OpenMP/MPI C codes provided for solving the time-dependent GP equation in a fully anisotropic

doi:10.1088/978-0-7503-5447-9ch13

trap [5] address 3D anisotropic traps, achieving near-linear speedup for large grids. These codes compute chemical potentials, root-mean-square sizes, and density profiles.

A.1.2 Dipolar BECs

For the case of dipolar BECs, involving long-range interactions and addressed by integro-differential GP equations, reference [6] provides Fortran and C codes for solving the time-dependent dipolar GP equation in 1D (along x or z axes), 2D (in x–y or x–z planes), and 3D traps, using Fourier transforms and the split-step Crank–Nicolson method. CUDA-accelerated versions of these codes [7] leverage cuFFT, achieving 12–25 times speedup. Parallelized versions, including OpenMP, OpenMP/MPI, and CUDA/MPI C programs [8], offer OpenMP and MPI support, with speedups of 11.5–16.5 on computer clusters. Recent OpenMP Fortran codes [9] reduce execution times on multicore processors. These codes compute energy, density, and vortex dynamics, validated against variational and Thomas–Fermi approximations.

A.1.3 Spin–orbit coupled and spinor BECs

Computational tools for solving coupled GP equations in reference [10] provide Fortran codes for solving the coupled GP equations for a harmonically trapped three-component spin-1 spinor BEC in 1D and 2D with or without SO and Rabi couplings using the split-step Crank–Nicolson scheme. Reference [11] provides an OpenMP-extended version of the Fortran codes for rotating spin-1 SO and Rabi-coupled BECs. Furthermore, the tools available in FORTRESS [12] could help study static and dynamic properties of spin-1 BECs in three-, quasi-two-, and quasi-one-dimensional settings. For spin-1 and spin-2 BECs, reference [13] uses a time-splitting Fourier spectral method for solving the three- and five-coupled GP equations.

References

[1] Muruganandam P and Adhikari S K 2009 Fortran programs for the time-dependent Gross-Pitaevskii equation in a fully anisotropic trap *Comput. Phys. Commun.* **180** 1888–912

[2] Vudragović D, Vidanović I, Balaž A, Muruganandam P and Adhikari S K 2012 C programs for solving the time-dependent Gross-Pitaevskii equation in a fully anisotropic trap *Comput. Phys. Commun.* **183** 2021–2025

[3] Young S L E, Vudragović D, Muruganandam P, Adhikari S K and Balaž A 2016 OpenMP Fortran and C programs for solving the time-dependent Gross-Pitaevskii equation in an anisotropic trap *Comput. Phys. Commun.* **204** 209–13

[4] Young S L E, Muruganandam P, Adhikari S K, Lončar V, Vudragović D and Balaž A 2017 OpenMP GNU and Intel Fortran programs for solving the time-dependent Gross-Pitaevskii equation *Comput. Phys. Commun.* **220** 503–6

[5] Satarić B, Slavnić V, Belić A, Balaž A, Muruganandam P and Adhikari S K 2016 Hybrid OpenMP/MPI programs for solving the time-dependent Gross-Pitaevskii equation in a fully anisotropic trap *Comput. Phys. Commun.* **200** 411–7

[6] Kumar R K, Young S L E, Vudragović D, Balaž A, Muruganandam P and Adhikari S K 2015 Fortran and C programs for the time-dependent dipolar Gross-Pitaevskii equation in an anisotropic trap *Comput. Phys. Commun.* **195** 117–28

[7] Lončar V, Balaž A, Bogojević A, Škrbić S, Muruganandam P and Adhikari S K 2016 CUDA programs for solving the time-dependent dipolar Gross-Pitaevskii equation in an anisotropic trap *Comput. Phys. Commun.* **200** 406–10

[8] Lončar V, Young S L E, Škrbić S, Muruganandam P, Adhikari S K and Balaž A 2016 OpenMP, OpenMP/MPI, and CUDA/MPI C programs for solving the time-dependent dipolar Gross-Pitaevskii equation *Comput. Phys. Commun.* **209** 190–6

[9] Young S L E, Muruganandam P, Balaž A and Adhikari S K 2023 OpenMP Fortran programs for solving the time-dependent dipolar Gross-Pitaevskii equation *Comput. Phys. Commun.* **286** 108669

[10] Ravisankar R, Vudragović D, Muruganandam P, Balaž A and Adhikari S K 2021 Spin-1 spin-orbit- and Rabi-coupled Bose-Einstein condensate solver *Comput. Phys. Commun.* **259** 107657

[11] Muruganandam P, Balaž A and Adhikari S K 2021 OpenMP solver for rotating spin-1 spin-orbit- and Rabi-coupled Bose-Einstein condensates *Comput. Phys. Commun.* **264** 107926

[12] Kaur P, Roy A and Gautam S 2021 FORTRESS: FORTRAN programs for solving coupled Gross-Pitaevskii equations for spin-orbit coupled spin-1 Bose-Einstein condensate *Comput. Phys. Commun.* **259** 107671

[13] Banger P, Kaur P, Roy A and Gautam S 2022 FORTRESS: FORTRAN programs to solve coupled Gross-Pitaevskii equations for spin-orbit coupled spin-f Bose-Einstein condensate with spin f=1 or 2 *Comput. Phys. Commun.* **279** 108442

IOP Publishing

An Introduction to Ultracold Atoms with Analytical and Numerical Methods

Paulsamy Muruganandam and Ramaswamy Radha

Appendix B

Raman-induced spin–orbit coupling in a spin-1 Bose–Einstein condensate

Consider a spin-1 ($F = 1$) atomic Bose–Einstein condensate of ^{87}Rb subjected to a bias magnetic field along the $\hat{y}$-direction, located at the intersection of two Raman laser beams propagating along $\hat{y} \pm \hat{x}$ with angular frequencies ω_L and $\omega_L + \Delta\omega_L$, respectively [1]. The lasers induce a two-photon Raman coupling (Ω_{R}) between adjacent ground-state Zeeman sublevels, far-detuned from excited states. This configuration generates an effective gauge field along $\hat{z}$, described by the Hamiltonian:

$$H_{\mathrm{R}} = \Omega_{\mathrm{R}} \hat{\sigma}_{3,z} \cos(2k_{\mathrm{L}}\hat{x} + \Delta\omega_L t), \tag{B.1}$$

where $\hat{\sigma}_{3,z}$ is the 3×3 spin-1 matrix, $k_{\mathrm{L}} = \sqrt{2\pi}/\lambda$ is the laser wavevector (λ being the wavelength), and $E_{\mathrm{L}} = \hbar^2 k_{\mathrm{L}}^2/2m$ is the photon recoil energy. After adiabatic elimination of excited states, the total Hamiltonian becomes

$$H_3 = \frac{\hbar^2 k^2}{2m}\hat{I}_3 + \begin{pmatrix} E_+ & 0 & 0 \\ 0 & E_0 & 0 \\ 0 & 0 & E_- \end{pmatrix} + \frac{\Omega_{\mathrm{R}}}{\sqrt{2}} \cos(2k_{\mathrm{L}}\hat{x} + \Delta\omega_L t) \begin{pmatrix} 1 & 0 & 0 \\ 0 & 0 & 0 \\ 0 & 0 & -1 \end{pmatrix}, \tag{B.2}$$

where E_+, E_0, and E_- represent the Zeeman energies of the $m_F = +1, 0, -1$ states, respectively. Under the rotating wave approximation (RWA) [2], $\hat{\sigma}_{3,z} \to \hat{\sigma}_{3,x}$:

$$\begin{pmatrix} 1 & 0 & 0 \\ 0 & 0 & 0 \\ 0 & 0 & -1 \end{pmatrix} \Rightarrow \begin{pmatrix} 0 & 1 & 0 \\ 1 & 0 & 1 \\ 0 & 1 & 0 \end{pmatrix}. \tag{B.3}$$

Defining $\theta_1 = 2k_{\mathrm{L}}\hat{x} + \Delta\omega_L t$, the Hamiltonian transforms to the following form

$$H_3 = \frac{\hbar^2 k^2}{2m}\hat{I}_3 + \begin{pmatrix} E_+ & 0 & 0 \\ 0 & E_0 & 0 \\ 0 & 0 & E_- \end{pmatrix} + \frac{\Omega_{\mathrm{R}}}{\sqrt{2}} \cos\theta_1 \begin{pmatrix} 0 & 1 & 0 \\ 1 & 0 & 1 \\ 0 & 1 & 0 \end{pmatrix}. \tag{B.4}$$

doi:10.1088/978-0-7503-5447-9ch14

B.1 Rotating frame transformation

Applying a unitary transformation U rotating at frequency $\Delta\omega_L$

$$U = \begin{pmatrix} e^{i\Delta\omega_L t} & 0 & 0 \\ 0 & 1 & 0 \\ 0 & 0 & e^{-i\Delta\omega_L t} \end{pmatrix}, \quad U^\dagger = U^{-1}, \tag{B.5}$$

the Raman Hamiltonian transforms as

$$UH_R U^\dagger = \frac{\Omega_R}{\sqrt{2}} \begin{pmatrix} 0 & e^{i\Delta\omega_L t}\cos\theta_1 & 0 \\ e^{-i\Delta\omega_L t}\cos\theta_1 & 0 & e^{i\Delta\omega_L t}\cos\theta_1 \\ 0 & e^{-i\Delta\omega_L t}\cos\theta_1 & 0 \end{pmatrix}. \tag{B.6}$$

Introducing $\theta_2 = \Delta\omega_L t$ and applying trigonometric identities yields

$$\begin{aligned} UH_R U^\dagger = {} & \frac{\Omega_R}{\sqrt{2}}\left[\cos(\theta_1 + \theta_2) + \cos(\theta_1 - \theta_2)\right]\hat{\sigma}_{3,x} + \frac{\Omega_R}{\sqrt{2}} \\ & \left[\sin(\theta_1 + \theta_2) + \sin(\theta_1 - \theta_2)\right](-\hat{\sigma}_{3,y}). \end{aligned} \tag{B.7}$$

The RWA neglects rapidly oscillating terms ($\theta_1 + \theta_2 = 2k_L\hat{x} + 2\Delta\omega_L t$), retaining only near-resonant terms ($\theta_1 - \theta_2 = 2k_L\hat{x}$)

$$H_R = \frac{\Omega_R}{\sqrt{2}}\cos(2k_L\hat{x})\hat{\sigma}_{3,x} - \frac{\Omega_R}{\sqrt{2}}\sin(2k_L\hat{x})\hat{\sigma}_{3,y}. \tag{B.8}$$

B.2 Effective two-level system

For large quadratic Zeeman shift $\hbar\omega_q$, the $|m_F = +1\rangle$ state decouples, yielding an effective two-level Hamiltonian as

$$H_2 = \frac{\hbar^2 k^2}{2m}\hat{I}_2 + \frac{\delta}{2}\hat{\sigma}_z + \frac{\Omega_R}{\sqrt{2}}\cos\theta\hat{\sigma}_x - \frac{\Omega_R}{\sqrt{2}}\sin\theta\hat{\sigma}_y, \tag{B.9}$$

where $\theta = 2k_L\hat{x}$. A pseudo-spin rotation about $\hat{z}$ followed by a global spin transformation simplifies this to

$$H_2 = \frac{\hbar^2 k^2}{2m}\hat{I}_2 + \frac{\delta}{2}\hat{\sigma}_z + \frac{\Omega_R}{2}\hat{\sigma}_x. \tag{B.10}$$

Rewriting in the dressed-state basis $\{|\uparrow, k_x + k_L\rangle, |\downarrow, k_x - k_L\rangle\}$ [3]:

$$H_2 = \frac{\hbar^2}{2m}\begin{pmatrix} (k_x + k_L)^2 & 0 \\ 0 & (k_x - k_L)^2 \end{pmatrix} + \frac{\hbar^2(k_y^2 + k_z^2)}{2m}\hat{I}_2 + \frac{\delta}{2}\hat{\sigma}_z + \frac{\Omega_R}{2}\hat{\sigma}_x. \tag{B.11}$$

Expanding the kinetic terms reveals the spin–orbit coupling

$$H_2 = \frac{\hbar^2 \mathbf{k}^2}{2m}\hat{I}_2 + \frac{2\hbar^2 k_x k_L}{2m}\hat{\sigma}_y + \frac{\Omega_R}{2}\hat{\sigma}_z + E_L \hat{I}_2 + \frac{\delta}{2}\hat{\sigma}_y, \tag{B.12}$$

where the unitary transformation U satisfies $U^\dagger H U - iU^\dagger \partial_t U$, with the second term neglected in this treatment.

References

[1] Lin Y -J, Jiménez-García K and Spielman I B 2011 Spin-orbit-coupled Bose-Einstein condensates *Nature* **471** 83–6

[2] Campbell D L, Juzeliūnas G and Spielman I B 2011 Realistic Rashba and Dresselhaus spin-orbit coupling for neutral atoms *Phys. Rev.* A **84** 025602

[3] Galitski V and Spielman I B 2013 Spin-orbit coupling in quantum gases *Nature* **494** 49–54

Appendix C

Expression for the dipole–dipole interaction potential in the Fourier domain

The dipole–dipole interaction integral can be significantly simplified through transformation to Fourier space by exploiting the convolution theorem. The spatial convolution integral takes the form

$$\int \mathrm{d}\mathbf{r}' \; V_{\mathrm{dd}}(\mathbf{r} - \mathbf{r}')|\phi(\mathbf{r}')|^2 = \int \frac{\mathrm{d}\mathbf{k}}{(2\pi)^3} \, \mathrm{e}^{-\mathrm{i}\mathbf{k}\cdot\mathbf{r}} \; V_{\mathrm{dd}}(\mathbf{k}) \; n(\mathbf{k}), \tag{C.1}$$

where $n(\mathbf{r}) = |\phi(\mathbf{r})|^2$ represents the particle density distribution. The Fourier transform pair is defined by the symmetric relations

$$A(\mathbf{k}) = \int \mathrm{d}\mathbf{r} \; B(\mathbf{r})\exp(\mathrm{i}\mathbf{k} \cdot \mathbf{r}), \tag{C.2}$$

$$B(\mathbf{r}) = \frac{1}{(2\pi)^3} \int \mathrm{d}\mathbf{k} \; A(\mathbf{k})\exp(-\mathrm{i}\mathbf{k} \cdot \mathbf{r}). \tag{C.3}$$

For the dipole potential, an analytical expression for $V_{\mathrm{dd}}(\mathbf{k})$ can be derived [1, 2]. Working in spherical polar coordinates (r, θ, ϕ) with the polar axis aligned along the momentum vector $\mathbf{k}$, and assuming that the dipole moment lies in the $y = 0$ plane at an orientation angle α relative to $\mathbf{k}$, the Fourier transform of the dipole interaction becomes

$$V_{\mathrm{dd}}(\mathbf{k}) = 3a_{\mathrm{dd}} \iiint \exp(-ikr \cos\theta)\left[1 - 3(\sin\alpha\sin\theta\cos\phi + \cos\alpha\cos\theta)^2\right] \frac{\sin\theta \; \mathrm{d}\theta \; \mathrm{d}\phi \; \mathrm{d}r}{r}. \tag{C.4}$$

Through the variable substitution $x = \cos\theta$ and subsequent integration over the azimuthal angle ϕ, this expression simplifies to

doi:10.1088/978-0-7503-5447-9ch15

$$V_{dd}(\mathbf{k}) = 3\pi a_{dd}(3\cos^2\alpha - 1)\int_b^\infty \frac{dr}{r}\int_{-1}^1 \exp(-ikrx)(1 - 3x^2)dx, \qquad (C.5)$$

where a short-distance cutoff b has been introduced to regularize the r-integration. Performing the x-integration yields

$$V_{dd}(\mathbf{k}) = 12\pi a_{dd}(1 - 3\cos^2\alpha)\int_{kb}^\infty \left[\frac{\sin u}{u^2} + \frac{3\cos u}{u^3} - \frac{3\sin u}{u^4}\right]du, \quad u = kr. \qquad (C.6)$$

The remaining integral over u can be evaluated through integration by parts, giving

$$\frac{kb\cos(kb) - \sin(kb)}{(kb)^3}, \qquad (C.7)$$

which in the limit $b \to 0$ approaches $-1/3$. This leads to the final expression for the Fourier-transformed dipole potential

$$V_{dd}(\mathbf{k}) = 4\pi a_{dd}(3\cos^2\alpha - 1) = 4\pi a_{dd}\left(\frac{3k_z^2}{|\mathbf{k}|^2} - 1\right). \qquad (C.8)$$

References

[1] Góral K and Santos L 2002 Ground state and elementary excitations of single and binary Bose-Einstein condensates of trapped dipolar gases *Phys. Rev.* A **66** 023613

[2] Lahaye T, Menotti C, Santos L, Lewenstein M and Pfau T 2009 The physics of dipolar bosonic quantum gases *Rep. Progr. Phys.* **72** 126401

www.ingramcontent.com/pod-product-compliance
Ingram Content Group UK Ltd.
Pitfield, Milton Keynes, MK11 3LW, UK
UKHW051905100926
8556IPUK00005B/769

9 780750 354455